ANNALS OF THE NEW YORK ACADEMY OF SCIENCES

Volume 422

EDITORIAL STAFF
Executive Editor
BILL BOLAND
Managing Editor
JOYCE HITCHCOCK
Associate Editor
SHEILA KATES TREITLER

The New York Academy of Sciences
2 East 63rd Street
New York, New York 10021

ELEVENTH TEXAS SYMPOSIUM ON RELATIVISTIC ASTROPHYSICS

ANNALS OF THE NEW YORK ACADEMY OF SCIENCES
Volume 422

ELEVENTH TEXAS SYMPOSIUM ON RELATIVISTIC ASTROPHYSICS

PROCEEDINGS

Edited by David S. Evans

The New York Academy of Sciences
New York, New York
1984

Library of Congress Cataloging in Publication Data

Texas Symposium on Relativistic Astrophysics (11th:1982: Austin, Tex.)
Eleventh Texas Symposium on Relativistic Astrophysics.

(Annals of the New York Academy of Sciences, ISSN 0077-8923; v. 422)
Includes bibliographical references and index.
1. Astrophysics—Congresses. 2. Relativity (Physics)—Congresses. I. Evans, David Stanley. II. Title. III. Series.
Q11.N5 vol. 422 [QB460] 500s [521] 84-4774
ISBN 0-89766-234-2
ISBN 0-89766-235-0 (pbk.)

SP
Printed in the United States of America
ISBN 0–89766–234–2 (Cloth)
ISBN 0–89766–235–0 (Paper)
ISSN 0077–8923

ANNALS OF THE NEW YORK ACADEMY OF SCIENCES

Volume 422
March 23, 1984

ELEVENTH TEXAS SYMPOSIUM ON RELATIVISTIC ASTROPHYSICS[a]

CONTENTS

[a]This volume is the result of a conference entitled Eleventh Texas Symposium on Relativistic Astrophysics held from December 12 to December 17, 1982 in Austin, Texas, sponsored by the American Institute of Physics, the American Astronomical Society, the Astronomical Society of the Pacific, the Committee on General Relativity, and The New York Academy of Sciences.

Part IX. Active Galaxies

Leon Mestel, *Chair*

Poster Papers

Financial assistance was received from:

- NATIONAL AERONAUTICS AND SPACE ADMINISTRATION
- NATIONAL SCIENCE FOUNDATION
- MRS. WINNIE SCHILD
- DR. HARLAN SMITH
- TRACOR INDUSTRIES
- UNITED STATES DEPARTMENT OF ENERGY
- UNIVERSITY OF TEXAS AT AUSTIN
- DR. STEVEN WEINBERG
- DR. JOHN A. WHEELER

Preface

DAVID S. EVANS AND J. CRAIG WHEELER

Department of Astronomy
University of Texas at Austin
Austin, Texas 78712

In 1982 the Texas Symposium on Relativistic Astrophysics returned to the state of its birth. More than 300 participants attended coming from the United States and some 30 foreign countries.

The symposium was opened by welcoming remarks from Dr. Gerhard J. Fonken, Vice President for Academic Affairs and Research at the University of Texas at Austin, and by Dr. John Archibald Wheeler, Jane and Roland Blumberg Professor in Physics at the university.

An innovation was the inclusion of two poster sessions which were enthusiastically supported. These are represented at the end of the volume by papers arranged in alphabetical order of the names of the first authors. In addition a contribution sent by Professor Bisnovatyi-Kogan is included.

The Council of the City of Austin declared the week of the symposium to be Relativistic Astrophysics week, and the declaration document was autographed by the participants. It was presented at the closing dinner to Mrs. Winnie Schild, whose husband, Professor Alfred Schild, was one of the prime movers in the original foundation of the symposium.

Mrs. Schild was hostess at a reception in the foyer at the Performing Arts Center, where the guests were entertained and vastly intrigued by the performance of the University of Texas Harp Ensemble.

A large share of the administrative work in connection with the symposium was undertaken by Mrs. Alice Herzog, and sterling work during registration was done by Mrs. Mae Collins, Mrs. Jane George and by Mr. Cecilio Martinez, who kept the financial arrangements in order.

The New Inflationary Universe[a]

ALAN H. GUTH

Center for Theoretical Physics
Laboratory for Nuclear Science and Department of Physics
Massachusetts Institute of Technology
Cambridge, Massachusetts 02139

INTRODUCTION

During the past three or four years I and a number of other people who have traditionally regarded themselves as particle theorists have begun to dabble in the study of the early universe. Some people think we were motivated mainly by jealousy of Carl Sagan, but I hope to convince you today that there is also a motivation that stems directly from recent advances in particle physics itself.

To explain this, let me begin by quickly summarizing the history of particle physics over the last decade or so. I was a graduate student in particle physics from 1968–1971, a period that now seems like the dark ages. The only interaction that we really understood at that time was electromagnetism—quantum electrodynamics—a theory that in fact had been fully developed about 20 years earlier. We had a defective theory of the weak interactions, and essentially no theory of the strong interactions at all. The idea of quarks had been proposed, but no one knew how quarks interacted or why we had not seen quarks, and few physicists took seriously the notion that quarks are really physical particles. As for a quantum theory of gravity, that was considered hopeless.

Then came a period of tremendous progress. The work of Weinberg, Salam, and Glashow was synthesized into an acceptable theory of the weak and electromagnetic interactions (for a review, see Reference 1). A theory of quark interactions was developed (for a review, see Reference 2) and dubbed quantum chromodynamics, or QCD, and it became the accepted model of the strong interactions. The combined theory of electromagnetic, weak, and strong interactions came to be known as the standard model. While only one of the four basic interactions of nature was understood when I was a graduate student, the standard model now provided an acceptable theory for three of the four—only gravity was omitted. The dark ages had come to an end.

In this atmosphere of enormous success, there was of course a drive to move further. In particular, there was a desire to construct a theory with greater unity. The standard model was a gauge theory, and it had three separate gauge groups: SU(3) for the strong interactions, and SU(2) and U(1) combined in a slightly complicated way to describe the weak and electromagnetic interactions. Then came the development of grand unified theories (for a review, see Reference 3), now known by their acronym—GUTs. The idea of a grand unified theory is to unite the three basic interactions of the standard model into a single gauge group—such a theory certainly seems more elegant than its predecessor.

[a]This work was supported in part through funds provided by the U.S. Department of Energy (DOE) under contract DE-AC02-76ER03069, and in part by an Alfred P. Sloan Fellowship.

Now I come at last to the important point. These grand unified theories make some predictions for physics at attainable energies, and these predictions seem to work. But the most dramatic predictions of grand unified theories occur only at the extraordinary energy scale of about 10^{14} GeV. To see how high this energy is, let me remind you that the most powerful beam available today has an energy per particle of about 500 GeV. If we were to try to build a 10^{14} GeV accelerator with present technology, we could in principle do it, more or less. It would be a linear accelerator with a length of about one light-year. Unfortunately, it seems quite unlikely that such an accelerator would be funded (at least during the Reagan administration). So, if we want to explore the dramatic, 10^{14} GeV consequences of grand unified theories, we have to turn to the only laboratory to which we have any access at all that has ever reached these energies. And that turns out to be the universe itself, in its very infancy. So that, to my mind, is the motivation for particle theorists to "dabble" in early universe studies.

Now, what I mainly want to talk about today is a scenario called the new inflationary universe. This is a scenario in which the universe supercools by many orders of magnitude below the critical temperature of a phase transition predicted by grand unified theories, and in the process it expands exponentially by many orders of magnitude[4]—hence the name "inflationary." The word "new" refers to a modification of my original proposal which was suggested independently by Linde[5] and by Albrecht and Steinhardt.[6] They suggested a new mechanism by which the phase transition could take place, solving some crucial problems[4,7,8] that were created by my original proposal.

The inflationary model is very attractive, because it can solve several very fundamental cosmological problems—which I will talk about later. If correct, it would also mean that grand unified theory mechanisms are responsible for the production of essentially all the matter, energy, and entropy in the observed universe. So the questions raised are very exciting—let me now get on with some of the details.[b]

Since I am a particle theorist, I will set $\hbar = c = k = 1$, and I will take the GeV as the fundamental unit for everything. Since I am not a general relativist, I will not set G equal to one. Instead I will set $G \equiv 1/M_P^2$, where $M_P = 1.22 \times 10^{19}$ GeV is the Planck mass. Note that 1 GeV $= 1.16 \times 10^{13}$ K $= 1.78 \times 10^{-24}$ g, and that 1 GeV$^{-1} = 1.97 \times 10^{-14}$ cm $= 6.58 \times 10^{-25}$ seconds.

GRAND UNIFIED THEORIES

Since so much of this work depends on GUTs, let me begin by outlining their basic properties.[3] I will place particular emphasis on those properties that are relevant to cosmology.

GUTs unite the SU(3), SU(2), and U(1) gauge fields associated with the standard model of the weak, electromagnetic, and strong interactions into a single gauge group. (The simplest model of this type was proposed by Georgi and Glashow, with SU(5) as the unifying group.)[10] The standard model contains three independent gauge coupling

[b]For the reader who would like a more detailed review of this subject, let me point out that Reference 9 contains reviews by both A. D. Linde and myself.

constants—one for each of the gauge groups. However, these coupling constants are not really constants—they vary with energy, in a calculable way. Using the values that are measured for the three coupling constants at low (i.e., attainable) energies, one finds that all three coupling constants are equal at an energy of about 10^{15} GeV.[11-13] Thus, at this energy scale the grand unified gauge group becomes an accurate symmetry. At lower energies the three interactions are distinguished by spontaneous symmetry breaking (also known as the Higgs mechanism), a mechanism that is already familiar from the Weinberg-Salam theory of electroweak interactions.

There are two important predictions of GUTs that lead me to believe that they have a reasonable chance of being valid. The first is the prediction of the equality in magnitude of the electron and proton charges. This equality is known to hold to an accuracy of one part in 10^{20},[14] but in the standard SU(3) $\times$ SU(2) $\times$ U(1) model it is achieved only by an *ad hoc* fine tuning of parameters. (Of course, one might easily imagine that other types of theories, such as composite models, would also require this equality.) The second prediction is the value of $\sin^2\theta_W$ (where θ_W is the Weinberg angle), which measures the relative strengths of the weak and electromagnetic interactions; it is calculated from the requirement that the three gauge coupling constants are all equal at some energy. The predicted value is 0.209 ± 0.005 (see Reference 15 and references therein), while the experimental value is 0.215 ± 0.012.

There are three consequences of GUTs that are of crucial importance to cosmology, and I would like to discuss these one at a time. The first is the existence of a phase transition (or possibly several phase transitions) with a critical temperature $T_c \approx 10^{14}$ GeV. To understand this phenomenon, note that the spontaneous symmetry breaking is accomplished by a scalar field Φ (called a Higgs field), which acquires a nonzero vacuum expectation value. The phase transition occurs because this nonzero expectation value is destroyed by thermal fluctuations at high temperatures[16-18] (for a review, see Reference 19), just as thermal fluctuations destroy the expectation value of the magnetic spin of a ferromagnet at temperatures above the Curie point.

The second consequence is the existence of magnetic monopoles in the particle spectrum. Any GUT in which a simple group is broken eventually to SU(3) $\times$ U(1) (i.e., QCD and electromagnetism) will necessarily (see for example reviews in References 20 and 21) contain magnetic monopoles of the 't Hooft-Polyakov type.[22,23] These monopoles are topologically stable knots in the Higgs field expectation value. (In this context one should recall that the Higgs field Φ is generally a multicomponent field, so the topology can be complicated.) The monopole masses are generically of order M_X/α, where M_X is the mass of the superheavy vector bosons and α is the fine-structure constant of the grand unified gauge interactions. In the minimal SU(5) GUT, $M_X \approx 4 \times 10^{14}$ GeV and $\alpha \approx 1/45$, so the monopole mass is of order 10^{16} GeV.

The third consequence is the nonconservation of baryon number. This means that the proton has a finite lifetime, estimated in the simplest models to lie in the range of 10^{30} to 10^{33} years. It also means that the net baryon number of the universe need not be taken as an initial condition, but can instead be generated dynamically (for a review, see Reference 24). Possibly even more important, it means that there may be no conserved quantity that distinguishes the observed universe from the vacuum. (This last point will be discussed further in the fifth section, The New Inflationary Universe.)

THE STANDARD SCENARIO

Now let me discuss what may be called the standard scenario of the very early universe.[c] The universe is assumed at the outset to be homogeneous and isotropic, and hence can be described in comoving coordinates by a Robertson-Walker metric. Whether it is open, closed, or flat, the curvature is negligible at very early times. Thus, the metric takes the simple form

$$ds^2 = -dt^2 + R^2(t)\, d\vec{x}^2 \tag{1}$$

The expansion of the universe is described by the scale factor $R(t)$, which is governed by the Einstein field equation

$$\left(\frac{\dot{R}}{R}\right)^2 = \frac{8\pi}{3} G\rho \tag{2}$$

where ρ is the energy density and the dot denotes differentiation with respect to the time t. The above equation would have an additional term if the curvature were not negligible. The quantity $H \equiv \dot{R}/R$ is the Hubble "constant" (which of course varies as the universe evolves). The mass density is dominated by the thermal radiation of effectively massless (i.e., $M << T$) particles at temperature T:

$$\rho = cT^4 \tag{3}$$

The constant c depends on the number of effectively massless particle species; for the minimal SU(5) GUT at the highest temperatures, c is about 50. The expansion is taken to be adiabatic, which implies (as long as the number of effectively massless particle species remains unchanged)

$$RT = \text{constant} \tag{4}$$

Putting these relations together, one finds

$$T^2 \approx \frac{M_P}{40t} \tag{5}$$

and

$$R \propto \sqrt{t} \tag{6}$$

To develop some feeling for the relevant numbers, consider $T = 10^{14}$ GeV $\approx T_c$. The equations then give $t \approx 1.9 \times 10^{-35}$ seconds, and $\rho \approx 1.2 \times 10^{75}$ g cm^{-3}. (For comparison, recall that in these units the density of an atomic nucleus is about 10^{15}.) Objects that are separated by 10^{10} light-years today were separated by only about 6 cm at this early time.

Finally, there must be a phase transition (or possibly several) when the temperature falls to T_c. In the standard scenario it is assumed that this phase transition occurs

[c]For a general background in cosmology, see Reference 25. At a less technical level, see References 26 and 27.

quickly, without any significant supercooling. The effects on the evolution of $R(t)$ are then negligible.

PROBLEMS OF THE STANDARD SCENARIO

The standard scenario suffers from three problems, which I will now describe. It is the existence of these problems that motivates the inflationary scenario.

The first is the monopole problem—that is, the standard scenario leads to a tremendous overproduction of magnetic monopoles.[28,29d] To see how this comes about, recall that the monopoles are in fact topologically stable knots in the Higgs field expectation value. If the Higgs field has a correlation length ξ, then one would expect a density of monopoles given roughly by the Kibble relation[31]

$$n_M \approx 1/\xi^3 \tag{7}$$

When the universe cools below the critical temperature $T_c \approx 10^{14}$ GeV of the GUT phase transition, it becomes thermodynamically favorable for the Higgs field to align uniformly over large distances. However, it takes time for these correlations to be established. Note that the horizon length, defined as the distance that a light pulse could have traveled since the initial singularity, is given in the radiation-dominated era by $l_H = 2t$, where t is the age of the universe. Thus, causality alone implies that[32–34]

$$\xi \lesssim 2t \tag{8}$$

Equations 7 and 8 can be used to obtain an approximate lower bound on n_M immediately after the phase transition. One can also calculate the entropy density s, and one finds

$$n_M/s \gtrsim 10^{-13} \tag{9}$$

At these densities, the rate of monopole-antimonopole annihilation can be shown to be insignificant.[28,29,35] Entropy is essentially conserved in this scenario, so the ratio n_M/s should be about the same today. One then finds that today one would have

$$\Omega \equiv \rho/\rho_c \gtrsim 3 \times 10^{11} \tag{10}$$

where $\rho_c = 3H^2/8\pi G$ is the critical mass density [which gives a precisely flat ($k = 0$) universe]. Such a huge value of Ω is clearly impossible. It would imply, for example, that the current age of the universe would be $\lesssim$ 30,000 years. (I am told that an age of this order is considered acceptable in some circles—but I have checked with my friends, and they are all confident that such an age can be ruled out.) Thus, some mechanism must be found to suppress the production of magnetic monopoles.

The second problem of the standard scenario is known as the horizon problem; it was first pointed out by Rindler[36] in 1956 (see also Reference 25). The observational basis for this problem is the uniformity of the cosmic background radiation, which is known to be isotropic to better than one part in 10^3. This fact is particularly difficult to

[d]For a more detailed review of this topic, see Reference 30 and the article by G. Lazarides in the same volume.

understand when one considers the existence of the horizon length, the maximum distance that light could have traveled since the beginning of time. Consider two microwave antennas pointed in opposite directions. Each is receiving radiation that is believed to have been emitted (or "decoupled") at the time of hydrogen recombination, when T was about 4000 K at $t \approx 10^5$ years. At the time of emission, these two sources were separated from each other by over 90 horizon lengths.[37] The problem is to understand how two regions over 90 horizon lengths apart came to be at the same temperature at the same time. Within the standard scenario this large-scale homogeneity cannot be explained; rather, it must be assumed as an initial condition.

The third problem of the standard scenario was pointed out by Dicke and Peebles in 1979,[38] and is known as the flatness problem. The basis of the problem is the fact that today the ratio between the actual mass density and the critical mass density (i.e., $\Omega \equiv \rho/\rho_c$) is conservatively known to lie in the range

$$0.01 < \Omega < 10 \quad \textbf{(11)}$$

No one is surprised by how narrowly this range brackets $\Omega = 1$. However, within the evolution of the standard model, the value $\Omega = 1$ is an *unstable* equilibrium point. Thus, to be near $\Omega = 1$ today, the universe must have been very near to $\Omega = 1$ in the past. When T was 1 MeV, Ω had to equal one to an accuracy of one part in 10^{15}. At the time of the GUT phase transition, when T was about 10^{14} GeV, Ω had to be equal to one to within one part in 10^{49}! In the standard model this precise fine tuning is simply put in as an initial condition. However, I feel that the model is believable only if a mechanism to create this fine tuning can be found.

Since the horizon and flatness problems are peculiarities of the initial conditions required by the standard model, it must be pointed out that both problems could conceivably be solved during the era of quantum gravity, when $T \gtrsim 10^{19}$ GeV. Since quantum gravity is only poorly understood, such a solution would be highly speculative at the present time. Inflation, on the other hand, relies only on rather well-accepted physical mechanisms. The fact that inflation can solve these problems, and the monopole problem as well, should (in my opinion) be considered strong evidence in its favor.

THE NEW INFLATIONARY UNIVERSE

I will now describe the new inflationary universe scenario. The scenario is cast in the context of the SU(5) GUT. In this theory the SU(5) gauge symmetry is broken to SU(3) × SU(2) × U(1) by means of a multicomponent Higgs field, which I will denote by Φ. I will use ϕ to denote that linear combination of components[e] that actually acquires the nonzero expectation value, normalized in the standard way for a real scalar field. The new ending proposed by Linde[5] and Albrecht and Steinhardt[6] requires that the parameters of the Higgs field potential energy function be chosen so that the effective potential $V(\phi)$ obeys the Coleman-Weinberg condition:[39]

$$\frac{\partial^2 V}{\partial \phi^2} = \frac{\partial^3 V}{\partial \phi^3} = 0 \text{ at } \phi = 0 \quad \textbf{(12)}$$

[e]More precisely, Φ denotes the adjoint representation Higgs field, represented as a traceless hermitian 5 × 5 matrix. I then set $\Phi = (2/15)^{1/2}\,\phi\,\mathrm{diag}[1,1,1-3/2,-3/2]$.

Including the one-loop gauge field quantum corrections (calculated in the flat-space approximation), the potential takes the form

$$V(\phi) = \frac{25}{16}\alpha^2 \left[\phi^4 \ln(\phi^2/\sigma^2) + \frac{1}{2}(\sigma^4 - \phi^4)\right] \tag{13}$$

where $\alpha \equiv g^2/4\pi \approx 1/45$ is the gauge coupling and $\sigma \approx 1.2 \times 10^{15}$ GeV. The form of this potential is shown in FIGURE 1. The minimum lies at $\phi = \sigma$, corresponding to the true vacuum. The point $\phi = 0$ is an equilibrium point which is just barely unstable. I will refer to the field configuration $\phi = 0$ as the false vacuum, even though this term is traditionally reserved for configurations that are classically stable. For any temperature $T > 0$, the false vacuum is stabilized by a bump in the finite temperature effective potential. For small T this bump has a height of order T^4, and a width of order T.

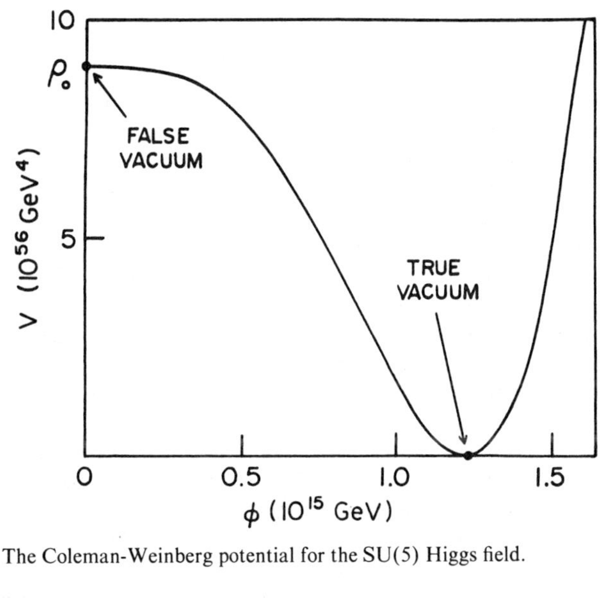

FIGURE 1. The Coleman-Weinberg potential for the SU(5) Higgs field.

There is a phase transition with critical temperature $T_c \approx 10^{14}$ GeV, above which the $\phi = 0$ configuration is the global minimum of the potential (and is hence the thermal equilibrium state).

The starting point of a cosmological scenario is somewhat a matter of taste and philosophical prejudice. Some physicists find it plausible that the universe began in some highly symmetrical state, such as a de Sitter space. I prefer to believe that the universe began in a highly chaotic state; one advantage of the inflationary scenario, from my point of view, is that it appears to allow a wide variety of starting configurations. I require only that the initial universe is hot ($T > 10^{14}$ GeV) in at least some places, and that at least some of these regions are expanding rapidly enough so that they will cool to T_c before gravitational effects reverse the expansion.

In these hot regions, thermal equilibrium would imply $\langle \Phi \rangle = 0$, where $\langle \Phi \rangle$ denotes the expectation value of Φ. (Actually, though, the universe has not had time at this point to thermalize.[40] Thus, I need to assume that there are some regions of high energy density with $\langle\Phi\rangle \approx 0$, and that some of these regions lose energy with Φ being trapped in the false vacuum.) These regions will cool to T_c, and nucleation rate calculations[41] indicate that they will continue to supercool well below T_c. The energy density ρ will approach $\rho_0 \equiv V(\phi = 0)$. Since this false vacuum state is Lorentz invariant, the energy-momentum tensor must then have the form

$$T_{\mu\nu} = \rho_0 g_{\mu\nu} \tag{14}$$

To the extent that Equations 1 and 2 are applicable, the universe will approach a space which can be described by the metric of Equation 1 with

$$R(t) = e^{\chi t} \tag{15}$$

where

$$\chi = \left(\frac{8\pi}{3} G\rho_0\right)^{1/2} \tag{16}$$

(For our parameters, $\chi \approx 10^{10}$ GeV.) Such a space is called a de Sitter space.[f] Recall, however, that the metric of Equation 1 embodies the assumption that the universe began in a state that is perfectly homogeneous and isotropic—precisely the assumption I am trying to avoid. Thus, I must consider perturbations about this metric. The behavior of these perturbations seems to be governed by a "cosmological no-hair theorem," which states that whenever the energy-momentum tensor is given by Equation 14, then any locally measurable perturbation about the de Sitter metric is damped exponentially on the time scale of χ^{-1}. The theorem has been demonstrated in the context of linearized perturbation theory,[43–46] and it is conjectured to hold even for large perturbations.[47,48] Thus, a smooth de Sitter metric arises naturally, without any need to fine tune the initial conditions. (Advocates of an initial de Sitter space are welcome to join the scenario at this point.)

As the space continues to supercool and exponentially expand, the energy density is fixed at ρ_0. Thus, the total matter energy (i.e., all energy other than gravitational) is increasing! If the inflationary model is right, this false vacuum energy is the source of essentially all the matter, energy, and entropy in the observed universe.

This seems to violate our naive notions of energy conservation, but we must remember that the gravitational field can exchange energy with the matter fields. The energy-momentum tensor for matter obeys a covariant conservation equation, which in the Robertson-Walker metric reduces to

$$\frac{d}{dt}(R^3\rho) = -p\frac{d}{dt}(R^3) \tag{17}$$

From Equation 14 it follows that the false vacuum has a large and negative pressure, $p = -\rho_0$; Equation 17 is then satisfied identically, with the energy of the expanding gas increasing due to the negative pressure. If the space were asymptotically Minkowskian,

[f] The properties of de Sitter space are well described in Reference 42.

it would be possible to define a conserved total energy (matter plus gravitational). However, in the Robertson-Walker metric, a global conservation law of this type does not exist, except perhaps in a trivial version in which the total energy vanishes identically.

Let me digress a moment to discuss the quantum numbers of the observed universe. Some approximate values are given as follows: baryon number $\approx 10^{78-79}$; matter energy $\approx 10^{78-80}$ GeV; electric charge ≈ 0; color (QCD) charge ≈ 0; and angular momentum ≈ 0.

I have just explained that matter energy is not conserved, but can be increased dramatically by exponential expansion. If baryon number is also nonconserved (as in GUTs), then the universe is, so far as we know, devoid of any conserved quantities (see Reference 49 and references therein). In that case, it is very tempting to believe that the universe began from nothing,[50] or from almost nothing. The inflationary model illustrates the latter possibility. Using the inflationary mechanism, it is possible to create the entire observed universe starting with a total matter energy of about 10 kilograms.[g]

Now let me return to my region of space, which is supercooling into a de Sitter phase. It is a striking fact about de Sitter space that quantum fluctuations mimic thermal effects at a Hawking temperature[47]

$$T_H = \chi/2\pi \tag{18}$$

In our case, $T_H \approx 10^9$ GeV. Calculations on the decay of the supercooling phase have been carried out mainly in Minkowski space, omitting the effects of gravitation (including the Hawking temperature). These calculations[41,5,6] indicate that the supercooling continues until $T \approx 10^{7-8}$ GeV, at which point the calculations break down. (They break down when the energy of a critical size bubble becomes comparable to T, resulting in a failure of the steepest descent approximation. The situation becomes even more difficult to analyze when T falls to $\sim 10^6$ GeV, at which point the SU(5) gauge coupling becomes strong.) Since gravitational effects prevent the temperature from ever falling below T_H,[48] these Minkowski space calculations are illustrative but inconclusive. The gravitational effects have been studied by several authors,[48,51–53] but important questions appear to be still unresolved.

It is clear that within the supercooling region, the Higgs field will be undergoing fluctuations due to thermal and/or quantum effects. One would expect that some fluctuations would begin to grow, and at some point these fluctuations will become large enough so that their subsequent evolution can be described by the classical equations of motion. I will use the term "fluctuation region" to denote any region in which this has occurred. The size and shape of the fluctuation regions are uncertain—they are presumably irregular,[6] and are presumably homogeneous on a length scale of order χ^{-1} (which appears to be the only relevant length scale in the problem). Note that this length scale of homogeneity is only about 10^{-11} proton diameters; the entire observed universe will evolve from a region of this size or smaller.

The Higgs field ϕ then "rolls" down the potential of FIGURE 1, obeying the classical equations of motion:

[g]To begin an inflationary universe with the minimal possible matter energy, one could start with a volume of about χ^{-3} of false vacuum; the total matter energy would be $\rho_0 \chi^{-3} \approx 10$ kg.

$$\ddot{\phi} + 3\frac{\dot{R}}{R}\dot{\phi} = -\frac{\partial V}{\partial \phi} \quad (19)$$

If the initial fluctuation is small, then the flatness of the potential for $\phi \approx 0$ will imply that the rolling begins very slowly. Note that the second term on the left-hand side of Equation 19 is a damping term, helping to further slow down the speed of rolling. As long as $\phi \approx 0$, the energy density ρ remains about equal to ρ_0, and the exponential expansion continues. The exponential expansion occurs on a time scale χ^{-1}, which is short compared to the time scale of the rolling.

For the scenario to work, it is necessary for the length scale of homogeneity to be stretched from χ^{-1} to at least about 10 cm before the Higgs field ϕ rolls off the plateau in FIGURE 1. This corresponds to an expansion factor of about 10^{25}, which requires about 58 times constants (χ^{-1}) of expansion. It can be shown that such an expansion would result if the initial fluctuation in ϕ were less than about 1.4 T_H (with no initial fluctuation $\dot{\phi}$).[54] One would expect fluctuations of roughly this magnitude, but a more detailed understanding is required to determine if the initial fluctuations are small enough or not.

When the ϕ field reaches the steep part of the potential, it falls quickly to the botton and oscillates about the minimum. The time scale of this motion is a typical GUT time of $(10^{14}\ \mathrm{GeV})^{-1}$, which is very fast compared to the expansion rate. The Higgs field oscillations are then quickly damped by the couplings to the other fields, and the energy is rapidly thermalized.[55–57] (The Higgs field oscillations correspond to a coherent state of Higgs particles; the damping is simply the decay into other species.) The release of this energy (which is just the latent heat of the phase transition) raises the temperature to the order of 10^{14} GeV. (More precisely, the reheating temperature is about one-sixth of the superheavy vector boson mass.)

From here on the standard scenario ensues, including the production of a net baryon number. The length scale of homogeneity increases to $\gtrsim 10^{10}$ light-years by the time T falls to 2.7 K.

SOLUTIONS TO THE COSMOLOGICAL PROBLEMS

Let me now explain how the three problems of the standard cosmological scenario discussed in the fourth section are avoided in the inflationary scenario.

First, let us consider the monopole problem. It is easy to see that the Kibble production mechanism, which produced the tremendous excess of monopoles in the standard scenario, is totally ineffective here. The Higgs field is correlated throughout the initial fluctuation region, and the exponential expansion increases the Higgs correlation length ξ so that it is greater than 10^{10} light-years today. Thus, this mechanism would produce $\lesssim 1$ monopole in the observable universe. Some monopoles would still be produced by thermal fluctuations after reheating, but this number would be negligible in the SU(5) model[58,30] and presumably in most other models as well.[h]

The horizon problem is clearly avoided in this scenario, since the entire observed

[h]A. S. Goldhaber and I have begun to study the production of monopoles by the quantum fluctuations in the Higgs field as it begins to roll off the plateau. While we expect this mechanism to produce a monopole density consistent with known bounds, we nonetheless believe that the density is much larger than the estimates for other mechanisms, and perhaps even detectable.

universe evolves from a single fluctuation region. This region had a size of order χ^{-1} at the time when the fluctuation began to grow classically—it was thus causally connected. As I explained earlier, the Higgs field is expected to be homogeneous on this length scale. The exponential expansion causes this very small region of homogeneity to grow to be large enough to encompass the observed universe.

The flatness problem is avoided by the dynamics of the exponential expansion of the fluctuation region. As ϕ begins to roll very slowly down the potential, the evolution of the metric is governed by the energy density ρ_0. Assuming that the fluctuation region (or a small piece of it) can be approximated locally by a Robertson-Walker metric, then the scale factor evolves according to the standard equation:

$$\left(\frac{\dot{R}}{R}\right)^2 = \frac{8\pi}{3} G\rho_0 - \frac{k}{R^2} \tag{20}$$

where $k = +1, -1$, or 0 depending on whether the region approximates a closed, open, or flat universe, respectively. (There could also be perturbations, but these would die out quickly, as discussed following Equation 16.) In this language, the flatness problem is the problem of understanding why the second term on the right-hand side is so extraordinarily small. But as the fluctuation region expands exponentially, the energy density ρ remains very nearly constant at ρ_0, while the k/R^2 term falls off as the square of the exponential factor. Thus the k/R^2 term is suppressed by at least a factor of 10^{50}, which provides a "natural" explanation of why its value immediately after the phase transition is less than 10^{-49} times the value of the other terms.

Except for a very narrow range of parameters, this suppression of the curvature term will vastly exceed that required by present observations. For example, the known fact that $\Omega \sim 1$ today can be explained if the initial fluctuations in ϕ are about 1.4 T_H. Suppose, however, that the initial fluctuations are just 10% smaller. It then turns out that today Ω would be equal to one to within one part in 10^{10}.[54] This leads to the prediction that the value of Ω today is expected to be equal to one to within an extraordinary degree of accuracy.

HISTORICAL SUMMARY AND UPDATE

The history of the inflationary universe has been an alternating sequence of "good news" and "bad news" events. The history began with the "good news" discovery[4] that inflation can solve the monopole, horizon, and flatness problems, and explain the origin of essentially all the matter, energy, and entropy in the observed universe. Then came the bad news. In this original scenario, the phase transition was assumed to take place by the random nucleation of bubbles of the new phase in the background of the old phase. The bad news was that the randomness of the nucleation would prevent a smooth coalescence of bubbles, resulting in gross inhomogeneities. This bad news was discovered[4] within months of the discovery of the original scenario, although the complete details were not published until somewhat later.[7,8] The inflationary universe remained tantalizing, but it was obviously in error.

I was of course delighted at Christmas of 1981 when I received Linde's preprint announcing the new inflationary universe[5]—by a special choice of the parameters in the Higgs potential, it was possible to have the entire observed universe evolve from a single bubble or fluctuation. The problem of bubble coalescence became irrelevant, yet all the successes of the original model were preserved. I soon learned that Albrecht and Steinhardt had independently discovered the same scenario.[6]

Now, unfortunately, comes some more bad news. It involves the calculation of density fluctuations in the new inflationary universe, which was a popular sport (see References 59–62 and also the contributions of these authors in Reference 9) at the Nuffield Workshop on the Very Early Universe, held last summer in Cambridge, England. The goal was to calculate $\delta\rho/\rho$ on a given comoving length scale, at the time when the physical length is comparable to the Hubble length H^{-1}. At the outset, all parties agreed that the value of $\delta\rho/\rho$ was roughly independent of length scale, in accord with the spectrum proposed by Harrison and Zeldovich to account for galaxy production.[63,64] (That part was considered good news.) However, the initial estimates of the amplitude ranged from 10^{-16} to 10^{2}, while the desired value is 10^{-4}–10^{-2}. By the end of the workshop, all of the groups converged to a value near 10^{2}.[i] The fluctuations arise mainly from the quantum fluctuations in the Higgs field ϕ, which cause it to fall off the plateau of FIGURE 1 and release the false vacuum energy density ρ_0 at a time that varies from one position to another.

It was of course disappointing to discover that the new inflationary universe could not correctly account for the density fluctuations in the universe. However, I would like to argue that this result should be viewed not as a discouraging failure, but rather as an encouraging near miss. First, one should remember that this is the first time we have ever had a scenario with enough detail to even allow a calculation of $\delta\rho/\rho$ from first principles. Traditionally, an *ad hoc* initial spectrum of fluctuations has simply been assumed. Second, I claim that the new inflationary universe does quite a bit better than a toy model that I would regard as a reasonable elaboration of the standard model. In this toy model, I will assume that the mysterious effects of quantum gravity leave the universe in thermal equilibrium at a perfectly uniform temperature $T \approx M_P$. However, I will insist that the gas that fills the universe exhibit the usual fluctuations of Poisson statistics. Due to the incredible instabilities of the early universe, it turns out that the fluctuations on galactic scales then grow to be 10^{12} times larger than those predicted by the new inflationary universe.[54] And third, recall that the fluctuation results are for the particular case of an underlying SU(5) model. The result is model dependent, and in a certain class of supersymmetric models the amplitude of the fluctuations is about right (see the contributions by P. J. Steinhardt and M. S. Turner in Reference 9). (Unfortunately, these supersymmetric models do not produce enough reheating.) Thus, it is possible that a model will yet be found that will solve all of these problems.

In conclusion, I want to say that the basic idea of inflation—the idea that the universe went through a period during which it expanded exponentially while trapped in a false vacuum—appears to me to be probably correct. It is a very simple and natural idea in the context of spontaneously broken gauge theories, and it seems to solve some very fundamental cosmological problems. On the other hand, we clearly do not yet have the details straight—the density fluctuations in the SU(5) new inflationary

[i]However, at this conference, Hawking and Moss circulated a preliminary preprint which contradicted the earlier answer, obtaining instead an answer that was acceptably small. They did, however, express some doubts about the validity of this new calculation. At this point the issues are not fully resolved, but to my knowledge none of the other physicists working on this problem have accepted this new result. [**Note added in proof:** Hawking and Moss have now revised their result. Their latest answer is smaller than that of other authors, but still unacceptably large.]

scenario turn out to be much too large. I would like to think that the cosmology is not working out right because the underlying particle physics is not yet correct in detail. As I said in my introduction, we would like to be able to use the very early universe as a GUT energy scale particle physics laboratory. So far, I think it is fair to say that we have not yet found a particle theory that fits these "experiments."

Finally, let me say again that probably the most dramatic recent discovery in cosmology is the realization that the universe may be completely devoid of conserved quantum numbers. If so, then even if we do not understand the precise scenario, it becomes very plausible that our observed universe emerged from nothing or from almost nothing. I have often heard it said that there is no such thing as a free lunch. It now appears possible that the universe itself is a free lunch.

REFERENCES

1. ABERS, E. S. & B. W. LEE. 1973. Phys. Rep. **9C:** 1.
2. MARCIANO, W. J. & H. PAGELS. 1978. Phys. Rep. **36C:** 137.
3. LANGACKER, P. 1981. Phys. Rep. **72C:** 185.
4. GUTH, A. H. 1981. Phys. Rev. **D23:** 347.
5. LINDE, A. D. 1982. Phys. Lett. **108B:** 389.
6. ALBRECHT, A. & P. J. STEINHARDT. 1982. Phys. Rev. Lett. **48:** 1220.
7. GUTH, A. H. & E. J. WEINBERG. 1983. Nucl. Phys. **B212:** 321.
8. HAWKING, S. W., I. G. MOSS & J. M. STEWART. 1982. Phys. Rev. **D26:** 2681.
9. GIBBONS, G. W., S. W. HAWKING & S. T. C. SIKLOS, Eds. 1983. The Very Early Universe. Cambridge University Press. Cambridge, England. (Proceedings of the Nuffield Workshop.)
10. GEORGI, H. & S. L. GLASHOW. 1974. Phys. Rev. Lett. **32:** 438.
11. GEORGI, H., H. R. QUINN & S. WEINBERG. 1974. Phys. Rev. Lett. **33:** 451.
12. GOLDMAN, T. J. & D. A. ROSS. 1979. Phys. Lett. **84B:** 208.
13. GOLDMAN, T. J. & D. A. ROSS. 1980. Nucl. Phys. **B171:** 273.
14. KING, J. G. 1960. Phys. Rev. Lett. **5:** 562.
15. GASIOROWICZ, S. & J. L. ROSNER. 1981. Am. J. Phys. **49:** 954.
16. KIRZHNITS, D. A. & A. D. LINDE. 1972. Phys. Lett. **42B:** 471.
17. WEINBERG, S. 1974. Phys. Rev. **D9:** 3357.
18. DOLAN, L. & R. JACKIW. 1974. Phys. Rev. **D9:** 3320.
19. LINDE, A. D. 1979. Rep. Prog. Phys. **42:** 389.
20. GODDARD, P. & D. I. OLIVE. 1978. Rep. Prog. Phys. **41:** 1357.
21. COLEMAN, S. 1983. *In* The Unity of the Fundamental Interactions. A. Zichichi, Ed.: 21–118. Plenum Press. New York, N.Y.
22. 'T HOOFT, G. 1974. Nucl. Phys. **B79:** 276.
23. POLYAKOV, A. M. 1974. Pis'ma Zh. Eksp. Teor. Fiz. **20:** 430 (1974. JETP Lett. **20:** 194).
24. YOSHIMURA, M. 1981. *In* Grand Unified Theories and Related Topics: Proceedings of the 4th Kyoto Summer Institute. M. Konuma & T. Maskawa, Eds.: 235–288. World Scientific. Singapore.
25. WEINBERG, S. 1972. Gravitation and Cosmology. John Wiley & Sons. New York, N.Y.
26. SILK, J. 1980. The Big Bang. W. H. Freeman & Co. San Francisco, Calif.
27. WEINBERG, S. 1977. The First Three Minutes. Bantam Books. New York, N.Y.
28. PRESKILL, J. P. 1979. Phys. Rev. Lett. **43:** 1365.
29. ZELDOVICH, YA. B. & M. Y. KHLOPOV. 1978. Phys. Lett. **79B:** 239.
30. GUTH, A. H. 1983. *In* Magnetic Monopoles. R. A. Carrigan & W. P. Trower, Eds.: 81–96. Plenum Press. New York, N.Y.
31. KIBBLE, T. W. B. 1976. J. Phys. **A9:** 1387.
32. PRESKILL, J. P. Private communication.
33. GUTH, A. H. & S.-H. TYE. 1980. Phys. Rev. Lett. **44:** 631, 963.
34. EINHORN, M. B., D. L. STEIN & D. TOUSSAINT. 1980. Phys. Rev. **D21:** 3295.

35. GOLDMAN, T., E. W. KOLB & D. TOUSSAINT. 1981. Phys. Rev. **D23:** 867.
36. RINDLER, W. 1956. Mon. Not. R. Astron. Soc. **116:** 663.
37. GUTH, A. H. 1983. *In* Asymptotic Realms of Physics: Essays in Honor of Francis E. Low. A. H. Guth, K. Huang & R. L. Jaffe, Eds.: 199–222. MIT Press. Cambridge, Mass.
38. DICKE, R. H. & P. J. E. PEEBLES. 1979. *In* General Relativity: An Einstein Centenary Survey. S. W. Hawking & W. Israel, Eds.: 504–517. Cambridge University Press. Cambridge, England.
39. COLEMAN, S. & E. J. WEINBERG. 1973. Phys. Rev. **D7:** 1888.
40. STEIGMAN, G. 1983. *In* Unification of the Fundamental Interactions II. J. Ellis & S. Ferrara, Eds. Plenum Press. New York, N.Y.
41. SHER, M. 1981. Phys. Rev. **D24:** 1699.
42. HAWKING, S. W. & G. F. R. ELLIS. 1973. The Large Scale Structure of Space-Time. Cambridge University Press. Cambridge, England.
43. FRIEMAN, J. A. & C. M. WILL. 1982. Astrophys. J. **259:** 437.
44. BARROW, J. D. (In Reference 9).
45. BOUCHER, W. & G. W. GIBBONS. (In Reference 9.)
46. GINSPARG, P. & M. J. PERRY. 1983. Nucl. Phys. **B222:** 245.
47. GIBBONS, G. W. & S. W. HAWKING. 1977. Phys. Rev. **D15:** 2738.
48. HAWKING, S. W. & I. G. MOSS. 1982. Phys. Lett. **110B:** 35.
49. ATKATZ, D. & H. PAGELS. 1982. Phys. Rev. **D25:** 2065.
50. VILENKIN, A. 1982. Phys. Lett. **117B:** 25; 1983. Phys. Rev. **D27:** 2848.
51. VILENKIN, A. & L. H. FORD. 1982. Phys. Rev. **D26:** 1231.
52. VILENKIN, A. 1982. Phase Transitions in de Sitter Space. Preprint TUTP-82-10. Tufts University. Medford, Mass.
53. HUT, P. & F. R. KLINKHAMER. 1981. Phys. Lett. **104B:** 439.
54. GUTH, A. H. (In Reference 9.)
55. ALBRECHT, A., P. J. STEINHARDT, M. S. TURNER & F. WILCZEK. 1982. Phys. Rev. Lett. **48:** 1437.
56. ABBOTT, L. F., E. FARHI & M. B. WISE. 1982. Phys. Lett. **117B:** 29.
57. DOLGOV, A. D. & A. D. LINDE. 1982. Phys. Lett. **116B:** 329.
58. LAZARIDES, G., Q. SHAFI & W. P. TROWER. 1982. Phys. Rev. Lett. **49:** 1756.
59. STAROBINSKY, A. A. 1982. Phys. Lett. **B117:** 175.
60. GUTH, A. H. & S.-Y. PI. 1982. Phys. Rev. Lett. **49:** 1110.
61. HAWKING, S. W. 1982. Phys. Lett. **115B:** 295.
62. BARDEEN, J. M., P. J. STEINHARDT & M. S. TURNER. 1982. Spontaneous Creation of Almost Scale-Free Density Perturbations in an Inflationary Universe. Preprint UPR-0202T. University of Pennsylvania. Philadelphia, Pa. (To be published in Phys. Rev. **D.**)
63. HARRISON, E. R. 1970. Phys. Rev. **D1:** 2726.
64. ZELDOVICH, YA. B. 1972. Mon. Not. R. Astron. Soc. **160:** 1P.

Microcosmology

New Particles and Cosmology

HEINZ R. PAGELS

The Rockefeller University
New York, New York 10021

We review the role of cosmology and astrophysics in providing information on the properties of new hypothetical particles that emerge as field quanta from unified relativistic quantum field theories. Our emphasis is on supersymmetric field theories and the gravitino and photino, but we also discuss the axion and majoron. Other topics include the role of heavy quanta and the galaxy formation and origin problems.

This article will review the role that some of the known and hypothetical particles of high-energy physics play in astrophysics and cosmology. Such a review is necessarily speculative. The common ground of high-energy physicists and the relativistic cosmologists is the early universe—a region of energy and momentum, time and space, for which they have the least confidence about their knowledge of what is going on. Yet such a circumstance is what creates the intellectual challenge.

Particle physicists have discovered that even some of their wildest models of the quantum particles, models that extrapolate by factors of 10^{16} or more beyond the energy range examined by terrestrial accelerators, can be tested by comparison to the evidence about the state of the very early universe and its subsequent evolution. We have already entered the era of postaccelerator physics, where the entire history of the universe becomes the proving ground for fundamental physics. Particle theorists, accustomed to comparing the numerical predictions of their theories against numbers provided by their experimental colleagues, now compare the *logarithms* of such numbers against the numbers provided by cosmologists and astrophysicists. Even with such numerical latitude no current model of high-energy phenomena can be said to fit all the facts of the very early universe.

From another viewpoint, cosmologists and astrophysicists find that the new unified field theories and the particles they predict may provide answers to some of the outstanding puzzles they confront—the origin of the universe, the distribution of possible dark matter, the structure and origin of galaxies, to name but a few. Yet some cosmologists feel there is an arbitrariness in the new field theory models and that clever particle theorists can merely conjure up any particle out of theory with properties made to order. Such feelings reflect the fact that the sense of necessity is not universally shared.

In this brief review we will examine one aspect of the interplay of cosmology and high-energy physics—the role of new quanta with special emphasis on supersymmetry. Much of this work is due to that tribe of physicists who feel equally at home in cosmology, astrophysics and fundamental particle physics. It is a tribe whose numbers are growing.

First we will outline a few of the predominant ideas about unified field theories with emphasis on the spectrum of particles. This is followed by a review of the role of new heavy particles and then possible light particles in cosmology. The reader should consult the numerous references for details.

SUPERSYMMETRY

The basic idea of supersymmetry is that the fundamental field theory Lagrangian possesses equal numbers of bose and fermi degrees of freedom and that the fields that represent those degrees of freedom can be transformed into one another by a supersymmetry transformation (supersymmetry has been reviewed in Reference 1). The transformation can either be a global one or local (depending on the space-time point x). We will discuss these possibilities in turn. But first a few words about why high-energy theoretical physicists are at all interested in supersymmetry.

There is not a single piece of experimental evidence that supports supersymmetry. Give up supersymmetry and you give up nothing as far as observed phenomena are concerned. Why then is there such interest in supersymmetry?

The attraction arises mainly from the theorist's ambition to build a unified field theory of all quanta. One of the profound discoveries of the last two decades is that gauge symmetries imply interacting fields with associated quanta—the photon, the colored gluons of the quark binding interaction, and the weak bosons. These gauge symmetries are the basis for the unification of electromagnetic, strong, and weak interactions. Yet in spite of the unified field theories that emerged from the gauge symmetries, fermi fields like those associated with electrons, neutrinos, or quarks were not incorporated in a fundamental way. They seemed superfluous. In supersymmetric gauge theories the fermions are treated on an equal footing with the bosonic gauge fields. They, too, are a consequence of symmetry and this seems to many people a correct step toward unification.

Another attractive feature of supersymmetry is that it may be part of the solution to the notorious "hierarchy problem." This problem is the modern version of the old problem of understanding the magnitude of the Dirac-Eddington numbers. The problem is easy to state. The mass scales associated with various quantum interactions have the values

Gravitational	$M_{\text{Planck}} =$	1.3×10^{19} GeV
The scale of grand unified theories (GUTs) of electromagnetic, weak, and strong interaction	$M_{\text{GUT}} \simeq$	10^{14} GeV
Mass scale of weak interactions	$M_{\text{Weak}} \simeq$	10^{2} GeV
Mass scale of strong interactions	$M_{\text{Strong}} \simeq$	1 GeV

and have the ratios

$$M_{\text{Planck}}/M_{\text{GUT}}/M_{\text{Weak}}/M_{\text{Strong}} \simeq 10^{19}/10^{14}/10^{2}/1$$

The hierarchy problem is to understand these ratios on the basis of a fundamental theory (and not just put them into the theory "by hand"). The presumption is that these ratios can be understood as manifestations of a broken symmetry. There are suggestions that supersymmetry might help solve this problem.[2-4]

A further outstanding problem is the "cosmological term problem"—why the cosmological term is so small, if not zero. Supersymmetry is the only known symmetry that is sensitive to the presence of a vacuum energy or cosmological term and this seems to hint at a solution to this old problem. Yet it remains unsolved.[5-7]

Although supersymmetry is attractive from the viewpoint of unification it has so far failed to solve any fundamental theoretical problem and no realistic model of quantum interactions that incorporates supersymmetry exists.[8-11] Many theorists feel some crucial idea is still lacking and when that idea is found a major step in the attempt to unify physics will have been taken.

Global Supersymmetry

The generators of global supersymmetry transformations are a set of spinor charges Q^a, $a = 1,2,3,4$, which are elements of a Grassmann algebra. They obey the fundamental anticommutation rule

$$[Q^a, \overline{Q}^b]_+ = 2\gamma_\mu^{ab} P_\mu \tag{1}$$

where $\overline{Q} = Q^{a\dagger} \gamma_0^{ab}$, γ_μ^{ab} with $\mu = 0,1,2,3$ are the Dirac matrices and P_μ is the generator of space-time translations. This relation, along with the more familiar commutation rules of P_μ and $J_{\mu\nu}$, the generator of Lorentz transformations, comprises the graded-Poincaré algebra. It follows from Equation 1 that the Hamiltonian

$$P_0 \equiv H = \frac{1}{4} \sum_{a=1}^{4} (Q^a)^2 \geq 0 \tag{2}$$

is positive and the supercharges are conserved

$$\dot{Q}^a = i\,[H, Q^a] = 0 \tag{3}$$

If we assume supersymmetry, i.e., Equation 1, then there are two possibilities—either the vacuum state $|0\rangle$ is supersymmetric or it is not. If the vacuum is supersymmetric so the supercharges Q^a annihilate it

$$Q^a|0\rangle = 0 \tag{4}$$

then Coleman's theorem—"a symmetry of the vacuum is a symmetry of the world"—requires that the spectrum of states be supersymmetric. This means that for every bosonic state there is a degenerate fermionic state and vice versa. The proton, for example, would have to have a bosonic partner.

Clearly this is in complete conflict with the observed particle spectrum so it is generally assumed that the vacuum is not supersymmetric,

$$Q^a|0\rangle \neq 0 \tag{5}$$

so supersymmetry is spontaneously broken in the vacuum. The order parameter for this spontaneous breaking is just the vacuum energy associated with the energy-momentum tensor $\theta_{\mu\nu}$

$$\langle \theta_{\mu\nu} \rangle_0 = \frac{1}{2} F^2 g_{\mu\nu} \quad (6)$$

so that

$$\langle H \rangle_0 = \frac{1}{2} F^2 \, \delta^3(0) \quad (7)$$

That F is a suitable order parameter follows from the fact that if the vacuum is supersymmetric, then using Equations 2, 4, and 7 one concludes $F = 0$, and if $F \neq 0$ then we must have vacuum breaking, Equation 5.

We assume $F \neq 0$ to avoid conflict with the degeneracy problem. Then the mass splitting of ordinary fermions and bosons from their superpartners is given by formulas of the form

$$\Delta m^2 = eF \quad (8)$$

where e is some coupling constant and F is the order parameter. If e is not too small and F is large, the splitting can be made large. No superpartner of any observed particle has been observed, and this fact places strong constraints on models of spontaneously broken supersymmetry and the F parameter. The simplest models do not work. It has been argued that no spontaneously broken globally supersymmetric model works[8–11] and that one must incorporate gravity (which gets us into local supersymmetry). This results in an effective low-energy theory in which supersymmetry is explicitly broken.

Although the assumption that supersymmetry is spontaneously broken solves the degeneracy problem it also created a new problem: the existence of a massless spin = ½ state called the Goldstino. As is well known (according to the Goldstone theorem), spontaneous symmetry breaking implies massless particles and supersymmetry is no exception. The Goldstino can not be identified with the neutrino.[12,13] This unwanted massless state can be eliminated if we now turn to considering local supersymmetry—at each point in space-time one requires supersymmetry.

Local Supersymmetry

Local supersymmetry in its simplest version implies supergravity—the theory of interacting bosonic gravity with spin 3/2 gravitinos—the superpartner of the spin 2 graviton (reviewed in Reference 14). If one quantizes the theory in the background of flat space-time, both the graviton and gravitino are massless. Alternatively one can consider supergravity with a cosmological term $-3\lambda^2$ corresponding to anti–de Sitter space, in which case supersymmetry requires a gravitino mass term[5–7]

$$M_{g3/2} = \kappa\lambda \quad (9)$$

where $\kappa = (8\pi G)^{1/2}$ with G the Newton-Cavendish constant. (One considers anti–de Sitter space so this mass is real.) If one adds to the Lagrangian of this supergravity theory the Lagrangian of spontaneously broken matter that has a vacuum energy $\frac{1}{2} F^2$, the effective cosmological term in the action is the sum of this vacuum energy and the anti–de Sitter term

$$\Lambda^2 = \frac{1}{2}F^2 - 3\lambda^2 \tag{10}$$

Requiring this cosmological term to vanish, $\Lambda = 0$, in agreement with the observation that it is very small, we find from Equations 9 and 10 for the gravitino mass the formula

$$M_{g_{3/2}} = \frac{\kappa F}{\sqrt{6}} \tag{11}$$

which is a true mass, since the background metric can be taken as flat. Furthermore one can show in complete analogy to the Higgs effect in conventional gauge theories that the two spin ½ degrees of freedom of the (unwanted) Goldstino state have combined with the two spin ³⁄₂ states of the massless gravitino to produce a massive gravitino with four helicity states. This is the super Higgs mechanism,[15] and it solves the problem of removing the massless Goldstino state from the spectrum (although for many purposes, to a good approximation, the Goldstino does not effectively decouple).[16,17] There remains the puzzle of why the effective cosmological term (**10**) is zero or nearly so.

What kind of spectrum results from a spontaneously broken local supersymmetry? In general the answer depends on the model and can be quite complicated. A minimal answer would be that each of the fundamental fields in the conventional model now has a superpartner.

Supersymmetric theories possess a discrete symmetry called R invariance such that $R^2 = 1$, so $R = \pm 1$. Ordinary particles are classified as R even $(+1)$ and their superpartners as R odd (-1). In FIGURE 1 are indicated the quanta of the standard model based on the gauge group $SU(3) \times SU_L(2) \times U(1)$ plus the graviton as R even states and their associated R odd superpartners. This figure primarily serves the purpose of giving the nomenclature for the superpartners (which is not generally agreed upon). It is known that the supersymmetric version of the standard model fails phenomenologically and that the gauge symmetry must be extended to include at least another U(1) group.[16,17] Even this extension may fail.[8–11]

In FIGURE 2 we indicate the possible role of supersymmetry in the unification of physics. It is known that as the temperature increases (appropriate for the description of the very early universe), the broken symmetries that distinguish the various interactions—strong, weak, electromagnetic, and possibly gravity—are restored. Broken supersymmetry is an exception—at high temperature, supersymmetry breaking increases and the order parameter gets larger.

The reason for this exception is not difficult to understand. At finite temperature bose and fermi statistics are different and this difference is exaggerated as the temperature increases. Hence bosons and fermions are treated differently at even the highest temperatures and consequently supersymmetry is broken for all temperatures.

At the bottom left of FIGURE 2 is the world of zero temperature corresponding to the standard model of interacting quarks and leptons and their associated gauge fields. Included in this scheme are possible topological defects—monopoles—which are stable

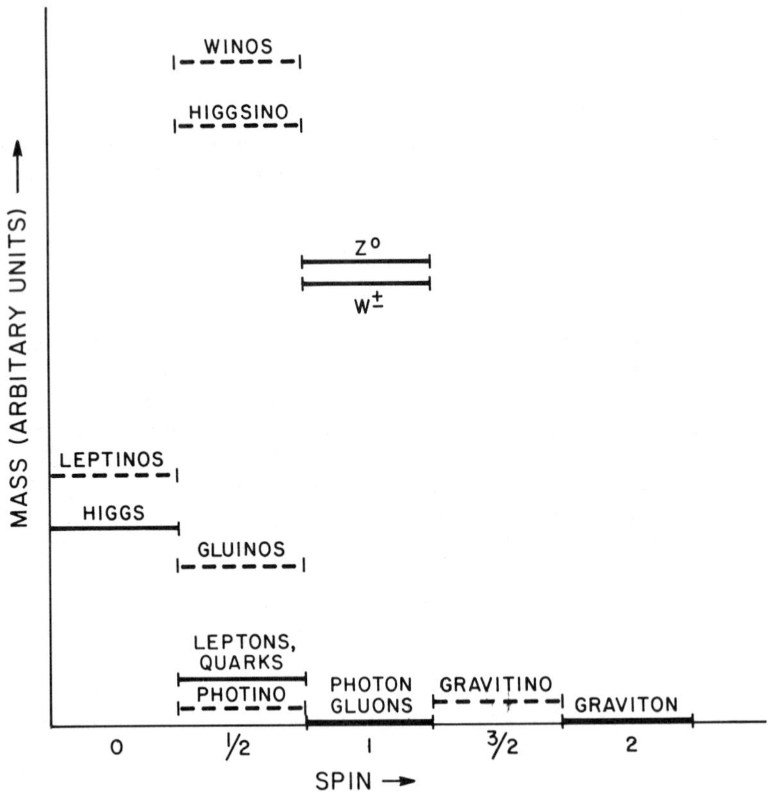

FIGURE 1. Super spectrum. States of the standard model (solid lines) and their superpartners (dashed lines). The indicated level spectrum of the superpartners is quite arbitrary; mostly it is intended to give the names of these states.

and occur in most gauge models. To the right is the standard model reflected in the "supermirror," which interchanges bosons and fermions. None of these objects, such as *R*-hadrons (the superpartners of ordinary hadrons), leptinos, photinos, quarkinos, have been detected.

At higher temperatures the distinctions between the various low-energy interactions are lost and the world is described by a GUT (grand unified theory) that at least contains SU(5) as a subgroup (GUTs are reviewed in Reference 18). At these temperatures the supersymmetric extensions of GUTs called SUSY GUTs (supersymmetric GUTs) may play a role.[19,20] Finally at higher temperatures corresponding to the Planck temperature gravity is presumably unified with the other interactions. The goal of the model builders is to find a TUT (totally unified theory) that unifies everything. Some people think that $N = 8$ supergravity is a candidate.[21,22]

While it is important to have such a unifying framework of all physics before us it is also important to bear in mind how little of this picture is actually verified. Particle theorists realize that the only hope of verifying some of these ideas is to turn to cosmology and astrophysics, and that is what we will do here. An important question that confronts the supersymmetric model builder is what is the value of $F^{1/2}$? Is it of the order MeV or less? (This is ruled out by the spectrum.) Is it of the order GeV. (Not likely.) Is it of the order TeV? (Possible.) Or does one have to go all the way to the Planck mass M_p? Before addressing these and other questions we will first examine the role of heavy particles in cosmology.

HEAVY PARTICLES AND COSMOLOGY (MASS $\gtrsim$ TeV)

Since all heavy particles have decayed away long ago, unless they carry some new conserved or almost conserved charge they are not present today. Therefore we have to look at the early universe when the temperature was very high or the very old universe

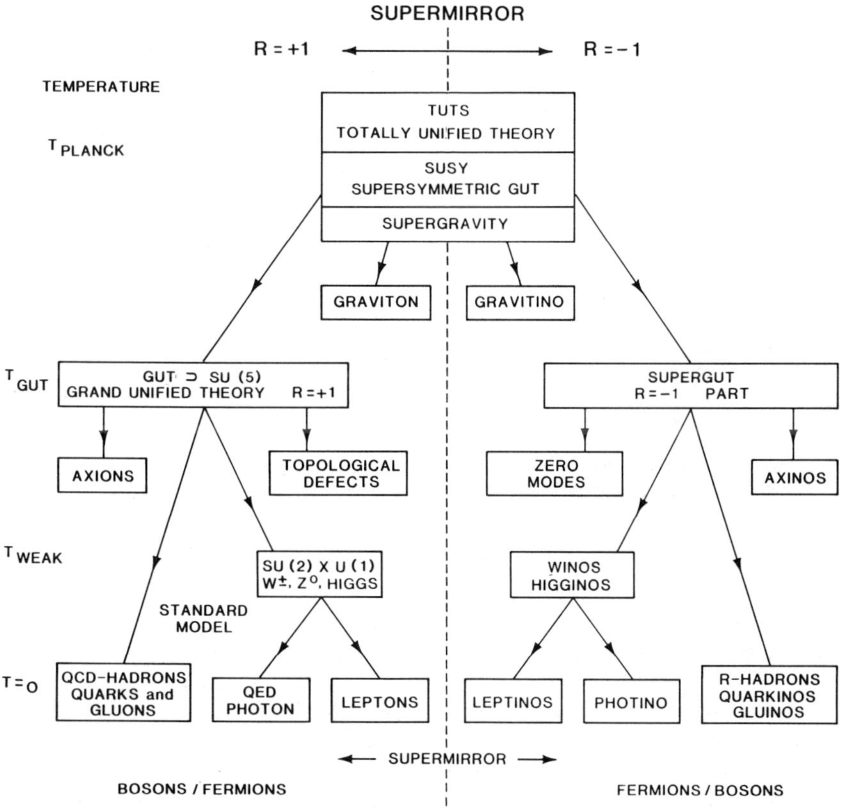

FIGURE 2. Through the looking glass: the unification of physics, including supersymmetry (ca. 1982). A consequence of supersymmetry is a doubling of the spectrum—each state of the GUT model is reflected into its corresponding superpartner. The unification proceeds as the temperature increases, but supersymmetry is never restored.

to see the cosmological effects of heavy particles. There are two important ways that heavy particles influence cosmology.

Degrees of Freedom in the Early Universe

In building models of the very early universe it is important to know the effective number of degrees of freedom $g(T)$ as a function of temperature T. As the temperature increases from zero to several GeV, more particles go into equilibrium as interaction thresholds are crossed. The number of massive particles thus directly influences the number of degrees of freedom. The possibilities are shown in FIGURE 3.

The standard model with three families of quarks and leptons has about 10^2 degrees of freedom above 10^2 GeV. In GUTs, this might increase by a factor of 2 or 3, and in

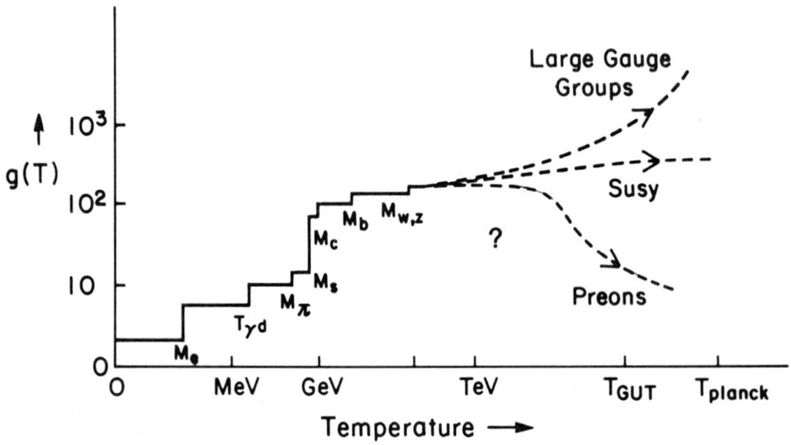

FIGURE 3. Quantum degrees of freedom vs. temperature. In the standard model there are about 100 degrees of freedom at temperature 100 GeV. Depending on the dynamical model, this number can increase or decrease for very high temperatures.

SUSYs with equal bosonic and fermi degrees of freedom by another factor of 2. Most models considered in the literature predict that at infinite temperature or effectively the Planck temperature there are $(5 \pm 3) \times 10^2$ degrees of freedom.

Conceivably very large gauge groups[23] come into play as the temperature becomes very large and the number of degrees of freedom might be as high as 10^3 or 10^4. An interesting (but unexamined) possibility is that the number of degrees of freedom becomes infinite at a large finite temperature.

Alternatively the number of effective degrees of freedom can fall for temperatures exceeding a TeV. This would be the case if the quarks, leptons, and gauge fields are bound states of more fundamental entities such as preons (reviewed in Reference 24) and melted into them at sufficiently high temperature—a phase transition.

There is a lot of freedom in picking $g(T)$ at high temperature. Field theory models specify $g(T)$, and it would be of interest for cosmologists to set bounds in this function, even crude bounds, to restrict the models.

Rare Decays at Low Energy

Heavy particles carrying exotic quantum numbers can influence low-energy physics by causing otherwise stable particles to decay. An old example is β-decay, and current interest is in possible proton decay. The heavy gauge fields in GUT models can violate baryon conservation causing proton decay and conversely proton (baryon number) synthesis. Baryon synthesis may occur in the very early universe, while proton decay is important for understanding the late open universe. Recently preliminary experiments on the lifetime of the proton imply the bound[25]

$$\tau_p > 6 \times 10^{31} \text{ years} \tag{12}$$

Two important results of GUTs are that they give a value for the proton lifetime and the weak interaction angle θ_W, which is known experimentally. The minimal SU(5) GUT implies $\tau_p \sim 2 \times 10^{29 \pm 1.7}$ years and a value of θ_W in agreement with experiment.[26] This model is already ruled out. The results of proton decay experiments currently in progress will improve by at least an order of magnitude the bound (**12**).

Modifications of the minimal SU(5) GUT have been proposed.[26] In a SUSY model with two Higgs doublets $\tau_p \sim 10^{35}$ years (out of reach for the experiments) but θ_W is too big. If one considers a SUSY with four Higgs doublets θ_W is alright but $\tau_p \sim 10^{30}$ years, which is already ruled out.

A further possibility is that the usual massive gauge field exchange process does not dominate the proton decay process, and one must examine dimension five loop induced operators. If this is the case proton decay depends on unknown Yukawa coupling constants and θ_W is likewise undetermined—little can be precisely predicted.

LIGHT PARTICLES AND COSMOLOGY (MASS $\lesssim$ keV)

The principal light particles in cosmology are the photon and the neutrino (currently only the experiment of Lubimov *et al.* finds a value of 35 eV for the electron neutrino mass).[27] The role of these observed quanta in cosmology has been extensively described,[28,29] and we will not be reconsidering it here. Instead our focus will be on the unobserved quanta that current supersymmetric models require—the gravitino and photino and a few other exotica not related to supersymmetry, axions and majorons.

What are the cosmological data that constrain the mass and interactions of quantum particles? There are several constraints.

1. A fundamental constraint on dissipationless light neutral particles is the bound on the mass density of the universe

$$\rho < \rho_c$$

where ρ_c is the critical density (or conservatively, $10\,\rho_c$) of matter in the F-R-W model. The density is $\rho = mn$ where m is the mass of the particle and n the number density which can be computed in the F-R-W model from the number of degrees of freedom of the particle and the number of degrees of freedom, $g(T^*)$ when it dropped out of equilibrium (this is discussed in Reference 30). This bound, for example, places a restriction on the sum of the masses of neutrinos to be $\leqq 100$ eV (see Reference 31 for a recent discussion; also see References 28 and 29).

2. Dissipationless particles cluster gravitationally on mass scales of the order of the maximum Jeans mass

$$M_{\text{Jeans}}^{\text{Max}} \cong \frac{M_{\text{pl}}^3}{m^2}$$

One may identify this Jeans mass with the invisible mass bound to galaxies or clusters of galaxies. There is good evidence that galaxies have invisible halos which dominate their mass.[32] Furthermore, there is good evidence that galaxies cluster (see References 33 and 34). The mass of galaxies, including the halo, is $\sim 10^{12}\ M_{\odot}$ and identifying this with the Jeans mass implies $m \sim 1$ keV. For large clusters the mass can be as large as $10^{15}\ m_{\odot}$ which implies $m \sim 30$ eV. Both of these are suggestive masses for the particle physicists.

3. Nucleosynthesis, especially the synthesis of helium, provides an important cosmological constraint.[35] The amount of helium produced in the big bang depends sensitively on the expansion rate and the interactions of light particles. For example this constraint requires that the number of neutrino species be $\leqq 4$ (this is reviewed in Reference 36).

4. Stellar evolution during its late stages provides other constraints—the emission rates for light particles may increase rapidly with temperature. The modeling of red giant stars requires that the incoherent production of neutrals (x) in the process

$$\gamma + e \rightarrow x + e$$

is bounded (a discussion is in Reference 37; see also References 38 and 40)

$$\sigma_x \lesssim 10^{-23}\ \text{GeV}^{-2}$$

This leads to a useful constraint on the axion and majoron couplings. One may also consider carbon burning stars or the cooling rates for neutron stars.[38]

In what follows we apply these constraints to the properties of hypothetical particles. It is important to develop a large list of constraints such as those above and others (like the constraint on the monopole flux and baryon density). Such cosmological constraints are today the major restraint on the imaginations of field theory model builders.

The Gravitino

The gravitino is the massive spin 3/2 superpartner of the graviton. Its primary property is that it decoupled very early in the thermal history of the universe. The mass of the gravitino is directly related to the order parameter F of supersymmetry breaking by $m_{g_{3/2}} = \kappa F/\sqrt{6}$. The energy $\sqrt{F}$ is approximately the energy at which most superpartners of ordinary particles will make their appearance in the spectrum.

There are two open windows on the allowed mass of the gravitino: a light one and a very heavy one. The heavy gravitino is unstable and decayed away long ago, while the light one is stable or quasi-stable and, if it exists at all, exists today. The basis for the bounds shown in FIGURE 4 is the following.

1a. Assuming that the mass of the photino and gravitino obey $M_{\lambda_{1/2}} + M_{g_{3/2}} < m_\psi = 3.1$ GeV where ψ is the ground state of charmonium, then from the known branching ratio of ψ into unobserved neutrals such as $\psi \rightarrow \overline{\gamma}_{1/2} + g_{3/2}$ one finds[39] $m_{g_{3/2}} > 1.5 \times 10^{-8}$ eV.

1b. This bound can be improved by an order of magnitude by considering neutrino emission from white dwarfs:[40] $M_{g_{3/2}} > 4.3 \times 10^{-7}$ eV.

1c. If the photino interacts weakly like an extra neutrino so that there are effectively four neutrino species (a strong assumption), then there is no room for gravitinos, which can spoil the He abundance, and one obtains a vast improvement:[41] $M_{g_{3/2}} > 10^{-2}$ eV.

2. Assuming the gravitino is stable—a light gravitino—the cosmological bound on the matter density implies $M_{g_{3/2}} \lesssim 1$ keV.[24] If this bound is saturated we are living in a gravitino-dominated universe and, interestingly, the mass value of 1 keV corresponds to a maximum mass corresponding to the mass of galaxies.

3. Heavy gravitinos decay into other R odd particles. This decay must destroy any produced He and restore the neutron-proton ratio to its equilibrium value. If the heavy gravitino is too light, this will not occur and one finds a model-dependent lower bound $M_{g_{3/2}} > 10^{5\text{-}13}$ GeV.[43] It is certainly possible for the gravitino to have a Planck mass.

We note that this latter bound eliminates geometric hierarchy models.[44] These models generate the weak mass scale $M_w \sim 10^2$ GeV as a radiative correction to a high mass scale $M_x \lesssim 10^{17}$ GeV. The typical relation is $M_w \sim (\alpha/\pi)\, F/M_x$ where α is on the order of the fine structure constant $\sim 10^{-2}$. This implies using Equation 11 $M_{g_{3/2}} \cong 10^5$ GeV (M_x/M_{pl}). The bound on the heavy gravitino mass $M_{g_{3/2}} > 10^5$ GeV then requires $M_x > M_{pl}$ which is inconsistent with the requirements of the model.

The Photino

The photino is the spin ½ fermionic partner of the photon. All the bounds we have discussed for the gravitino also apply to the photino. Like the gravitino the photino can

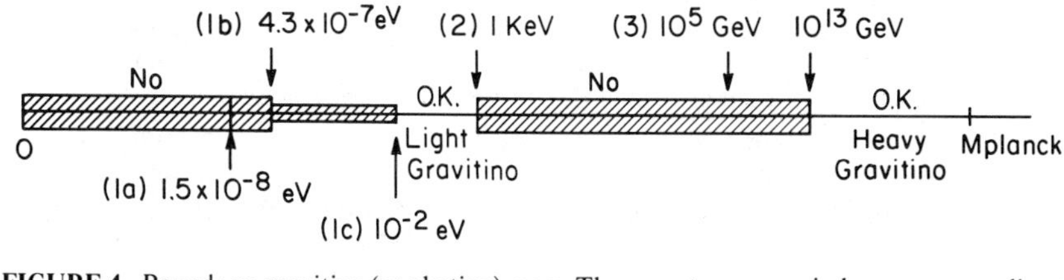

FIGURE 4. Bounds on gravitino (or photino) mass. There are two open windows corresponding to a light gravitino (stable or quasi-stable) and a heavy one (unstable). The light gravitino or photino could be cosmologically important.

be either light or very heavy. If one makes the extra assumption that the photino couples with about the same weak interaction strength as ordinary neutrinos then the upper bound on the light photino is reduced to $M_{\gamma_{1/2}} < 100$ eV instead of 1 keV.[41]

If supersymmetry is at all relevant to physics, then photinos and gravitinos may play an important role in cosmology. But which of these two particles is relevant? If both particles are heavy then they are not of interest except in the very early universe with temperature $T > 10^5$ GeV. If one is heavy and the other is light, the heavy one will decay into the light one which is the lightest R odd state. The light particle could then dominate the mass density of the universe and play an important role in galaxy formation.

An interesting possibility is if they are both light. Then the heavier one decays according to

$$g_{3/2} \rightarrow \gamma_{1/2} + \gamma$$

or

$$\gamma_{1/2} \rightarrow g_{3/2} + \gamma$$

The resulting monochromatic photon can contribute significantly to the x-ray background.[45] There is (disputed) evidence for such photons. The lifetime of the unstable one can be calculated with the result[46]

$$\frac{\tau}{\tau_0} \cong \left(\frac{M_{g_{3/2}}}{1\ \text{keV}}\right)^2 \left(\frac{1\ \text{MeV}}{M_{\gamma_{1/2}}}\right)^5$$

where τ_0 = present age of universe $\sim 10^{10}$ years. If the lower bound $M > 10^{-2}$ eV is valid, both these quanta must exist today since $\tau/\tau_0 > 10^5$; but if we take only the weaker lower bound $M > 10^{-7}$ eV, then the heavier particle may have decayed away long ago.

The Axion

Unlike the quanta we discussed above the axion has nothing to do with supersymmetry. Instead it was introduced into particle physics to render the current theory of the strong interaction—quantum chromodynamics (QCD)—automatically CP or time reversal invariant.

An unfortunate feature of QCD with massive quarks (all the quarks seem to possess a small intrinsic mass) is that the most general form of the field theory consistent with renormalizability violates time reversal invariance. This is intolerable experimentally unless for some reason the amplitude is extremely small. It turns out that by introducing a new complex pseudoscalar field into QCD, the time reversal violating phases can be removed by rotating them into the complex phase of the new field.[47,48] While this rids QCD of potential strong CP violations the price of this solution is the existence of a new field—the axion.[49,50]

If the axion is too massive it would have been "visible" and shown up in beam dump experiments. From a study of stellar evolution and the properties of red giants it is known that the axion mass M_A is bounded above[51–53]

$$M_A \lesssim 10^{-2}\text{ eV}$$

Axions satisfying this bound were thought to be "invisible." Recently a lower bound has been established on the assumption that axions dominate the matter density of the universe. This bound is established as follows.[54–56]

The mass of the axion can be shown to be given by

$$M_A = M_\pi f_\pi / \text{v}$$

where $M_\pi = 140$ MeV is the pion mass, $f_\pi \sim 90$ MeV is the pion decay constant, and v is the vacuum expectation value of the axion field

$$\langle \phi_A \rangle_0 = \text{v}e^{i\theta}$$

where θ is the CP violating phase parameter. The axion field $\phi_A(t)$ behaves as a coherent field with an approximate time dependence

$$\phi_A(t) \simeq A\text{v} \cos (M_A t)$$

where A is the amplitude. Assuming that this coherent field does not dominate the mass density of the universe so $\rho < 10\,\rho_c$ implies that today $A^\circ \leqq 10^{-21}$. Furthermore

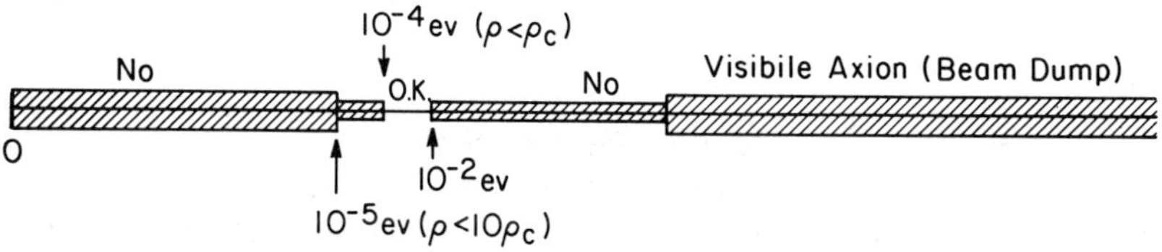

FIGURE 5. Bounds on the axion mass. Only a small window is open. Is it possible to close the window?

from the standard Freedmann cosmology and QCD one can show

$$A^\circ = 10^{-21}\left[\frac{\text{v}}{10^{12}\text{ GeV}}\right]^{7/12}$$

and hence $\text{v} < 10^{12}$ GeV which implies

$$M_A > 10^{-5}\text{ eV}$$

very close to the lower bound. These bounds are shown in FIGURE 5. If we are not so conservative and assume $\rho < \rho_c$ then

$$M_A > 10^{-4}\text{ eV}$$

These cosmological bounds narrow the range of allowed axion masses to the narrow window $10^{-2}\text{ eV} > m_A > 10^{-4}$ eV. Is it possible to close the window?

Majorons

If neutrinos are massive majorana particles[57,58] (rather than massive Dirac particles) then lepton number is not conserved. There are three possibilities for the origin of this nonconservation.

1. A majoron mass term put in "by hand."
2. A spontaneously broken local symmetry resulting in a massive vector boson.
3. A spontaneously broken global symmetry resulting in a massless Goldstone particle (the majorons).

While one may or may not consider the actual existence of majorans reasonable there certainly is not any compelling theoretical necessity for their existence. Nonetheless they raise the interesting and recurrent question, What are the cosmological bounds on the properties of massless particles or nearly massless particles?

Majorons like axions are pseudoscalar particles and couple to electrons with coupling constant g_ϕ. The electronic mass generated by the spontaneous symmetry breaking is

$$M_e = g_\phi u^2/2\text{v}$$

where $u = 250$ GeV and v is the vacuum value of the majoron field. The best bounds on g_ϕ come from the allowed emission rates for red giants, which imply $g_\phi^2/4\pi \lesssim 3 \times 10^{-26}$ or $\text{v} < 40$ keV.[38,40] While carbon-burning stars and neutron stars also imply bounds, the uncertainties are numerous.[38]

GALAXIES AND INOS

One of the interesting developments in microcosmology is the implication of the new particles for the old problem of the origin and formation of galaxies. The growth of fluctuations from dissipationless particles has been extensively studied especially since the time a small neutrino mass became theoretically fashionable (with some experimental support.) The basic idea is rather simple. One assumes the mass of the universe is dominated by some light neutral particle which decoupled early in the thermal history of the universe. Some initial spectrum of fluctuations in the gas of these particles is presumed. The fluctuations exhibit growing modes, and these can be calculated in the standard model. At some time the density contrast $\delta\rho/\rho$ of the growing fluctuations becomes unity and complicated nonlinear effects take over. Baryonic matter is gravitationally coupled to these fluctuations and after lumps have formed the infalling baryonic matter dissipates, initiating complicated processes eventually leading to true galaxy formation.

Growth of Fluctuations from Dissipationless Particles

Currently there are two sets of dissipationless particles being considered by theorists in the galaxy formation problem: (1) the (massive) neutrino with $m \sim 30$–100 eV;[28,29] (2) the gravitino (or photino) with $m \sim 1$ keV.[42,59] While we may consider the possibility that both components are present together, we will not do so here. Instead, the contrast between these possibilities is instructive as it reveals rather different pictures of galaxy formation, which can be observationally distinguished.

The clustering mass for massive neutrinos is $\sim 10^{15}\ M_{\odot}$, about the mass of a large cluster of galaxies, while for gravitinos it is $\sim 10^{12}\ M_{\odot}$, the mass of a galaxy if we include the nonluminous halo (which is the gravitino gas in this picture). Observationally, both clusters of galaxies and galaxies exist, so that on these grounds alone there is no way of distinguishing between the two pictures. However, in the neutrino picture galaxies can only form by dissipational fragmentation of a large pancake cluster while in the gravitino picture the cluster forms dissipationlessly only after galaxy formation. These different mechanisms have observational consequences (our discussion follows Reference 60).

First, there is no evidence for excessive flattening in superclusters as would be expected in the neutrino-pancake model. Second, the absence of Ly_{α} absorption lines in quasar spectra is consistent with the gravitino picture. Third, if galaxies have large nonluminous halos and a mass-to-luminosity mass ratio $M/L \sim 50$, then there is no more than a factor of two increase in the total-to-baryonic mass ratio in going from the scale of galaxies to that of clusters. This is consistent with the suggestion from data that this mass ratio is approximately constant for all scales larger than galaxies, as is expected in the gravitino picture but not the neutrino picture. Finally, recent observations on the Draco dwarf galaxy indicate a ratio $M/L \sim 80$.[61] The phase-space constraints are very strong for dwarf galaxies and an analysis suggests that the nonluminous particles in the Draco halo have a mass $\gtrsim 600$ eV.[62] This supports the gravitino picture.

While it is too soon to tell, some evidence supports the existence of dissipationless particles of mass ~ 1 keV but there is little to support a neutrino of mass ~ 30 eV. Exacting experiments to determine or bound the neutrino mass are currently under way and ought to settle this issue of a cosmologically significant (electron) neutrino.

Origin of Galaxies in Inflationary Universes

While the growth of fluctuations can be examined in detail, such a study leaves unanswered the question of the actual origin of these fluctuations and their spectrum. This is the problem of the origin of galaxies and we will discuss it in the context of the inflationary universe. If there is a large inflation then all topological defects—monopoles or strings—are inflated away and cannot play a role in creating the fluctuations. However, there remains the possibility of quantum and thermal fluctuations. Such fluctuations give rise to a density constant $\delta\rho/\rho$. In a model, this quantity—evaluated as a fluctuation enters the horizon—must have the magnitude

$$\delta\rho/\rho)_{H} \sim 10^{-4}$$

or galaxies will fail to form.[63,64]

Quantum fluctuations in the standard model cannot give rise to a density contrast as large as this. Typically quantum fluctuations go as $(N)^{-1/2}$ where N is the number of particles within the horizon $\sim 10^{60}$, so $\delta\rho/\rho)_{H} \sim 10^{-30}$ is many orders of magnitude too small.

One might think that any quantum fluctuations in the inflationary universe would get inflated away. But this is not the case. A gauge invariant combination of amplitudes has the property that as the fluctuation leaves the horizon it gets "frozen out" so that when it reenters it has the same amplitude and a scale invariant spectrum.[65,66] Evidently some quantum fluctuations can survive the inflation and a calculation indicates[67]

$$\delta\rho/\rho)_H \sim 42N\left(\frac{T_{GUT}}{T_{pl}}\right)^3 \sim 10^{-4}$$

where N is the number of degrees of freedom $\sim 10^3$, T_{GUT} is the inflation temperature 1.6×10^{16} GeV, and $T_{pl} \sim 1.3 \times 10^{19}$ GeV. So quantum fluctuations could give rise to galaxies.

Thermal fluctuations in the scalar field that describes the matter in the inflationary model also contribute to $\delta\rho/\rho)_H$. These fluctuations are large during the phase transition from the inflationary phase to the *FWR* phase, essentially due to the lack of simultaneity in the phase transition (it does not happen everywhere all at once). They have been estimated to be[68–71]

$$\delta\rho/\rho)_H \sim \lambda^{1/2}\,(n_{ef})^{3/2} \sim 50$$

where $\lambda \sim 0(1)$ is the self-coupling of the scalar field and $n_{ef} \sim 60$ is the number of e foldings of the inflation. Significantly, this spectrum is scale invariant.

Clearly this number is far too large and can be a serious problem for the inflationary models. One can reduce its magnitude by making the phase transition very gentle, but then the universe does not reheat. Should this problem be solved and someone succeeds in making the scalar fluctuations somehow smaller, then the quantum fluctuations (which have the right magnitude) might compete with or even dominate the scalar fluctuations. We have for the first time the problem of reducing $\delta\rho/\rho$ rather than making it larger. This seems like progress.

How many of these ideas will survive the year or even the decade is not clear. What is clear is that cosmology and astrophysics are providing the major proving ground for current ideas in fundamental field theory.

ACKNOWLEDGMENT

It is my pleasure to thank my collaborator and superpartner Prof. Joel Primack of the University of California at Santa Cruz for many enlightening discussions.

REFERENCES

1. Fayet, P. & S. Ferrara. 1977. Phys. Rep. **32C:** 249.
2. Witten, E. 1981. Nucl. Phys. **B188:** 513.
3. Dimopoulis, S. & H. Georgi. 1981. Nucl. Phys. **B193:** 150.
4. Sakai, N. 1982. Z. Phys. **C11:** 153.
5. Pagels, H. 1981. *In* Gauge Theories, Massive Neutrinos and Proton Decay. A. Perlmutter, Ed.: 121. Plenum Press. New York, N.Y.

6. DESER, S. & B. ZUMINO. 1977. Phys. Rev. Lett. **38:** 1433.
7. DAS, A. & D. Z. FREEDMAN. 1977. Nucl. Phys. **B120:** 221.
8. FARRAR, G. & S. WEINBERG. Report No. RU-82-38. Rutgers University–Texas University. Rutgers, N.J. (To be published.)
9. LAZARIDES, G. 1982. Cosmological Constraints on GUTs. Report HO RU82/B/40. Rockefeller University. New York, N.Y.
10. LAZARIDES, G. & Q. SHAFI. 1982. Phys. Lett. **115B:** 21.
11. ALBRECHT, A., S. DIMOPOULIS, W. FISCHLER, E. W. KOLB, S. RABY & P. J. STEINHART. 1982. Cosmology, Inflation and Supersymmetry. Preprint LA-UR-82 2947. Los Alamos Scientific Laboratory. Los Alamos, N. Mex. (To be published in Nucl. Phys. B.)
12. BARDEEN, W. A. Unpublished notes.
13. DE WITT, B. & D. Z. FREEDMAN. 1975. Phys. Rev. Lett. **35:** 827.
14. VAN NIEUWENHUIZEN, P. 1981. Phys. Rep. **68:** 189.
15. VOLKOV, D. V. & V. A. SOROKA. 1973. Pisma Zh. Eksp. Teor. Fiz. **18:** 529. (1973. JETP Lett. **18:** 312.)
16. FAYET, P. 1980. Supersymmetry particle physics and gravitation. *In* Unification of the Fundamental Particle Interactions. S. Ferrara, J. Ellis & P. van Nieuwenhuizen, Eds. Plenum Press. New York, N.Y. (and 1981. Ecole Normale Supérieure preprint LPTENS 81/9).
17. FAYET, R. 1977. Phys. Lett. **70B:** 461.
18. LANGACKER, P. 1981. Phys. Rep. **72:** 185.
19. GEORGI, H. & S. DIMOPOULIS. 1981. Nucl. Phys. **B193:** 150.
20. SAKAI, N. 1982. Z. Phys. **C11:** 153.
21. ELLIS, J., M. K. GAILLARD, L. MAIANI & B. ZUMINO. 1980. *In* Unification of the Fundamental Particle Interactions. S. Ferrara, J. Ellis & P. van Nieuwenhuizen, Eds.: 69. Plenum Press. New York, N.Y.
22. ELLIS, J., M. K. GAILLARD & B. ZUMINO. 1980. Phys. Lett. **94B:** 349.
23. HARVEY, J. A., E. W. KOLB & S. WOLFMAN. 1983. Phys. Rev. **D27:** 315.
24. BEG, M. A. B. 1981. Paper presented at the International Conference on High Energy Physics, Lisbon, Portugal. Preprint No RU81/B/9. Rockefeller University. New York, N.Y.
25. GOLDHABER, M. 1983. Paper presented at Orbis Scientiae Conference, Coral Gables, Fla.
26. MARCIANO, W. 1982. Proton decay 1982. *In* Orbis Scientiae 1982. Plenum Press. New York, N.Y.
27. LUBIMOV, V. A., *et al.* 1980. Phys. Lett. **B94:** 266.
28. DOROSHKEVICH, A. G., M. YU. KHLOPOV, R. A. SUNYAEV, A. S. SZALAY & YA. ZELDOVICH. 1981. Cosmological impact of the neutrino rest mass. Ann. N.Y. Acad. Sci. **375:** 32–42.
29. FREESE, K., E. W. KOLB & M. S. TURNER. 1982. Massive, Degenerate Neutrinos and Cosmology. Los Alamos Scientific Laboratory Report LA-UR 82-3068. OLIVE, K. & M. TURNER. 1982. Preprint NSF-ITP-81-51. National Science Foundation. Washington, D.C.
30. PRIMACK, J. 1981. Supersymmetry and cosmology. Presented at Banff Summer Institute on Particles and Fields.
31. SCHRAMM, D. N. & G. STEIGMAN. 1981. Astrophys. J. **243:** 1.
32. FABER, S. M. & J. S. GALLAGHER. 1979. Annu. Rev. Astron. Astrophys. **17:** 135.
33. FABER, S. M. *In* Proceedings, Vatican Conference on Cosmology and Fundamental Physics. (In press.)
34. SILK, J. 1981. Fundamental tests of galaxy formation theory *In* Proceedings, Vatican Conference on Cosmology.
35. WAGONER, R. V., W. A. FOWLER & F. HOYLE. 1967. Astrophys. J. **148:** 3.
36. STEIGMAN, G. 1981. Cosmology and neutrino physics. *In* Particles and Fields 1981: Testing the Standard Model. Santa Cruz, Calif.
37. BOUQUET, A. & C. E. VAYONAKIS. 1982. Cosmological and Astrophysical Bounds on the Scale of Supersymmetry Breaking. Report LPTHE 82-7. Paris, France.
38. ELLIS, J. & K. A. OLIVE. 1982. Constraints on Light Particles from Stellar Evolution. Report TH 3328. CERN. Geneva, Switzerland.
39. FAYET, P. 1979. Phys. Lett. **B84:** 421.

40. Fukugita, M. & N. Sakai. 1982. Phys. Lett. **B114:** 23.
41. Fayet, R. 1982. Phenomenology of supersymmetry. Talk given at the XVIIth Rencontre de Moriond, March, LPTENS 82/10.
42. Pagels, H. & J. R. Primack. 1982. Phys. Rev. Lett. **48:** 223.
43. Weinberg, S. 1982. Cosmological Constraints on the Scale of Supersymmetry Breaking. University of Texas Preprint. Austin, Texas.
44. Dimopoulos, S. & S. Raby. 1982. Preprint LA-UR-82-1282. Los Alamos Scientific Laboratory. Los Alamos, N. Mex.
45. Sciama, D. 1982. IAAS Trieste preprint 18/82/A.
46. Cabibbo, N., G. R. Farrar & L. Maiani. 1981. Phys. Lett. **105B:** 155.
47. Pecci, R. D. & H. R. Quinn. 1977. Phys. Rev. Lett. **38:** 1440.
48. Pecci, R. D. & H. R. Quinn. 1977. Phys. Rev. **D16:** 179.
49. Weinberg, S. 1978. Phys. Rev. Lett. **40:** 223.
50. Wilczek, F. 1978. Phys. Rev. Lett. **40:** 279.
51. Dicus, D. A., E. W. Kolb, V. L. Teplitz & R. V. Wagoner. 1978. Phys. Rev. **D18:** 1829.
52. Sato, K. & H. Sato. 1975. Progr. Theor. Phys. **54:** 1564.
53. Vysotsskii, M. I., Ya. B. Zeldovich, M. Yu. Khlopov & V. M. Chechetkin. 1978. JETP Lett. **27:** 502.
54. Preskill J., M. B. Wise & F. Wilczek. 1983. Phys. Lett. **120B:** 127.
55. Abbott, L. F. & P. Sikivie. 1983. Phys. Lett. **120B:** 133.
56. Dine, M. & W. Fischler. 1983. Phys. Lett. **120B:** 137.
57. Gelmini, G. B. & M. Roncadelli. 1981. Phys. Lett. **99B:** 411.
58. Chikashige, Y., R. N. Mohapatra, & R. D. Peccei. 1981. Phys. Lett. **98B:** 265.
59. Bond, J. R., A. S. Szalay & M. Turner. Phys. Rev. Lett. (In press.)
60. Blumenthal, G. R., H. Pagels & J. R. Primack. 1982. Nature **229:** 37.
61. Aaronson, M. Astrophys. J. Lett. (Submitted for publication.)
62. Lin, D. N. C. & S. M. Faber. 1982. Some implications of nonluminous matter in dwarf spheroidal galaxies and Is there non-luminous matter in dwarf spheroidal galaxies? *In* Lick Observatory Bulletin. University of California. Santa Cruz, Calif.
63. Zeldovich, Y. B. 1972. Mon. Not. R. Astron. Soc. **160:** 1P.
64. Gott, J. R. & M. J. Rees. 1975. Astron. Astrophys. **45:** 365.
65. Lukash, V. N. 1981. JETP Lett. **31:** 11.
66. Chibisov, C. V. & V. F. Mukhanov. 1980. Galaxy Formation and Phonons. Preprint No. 162. Moscow, USSR.
67. Pagels, H. 1983. Astrophys. J. Lett. **23:** 151.
68. Hawking, S. W. 1982. Phys. Lett. **115B:** 295.
69. Starobiniski, A. 1982. Phys. Lett. **117B:** 175.
70. Guth, A. & S.-Y. Pi. 1982. Phys. Rev. Lett. **49:** 1110.
71. Bardeen, J. M., P. J. Steinhardt & M. S. Turner. 1982. Phys. Rev. **D28:** 679.

Massive Magnetic Monopoles in Cosmology and Astrophysics

EDWARD W. KOLB[a]

Theoretical Division
Los Alamos National Laboratory
Los Alamos, New Mexico 87545

INTRODUCTION

Just over 50 years ago, P. A. M. Dirac showed that the existence of magnetic monopoles implies that the electric charge, e, and the magnetic charge, g, are quantized:[1]

$$eg = n/2 \quad (n, \text{integer}) \tag{1}$$

This would offer an explanation for the fact that the magnitude of the proton charge is exactly equal to the magnitude of the electron charge (rather than say $3\pi/17$ times the electron charge). However, it was realized that the theory had the shortcoming that monopoles had to be introduced as new dynamical degrees of freedom, and that monopole properties such as mass, spin, etc., were free parameters.

This situation changed in 1974, when 't Hooft and Polyakov demonstrated that monopoles *must* occur in certain classes of gauge theories.[2,3] Among the classes of gauge theories with monopoles are the grand unified theories (GUTs), which attempt to unify the strong and electroweak interactions. In these theories, all the monopole properties are calculable. In this paper, I will consider only the monopoles in grand unified models based on SU(5). The specialization to SU(5) is for convenience only; the same conclusions would obtain in any GUT.

In order to discuss the astrophysical and cosmological consequences of these monopoles, it is necessary to understand their origin and their structure. Monopoles in GUTs arise from "knots" in the Higgs field. The Higgs field is necessary to break the SU(5) symmetry down to the unbroken symmetry observed today, $SU(3)_C \times U(1)_{em}$. The $SU(3)_C$ is the color symmetry of the strong interactions, and $U(1)_{em}$ is quantum electrodynamics. This symmetry breaking occurs in two stages. The first stage of symmetry breaking occurs at a mass scale of about 10^{15} GeV, and breaks SU(5) down to $SU(3)_C \times SU(2)_L \times U(1)_Y$, where $SU(2)_L \times U(1)_Y$ is the symmetry of the electroweak interactions. The second stage of symmetry breaking occurs at a mass scale of about 10 GeV, and breaks $SU(2)_L \times U(1)_Y$ to $U(1)_{em}$. The monopoles appear at the first stage of symmetry breaking; in general, they appear whenever a simple group, G, breaks to a symmetry that contains a U(1) factor, $G \rightarrow G' \times U(1)$. The Higgs field responsible for the breaking is, in general, a tensor in group space, and the "direction" of the symmetry breaking is not determined. In a gauge theory, the Higgs

[a]Supported in part by the Department of Energy. Present affiliation: Theoretical Astrophysics Group MS 209, Fermi National Accelerator Laboratory, P.O. Box 500, Batavia, Ill. 60510.

field may be locally gauge rotated in any direction in the group space. However, the topology of mappings of manifolds in real space onto the group is gauge invariant, and topologically nontrivial defects cannot be eliminated by nonsingular gauge transformations. These topological defects, or knots, correspond to magnetic monopoles.

The GUT monopole shown in FIGURE 1 has an onion-skin structure that reflects the pattern of symmetry breaking. At small distances, less than 10^{-29} cm, the monopole has many degrees of freedom excited, gluons (g), photons (γ), and the "end of the alphabet" gauge fields including the weak bosons (W, Z) and the baryon numbers violating bosons of SU(5) (X, Y). There are also the supermassive Higgs bosons of SU(5) (S), and the Higgs bosons of the Weinberg-Salam model (ϕ). At distances greater than m_X^{-1} ($\simeq 10^{-29}$ cm), the X, Y, and S fields are no longer excited. At distances greater than m_W^{-1} ($\sim 10^{-16}$ cm), the W, Z, and ϕ fields are no longer excited. At distances greater than $\Lambda_{\rm QED}$ ($\sim 10^{-14}$ cm), the gluon fields are screened. Only at distances greater than 10^{-14} cm does the GUT monopole resemble a classical Dirac monopole.

The mass of the GUT monopole is

$$m_M \simeq \langle \phi \rangle \simeq m_X/\alpha \gtrsim 10^{16}\ \text{GeV} \tag{2}$$

where $\langle \phi \rangle$ is the vacuum expectation value of the Higgs field responsible for the SU(5) breaking. In GUTs, the monopole mass must be greater than about 10^{16} GeV, since the limit on the proton lifetime requires m_X to be greater than about 10^{14} GeV. This enormous mass for the monopole ($m_M \gtrsim 10^{16}$ GeV $\simeq 10^{-8}$ g) prevents their creation in accelerators. Only in the early universe can the energy be found to create the monopoles.

MONOPOLES FROM THE EARLY UNIVERSE

The enormous mass of GUT monopoles leads one naturally to the early universe for production of monopoles.[4,5] There are two sources of monopole creation in the early universe: thermal production, and spontaneous production in the phase transition. As

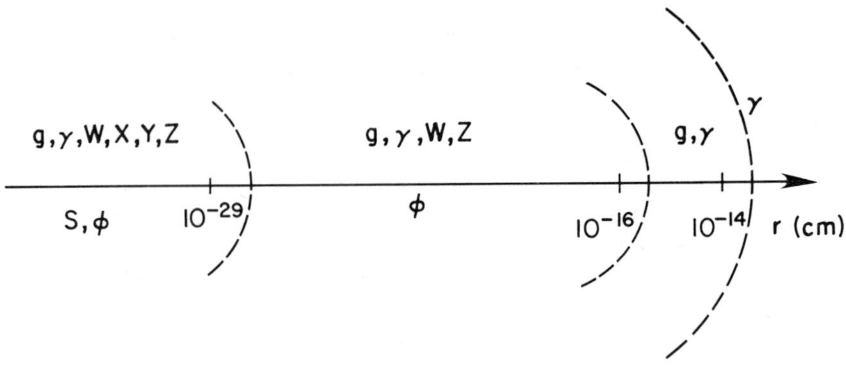

FIGURE 1. The onion-skin structure of the monopole in grand unified theories.

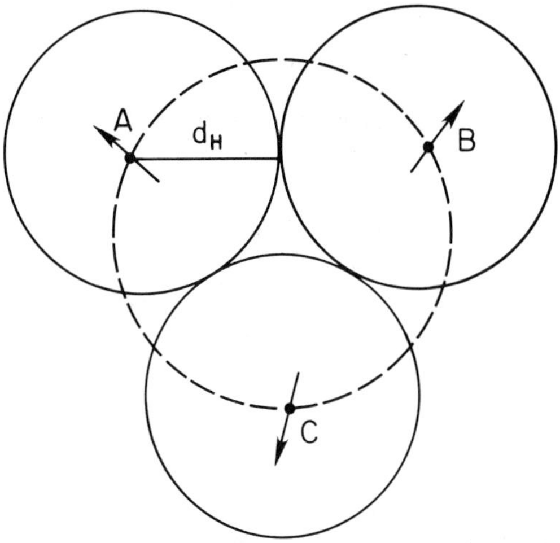

FIGURE 2. Monopole production due to the domain structure in the early universe; d_H is the causal horizon distance.

spontaneous production dominates in the standard cosmology, I will ignore thermal production.

The key ingredient in spontaneous production is the fact that at high temperatures, thermal effects of the ambient background gas should have restored the symmetry. That is, at temperatures much greater than 10^{15} GeV, $\langle\phi\rangle = 0$, where $\langle\phi\rangle$ is the vacuum expectation value of the Higgs field responsible for the breaking, SU(5) $\rightarrow$ SU(3) $\times$ SU(2) $\times$ U(1). Since monopoles exist as knots in $\langle\phi\rangle$, in the symmetric phase ($\langle\phi\rangle = 0$) there are no monopoles. At some critical temperature, $T_c \simeq 10^{15}$ GeV, spontaneous symmetry breaking occurs and monopoles will be formed. The number of monopoles formed depends on the distance over which the Higgs field can be correlated, which in turn, depends upon the size of the causal horizon.

The causal horizon, d_H, in the standard big bang model is given by

$$d_H = 2t \simeq m_{Pl}/10T^2 \tag{3}$$

where t is the time since the bang, which can be expressed in terms of temperature (T) and the Planck mass (m_{Pl}) in the standard model. The Higgs field cannot be correlated over distances larger than d_H, and the finiteness of the correlation length leads to the production of monopoles. The number of monopoles produced may be calculated by considering three points, A, B, and C, separated by a distance greater than d_H (see FIGURE 2). At each point, we can represent the orientation of $\langle\phi\rangle$ in group space by a vector. As the distances $\overline{AB}$, $\overline{AC}$, $\overline{BC}$ are greater than the causal horizon, the direction in group space of $\langle\phi\rangle$ at the three points is independent. There exists a finite probability, about 10%, that any given orientation of the vectors as shown in FIGURE 2 will correspond to a nontrivial topology, i.e., correspond to a monopole. Therefore, the expectation is that on average one monopole per 10 horizon volumes should be created (and an equal number of antimonopoles). This leads to the creation of a number density of monopole-antimonopole pairs of

$$n_M = n_{\overline{M}} \simeq 0.1\, d_H^{-3} \simeq 10^3\, T_c^6/m_{Pl}^3 \tag{4}$$

It is convenient to compare this number density to the photon number density at $T = T_c$, $n_\gamma \simeq T_c^3$

$$\frac{n_M}{n_\gamma} = \frac{n_{\overline{M}}}{n_\gamma} \simeq 10^3 \frac{T_c^3}{m_{Pl}^3} \simeq 10^{-9} \qquad (T_c = 10^{15}\text{ GeV}) \tag{5}$$

After creation of the monopoles, monopole-antimonopole pairs can annihilate. This annihilation causes the monopole density to decrease as (ignoring monopole-antimonopole creation in collisions)

$$\frac{dn_M}{dt} = \frac{dn_{\overline{M}}}{dt} = -\langle n_M n_{\overline{M}} \sigma_A |v| \rangle - 3\frac{\dot{R}}{R} n_M \tag{6}$$

where σ_A is the $M\overline{M}$ annihilation cross section. The classical calculation of σ_A involves monopole-antimonopole capture into a bound state, with subsequent emission of radiation until the eventual annihilation.[5,6] The solution to Equation 6 results in a "freeze out" density of monopoles of about

$$\frac{n_{\overline{M}}}{n_\gamma} = \frac{n_M}{n_\gamma} \simeq 10^{-10} \tag{7}$$

This freeze out occurs because as the monopole-antimonopole density becomes too small, a monopole cannot find an antimonopole to annihilate.

The ratio in Equation 7 is roughly constant during the subsequent evolution of the universe. Therefore, today the monopoles would contribute an energy density of

$$\rho_{M\overline{M}} = m_M n_M \simeq 10^{-10}\, m_M\, n_\gamma \tag{8}$$

There are about 400 cm^{-3} photons in the universe today, and if the monopole mass is 10^{16} GeV (10^{-8} g), the monopoles would contribute an energy density of

$$\rho_{M\overline{M}} \simeq 10^{-15}\text{ g cm}^{-3} \simeq 10^{14}\, \rho_c \tag{9}$$

where ρ_c is the critical density. Since observationally we know the total energy density of the universe must be less than $2\rho_c$, the energy density in Equation 9 is about 14 orders of magnitude too large. Therefore, there must be some way either to avoid making so many monopoles, or to efficiently annihilate the monopoles.

The obvious lesson is that we are missing something very fundamental in the simple picture of the early universe we have just discussed. Below we discuss some proposals that have been made to alleviate the monopole problem.

Prevent Symmetry Restoration

The monopoles are formed in the phase transition; if one could avoid the phase transition, monopoles would not be produced. One possible way to avoid the phase transition is to assume that the universe had a maximum temperature, T_{MAX}, less than the critical temperature, T_c. This assumption has the unwanted feature that baryon production is difficult, since temperatures of the order $T_c \simeq m_X$ are necessary to make baryons. There may be a narrow range for T_{MAX}, where $T_{MAX} \leq T_c$, and T_{MAX} is still large enough to produce baryons; but unless there is some physical reason for T_{MAX} to be in that narrow range, this solution cannot be taken seriously. One possible physical reason for $T_{MAX} \lesssim T_c$ can arise in grand unified models that are not asymptotically free.[7] Such asymptotically nonfree models are considered ugly on esthetic grounds, but quite naturally arise in complicated GUTs.[7]

Another possible way to prevent the universe from being in the symmetric state is to assume that there is a large fermion asymmetry that prevents high-temperature restoration of symmetry.[8] Such an initial large fermion member may have further implications for cosmology, e.g., for the neutrino number of the universe.[9]

Enhance Annihilation

Let us suppose that spontaneous production of monopoles proceeds as discussed above, and try to find a way to get rid of them by annihilation. To do this, it is necessary that Equation 6 be insufficient to describe the annihilation.

One possibility in enhancing annihilation is to assume that the monopoles are clumped, and n_M appearing in Equation 6 is larger than the value obtained in uniform expansion. The gravitational attraction of monopoles provides just such a mechanism. The possibility that monopoles form bound objects and annihilate was suggested by Preskill,[5] and examined in detail by several groups.[6,10,11] All have concluded that the mechanism of gravitational clumping cannot solve the monopole problem.

Another possibility is that when the monopoles and antimonopoles form, they are connected by strings, and are confined.[12] A basic assumption in Equation 6 is that the initial states are uncorrelated in the scattering. If monopoles are connected to antimonopoles by strings, the initial state is correlated, and the annihilation rate will be much enhanced. Although the physics behind the motivation is somewhat uncertain, there is at least some reason to believe in monopole confinement since monopole confinement could be an artifact of the non-Abelian nature of the monopole. In QED, the electric mass of the photon, m_γ^{el}, and the magnetic mass of the photon, m_γ^{mg}, are given by[13]

$$\begin{aligned}(m_\gamma^{el})^2 &= \Pi_{00}(\omega = 0, \vec{k} = 0) \simeq g^2T^2 \\ (m_\gamma^{mg})^2 &= \Pi_{ij}(\omega = 0, \vec{k} = 0) = 0\end{aligned} \qquad \mathbf{(10)}$$

where $\Pi_{\mu\nu}$ is the photon self-energy. The vanishing of the magnetic photon mass persists to all orders in perturbation theory, and means that the magnetic field is unscreened while the electric field is screened. To one loop, the electric and magnetic gluon masses are given by Equation 10. The fact that the electric mass of the gluon is nonzero means that chromoelectric fields are screened, and quarks are free at high temperatures. Although to one loop the gluon magnetic mass is zero, it is possible that higher-order calculations can lead to a nonzero magnetic gluon mass. In this case the plasma would not be a magnetic conductor, but rather the plasma would be a magnetic superconductor with the magnetic flux lines confined to flux tubes. Monopole-antimonopole pairs would be connected by these flux tubes, and confined.

Postpone the Phase Transition

The number of monopoles created in the transition is proportional to T_c^3/m_{Pl}^3. If the critical temperature is less than about 10^{10} GeV, an acceptable monopole density would result. One method of postponing the phase transition involves supercooling,[14,15] and was the original motivation of the inflationary universe. Another possibility is an intermediate symmetry as proposed by Langacher and Pi,[16] where the U(1) appearance is postponed until $T \leqq 10^{10}$ GeV.

In conclusion, the dearth of monopoles in the universe gives us very important information about the early universe. Unfortunately, it is not clear how to interpret the information. Perhaps one of the scenarios for monopole suppression discussed above is the solution to the "monopole problem," but it is just as likely that the solution has not yet been found. Some of the solutions discussed above get rid of all the monopoles. Some of the solutions get rid of some of the monopoles. The obvious next step is to look for monopoles today. The next section discusses limits on the present monopole flux, both astrophysical limits and experimental limits.

GALACTIC MAGNETIC MONOPOLES

So far we have seen that massive magnetic monopoles are predicted to exist in grand unified theories, and should have been copiously produced in the big bang. In fact, the predicted density of monopoles is so large that some mechanism must either suppress their production or efficiently reduce their number after creation. If monopoles could be detected, knowledge of the present monopole flux would help in determining the suppression mechanism.

Before discussing terrestrial experiments, it is useful to discuss some astrophysical limits. The first limit comes from the mass density of the universe. The "critical density" of the universe is (H_0 is the Hubble constant)

$$\rho_c \equiv \frac{3H_0^2}{8\pi G} = 2 \times 10^{-29}\, H_{100}^2 \text{ g cm}^{-3} \tag{11}$$

where $H_{100} = H_0/100$ km s^{-1} Mpc^{-1}. We know observationally that the energy density of the universe is less than $2\rho_c$. This implies that the average number density of monopoles, $\langle n_M \rangle$, is limited by requiring that the energy density of the monopoles is less than $2\rho_c$

$$\rho_M = \langle n_M \rangle m_M \leqq 2\rho_c$$

$$\langle n_M \rangle \lesssim 4 \times 10^{-21} \frac{10^{16}\text{ GeV}}{m_M} \text{cm}^{-3} \tag{12}$$

In order to compare with experiments, we must form a flux from the density. If we assume the monopoles are completely uniform throughout the universe, they would be cold, and the relevant velocity would be the velocity of our galaxy relative to the cold monopoles. This velocity is our peculiar velocity relative to the 3K background radiation,[17-19] $v_M \cong 10^{-3}\,c$. Therefore, the flux of monopoles in the universe must be

$$\begin{aligned} \langle F_M \rangle_U &\lesssim \langle n_M \rangle_U v_M \\ &\lesssim 10^{-14}\,\frac{10^{16}\ \text{GeV}}{m_M}\,\frac{v_M}{10^{-3}}\ \text{cm}^{-2}\,\text{s}^{-1}\,\text{sr}^{-1} \end{aligned} \tag{13}$$

It is perhaps more reasonable to assume that the monopoles are clumped, and that their density in our galaxy is enhanced over the average density throughout the universe, as are the protons. In that case, the relevant mass density limit is roughly $10^6\ \rho_c$, and the relevant velocity would be the galactic virial velocity, again about $10^{-3}\,c$. The limit in this case is

$$\langle F_M \rangle_G \lesssim 10^{-9}\,\frac{10^{16}\ \text{GeV}}{m_M}\,\frac{v_M}{10^{-3}}\ \text{cm}^{-2}\,\text{s}^{-1}\,\text{sr}^{-1} \tag{14}$$

The best astrophysical limit comes from the observation of large-scale galactic magnetic fields. The original argument is due to Parker.[20-22] He argued that if the galactic magnetic fields are due to currents, then monopoles drain energy from the field as they are accelerated. The rate of extracting energy from the magnetic field is $F_M B_G$, where B_G is the galactic magnetic field ($B_G \cong 10^{-6}$ gauss). The energy in the magnetic field is $B_G^2/8\pi$, and this energy will be drained on a time scale

$$\tau_B = \frac{B^2/8\pi}{4\pi\, F_M B} \tag{15}$$

In order for magnetic fields to survive, this time scale must be less than the time scale for regeneration of the magnetic field, which is the order of the galactic rotation time, about 10^8 years.[20-22] The requirement $\tau_B \gtrsim 10^8$ years requires

$$F_M \lesssim 10^{-16}\ \text{cm}^{-2}\,\text{s}^{-1}\,\text{sr}^{-1} \tag{16}$$

Equation 16 has become known as the "Parker limit." More detailed studies have closed several loopholes in the simple argument given above,[23] and have concluded that the limit (**16**) is probably good to an order of magnitude for reasonable values of the monopole mass. It is possible to evade the Parker limit if the monopoles themselves are the sources of the magnetic field.[23,24] This scenario is hard to imagine, since it needs large magnetic charge fluctuations. However, it potentially weakens the Parker limit by about 5 orders of magnitude.

The uncertainty in the astrophysical limits is easily matched by the uncertainties in terrestrial searches. The largest source of uncertainty comes from the difficulty in calculating the energy loss from slowly moving monopoles. For a discussion of the difficulties of such a calculation, see Ahlen and Kinoshita[25] (see also Reference 26). They find an energy loss for massive, slowly moving monopoles in silicon of (for $\beta \leq 0.01$)

$$\frac{dE}{dx} = \cong 50\,\beta\ \text{GeV cm}^{-1} \tag{17}$$

where β is the monopole velocity. There are several things to note about this result.

1. It is linear in β.
2. It is at least as large as the energy loss for slowly moving protons.
3. It agrees with previous calculations.

All reported experiments that detect monopoles by ionization depend on calculations of the monopole ionization. Although different experiments use slightly different values for the ionization, it is still interesting to compare some quoted limits:

$$F_M \lesssim 10^{-10}\ \text{cm}^{-2}\,\text{s}^{-1}\,\text{sr}^{-1} \qquad (\beta \gtrsim 3 \times 10^{-3}) \tag{18a}$$

$$F_M \lesssim 3 \times 10^{-11}\ \text{cm}^{-2}\,\text{s}^{-1}\,\text{sr}^{-1} \qquad (3 \times 10^{-4} \leqq \beta \leqq 10^{-3}) \tag{18b}$$

$$F_M \lesssim 10^{-11}\ \text{cm}^{-2}\,\text{s}^{-1}\,\text{sr}^{-1} \qquad (2 \times 10^{-4} \leqq \beta \leqq 3 \times 10^{-2}) \tag{18c}$$

$$F_M \lesssim 2 \times 10^{-12}\ \text{cm}^{-2}\,\text{s}^{-1}\,\text{sr}^{-1} \qquad (10^{-2} \lesssim \beta) \tag{18d}$$

$$F_M \lesssim 3 \times 10^{-13}\ \text{cm}^{-2}\,\text{s}^{-1}\,\text{sr}^{-1} \qquad (10^{-3} \leqq \beta \leqq 10^{-2}) \tag{18e}$$

where the limits quoted in (**18a**)–(**18e**) are given in References 27–31 respectively.

The cleanest way to look for monopoles is the superconducting search done by Cabrera.[32] This method is based on the change in the macroscopic quantum state of a superconducting ring when a magnetic charge passes through the ring. The magnetic flux through the superconducting ring changes the current, which is detected by a SQUID. This method is clean, since it is independent of the monopole mass, monopole velocity, monopole electric charge, and energy loss calculations. Cabrera has reported a limit[32]

$$F_M \lesssim 6 \times 10^{-10}\ \text{cm}^{-2}\,\text{s}^{-1}\,\text{sr}^{-1} \tag{19}$$

and a candidate event. If we assume the candidate event was the actual detection of a magnetic monopole, the flux of $6 \times 10^{-10}\ \text{cm}^{-2}\,\text{s}^{-1}\,\text{sr}^{-1}$ represents the actual monopole flux.

Although one event does not give unequivocal evidence for the existence of magnetic monopoles, it is interesting to assume this event is the result of a flux given in Equation 18 and examine the consequences. First, it is obviously in conflict with the ionization searches discussed above. This would mean that either the monopoles are very slow ($\beta \ll$ galactic virial velocity) or the ionization calculations are in serious error. The Cabrera flux is also in serious disagreement with the Parker limit. Either we do not understand galactic magnetic fields, or the monopole flux on earth is much different than the average galactic flux. The latter possibility has been examined by Dimopoulos, Glashow, Prucell, and Wilczek.[33] They suggested that there is a "cloud" of monopoles orbiting the sun, which enhances the monopole flux at the earth by a factor of 10^5 relative to the average galactic flux. In this manner it may be possible to reconcile the Cabrera flux with the Parker limit.

In conclusion, there is one hint that monopoles exist in our galaxy. If this hint proves correct, it may have drastic implications for our understanding of galactic magnetic fields.

MONOPOLE-INDUCED PROTON DECAY

Two dramatic predictions of GUTs are proton decay, and the existence of magnetic monopoles. Recent model calculations by Rubakov,[34,35] Callan,[36–38] and Wilczek[39] have suggested that proton decay in the presence of a monopole occurs at a much faster rate than one might guess on dimensional grounds.

A naive dimensional estimate of the cross section $pM \longrightarrow MX$, where X denotes the proton decay products, might be $\sigma_{\Delta B} = m_M^{-2}$, where m_M is the monopole mass. However Rubakov and Callan have argued that the cross section is *independent* of the mass of the monopole, and it is related instead to the mass scale of color confinement, about 1 GeV. We shall parameterize the cross section for monopole-induced proton decay by a mass scale Λ:

$$\sigma_{\Delta B} = \frac{\pi}{\Lambda^2} \cong \frac{10^{-27}}{\Lambda_{\text{GeV}}^2}\ \text{cm}^2 \tag{20}$$

where $\Lambda_{\text{GeV}} = \Lambda/1$ GeV. If the conjecture of Callan and Rubakov is correct, then $\Lambda_{\text{GeV}} \cong 1$.

There are two obvious questions to be asked about monopole-induced proton decay. First, does it open up new methods for detecting cosmic monopoles? Second, does monopole-induced proton decay have any important astrophysical consequences? The answer to both questions is yes!

If monopoles induce proton decay while traversing a proton-decay detector, then proton-decay detectors would make the most sensitive monopole telescopes. A preliminary study by the Irvine-Michigan-Brookhaven proton decay collaboration has concluded that with minor modifications, the detector would have a sensitivity to a monopole flux of about $10^{-15}\ \text{cm}^{-2}\ \text{s}^{-1}\ \text{sr}^{-1}$, if $\sigma_{\Delta B} \cong 10^{-27}\ \text{cm}^2$.[40] This is several orders of magnitude better than the limits in (**18**).

If monopoles induce proton decay, they would have a drastic effect on the x-ray luminosity of neutron stars.[41,42] By using the limit on the neutron star luminosity, it is possible to place a limit on the product of the galactic monopole flux and baryon-violating cross section, which suggests that the phenomenon of monopole-induced proton decay is not likely to be detected terrestrially.

Neutron stars are believed to be plentiful in our galaxy,[43–45] and to be as old as the galaxy itself. During the lifetime of the neutron star, the neutron star would have been subjected to a galactic monopole flux F_M. This resulted in a number of neutron star/monopole collisions given by

$$N_M = \frac{2\pi}{3} F_M A_c t_{\text{NS}} \tag{21}$$

where t_{NS} is the lifetime of the neutron star, and A_c is the "capture" area given by

$$A_c = R_{\text{NS}}^2 \frac{1 + 2M_{\text{NS}} G/v_M^2 R_{\text{NS}}}{1 - R_S/R_{\text{NS}}} \tag{22}$$

In Equation 22, M_{NS} and R_{NS} are the mass and radius of the neutron star, v_M is the galactic velocity of the monopole, and R_S is the Schwarzschild radius of the neutron star. Putting in values of $M_{NS} = 1\ M_\odot$, $R_{NS} = 10^6$ cm

$$N_M = 10^{36}\ F_M(\text{cm}^{-2}\ \text{s}^{-1}\ \text{sr}^{-1}) \tag{23}$$

The monopole will be trapped in the neutron star, mainly through energy loss in scattering with electrons.[25,26,42] This results in an energy loss of

$$\frac{dE}{dx} \simeq \beta\ 10^{11}\ \text{GeV cm}^{-1} \tag{24}$$

where β is the monopole velocity in the star ($\beta \simeq 10^{-1}$ is escape velocity). If $\beta \gtrsim 10^{-1}$, the monopole will lose about 10^{16} GeV traversing the star (if $\beta \lesssim 10^{-1}$, the monopole must be trapped). The initial energy of the monopole in the galaxy was

$$\begin{aligned} E_0 &= \frac{1}{2} m_M v_M^2 \\ &\simeq 5 \times 10^9 \frac{m_M}{10^{16}\ \text{GeV}} \left[\frac{v_M}{10^{-3}}\right]^2 \text{GeV} \end{aligned} \tag{25}$$

Therefore, if the monopoles have typical virial velocities of 10^{-3}, monopoles of mass less than 10^{22} GeV ($10^3\ m_{Pl}$!) will be trapped.

The monopoles trapped in the star catalyze nucleon decay, and release energy at a rate

$$\begin{aligned} L_M &= m_N n_N \sigma_{\Delta B} |v| \\ &\simeq 8.5 \times 10^{17}\ \Lambda_{\text{GeV}}^{-2}\ \text{erg s}^{-1}\ \text{monopole}^{-1} \end{aligned} \tag{26}$$

where m_N is the nucleon mass, n_N is the nucleon density, and we have used $|v| \simeq 10^{-1}$, the nucleon Fermi velocity.

Combining (**23**) and (**26**), the energy produced in the neutron star by monopole-induced nucleon decay is

$$L = 2 \times 10^{54}\ \Lambda_{\text{GeV}}^{-2}\ F_M(\text{cm}^{-2}\ \text{s}^{-1}\ \text{sr}^{-1})\ \text{erg s}^{-1} \tag{27}$$

This luminosity is the *total* luminosity, which in general may be emitted as photons and neutrinos. The best limit on the photon luminosity comes from recent surveys for serendipitous x-ray sources.[46,47] These surveys could detect an x-ray luminosity of $L_\gamma = 10^{31}$ erg s^{-1} at a distance of 1 kpc.[46,47] Estimates based on the current pulsar birth rate in the solar neighborhood indicate that there should be an old neutron star density of $n_{NS} \gtrsim 4 \times 10^{-3}$ pc^{-3}. Therefore, it is reasonable as an extremely conservative estimate to take 10^{31} erg s^{-1} as the maximum luminosity of old neutron stars.

The relationship between the x-ray luminosity and the total luminosity depends on the structure of the neutron star. The smallest L^γ/L^ν ratio results for zero surface magnetic field, and for pion condensation in the neutron star interior. In this case, $L^\gamma \lesssim 10^{31}$ erg s^{-1} corresponds to a total luminosity of about 10^{33} erg s^{-1}.[48] It is also important to note that a photon luminosity of 10^{31} erg s^{-1} corresponds to a surface temperature of

$T_S = 0.04$ keV. About 80% of the energy radiated at this temperature will be in the window of the Einstein satellite.

A maximum total luminosity of 10^{33} erg s^{-1} requires

$$F_M(\mathrm{cm}^{-2}\,\mathrm{s}^{-1}\,\mathrm{sr}^{-1})\,\Lambda_{\mathrm{GeV}}^{-2} < 5 \times 10^{-22} \qquad \textbf{(28)}$$

Therefore, if monopoles catalyze proton decay at a strong rate ($\Lambda_{\mathrm{GeV}} \simeq 1$), the flux limit of (**28**) is about 6 orders of magnitude better than the Parker limit discussed in the previous section. This limit is also 6 orders of magnitude below the possible limit for detection reported by the IMB group.

Conversely, if proton decay detectors do see monopoles, something very basic is very wrong in our understanding of neutron stars.

CONCLUSIONS

Not only do magnetic monopoles provide a natural explanation for charge quantization, but they must exist in grand unified theories. In such theories, the monopole is very massive, and the early universe is probably the only source for production.

The early universe should have provided a plentiful supply of monopoles. In fact, the simplest prediction one might make results in a monopole density 10^{14} orders of magnitude too large. This important cosmological probe of the early universe tells us that either the simple picture of unification or the simple picture of the early universe is seriously wrong. There have been several proposals to solve the monopole problem, they all have important implications for cosmology, and could potentially alter our picture of the early universe. Detection of a monopole would be strong evidence that the universe was once at a temperature greater than 10^{15} GeV.

There have been several recent terrestrial searches for massive magnetic monopoles. One experiment can be interpreted as detection of a monopole.[32] If this experiment is confirmed, then the flux of magnetic monopoles would be many orders of magnitude too large for us to understand the existence of galactic magnetic fields.

There are theoretical indications that the cross section for monopole-induced proton decay is a strong cross section. If this is true, then a galactic monopole flux greater than 10^{-22} cm^{-2} s^{-1} sr^{-1} would result in an enormous x-ray luminosity for neutron stars. Therefore, a detection of monopoles on earth would not only mean that galactic magnetic fields do not exist, but that neutron stars do not exist.

Not only do magnetic monopoles provide a probe of particle physics, but potentially are a great source of information in astrophysics.

REFERENCES

1. DIRAC, P. A. M. 1931. Proc. R. Soc. **133:** 60.
2. 'T HOOFT, G. 1974. Nucl. Phys. **B79:** 276.
3. POLYAKOV, A. M. 1974. JETP Lett. **20:** 194.
4. ZEL'DOVICH, YA. B. & M. YU. KHLOPOV. 1978. Phys. Lett. **79B:** 239.
5. PRESKILL, J. P. 1979. Phys. Rev. Lett. **43:** 1365.
6. DICUS, D. A., D. N. PAGE & V. L. TEPTLITZ. 1982. Phys. Rev. **D26:** 1306.
7. HARVEY, J. A., E. W. KOLB & S. WOLFRAM. 1983. Phys. Rev. **D27:** 315.

8. LINDE, A. 1976. Phys. Rev. **D14:** 3345.
9. HARVEY, J. A. & E. W. KOLB. 1981. Phys. Rev. **D24:** 2090.
10. GOLDMAN, T., E. W. KOLB & D. TOUSSAINT. 1981. Phys. Rev. **D23:** 867.
11. FRY, J. N. 1981. Astrophys. J. **246:** L93.
12. LAZARIDES, G. & Q. SHAFI. 1980. Phys. Lett. **94B:** 149.
13. GROSS, D. J., R. D. PISARSKI & L. G. YAFFE. 1981. Rev. Mod. Phys. **53:** 43.
14. GUTH, A. H. & S.-H. H. TYE. 1980. Phys. Rev. Lett. **44:** 631.
15. EINHORN, M. B., D. L. STEIN & D. TOUSSAINT. 1980. Phys. Rev. **D21:** 3295.
16. LANGACKER, P. & S.-Y. PI. 1980. Phys. Rev. Lett. **45:** 1.
17. SMOOT, G. F. & P. M. LUBIN. 1979. Astrophys. J. **234:** L83.
18. FABBRI, R., *et al.* 1980. Phys. Rev. Lett. **44:** 1563.
19. BOUGHN, S. P., E. S. CHENG, & D. T. WILKINSON. 1981. Astrophys. J. **243:** L113.
20. PARKER, E. N. 1970. Astrophys. J. **160:** 383.
21. PARKER, E. N. 1971. Astrophys. J. **163:** 225.
22. PARKER, E. N. 1971. Astrophys. J. **166:** 295.
23. TURNER, M. S., E. N. PARKER & T. J. BOGDAN. 1981. Phys. Rev. **D26:** 1296.
24. SALPETER, E. E., S. L. SHAPIRO & I. WASSERMAN. 1982. Phys. Rev. Lett. **49:** 1114.
25. AHLEN, S. P. & K. KINOSHITA. 1982. Phys. Rev. **26D:** 2347.
26. DRELL, S. D. 1983. Phys. Rev. Lett. **50:** 644.
27. SOKOLOWSKI, J. K. & L. R. SULAK. University of Michigan Report. Ann Arbor, Mich.
28. ULMAN, J. D. 1981. Phys. Rev. Lett. **47:** 289.
29. GROOM, D. E., *et al.* University of Utah Report. Salt Lake City, Utah.
30. BONARELLI, R., *et al.* 1982. Phys. Lett. **112B:** 102.
31. BARTELT, J., *et al.* 1983. Phys. Rev. Lett **50:** 655.
32. CABRERA, B. 1982. Phys. Rev. Lett. **48:** 1378.
33. DIMOPOULOS, S., S. L. GLASHOW, E. M. PURCELL & F. WILCZEK. 1982. Nature **298:** 824.
34. RUBAKOV, V. A. 1982. Nucl. Phys. **B203:** 311.
35. RUBAKOV, V. A. 1981. Zheft Pis'ma **33:** 658.
36. CALLAN, C. G. 1982. Phys. Rev. **D25:** 2141.
37. CALLAN, C. G. 1982. Phys. Rev. **D26:** 2058.
38. CALLAN, C. G. 1982. Nucl. Phys. **B204.**
39. WILCZEK, F. 1982. Phys. Rev. Lett. **48:** 1146.
40. ERREDE, S. Private communication.
41. KOLB, E. W., S. A. COLGATE & J. A. HARVEY. 1982. Phys. Rev. Lett. **49:** 1373.
42. DIMOPOULOS, S., J. P. PRESKILL & F. WILCZEK. 1982. Phy. Lett. **119B:** 320.
43. LAMB, D. Q., F. K. LAMB & D. PINES. 1973. Nature **246:** 52.
44. HILLS, J. G. 1978. Astrophys. J. **219:** 550.
45. HILLS, J. G. 1980. Astrophys. J. **240:** 242.
46. CÓRDOVA, F. S., K. O. MASON & J. E. NELSON. 1981. Astrophys. J. **245:** 609.
47. REICHERT, G. A., K. O. MASON, J. R. THORSTENSEN & S. BOWYER. In preparation.
48. VAN RIPER, K. A. & D. Q. LAMB. 1981. Astrophys. J. **244:** L13.

Energy in Gravitational Theories[a]

Definition, Positivity Theorems, and Stability

S. DESER

Physics Department
Brandeis University
Waltham, Massachusetts 02254

INTRODUCTION

This talk is intended to explain and summarize the positive energy and stability results that have recently been established in gravitational theories. Apart from its essential bearing on any field theoretical model, stability is obviously important to considerations on the early universe and cosmology, particularly since we shall see that it can now be applied to systems with nonvanishing cosmological constant Λ. Whatever the ultimate form, at trans-Planck scales, of the correct gravitational theory, it seems likely that these results will remain physically relevant to the present "effective" models, such as general relativity and supergravity.

Because it is often a subject of confusion, we will begin with a review of the physical significance and domain of applicability of the notion of energy. Then we will implement this physics with a more formal definition which is simple, invariant, and reflects the circumstances under which energy makes sense. The known positive energy and stability theorems will then be given for the various models of current interest. I will also try to show explicitly how they can be derived from simple supersymmetry ideas. Semiclassical stability, Kaluza-Klein models, and some other theories are also briefly considered.

ENERGY DEFINITION

The meaning, definition, and positivity of energy in generally covariant theories have given rise to a vast and often confusing literature ever since general relativity began. A clear, consistent, and unique picture of gravitational energy for $\Lambda = 0$ emerged about 20 years ago. It is physically quite similar to that for special relativistic systems, on which it is based; the recent extension to $\Lambda \neq 0$ models is surprisingly straightforward. Because of the nonlinearities of these theories, general positivity and stability theorems were only obtained in the past few years; here supergravity has played a basic role.

The early confusion about energy had at least two causes. First, general covariance means there is no absolute time with respect to which translations (i.e., a Hamiltonian) can be defined; nor is there an absolute space, so the notion of local energy density is also coordinate dependent. We will see that these objections are correct but irrelevant.

[a]Supported in part by National Science Foundation Grant PH 82-01094.

The time problem is overcome by a proper physical understanding of what an isolated system means; the most important physical point here is that the solutions of the field equations provide their own criteria for defining energy. More explicitly, a theory is defined by boundary conditions as well as by field equations. One can compare (and only compare) sets of solutions with the same boundary conditions, and it is on the properties of the latter, namely, the symmetries of asymptotic solutions, that the possibility of defining an energy concept depends in a generally covariant theory. Because energy is a global quantity, the ambiguities in defining energy density become irrelevant, although in practice it is often useful (and approximately correct) to speak of local energy in the usual flat space sense.

A second cause of confusion is that energy (which is also frequently called mass in this context) plays several roles in general relativity. It is both "inertial" and "gravitational," as well as the source ("charge") of the self-coupled gravitational field. The gravitational mass aspect means that we expect to be able to read off the total mass of an isolated system from its asymptotic properties alone, e.g., by a Cavendish experiment performed far from its regions of excitation (using a test particle for example) and the same goes for the inertial mass notion, where we imagine a soft collision with another isolated system. From the gauge aspect of the theory, we also expect that energy, as its "charge," should be definable as a flux integral at spatial infinity (spatial, because if energy is to be interesting, it must be both defined in terms of the system's initial data and conserved off the initial surface). Note again that we have only mentioned the total mass rather than density, although integration over all space of the (locally ambiguous) energy density does of course yield the same invariant number as an asymptotic measurement.

Our usual definition of energy in special relativity rests on the existence of a timelike symmetry in flat space. In curved space, what remains (and suffices) is the global aspect: sets of solutions (geometries) that tend asymptotically to a common background solution can be labeled and compared by energy if (and only if) this background has a timelike Killing vector (the coordinate-free definition of a preferred time). Thus, the frustration of general covariance is avoided by the boundary conditions on physically interesting solutions. For $\Lambda = 0$, the obvious background is flat space, and is clearly what we have in mind when speaking of isolated systems, whose excitations (both matter and geometry) die off outside a finite spatial domain, although energy with respect to other backgrounds possessing a timelike Killing vector could in principle be considered. How does one proceed in the presence of a nonvanishing cosmological constant? Here flat space is not an asymptotic solution, but is replaced as the relevant vacuum state by (the maximally symmetric) de Sitter ($\Lambda > 0$) or anti–de Sitter ($\Lambda < 0$) space. The latter has a global timelike Killing vector; the former does so only inside its (built-in) event horizon, beyond which this vector becomes spacelike. Furthermore, the physics of isolated systems is quite similar to that discussed for $\Lambda = 0$, and as we shall see, these similarities suffice to generalize the energy idea to $\Lambda \neq 0$ models. We conclude this physical resumé with three general remarks: (1) Closed spaces have no asymptotic region and so neither require nor admit an energy definition (except in an approximate local way). (2) One cannot compare "energies" of two systems with different asymptotic geometries. A fortiori, solutions corresponding to two different theories, e.g., different values of Λ, cannot be so compared. (3) Scalar-tensor theories do not in general obey the equivalence principle, and the scalar's asymptotic behavior becomes separately relevant. This is also true of theories involving a potential depending on scalar fields, which acts like a "variable" cosmological constant.

Let us now turn to the formal implementation of the above physical demands for Einstein theory with arbitrary cosmological constant and matter sources $T^M_{\mu\nu}$.[1] To build in the role of the boundary conditions, we decompose the metric tensor into a background $\bar{g}_{\mu\nu}$ corresponding to the asymptotic geometry and a deviation tensor $h_{\mu\nu}$, which vanishes at spatial infinity,

$$g_{\mu\nu} = \bar{g}_{\mu\nu} + h_{\mu\nu}, \qquad h_{\mu\nu} \xrightarrow[r\to\infty]{} 0$$

We emphasize that this is a coordinate-invariant splitting and that no smallness of $h_{\mu\nu}$ is implied. The required spatial falloff of $h_{\mu\nu}$ can be given precisely and invariantly.[2–5] Essentially, it must decay as $1/r$; the constraint equations themselves then fix the asymptotic behavior of the various components of $h_{\mu\nu}$. The field equations are

$$\mathcal{G}_{\mu\nu}(g) \equiv \sqrt{-g}\,[G_{\mu\nu}(g) + \Lambda g_{\mu\nu}] = \sqrt{-g}\,T^M_{\mu\nu} \tag{1a}$$

and the background is itself a solution of the source-free equations, $\mathcal{G}_{\mu\nu}(\bar{g}) = 0$. A rearrangement (not an approximation) of Equation 1a leads to the form

$$\mathcal{G}^L_{\mu\nu}(h;\bar{g}) = \sqrt{-g}\,T^M_{\mu\nu} - T^{\rm NL}_{\mu\nu}(h;\bar{g}) \equiv T_{\mu\nu} \tag{1b}$$

where $T^{\rm NL}_{\mu\nu}$ is the nonlinear part in h of $\mathcal{G}_{\mu\nu}$ and all quantities are tensors under common coordinate transformations of h and $\bar{g}$. The field equations (**1a**) imply that

$$\bar{D}_\mu\, T^{\mu\nu} = 0 \tag{2}$$

where $\bar{D}_\mu$ is the covariant derivative with respect to $\bar{g}_{\mu\nu}$. The covariant conservation (**2**) is still not enough to yield a conserved charge, since for that we need a vectorial law $\partial_\mu j^\mu = 0$ where j^μ is a vector density. This is where the background Killing vector ξ_μ enters: because it satisfies $\bar{D}_\mu\xi_\nu + \bar{D}_\nu\,\xi_\mu = 0$ (and $T^{\mu\nu}$ is symmetric), it defines a conserved vector current

$$\bar{D}_\mu\,(T^{\mu\nu}\xi_\nu) \equiv \partial_\mu\,(T^{\mu\nu}\xi_\nu) = 0 \tag{3a}$$

The corresponding conserved coordinate-independent spatial integral is

$$E(\xi) = 1/8\pi G \int d^3x T^{o\nu}\,\xi_\nu \tag{3b}$$

where we have restored conventional units. As we have argued, because this is a gauge theory, the volume integral form must also be expressible as a flux integral over a closed two-surface at infinity. Indeed, replacing $T^{o\nu}$ by its left side in Equation 1b, one obtains the asymptotic representation[1]

$$E(\xi) = 1/8\pi G \int dS_i\, K^{io}\,(h,\xi;\bar{g}) \tag{4}$$

where K^{io} is a simple expression linear in h, ξ, and $\bar{D}_j$. Note that this definition involves neither the nature of the matter sources nor the choices of coordinates, inside or out.[b] The form of Equation 4 is also independent of the value of Λ and of the time- or

[b]It reduces, for $\Lambda = 0$ and with the usual flat space gauge choice $\bar{g}_{\mu\nu} = \eta_{\mu\nu}$ where the timelike Killing vector becomes $\xi^\mu = \delta^\mu_0$, to the usual energy expression.[2–5] We mention that the above treatment has a close parallel in the definition of color charge in Yang-Mills theories.[6]

spacelike character of ξ. A system has as many conserved "charges" as there are background Killing vectors, but it is only timelike background isometries that lead to an energy concept. For flat space $\bar{g}$, there are 10 ξ^{μ}_{a}, which yield the 10 Poincaré generators $E(\xi_a)$; they satisfy the algebra $[E(\xi_a), E(\xi_b)] = E([\xi_a, \xi_b])$. This automatically guarantees that energy, momentum, and angular momentum all transform properly under rigid overall Poincaré transformations, in addition to being invariant under arbitrary interior changes. Correspondingly, for $\Lambda \neq 0$, the Es are the de Sitter algebra generators. Other equivalent invariant forms for E may be given; as we shall see, that used in Reference 7 replaces $\bar{g}$, ξ, and $\overline{D}h$ by the background vierbein, its associated Killing spinor ($\overline{D}_{\mu}\epsilon_{o} = 0$) and derivatives of a spinor obeying the Dirac equation in the interior geometry. But the general idea is the same: the energy is always given by an asymptotic surface integral at spatial infinity. The integral depends on the background geometry, its appropriate timelike symmetry generator, and linearly on a representative component of the interior geometry.

POSITIVITY THEOREMS

The physical basis for expecting that gravitational systems, including matter sources with positive flat space energy, will be stable can be illustrated as follows (for a history see References 8 and 9). First, for gravitational excitations about flat space, it is known rigorously that the energy is positive; this linearization stability[10,11] means essentially that weak gravity really looks like the free massless spin 2 field. Next it is also known that the energy as a functional of the excitations is stationary only at flat space or zero energy.[12,13] Third, a simple equivalence principle argument indicates that (in contrast with Newtonian theory) the attractive gravitational self-energy corrections at the nonlinear level will indeed lower a positive bare energy (including that of gravitons) but cannot bring it below zero. This is obvious from the suggestive equation

$$M = M_o - GM^2/r \geqq 0, \qquad \text{if } M_o \geqq 0$$

where it is the *total* mass M that self-gravitates, rather than the bare M_o. General theorems did not come until quite recently, however, because a frontal attack on the nonlinear Einstein constraint equations was so difficult and their supersymmetry was hidden. We now list the results, following which their derivation will be sketched.

For $\Lambda = 0$, and an asymptotically flat background, the energy corresponding to the timelike Killing vector is nonnegative, and vanishes only for the vacuum ($h_{\mu\nu} = 0$ everywhere). This theorem holds for arbitrary sources whose own energy is positive (essentially $T^M_{oo} \geqq 0$ or $\geqq |T^M_{oi}|$). The derivations are of two basic inspirations, from differential geometry[14–16] and from supergravity.[7,17] The geometrical proof is based on the fact that negative mass would violate the properties (dictated by the constraints) of the spatial curvature on the initial surface. Witten's proof[7] begins from a definition of energy in terms of a spinor obeying the Dirac equation at initial times, and provides a manifest positive energy form once the properties of the equation are related to the initial constraints; a rigorous version is given in References 18 and 19. However, it is somewhat mysterious as it stands, with no apparent connection with supersymmetry. Recently, it has been possible to deduce it from the classical supergravity, using the classical Rarita-Schwinger action for spin 3/2 in a simple way.[20,21] The original positivity proof, initially for supergravity itself,[17] was actually very similar in form to this last argument. It makes use of the fact that this theory, being locally supersymmetric, is also globally so. This enables one to use the very hallmark of supersymmetric theories, namely, the positivity relation between the Hamiltonian and the generator Q of supersymmetry transformations

$$H = \frac{1}{2}\sum Q_\alpha^2 \geqq 0, \qquad Q = 0 \rightarrow \text{vacuum} \tag{5}$$

which is valid for all $1 \leqq N \leqq 8$ supergravities. This formal operator statement can be converted to a classical result[22] because in the limit as $\hbar = 0$, the sector of supergravity with no external fermions describes tree level classical gravity and the operator relation (**5**) retains its form in this limit. Witten arrived at his formulation for classical gravity while seeking to short-circuit this process of taking the classical limit of quantum supergravity; we shall provide the direct route below.

Since the ideas and methods in the supersymmetry-based derivations are physically simple, instructive, and likely to have further important applications, we will sketch them here. (Implicit is that all the hard technical steps that turn the arguments into proofs are omitted!) We will follow the historic route in three steps: first the full quantized supergravity result, and its classical gravity limit; next, the Witten proof; and finally the derivation of the latter from purely classical supergravity, which avoids any quantum elements. For simplicity, we will stick to the $\Lambda = 0$ case here, and omit sources; generalization to $\Lambda \neq 0$ will be discussed later. As for sources, it is not difficult to show that matter with positive flat space energy density (more exactly such that $T^M_{00} \geqq |T^M_{0i}|$) or matter supermultiplets coupled to supergravity do not alter the results; nor is it surprising that negative energy matter could give negative total energy—minus five grams of a very dilute negative mass "dust" would clearly have total energy of (a bit less than) minus five grams.

Supergravity is a quantized system consisting of the Einstein field coupled minimally to a massless (fermionic) spin 3/2 field ψ_μ. The total action is invariant under local supersymmetry transformations that mix the two fields, with a local fermionic gauge parameter $\alpha(x)$. The relevant property of these transformations is that their product corresponds to a local translation. This does not yet get us to the desired form (**5**) because we must first go through the steps that lead from local invariance to existence of global conserved quantities and then show that the latter obey (**5**) as our operator relation. For energy, this has already been accomplished in our previous discussion, with $T^M_{\mu\nu}$ identified as the spin 3/2 field's stress tensor. We must therefore do the same for supercharge, the quantity associated to the supersymmetry invariance. The steps are very similar, because the latter is just the gauge generator for the spin 3/2 field whose field equations read

$$R^\mu(x) \equiv \epsilon^{\mu\nu\alpha\beta}\gamma_5\gamma_\nu D_\alpha \Psi_\beta = 0 \tag{6}$$

They obey the consistency (conservation) laws $D_\mu R^\mu = 0$ as a consequence of the Einstein equations. We now break up the field $\Psi_\mu = \overline{\Psi}_\mu + \psi_\mu$ where $\overline{\Psi}_\mu$ is the background solution and ψ_μ the deviation; we will be interested in the "flat" case $\overline{\Psi}_\mu$ a pure gauge, which we set to zero for convenience. It then follows that $\overline{D}_\mu R^\mu_L(\psi) = 0$; this is still a covariant conservation law because $R^\mu_L(\psi)$ is a vector-spinor rather than a vector. Next invoke the background Killing spinor $\epsilon(x)$ which satisfies $\bar{D}_\mu \epsilon = 0$ to convert this to the desired law $\partial_\mu(\bar{\epsilon} R^\mu_L) = 0$. The corresponding conserved charge is essentially $\bar{\epsilon}$ times the fermionic Q in (**5**):

$$Q(\epsilon) = \int d^3x\, \bar{\epsilon}\, R^o \tag{7}$$

Just as the Poincaré algebra $[E(\xi_1), E(\xi_2)] = E[(\xi_1, \xi_2)]$ followed in the bosonic sector, here we obtain the fermionic part

$$[Q(\epsilon_1), Q(\epsilon_2)] = E(\xi) \tag{8}$$

where the Killing vector is bilinear in the spinors. Since the Qs are Hermitian (Majorana) spinors and the Hilbert space has positive metric when all gauges have been fixed and constraints eliminated, the energy is positive just as in globally supersymmetric models (with positive Hilbert space metrics!), i.e., we are back at (**5**). The above is a sketch of the very first (1977) energy positivity proof[17] restated in the fancier language of Reference 1. Its only drawback is that it assumes that $N = 1$ supergravity exists! Since the latter is probably a nonrenormalizable theory, its (rigorous) meaning as a field theory is not clear. The same derivation also holds for all extended models; if, as is hoped, the maximal extension to $N = 8$ is finite, then there is no problem with existence there. One may also formally take the $\hbar \rightarrow 0$ classical limit of the above results.[22] Here, one simply notes that matrix elements of (**8**) between states without external fermions and with $\hbar \rightarrow 0$ just represent tree level supergravity (no closed loops) without fermions. But this is just ordinary classical Einstein theory, so its energy is also positive just because time translation is the "square" of another transformation, that of supersymmetry. We will return to a direct classical version of this argument in our third step. Note that in any case, we have nothing to say about quantum gravity because there is no way to limit from full supergravity to it—we cannot remove just loops due to the ψ_μ field without also losing the purely graviton loops. This is probably no great tragedy since that theory is not likely to make much sense by itself.

Next, we state Witten's proof, which is purely classical but appears rather mysterious, being neither purely metric nor in any manifest sense anchored in supergravity except that it too involves spinors (and was obtained by Witten in attempting to circumvent the latter's quantum character). The idea is amazingly simple. Consider, on an asymptotic flat $t = 0$ surface, where the Einstein constraints $G_{o\mu} = 0$ hold, a spinor field ϵ obeying

$$\gamma^i D_i \epsilon = 0 \tag{9}$$

where D_i is the projection of D_μ on the initial surface. Iterate, and use the commutation identities between covariant derivatives to obtain

$$0 = \gamma^i D_i \gamma^j D_j \epsilon + \gamma^o \gamma^\mu G_{o\mu} \epsilon = \nabla^2 \epsilon \tag{10}$$

Now multiply by ϵ^* and integrate over all space to obtain the positive expression

$$\int \epsilon^* \nabla_i \epsilon \cdot dS^i = \int d^3x |\nabla \epsilon|^2 \geqq 0 \tag{11}$$

If $\nabla\epsilon$ vanishes, it follows that the initial surface is flat and embedded in the four-space in such a way that the latter must be flat as a consequence of the dynamical $G_{ij} = 0$ equations. All that remains to be shown is that the left side is proportional to the energy. This comes about as follows: (**9**) has no solutions unless

$$\epsilon = \epsilon_o \left[1 + \frac{f(x)}{r} + \cdots \right]$$

where ϵ_o is a Killing spinor at infinity ($\overline{D}_i \epsilon_o = 0$), i.e., a (nonvanishing) constant. (This is the hard part, mathematically.)[18,19] It is then easy to check that the left side of (**11**) is just $\frac{1}{2} |\epsilon_o|^2 E$, which proves the theorem. [It is not surprising that $\epsilon_o \neq 0$ is required; there are no well-behaved harmonic ($\nabla^2 \psi = 0$) functions that vanish at infinity in a "normal" elliptic space.] As mentioned earlier, the left side of (**11**) provides an example of an alternative energy definition in terms of a Killing spinor ϵ_o and of the interior geometry as represented by the harmonic spinor $\epsilon(x)$.

We now come to the third step—deriving the Witten proof from "classical" supergravity.[20,21] Again, we will omit the fine points and emphasize the power available from knowing that a square root of coordinate translations exists if an auxiliary classical fermionic field ψ_μ is introduced (it obeys "anti"-Poisson bracket dynamics). The logic is as follows. First, we define a local supersymmetry transformation on classical metric fields by introducing an auxiliary variable ψ_μ which is just a classical field times a Grassmann parameter (for anticommutation). It looks just like the usual rule, with the parameter $\alpha(x)$ also a Grassmann element. It then follows that the result of two such transformations on the metric is a local translation, $[\delta_1, \delta_2] g_{\mu\nu} = K^\alpha_{12} \partial_\alpha g_{\mu\nu}$ where $K^\alpha_{12} = 2i\, \overline{\alpha}_1 \gamma^\alpha \alpha_2$, and we set $\psi_\mu = 0$ after transforming. Next we note that there exists an extension of classical gravity—classical supergravity—invariant under these transformations to first order in ψ_μ (we can drop torsion). The following step is to define the generators $Q(\alpha)$ of these transformations, since their square will correspond to the metric's Hamiltonian. In this process, we need not consider gravitational dynamics at all (since these would only enter in two successive transformations of a fermi field), and can restrict ourselves to an initial asymptotically flat surface $t = 0$. Here the dynamics reduce to the Einstein ($G_{o\mu} = 0$) and spin 3/2 ($R^o = 0$) constraints, since energy is defined solely in terms of initial data. The supersymmetry generators Q should be such that their Poisson bracket with the initial ψ_i have the required form

$$\{Q, \psi_i\} = 2D_i \alpha(x) \tag{12}$$

We reiterate that for our purpose, we need not worry about how Q affects metric variables. Once we have the physical subspace of dynamical ψ variables, the form of Q will be manifest. To obtain this space, one decomposes ψ_i into a gauge, constraint, and dynamical part according to a generalized breakup appropriate to the $G_{o\mu} = 0$ background. This turns out to be relatively simple, and the $R^o = 0$ constraint, which reads $\sigma^{ij} D_i \psi_j = 0$, implies that the constraint part of ψ_i vanishes. Then ψ_i is just the sum of a pure gauge part, $D_i \eta$ and a dynamical ψ_i^{Tt} part, such that $\gamma \cdot \psi^{Tt} = 0 = D \cdot \psi^{Tt}$. The pure gauge part (like the longitudinal vector potential) drops out of the dynamics automatically in this essentially abelian situation, so that we do not need to impose interior gauge conditions on ψ. Physically, it does not matter to the overall generators (E, Q) what choice of coordinates or ψ-gauge has been used in the interior. So the physical (Tt) subspace is characterized by ψ^{Tt} and its conjugate momentum π^{Tt} according to the (anti) Poisson bracket algebra

$$\{\pi^{Tt}(\mathbf{r})^{i}, \psi_j^{Tt}(\mathbf{o})\} = [\delta_j^i (\mathbf{r})]^{Tt} \tag{13}$$

Here there are no gauge ghosts. It is now clear from (**12**) and (**13**) that Q must have the form

$$Q = \int d^3x[\pi^i 2D_i\alpha + iD^i\alpha\psi_i] \tag{14}$$

However, since the field variables in (**14**) are Tt only, only the Tt parts of the vector $D_i\alpha$ enter; this means that only those transformations for which

$$\gamma^i D_i\alpha = 0 \tag{15}$$

(and therefore $D^2\alpha = 0$) is satisfied will be generated. But this is just right physically since we have seen that such α are entirely determined by their constant asymptotic values α_o, which correspond to the global supersymmetry transformations we are interested in. This is made manifest by the alternate form of (**14**) implicit in the (Tt) property:

$$Q = \int dS^i [\pi_i 2\alpha_o + i\alpha_o \psi_i] \tag{16}$$

We now complete the link to Witten's results: if we take $[Q, Q]$ according to (**13**) and (**14**),

$$\tfrac{1}{4}i[Q, Q] = \int d^3x |\nabla\alpha|^2 \geqq 0 \tag{17}$$

On the other hand, using (**14**) and (**16**) together, we have

$$\tfrac{1}{4}i[Q, Q] = \int dS^i \alpha_o \nabla_i\alpha = \tfrac{1}{2}|\alpha_o|^2 E \tag{18}$$

The system (**15**), (**17**), and (**18**) is just Witten's framework.

To summarize the foregoing, Witten's proof can be viewed as the implementation of a "virtual" supersymmetry of classical gravity. Here the energy is expressed as a positive functional of a spinor which is defined, through the surface Dirac equation (**15**), as the continuation in the interior of the asymptotic (global) supersymmetry parameter. The square of the latter is the constant infinitesimal global time translation parameter which goes with the Hamiltonian.

This completes our discussion of the positive energy theorems for quantum supergravity (with arbitrary $1 \leqq N \leqq 8$) and classical gravity (which is a supersymmetric "point" of classical supergravity), for $\Lambda = 0$.

Models with nonvanishing cosmological constant must be dealt with separately according to whether it is positive or negative. For negative Λ, one may prove that general relativity has positive timelike Killing energy, defined as in our earlier discussion. This is because there exists a locally supersymmetric extension corresponding to the graded global (anti) de Sitter algebra. This cosmological supergravity has positive energy[1] as does its classical limit. Here the vacuum state is the maximally symmetric analogue of flat space—anti–de Sitter space. Although this space has many peculiar geometrical aspects in its time evolution, it does possess a timelike Killing vector with respect to which energy is defined on a given spacelike surface. The energy here is the "rotation" generator J_{05} (which reduces to P_0 as $\Lambda \to 0$). The latter is the square of the corresponding supercharge Q_0, so the theory has positive energy. This then also implies positive classical gravity energy either by the $\hbar \to 0$ argument, or by an extension of our Witten argument discussion. Explicit study of small fluctuations about anti–de Sitter space corroborates this general result when due care is taken to identify the correct modes.

The case $\Lambda > 0$, however, is not amenable to grading (there are a number of physical and mathematical reasons for the asymmetry from $\Lambda < 0$; they may be traced to the fact that $\sqrt{-\Lambda}$ is the relevant parameter). In addition, its vacuum state, de Sitter space, only allows for a timelike Killing vector inside the event horizon present to any observer there. This vector becomes spacelike as it is followed through the horizon. The physical picture is somewhat less clear here. What is known is that small metric fluctuations (about de Sitter space) inside the horizon have positive energy, and we believe that this is true for general interior excitations. It now appears that stability may persist for small excitations in more global solutions,[23] at least classically.

SEMICLASSICAL STABILITY

We do not yet have a meaningful quantized Einstein gravity theory because of renormalizability problems, nor is it known to have positive energy even formally, since the quantum supergravity proof only assures us that classical relativity is stable. What about semiclassical stability? The generic source of instability at this level is the possibility of tunneling from a "false" vacuum to a more stable one through a finite barrier. Since energy is of course conserved, there is danger only if there exist "bounce" (instanton-like) solutions whose asymptotic behavior is the same as that of the false vacuum. Instability is therefore precluded when a positive energy theorem exists, because then there are no other zero energy states for vacuum to decay into.[24] For example, in $\Lambda = 0$ gravity with positive sources, we know that all solutions behave effectively as $g_{\mu\nu} \sim \eta_{\mu\nu} + m/r + 0(1/r^2)$ and asymptotic behavior with $m = 0$ is unique to flat space. However, this need not be the case for compactified higher dimensional (Kaluza-Klein) models. We refer here to the detailed discussions in References 7 and 25. Witten concludes that tunneling could occur there, but only between solutions of different topology.[c] The instabilities, for a compactified fifth dimension, can be pictured as due to a singularity that, as it expands, actually "pushes" all geometry out to infinity. A formally similar instability occurs for four-dimensional gravity at finite temperature, where it is black holes that grow by accreting thermal radiation: according to Reference 27 this may imply that at finite temperatures, relativity is not stable.

For the global $\Lambda > 0$ de Sitter case, there are no known bounce solutions (or general theorems forbidding them). One can explicitly exclude[1] some obvious candidates such as the Schwarzschild–de Sitter metric, where the black hole a priori could, but does not, exhibit the above unstable expansion.

A rather different problem concerns tensor-scalar theories, which also result from

[c]Topological instability can also exist in ordinary gravity,[26] although the issue seems rather academic there.

dimensionally reduced purely geometric Kaluza-Klein models. These involve a long-range scalar field which violates the equivalence principle and endangers the positivity proofs because it couples nonminimally ($\sim\phi R$) to curvature, resulting in a nonpositive stress tensor. The conformal-invariant coupling of scalars to gravity (with the well-known $\frac{1}{6} R\phi^2$ "improvement" term) is also of this type. However, it is possible (by rescaling the metric) to reduce these systems to minimal coupling form with positive effective scalar field stress-tensor. Stability in terms of the rescaled geometry is then regained,[28] at least if the scalar field amplitudes do not exceed the Planck mass. The physics here is not yet completely clear, although a conflict between matter conformal invariance and positive energy seems to be averted. A rather different question involves scalar fields and associated potentials $V(\phi)$ which occur in gauged $N \geqq 4$ supergravity and which act like "variable" cosmological constants; here too boundary conditions on the scalars play an essential role.[29,30] On the basis of the $\Lambda < 0$ results,[1] it has been possible to understand in detail how the apparently unbounded (from below) scalar field potentials in gauged higher N supergravities do not defeat stability.[29,30]

Finally, we mention the higher derivative gravity models, whose Lagrangian has the generic form $\kappa^{-2}R + \alpha R^2_{\mu\nu} + \beta R^2$.[d] Although these share the formal supersymmetrizability of pure gravity, and can be embedded in corresponding supergravities, this does not mean they are stable. Indeed, we know that they necessarily involve ghost excitations and expect a corresponding instability. The resolution of this apparent paradox involves the negative Hilbert space metric of a ghost sector:[31] the formal $H = Q^2$ relation does *not* entitle one to deduce positivity, since the "square" of the ghost parts of Q actually carries a minus sign. Classically, the right side corresponding to (**17**) will now also carry negative terms.

SUMMARY

The physical meaning of energy in generally covariant gravitational models is analyzed and shown to lead to a simple invariant formal definition. From this basis we state and sketch the derivation of the recent positive energy and stability theorems for Einstein and supergravities with arbitrary cosmological constant. Other models, such as tensor-scalar and higher derivative gravity, are briefly considered, as is semiclassical stability and Kaluza-Klein theory. We conclude that energy is unique, well defined, and has all the desirable stability properties for physically relevant solutions in most gravitational models with or without cosmological constant. It is positive-definite for all classical solution of Einstein's equations asymptotic to the appropriate vacuum—Minkowski or de Sitter spaces. Energy positivity also holds, at least formally, in all (quantized) supergravities with arbitrary N.

REFERENCES

1. ABBOTT, L. & S. DESER. 1982. Nucl. Phys. **B195:** 76.
2. ARNOWITT, R., S. DESER & C. W. MISNER. 1959. Phys. Rev. **116:** 1322.

[d]The case $\alpha = 0$ is degenerate and represents a tensor-scalar theory, as can be seen by writing it in the equivalent form $(\kappa^{-2} - 2\beta f)R - \beta f^2$ and rescaling the metric to remove the coefficient of R. The Lagrange multiplier f then becomes a scalar field variable.

3. Arnowitt, R., S. Deser & C. W. Misner. 1960. Phys. Rev. **117:** 1595.
4. Arnowitt, R., S. Deser & C. W. Misner. 1960. Phys. Rev. **118:** 1100.
5. Arnowitt, R., S. Deser & C. W. Misner. 1961. Phys. Rev. **122:** 997.
6. Abbott, L. & S. Deser. 1982. Phys. Lett. **116B:** 259.
7. Witten, E. 1981. Comm. Math Phys. **80:** 381.
8. Brill, D. R. & P. S. Jang. 1980. *In* General Relativity and Gravitation. A. Held, Ed. **1:** 173. Plenum Press. New York, N.Y.
9. York, J. 1980. *In* Essays in General Relativity. F. J. Tipler, Ed. Academic Press. New York, N.Y.
10. Choquet, Y. & S. Deser. 1973. Ann. Phys. **81:** 165.
11. Choquet, Y. & J. Marsden. 1976. Comm. Math Phys. **51:** 283.
12. Brill, D. R. & S. Deser. 1968. Ann. Phys. **50:** 548.
13. Brill, D. R., S. Deser & L. Faddeev. 1968. Phys. Lett. **26A:** 538.
14. Schoen, R. M. & S. T. Yau. 1981. Comm. Math. Phys. **79:** 231.
15. Schoen, R. M. & S. T. Yau. 1979. Comm. Math. Phys. **65:** 45.
16. Schoen, R. M. & S. T. Yau. 1979. Phys. Rev. Lett. **43:** 1457.
17. Deser, S. & C. Teitelboim. 1977. Phys. Rev. Lett. **39:** 249.
18. Parker, T. & C. H. Taubes. 1982. Comm. Math Phys. **84:** 223.
19. Reula, O. 1982. J. Math Phys. **23:** 810.
20. Horowitz, G. T. & A. Strominger. 1983. Phys. Rev. **D27:** 2793.
21. Deser, S. 1983. Phys. Rev. **D27:** 2805.
22. Grisaru, M. T. 1978. Phys. Lett. **73B:** 207.
23. Abbott, L. & C. Burges. (In preparation.)
24. Schoen, R. M. & S. T. Yau. 1979. Phys. Rev. Lett. **42:** 547.
25. Witten, E. 1982. Nucl. Phys. **B195:** 481.
26. Brill, D. R. & S. Deser. 1973. Comm. Math. Phys. **32:** 291.
27. Gross, D. J., M. J. Perry & L. G. Yaffe. 1982. Phys. Rev. **D25:** 330.
28. Deser, S. (In preparation.)
29. Breitenlohner, P. & D. Z. Freedman. 1982. Phys. Lett. **115B:** 197.
30. Breitenlohner, P. & D. Z. Freedman. 1982. Ann. Phys. **144:** 249.
31. Boulware, D., S. Deser, G. Gibbons & K. Stelle. (In preparation.)

General Relativistic Collapse of Rotating Stars

TAKASHI NAKAMURA

Department of Physics
Kyoto University
Kyoto 606, Japan

INTRODUCTION

It is my pleasure to talk on numerical relativity here in Texas. Texas is a special place for numerical relativity. I think it was born here in 1976 because the famous work on two black hole collision by Smarr, Cadez, De Witt, and Eppley appeared that year.[1] Today, I would like to review a part of it, that is, general relativistic collapse of rotating stars.

FIGURE 1 shows a standard viewpoint of gravitational collapse of a rotating star to a black hole.[2] When a rotating star begins to collapse, the gravity becomes so strong that there appears a region from which even a photon cannot escape. After the distortion of space-time is radiated as gravitational waves, a Kerr black hole is formed finally. One of the main goals for numerical relativity is to simulate the collapse of a rotating star (such as FIGURE 1) under realistic initial conditions. However, to know both the dynamics of matter and the propagation of gravitational radiation seems to be very difficult. Therefore we divide the problem into four stages.

Stage 1. We calculate the time evolution of pure gravitational waves in a two-dimensional (2D) code (2-dimension means the dimension of spatial coordinates is two). In this stage the question is, Is it possible to simulate the propagation of gravitational waves by a computer? The answer is yes. Smarr, Eppley, and Miyama have calculated this problem.[3–5]

Stage 2. We try to know the dynamics of a collapsing, rotating star in 2D code. In this stage, we do not try to estimate the amount of the gravitational waves. The relevant question to this stage is, Is it possible to simulate the collapse of a rotating star (or matter) up to the formation of an apparent horizon? The answer to this question will be given later in this paper.

Stage 3. Combining the techniques from stages 1 and 2, we simulate FIGURE 1 in 2D, that is, we try to know both the dynamics of matter and the propagation of gravitational waves generated by the nonspherical motion of matter. The difficulty of stage 3 is that we must treat both the strong gravity region and the weak gravity region where gravitational waves propagate at the same time.[6,7]

Stage 4. We simulate FIGURE 1 in 3D.

Here, I would like to show some numerical results from stage 2. Let us define a nondimensional quantity q by

$$q \equiv J/(M \cdot \frac{GM}{c^2} \cdot c)$$

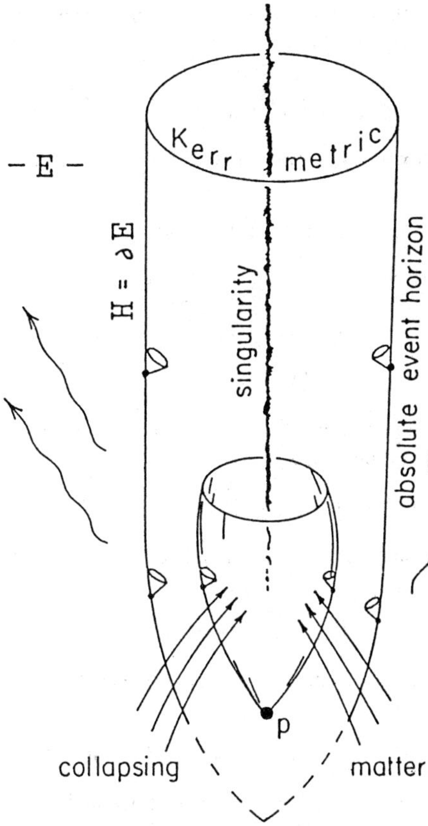

FIGURE 1. Standard viewpoint of gravitational collapse of a rotating star to a black hole. (Reproduced from Reference 2 with permission from the publisher.)

where J and M are the total angular momentum and the gravitational mass of the star. As q corresponds to a/M in Kerr metric, q should be smaller than unity if a Kerr black hole is finally formed as in FIGURE 1. However, q of a collapsing star can be greater than unity. For example, q of a massive star of mass $M \gtrsim 10M_\odot$ is believed to be greater than unity.[8] We note that q of the sun is 0.18 and q of a new fast pulsar[9] is 0.24 if $M = 1.4\ M_\odot$ and the moment of inertia is 10^{45} cgs.

Questions relevant to stage 2 are: (1) If the collapsing star has $q > 1$, what happens? (2) If $q < 1$, is a Kerr black hole really formed? Does no naked singularity as in Tomimatsu-Sato metrics appear?

[(2 + 1) + 1]-FORMALISM

In the axially symmetric space-time, there is a rotational Killing vector (ξ_μ). This implies we can divide out the Killing direction by using Geroch's three-dimensional formalism for space-time with one Killing vector.[10] Let us define λ, $h_{\mu\nu}$, ω_μ, and $\epsilon_{\mu\nu\rho}$ as

$$\lambda^2 = \xi_\mu \xi^\mu,\ h_{\mu\nu} = g_{\mu\nu} - \lambda^{-2}\xi_\mu\xi_\nu,$$

$$\omega_\mu = \epsilon_{\mu\nu\rho\sigma}\xi^\mu\nabla^\rho\xi^\sigma,\ \epsilon_{\mu\nu\rho} = \lambda^{-1}\xi^\sigma\epsilon_{\mu\nu\rho\sigma}$$

Then, the four-dimensional Einstein equations become the set of equations for $h_{\mu\nu}$, λ, and ω_μ;

$$^{(3)}R_{\mu\nu} = (2\lambda^4)^{-1}[\omega_\mu\omega_\nu - h_{\mu\nu}\,\omega^\rho\omega_\rho] + \lambda^{-1}D_\mu D_\nu\lambda + 8\pi[J_{\mu\nu} - h_{\mu\nu}(J^\rho_\rho + \lambda^{-2}J)/2] \qquad (1)$$

$$\lambda^{-1}D^\rho D_\rho\lambda = -(2\lambda^4)^{-1}\omega^\rho\omega_\rho - 4\pi(\lambda^{-2}J - J^\rho_\rho) \qquad (2)$$

$$D_{[\mu}\omega_{\nu]} = 8\pi\lambda\epsilon_{\mu\nu\rho}J^\rho \qquad (3)$$

$$D^\rho[\lambda^{-3}\omega_\rho] = 0 \qquad (4)$$

where D^μ and $^{(3)}R_{\mu\nu}$ are covariant derivative and three-dimensional Ricci tensor with respect to $h_{\mu\nu}$. J, J_ρ, and $J_{\rho\sigma}$ are defined by

$$J = T_{\mu\nu}\xi^\mu\xi^\nu,\ J_\rho = h_{\mu\rho}\xi_\nu T^{\mu\nu},\ J_{\rho\sigma} = h_{\rho\mu}h_{\sigma\nu}T^{\mu\nu}$$

where $T_{\mu\nu}$ is the energy-momentum tensor of matter. Next we define a projection tensor as

$$H_{\mathrm{ab}} = h_{\mathrm{ab}} + n_{\mathrm{a}}n_{\mathrm{b}}$$

where n_{a} is a unit normal of t = constant hypersurface. We define α (lapse function) and η^{A} (shift vector) as

$$ds^2 = h_{\mathrm{ab}}dx^{\mathrm{a}}dx^{\mathrm{b}} = -\alpha^2dt^2 + H_{\mathrm{AB}}(dx^{\mathrm{A}} + \eta^{\mathrm{A}}dt)(dx^{\mathrm{B}} + \eta^{\mathrm{B}}dt)$$

Let the extrinsic curvature be expressed by χ_{AB} (small Latin indices refer to the ranges 0, 1, 2 and capital Latin ones, 1, 2). Moreover, we define various quantities as

$$H = det\,(H_{\mathrm{AB}}),\ \chi = \chi^{\mathrm{A}}_{\mathrm{A}},\ \lambda K^\varphi_\varphi = -h^\mu\partial_\mu\lambda,\ E^{\mathrm{A}} = \epsilon^{\mathrm{AB}}H^{\mathrm{b}}_{\mathrm{B}}\omega_{\mathrm{b}}\lambda^{-2},$$

$$B_\varphi = n_{\mathrm{a}}\omega^{\mathrm{a}}\lambda^{-2},\ \epsilon_{\mathrm{AB}} = n^{\mathrm{c}}\epsilon_{\mathrm{CAB}},\ \rho_{\mathrm{H}} = n_{\mathrm{a}}n_{\mathrm{b}}J^{\mathrm{ab}},\ J_\varphi = -n_{\mathrm{a}}J^{\mathrm{a}},$$

$$J^{\mathrm{A}} = -n_{\mathrm{a}}H^{\mathrm{A}}_{\mathrm{b}}J^{\mathrm{ab}},\ S^{\mathrm{A}} = H^{\mathrm{A}}_{\mathrm{b}}J^{\mathrm{b}},\ S_{\mathrm{AB}} = H_{\mathrm{Aa}}H_{\mathrm{Bb}}J^{\mathrm{ab}}$$

If we carry out projection of all the tensors appeared in Equations 1–4 like the (3 + 1)-split, we obtain the [(2 + 1) + 1]-representation of the Einstein equations[11] as

$$\partial_{\mathrm{o}}H_{\mathrm{AB}} = -2\alpha\chi_{\mathrm{AB}} + \eta_{\mathrm{A}\|\mathrm{B}} + \eta_{\mathrm{B}\|\mathrm{A}} \qquad (5)$$

$$\begin{aligned}\partial_{\mathrm{o}}\chi_{\mathrm{AB}} = {} & \eta^{\mathrm{c}}\chi_{\mathrm{AB}\|\mathrm{C}} + \alpha[^{(2)}R_{\mathrm{AB}} + \chi\chi_{\mathrm{AB}}] - 2\alpha\chi^{\mathrm{C}}_{\mathrm{A}}\chi_{\mathrm{CB}} - \alpha_{\|\mathrm{A}\|\mathrm{B}} \\ & + (\chi_{\mathrm{AC}}\eta^{\mathrm{C}}_{\|\mathrm{B}} + \chi_{\mathrm{BC}}\eta^{\mathrm{C}}_{\|\mathrm{A}}) - \alpha\lambda^{-1}\lambda_{\|\mathrm{A}\|\mathrm{B}} + \alpha K^\varphi_\varphi\chi_{\mathrm{AB}} - \frac{\alpha}{2} \\ & [\epsilon_{\mathrm{CA}}\epsilon_{\mathrm{DB}}E^{\mathrm{C}}E^{\mathrm{D}} - H_{\mathrm{AB}}(E^{\mathrm{C}}E_{\mathrm{C}} - B^2_\varphi)] \\ & - 8\pi\alpha[S_{\mathrm{AB}} + \frac{1}{2}H_{\mathrm{AB}}\,(\rho_{\mathrm{H}} - S^{\mathrm{C}}_{\mathrm{C}} - \lambda^{-2}J)]\end{aligned} \qquad (6)$$

$$\partial_o \lambda = \eta^A \partial_A \lambda - \alpha \lambda K^\varphi_\varphi \tag{7}$$

$$\partial_o K^\varphi_\varphi = \eta^A \partial_A K^\varphi_\varphi + \alpha K^\varphi_\varphi (K^\varphi_\varphi + \chi) - H^{AD} \alpha_{\|A} \lambda_{\|B} \lambda^{-1} - \alpha^{(2)} \Delta \lambda \cdot \lambda^{-1}$$
$$- \alpha (E_A E^A - B^2_\varphi)/2 - 4\pi\alpha [\rho_H - S^A_A + \lambda^{-2} J] \tag{8}$$

$$\chi^2 - \chi^{AB}\chi_{AB} + {}^{(2)}R = 2\lambda^{-1}\,{}^{(2)}\Delta\lambda - 2\chi K^\varphi_\varphi + (E_A E^A + B^2_\varphi)/2 + 16\pi\rho_H \tag{9}$$

$$\lambda^{-1}(\lambda\chi^B_A)_{\|B} - \lambda^{-1}\lambda_{\|A} K^\varphi_\varphi - \partial_A(\chi + K^\varphi_\varphi) = 8\pi J_A - B_\varphi \epsilon_{CA} E^C/2 \tag{10}$$

$$\partial_o(\lambda^2 \sqrt{H} E^A) = (\eta^B(\lambda^2 E^A)_{\|B} + \epsilon^{AB}\eta^C{}_{\|B}\epsilon_{DC}\lambda^2 E^D)\sqrt{H}$$
$$+ \sqrt{H}\,\epsilon^{AB}\partial_B(\alpha\lambda^2 B_\varphi) - 16\pi\alpha\lambda S^A \sqrt{H} \tag{11}$$

$$\partial_o(\lambda^{-1}\sqrt{H} B_\varphi) = \partial_A(\eta^A \sqrt{H} B_\varphi \lambda^{-1}) + \partial_A(\alpha E_B \epsilon^{BA} \sqrt{H} \lambda^{-1}) \tag{12}$$

$$(\lambda^2 E^A)_{\|A} = 16\pi\lambda J_\varphi \tag{13}$$

In Equations 5–13, ${}^{(2)}\Delta$, $\|$, ${}^{(2)}R$, and ${}^{(2)}R_{AB}$ are Laplacian, covariant differentiation, scalar curvature, and Ricci tensor with respect to H_{AB}, respectively.

We assume the perfect fluid for $T_{\mu\nu}$,

$$T_{\mu\nu} = \rho(1 + \epsilon + p/\rho) u_\mu u_\nu + p g_{\mu\nu}$$

TABLE 1. Initial Parameters and Main Results of the Calculated Models[a]

Model	Case	q	U	J	Time Slice	Horizon?[b]	Remarks
A146	A	1.46	0.94	0.77	max	no	jet
A122	A	1.22	0.86	0.51	max	no	jet
A105	A	1.05	0.84	0.37	max	yes	
A93	A	0.93	0.82	0.29	max	yes	
A75	A	0.75	0.82	0.19	max	yes	
A50	A	0.50	0.81	0.08	max	yes	
B143	B	1.43	1.01	1.20	max	no	expanding disk
B121	B	1.21	0.88	0.76	max	no	expanding disk
B104	B	1.04	0.84	0.54	max	no	expanding disk
B92	B	0.92	0.82	0.42	max/hyper	yes	
B74	B	0.74	0.81	0.27	max/hyper	yes	
B51	B	0.51	0.81	0.12	max	yes	
C137	C	1.37	0.25	1.56	hyper	no	expanding ring
C109	C	1.09	0.23	0.79	hyper	no	expanding ring
C95	C	0.95	0.22	0.55	hyper	no	expanding ring
C86	C	0.86	0.21	0.45	hyper	yes	ring singularity
C80	C	0.80	0.21	0.36	hyper	yes	ring singularity
C64	C	0.64	0.20	0.22	hyper	yes	ring singularity
C56	C	0.56	0.20	0.17	hyper	yes	ring singularity
C48	C	0.48	0.20	0.12	hyper/max	yes	ring singularity
C32	C	0.32	0.20	0.05	hyper	yes	

[a]In the sixth column, max and hyper mean the maximal slicing and the hypergeometric slicing, respectively. In the eighth column, the important characteristic of each model is written.

[b]A method for identifying an apparent horizon is shown in Reference 22.

where ρ, ϵ, and p are proper mass density, internal energy per gram, and pressure, respectively. The hydrodynamics equations can be written by using J_A, J_ψ, ρ, and ρ_H as (1) energy equation:

$$\partial_o(\alpha u^o \sqrt{H}\lambda\rho\epsilon) + \partial_A(\alpha u^o \sqrt{H}\lambda U^A\rho\epsilon) = -p[\partial_o(\alpha u^o \sqrt{H}\lambda) + \partial_A(\alpha u^o \sqrt{H}\lambda U^A)] \quad \mathbf{(14)}$$

(2) Euler equations:

$$\partial_o(\lambda \sqrt{H}J_A) + \partial_B(U^B\lambda \sqrt{H}J_A) = -\alpha\lambda \sqrt{H}(\partial_A p + (p + \rho_H)\alpha^{-1}\partial_A\alpha)$$

$$+ \alpha\lambda \sqrt{H}(p + \rho_H)[(\partial_A H_{BC})V^B V^C/2 + \alpha^{-1}V_C\partial_A\eta^C]$$

$$+ \alpha\lambda \sqrt{H}\lambda^{-1}J_\varphi[E_A + \epsilon_{AC}(2B_\varphi V^C - \lambda^{-2}\epsilon^{CB}\lambda_{\|B}V^\varphi)] \quad \mathbf{(15)}$$

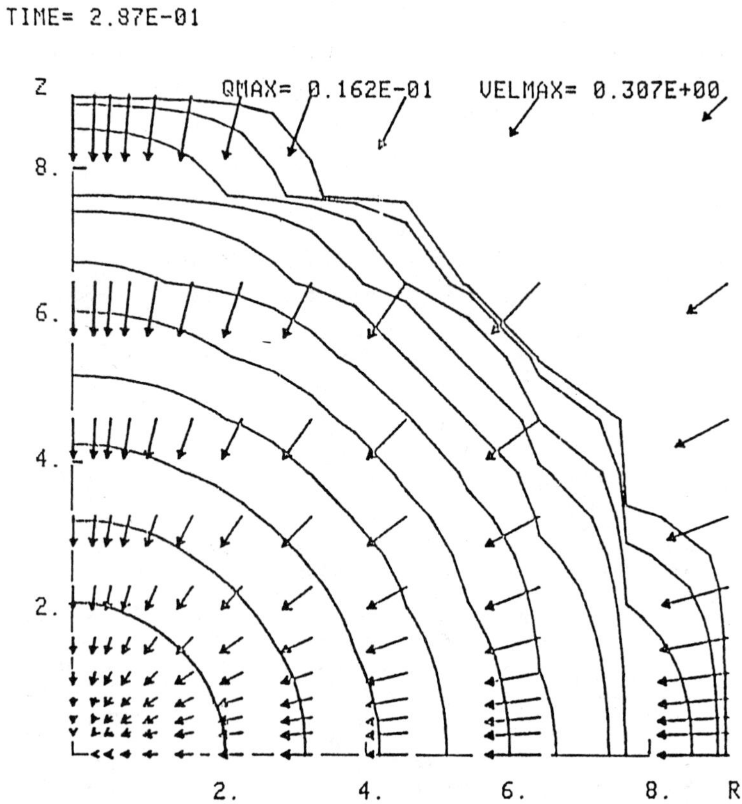

FIGURE 2a. Contour lines of Q_b at $t = 0.287$ for A50. The space integral of Q_b becomes M_B, that is, $2\pi \int_0^\infty Q_b R dR dZ = M_B$. The precise definition of Q_b can be found in Reference 14. Each line corresponds to $Q_b = Q_{MAX}\, 10^{-n/2}$ for $n = 1, 2, \ldots, 11$. Q_{MAX} is shown in the figure. Arrows show the vector $(J_R/Q_b, J_z/Q_b)$. The maximum of this vector is shown in the figure as VELMAX.

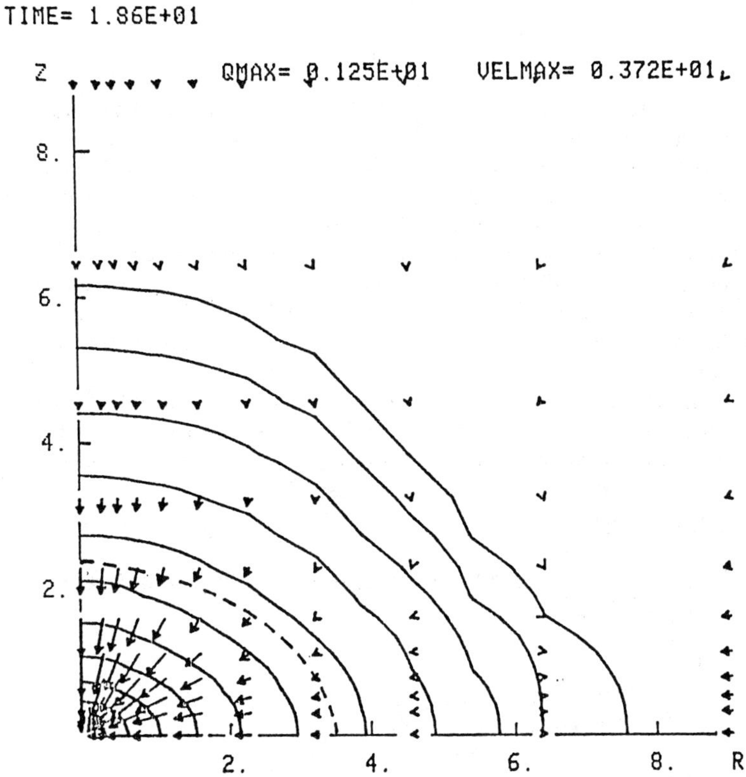

FIGURE 2b. The contour lines of Q_b for A50 at $t = 18.6$. The notations are the same as FIGURE 2a. The dashed line shows the apparent horizon.

(3) conservation of angular momentum:

$$\partial_0(\lambda\sqrt{H}J_\varphi) + \partial_A(U^A\lambda\sqrt{H}J_\varphi) = 0 \tag{16}$$

(4) conservation of baryon number:

$$\partial_0(\alpha u^0\sqrt{H}\lambda\rho) + \partial_A(U^A\alpha u^0\sqrt{H}\lambda\rho) = 0 \tag{17}$$

(5) equation of state:

$$p = p(\rho, \epsilon) \tag{18}$$

(6) normalization of four-velocity:

$$\alpha u^0 = 1/\sqrt{1 - V^B V_B - V^\varphi V_\varphi} \tag{19}$$

where

$$V^B = (p + \rho_H)^{-1}J^B, J^\varphi = (p + \rho_H)V^\varphi$$

and

$$U^{A} = \alpha V^{A} - \eta^{A} \tag{20}$$

One can see that the basic equations in the [(2 + 1) + 1]-formalism strongly resemble those of a nonrotating system with toroidal magnetic field B_φ and poloidal electric fields E^A. Angular momentum density J_φ behaves like electric charge density. Conservation of angular momentum corresponds to conservation of electric charge. Moreover, there are "Lorentz force"–like terms in the Euler equations (**15**).

Our method for evolving the basic quantities is called the free evolution.[12] We solve the constraint equations (**9, 10,** and **13**) only at $t = 0$ to determine the initial data. For $t > 0$ we solve Equations 5–8, 11, and 12 as well as the hydrodynamics equations (**14–20**). We use the constraint equations for $t > 0$ to see the accuracy of the numerical calculations.[14]

We adopt the cylindrical coordinates R, Z, and φ; we assume that the system has reflection symmetry about $z = 0$ plane. In this case, the regularity conditions at $R = 0$ and the reflection symmetry enable us to use the following basic variables instead of

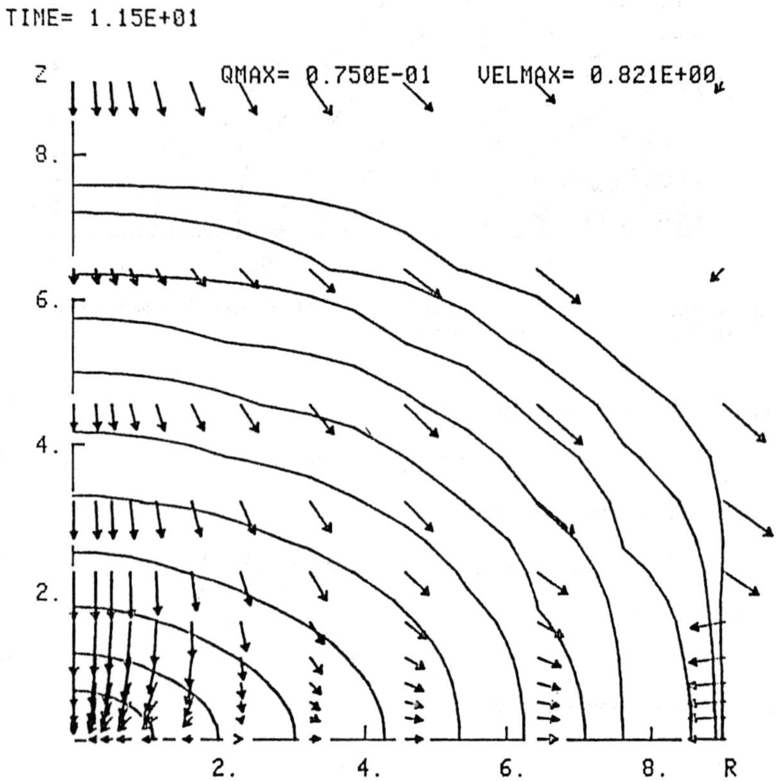

FIGURE 3a. The contour lines of Q_b for A105 at $t = 11.5$. The notations are the same as FIGURE 2a.

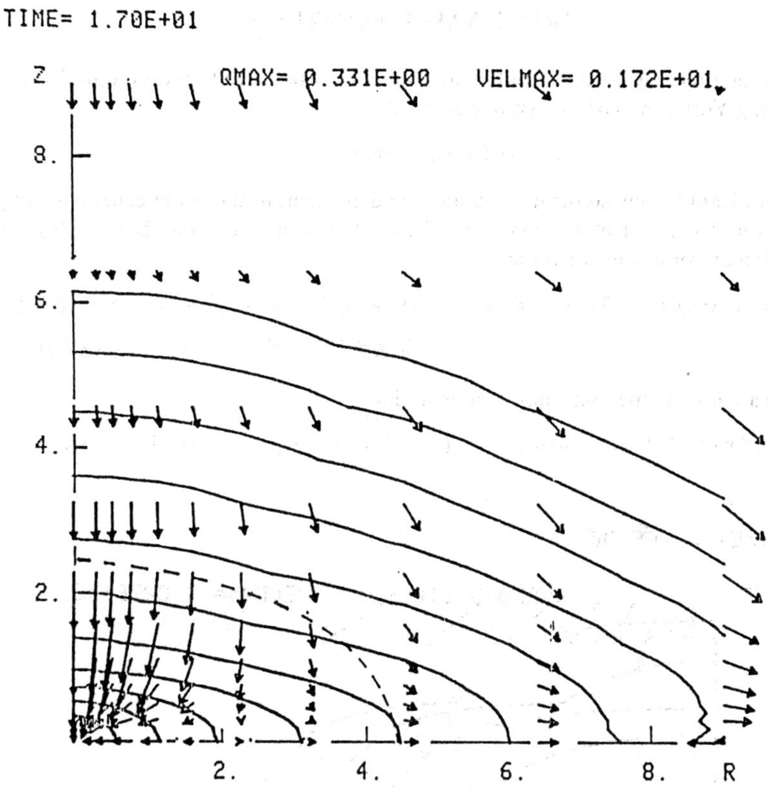

FIGURE 3b. The contour lines of Q_b for A105 at $t = 17.0$. The notations are the same as FIGURE 2b.

the original variables appeared in Equations 5–20:

$$B \equiv \lambda/R,\ a \equiv (\sqrt{H}_{RR} - B)/R^2,\ C \equiv H_{RZ}/R/Z,\ F \equiv \sqrt{H}_{ZZ},$$

$$k^R_R \equiv (\chi^R_R - K^\varphi_\varphi)/R^2,\ k^Z_R \equiv \chi^Z_R/R/Z,\ K^Z_Z \equiv \chi^Z_Z,$$

$$e^R \equiv E^R/R^2,\ e^Z \equiv E^Z/R/Z,\ b_\varphi \equiv B_\varphi/R^2/Z,\ \tilde{\eta}^R \equiv \eta^R/R,\ \tilde{\eta}^Z \equiv \eta^Z/Z,$$

$$Q_b \equiv \alpha u^0 B\sqrt{H}\rho,\ h = 1 + \epsilon + p/\rho,\ \alpha u^0 = B\sqrt{H}(\rho_H + \rho)/h/Q_b,$$

$$Q_R \equiv B\sqrt{H}J_R/Q_b\,R,\ Q_Z \equiv B\sqrt{H}J_Z/Q_b/Z,$$

$$\Omega \equiv B\sqrt{H}J_\varphi/Q_b/R^2,\ \tilde{V}^R \equiv V^R/R,\ \tilde{V}^Z \equiv V^Z/Z,$$

$$\tilde{U}^R \equiv \alpha\tilde{V}^R - \tilde{\eta}^R, \quad \text{and} \quad \tilde{U}^Z \equiv \alpha\tilde{V}^Z - \tilde{\eta}^Z$$

By this choice of the basic variables, the regularity at $R = 0$ is automatically satisfied. In general, any quantity Q is a function of $x(\equiv R^2)$ and $y(\equiv Z^2)$.

INITIAL VALUE EQUATIONS

The method for solving the initial value equations is beautifully reviewed by York.[13] Following York's method, we assume at $t = 0$,

$$\gamma_{ij} = \phi^4 (\gamma_{ij})_{\text{flat}} \quad \text{and} \quad K_{ij}^{TT} = 0$$

We solve Hamiltonian constraint equations and momentum constraint equations using Robin's and vector Robin's condition. The initial value equations become (a) the Hamiltonian constraint equation:

$$4(\partial_x \phi + x\partial_{xx}\phi) + 2\partial_y \phi + 4y\partial_{yy}\phi = -2\pi(\rho_H \phi^6)/\phi - \phi^5 x[x(\partial_x W^\varphi)^2 + y(\partial_y W^\varphi)^2]$$
$$- \phi^5[(K_R^R)^2 + (K_Z^Z)^2 + (K_\varphi^\varphi) + 2xy(k_R^Z)^2]/8$$

(b) the angular momentum constraint equation:

$$(8 + 24x\partial_x \ln \phi)\partial_x W\varphi + 4x\partial_{xx} W^\varphi + (2 + 2xy\partial_y \ln \phi)\partial_y W^\varphi + 4y\partial_{yy} W\varphi = 8nJ_\varphi x^{-1}$$

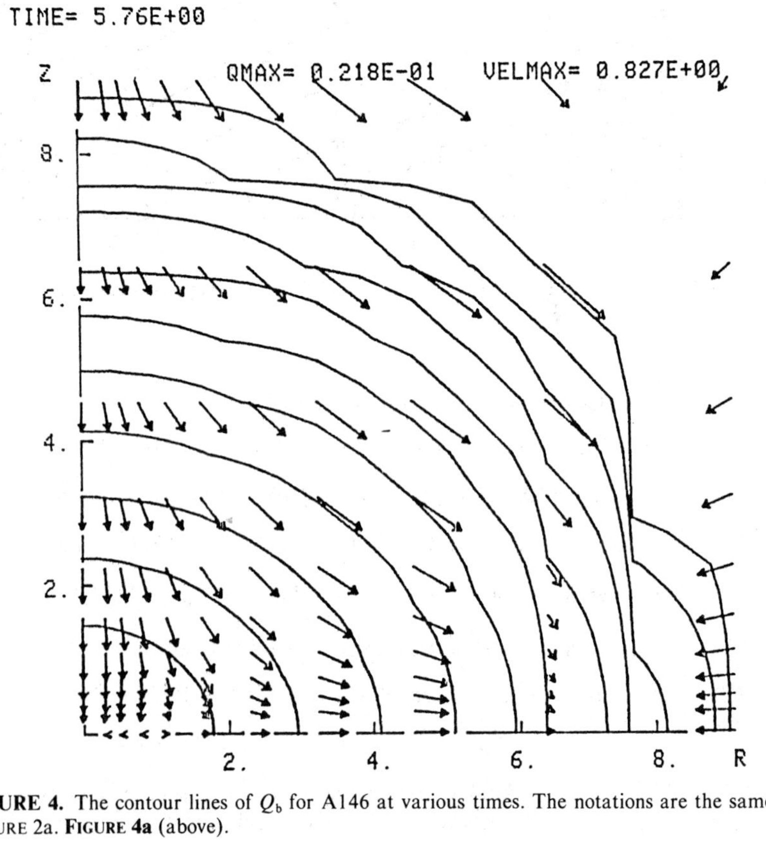

FIGURE 4. The contour lines of Q_b for A146 at various times. The notations are the same as FIGURE 2a. **FIGURE 4a** (above).

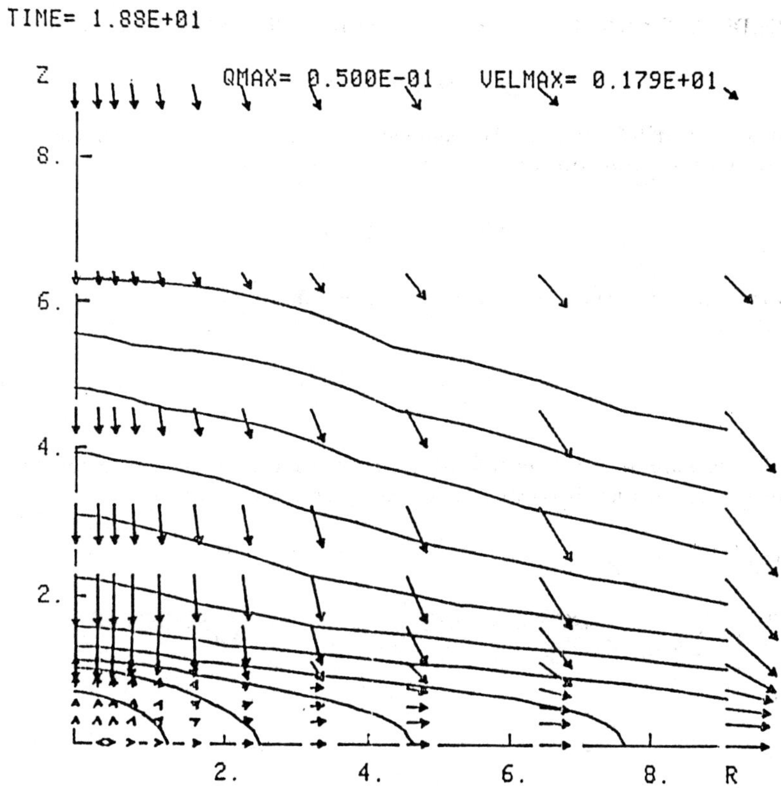

FIGURE 4b

(c) the momentum constraint equation:

$$16x/3\partial_{xx}w^R + 4y\partial_{yy}w^R + (32/3 + 32x\partial_x \ln\phi)\partial_x w^R$$

$$+ (2 + 24y\partial_y \ln\phi)\partial_y w^R + 8\partial_x \ln\phi w^R + 4/3y\partial_{xy}w^Z$$

$$+ (2/3 + 24y\partial_y \ln\phi)\partial_x w^Z - 8y\partial_x \ln\phi\partial_y w^Z - 8\partial_x \ln\phi w^Z = 8\pi J_R/R,$$

$$4x\partial_{xx}w^Z + 16/3y\partial_{yy}w^Z + (4 + 24x\partial_x \ln\phi)\partial_x w^Z$$

$$(8 + 32y\partial_y \ln\phi)\partial_x w^Z + 16\partial_y \ln\phi w^Z + 4/3x\partial_{xy}w^R$$

$$- 16x\partial_y \ln\phi\partial_x w^R + (4/3 + 24x\partial_x \ln\phi)\partial_y w^R - 16\phi^{-1}\partial_y\phi w^R = 8\pi J_Z/Z$$

where (Rw^R, Zw^Z, W^φ) is the longitudinal part of K_{ij}. The asymptotic behavior of ϕ, w^R, w^Z and W^φ for $r \to \infty$ is given by

$$\phi = 1 + C_1/r,\ w^\varphi = C_2/r^3$$

$$w^R = C_3/r^3 - C_4 3R^2/r^5/2 \quad \text{and} \quad w^Z = C_3/r^3 + C_4(2Z^2 + R^2)/2r^5$$

COORDINATE CONDITIONS AND METHOD FOR FINITE DIFFERENCE

Coordinate Conditions

We use zero shift vector, i.e., the coordinate lines agree with the normal lines. For the lapse function we use two kinds of slices: (1) maximal slicing, i.e.,

$$x_R^R + x_Z^Z + K_\varphi^\varphi = 0$$

(2) hyper geometric slicing.[14] In this slice, α is determined by

$$r^{-2}\frac{d}{dr}r^2\frac{d}{dr}\alpha = V_0\,\mathrm{sech}^2(dr)\,\alpha \tag{21}$$

where V_0 and d are two constants that are determined at each time step appropriately. The solution of Equation 21 is expressed by the hypergeometric function.

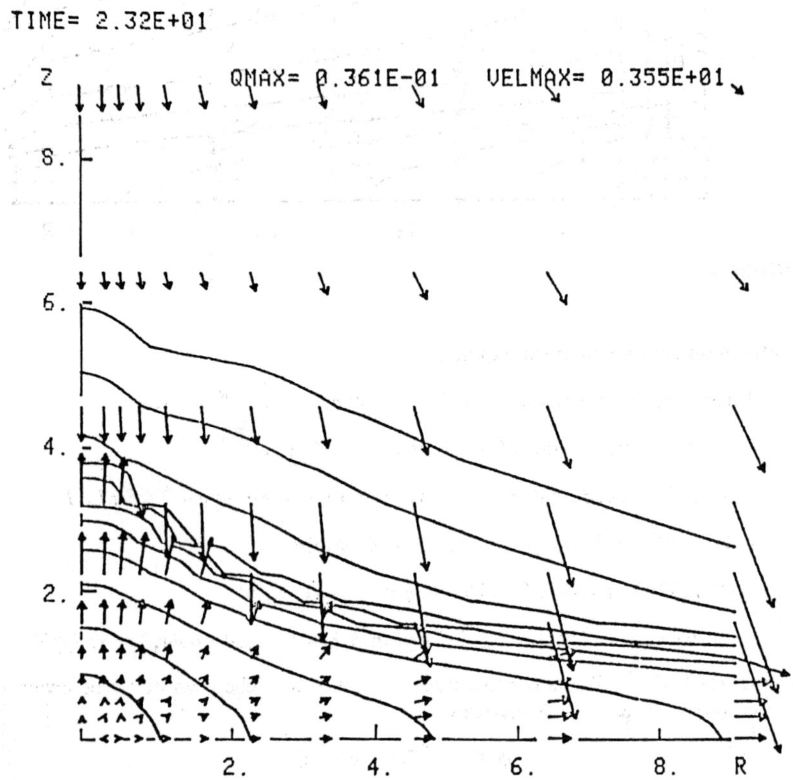

FIGURE 4c

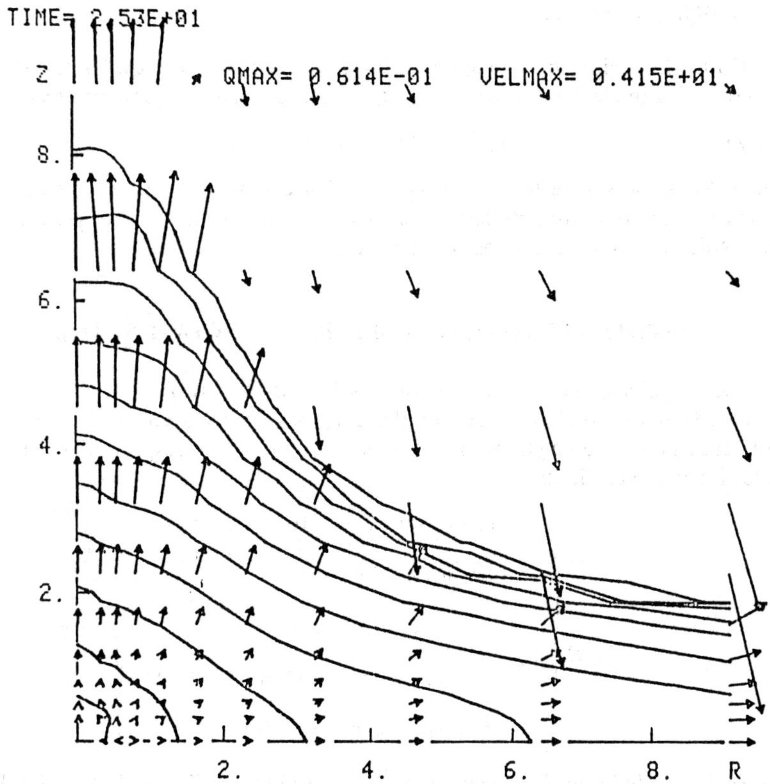

FIGURE 4d

Method for Finite Difference

Grids

A grid point is determined by

$$\left.\begin{aligned} x_{i+1} - x_i &= 1.4(x_i - x_{i-1}) \\ z_{i+1} - y_i &= 1.4(y_i - y_{i-1}) \end{aligned}\right\} i = 2, 3, \ldots$$

The number of grids is typically 28×28. Some of the models are calculated by using finer grids (42×42). The outermost grid point has a coordinate [25 (GM/c^2), 25 (GM/c^2)]. This is far enough for the gravitational potential. For example $\phi \sim 1.04$. But it is not far enough to estimate the gravitational wave whose wavelength is expected to be 16 (GM/c^2). This is the reason why we do not estimate the amount of gravitational waves in stage 2.

Finite Difference Method

For the hydrodynamics equations, a donor cell type method is used unless $R = 0$ or $Z = 0$. For example, for $R = 0$ and $Z = 0$, the hydrodynamics equations become

$$\partial_0 Q + (2\,\tilde{U}^R + \tilde{U}^Z)Q = S_Q$$

where S_Q is the source term. For the evolution equations of space-time geometry, Friedrichs-Lax type viscosity terms are added to make the numerical systems stable. (See Reference 14 for details on this method.)

GENERAL RELATIVISTIC COLLAPSE OF ROTATING STARS[a]

We use the units of mass, length, and time as $M = M_b$, $L = GM_b/c^2$, and $T = GM_b/c^3$ where M_b is the total rest mass of the baryon. Let $\rho_3(r)$ and r_0 be the density distribution of $N = 3$ polytrope and an initial radius of a star, respectively. We use the following initial conditions:

$$\rho_H\phi^6 = \begin{cases} \rho_3(r) & \text{for } r \leqq \rho_3^{-1}(10^{-6}\rho_3(0)) \equiv r^* \\ 10^{-6}\rho_3(0) & \text{for } r > r^* \end{cases} \tag{22}$$

$$J_R/R = J_Z/Z = \begin{cases} \rho_H c_v & \text{for } r \leqq r_0 \\ \rho_H c_v \exp\,(1 - (r/r_0)^2) & \text{for } r > r_0 \end{cases} \tag{23}$$

$$J_\varphi = x\rho_H\Omega_o \exp\,(-\,c_\varphi x/r_0^2) \tag{24}$$

where c_v, Ω_0, and c_φ are the constants that determine the initial velocity, the initial central angular velocity, and the rotation law, respectively.

As an equation of state, we use

$$p = \rho\epsilon/3$$

Initial distribution of ϵ is taken as

$$\epsilon = K\,\rho^{1/3}$$

Now r_0, c_v, Ω_0, c_φ, and K determine the initial conditions uniquely. In all the calculated models, r_0 is 10.5 and c_v is chosen so that the infall velocity at $r = r_0$ becomes the free-fall velocity. Instead of Ω_0 and K, we use $\tilde{J}$ and U defined by $\tilde{J} = E_{rot}/|E_{grav}|$ and $U = E_{int}/|E_{grav}|$ where E_{rot}, E_{int}, and E_{grav} are the rotational energy, the internal energy, and the gravitational energy of a star, respectively.

As the rotation laws, we use two c_φ's as (a) rotation law A; $c_\varphi = 2$ (almost rigidly rotating case); and (b) rotation law B; $c_\varphi = 10$ (strongly differential rotation case).

In TABLE 1, the initial parameters of each model are shown. Since the models are characterized mainly by q and the rotation law, we use them as a name of each model. For example A146 means the collapse of the rotating supermassive star with $q = 1.46$

[a] For details, see References 14–21.

and the rotation law A. Case C corresponds to the collapse of a star of $10 M_\odot$,[14] with

$$J_R = J_Z = 0, \rho = 3 \times 10^{13} \exp\{-(x + y)/4.5\}\ \mathrm{g/cm^3}, \qquad \Omega = \Omega_0 \exp(-x/4.5)$$

$$P = \begin{cases} \rho\epsilon/3 & \text{for } \rho \leqq \rho^* \equiv 3 \times 10^{14}\mathrm{g/cm^3} \\ (\rho - \rho^*)\epsilon + \rho^*\epsilon/3 & \text{for } \rho > \rho^* \end{cases}$$

where Ω is the initial angular frequency.

Rotation Law A

In this rotation law, the centrifugal force has a maximum value at $R = 0.35\ r_0$ where $\rho_3(R)$ is less than $0.1\ \rho_3(0)$. This means that if the mass shedding occurs, it will occur from the outer part of the star. In the following, we show the details of the numerical results of the three typical models.

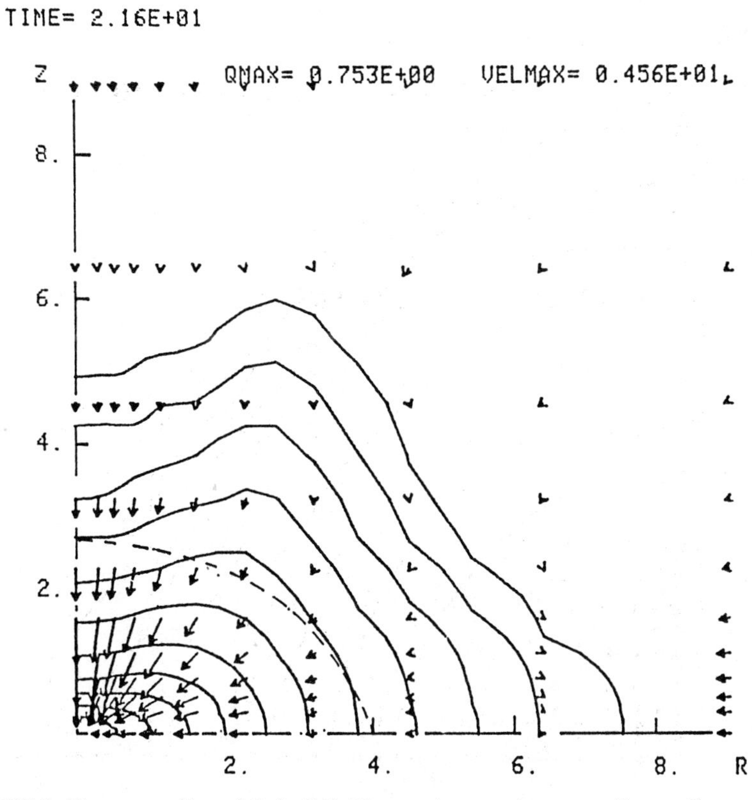

FIGURE 5. The contour lines of Q_b for B51. The notations are the same as FIGURE 2b.

Model A50

In this model, the rotation is very slow. At $t = 0.287$, the matter distribution is almost spherical and the velocity pattern shows the spherical collapse (FIGURE 2a). This feature is kept through the entire time up to when the apparent horizon (dashed line) is formed. (FIGURE 2b). Although the matter distribution becomes slightly oblate by the effect of rotation, all the matter will be swallowed into the slowly rotating black hole.

Model A105

In this model, the star is rotating rather rapidly. At $t = 11.5$ (FIGURE 3a), the matter falls vertically for $R \lesssim 2$. The collapse in the equatorial plane is considerably suppressed by the effect of rotation. For $2.4 \lesssim R \lesssim 7.7$, the outflow velocity reaches up to 0.3 c. Finally, the oblate-shaped core is formed in the central region and an apparent horizon is formed outside this core (FIGURE 3b). The outer envelope expands along the lateral direction with relativistic velocity. We can expect that some part of this

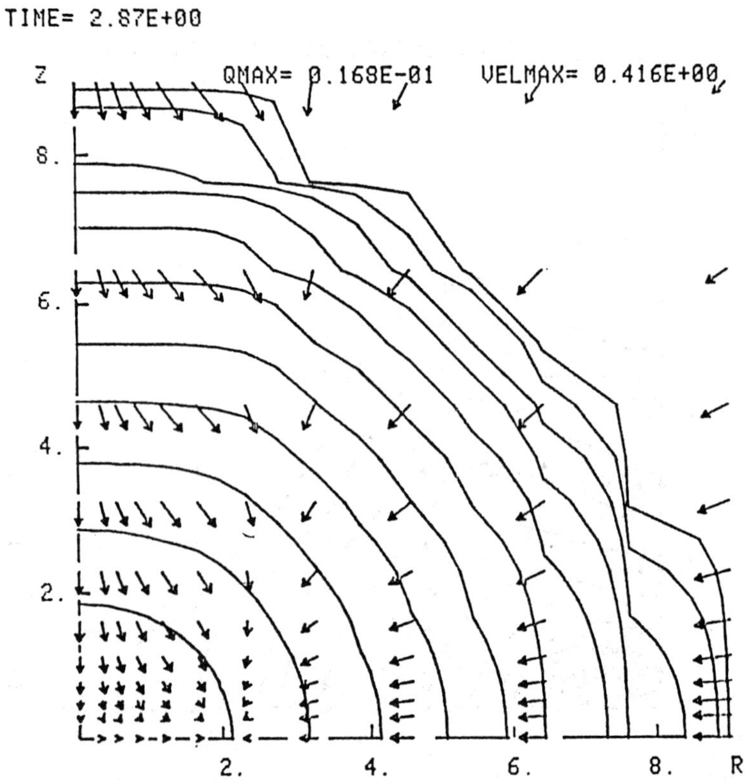

FIGURE 6. The contour lines of Q_b for B92. **FIGURE 6a** (above).

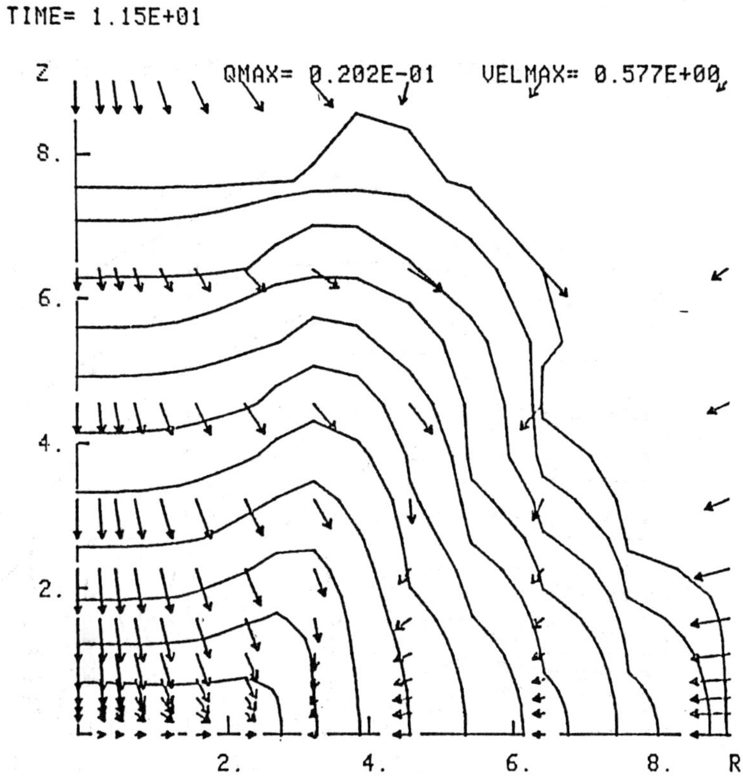

FIGURE 6b

envelope will return to the central black hole and the other part will expand to infinity. Thus the ultimate fate of the collapsing supermassive star in this model is completely different from that in model A50.

Model A146

At $t = 5.76$, the matter in the central part falls almost vertically. For $2 \lesssim R \lesssim 6.5$, we can see a strong outflow with the velocity up to 0.5 (FIGURE 4a). At $t = 18.8$, the central core bounces and a shock wave is formed. Near the equatorial plane the outflow extends up to $R = 9$. The outer thin envelope falls vertically to this outflow and the shock front is formed (FIGURE 4b). At $t = 23.2$, we can see the strong jet along the rotational axis. The central core has almost stopped moving and the rather dense envelope expands both in the lateral direction and in the Z-direction (FIGURE 4c). Finally the strong jet reaches $Z = 9$ (FIGURE 4d). The kinetic and internal energies of this jet are 2×10^{-3} and 1.5×10^{-3} in our units, respectively. The total mass of the jet is 5×10^{-3}. Thus the mean kinetic energy per gram of this jet becomes 0.4 (3.6×10^{20} ergs/g). The total mass and the angular momentum of the relaxed core are 0.21 and

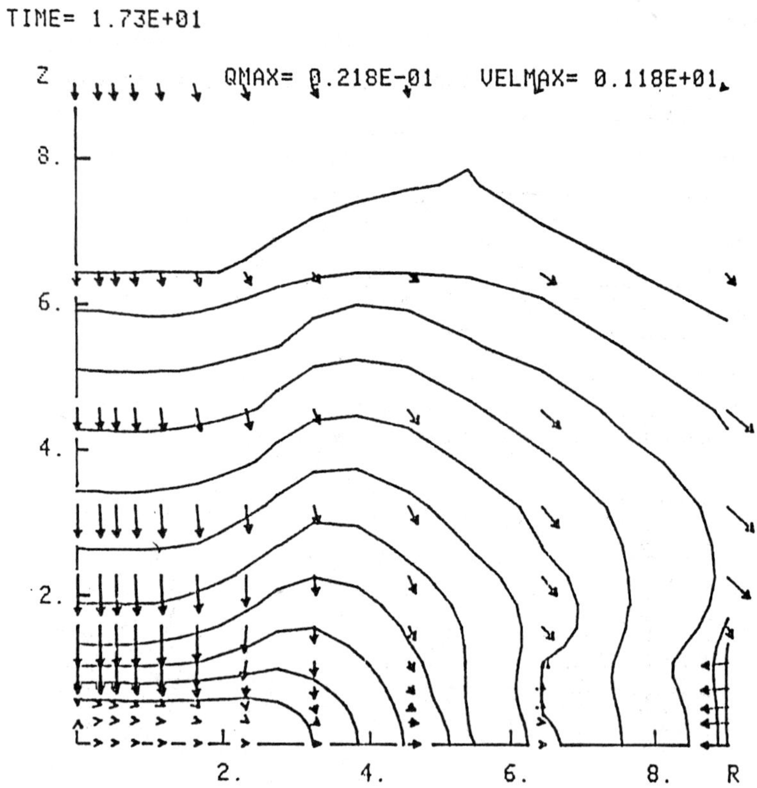

FIGURE 6c

4×10^{-2}, respectively. As the core has a rather small q value (≈ 1.0), it may recollapse eventually and a black hole may be formed after all.

Rotation Law B

In this rotation law, the centrifugal force is more effective for small R. This means that if the mass shedding occurs, it will occur from the central region. In the following, we show the details of the numerical results of the three typical models.

Model B51

In FIGURE 5, we show the density contours and the flow pattern at $t = 21.6$. Although A50 and B51 have almost the same angular momentum, we can see that the central core is deformed rather strongly in B51. Of course, this is due to a rapidly rotating core because of the differential rotation of rotation law B.

Model B92

In this model, the centrifugal force near the center is greater than the gravitational force at $t = 0$. This fact causes the outflow of the matter from a small R region. As the rotation is very slow for large R, the matter in the outer part falls almost spherically (FIGURE 6a). At $t = 11.5$, the inflow in the Z-direction and the outflow in the lateral direction form a disk in the central region (FIGURE 6b). At $t = 17.3$, the outgoing velocity of the disk is considerably decelerated. The outer envelope falls into this disk vertically and forms the almost steady shock. For large R, a very thin envelope expands in the lateral direction (FIGURE 6c). In this model, we have tried to identify an apparent horizon, but in vain. We have recalculated this model using hypergeometric slicing. In this slicing, an apparent horizon is identified. This is due to the fact that, if one uses the maximal slicing as a time slice, the proper time of the comoving observer, whose four velocity is that of the matter, stops increasing too soon after the density distribution becomes disklike.

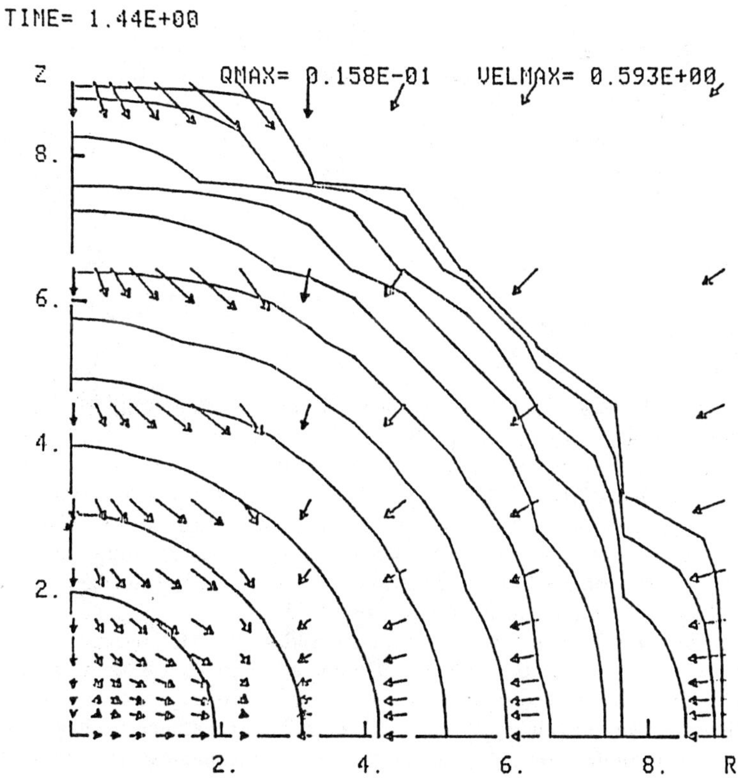

FIGURE 7. The contour lines of Q_b for B143. **FIGURE 7a** (above).

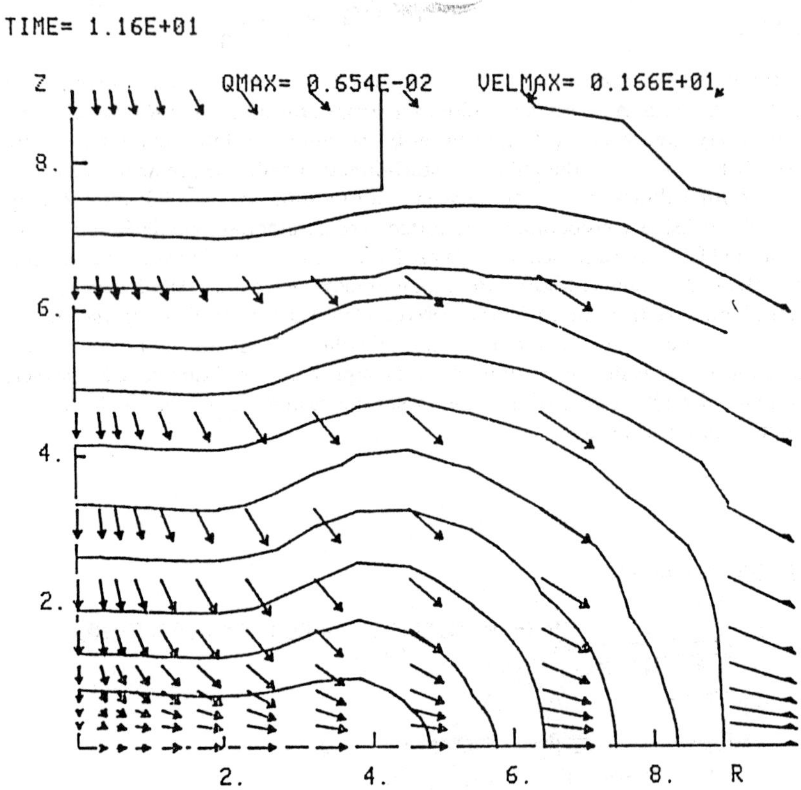

FIGURE 7b

Model B143

In this model, we can see the outflow for small R even at a rather early time (FIGURE 7a). At $t = 11.6$, an expanding disk is clearly formed (FIGURE 7b). At $t = 17.5$, the expansion velocity of the disk is decelerated in the central part and it is the fastest at the edge of the disk (FIGURE 7c). At $t = 20.8$, the central part of the disk is almost stopped. The outer thin envelope falls into the disk continuously and forms the almost steady shock front. As the expansion velocity is very large in this model, all the matter except the central part will go away from the system. In this model, no jet is formed, in contrast to the model A146 (FIGURE 7d).

From the above numerical results we can conclude that, first, the apparent horizons are formed for $q \lesssim 1.05$ in case A, $q \lesssim 0.92$ in case B, and $q \lesssim 0.86$ in case C. These models are different from each other in their equation of state, initial density distribution, and rotation law. However, they have almost the same critical value of q (q_c) for the formation of the apparent horizon. From the numerical experiments,

$$q_c = 0.94 \pm 0.1$$

In all the models in which the apparent horizon is formed, nothing peculiar has been found outside the apparent horizon. If we remember that the singularity of the Kerr black hole is hidden by the event horizon for $q < 1$, our numerical results suggest that the Kerr black holes may be formed for wide ranges of the initial conditions provided that the initial density and the angular velocity decrease with the radius.

Second for $q > 1$, the dynamics of matter strongly depend on the initial conditions and the equation state. Jets (case A), an expanding disk (case B), or an expanding ring (case C) appears. Anyway for $q > 1$, all the matter will not be swallowed into a black hole. The matter with large specific angular momentum is ejected as jets, an expanding disk, or an expanding ring, and the matter with $q < 1$ may recollapse to a black hole as in model A146.

ACCRETING NEUTRON STAR WITH ROTATION

There have been many calculations on type II supernova problems. However, models based on the massive initial iron-nickel cores are unlikely to give explosions generated

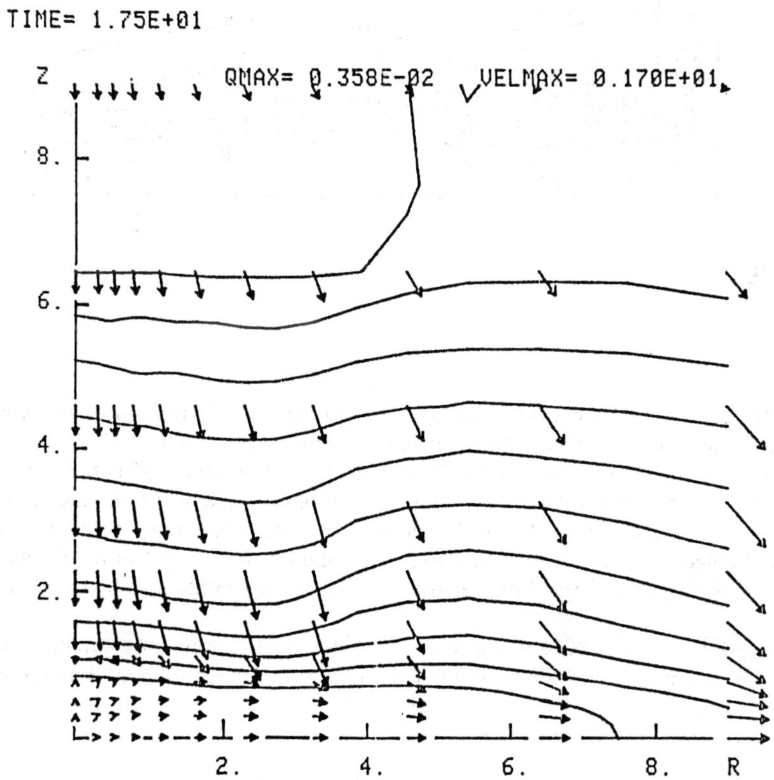

FIGURE 7c

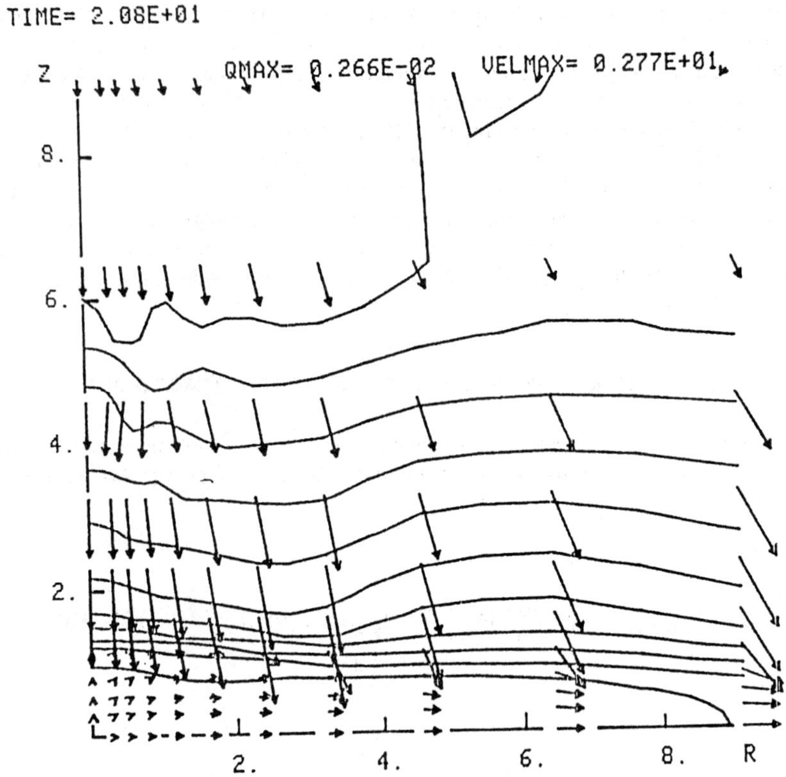

FIGURE 7d

directly from core bounce. This is due to the energy losses from the shock mainly by nuclear dissociation. Hillebrandt claimed in this symposium that a star of ($8M_\odot \lesssim M \lesssim 10M_\odot$) may result in supernova explosion,[23] but the outer envelope of a star of $M \gtrsim 10M_\odot$ will not be ejected. If so, the envelope of a massive star will fall back onto a proto–neutron star of mass $\approx 0.8M_\odot$. In nonrotating cases, a Schwarzschild black hole will be formed. However in rotating cases, what happens for $q > 1$? In the previous section, it was concluded that all the matter will not be swallowed into a black hole for $q > 1$.

Considering the above situation concerning the type II supernova problem, we adopt a very simple model for the infall of an unejected envelope of a massive star of mass $\gtrsim 10M_\odot$.

Units of mass, length, and time

$$M = 1.4\ M_\odot,\ L = \frac{GM}{C^2}\ ,\ T = \frac{GM}{C^3}$$

As an equation of state we use

$$p = \begin{cases} K\rho^{4/3} & \text{for } \rho \leqq \rho^* \equiv 3 \times 10^{14}\ \text{g/cm}^3 \\ \left(\rho - \dfrac{2}{3}\rho^*\right)\epsilon & \text{for } \rho \geqq \rho^* \end{cases}$$

where

$$\epsilon = 3k\rho/(\rho^*)^{2/3} \text{ and } k = \frac{2}{3}(1.4\ M_\odot/4\pi\ 2.018)^{2/3}\pi G$$

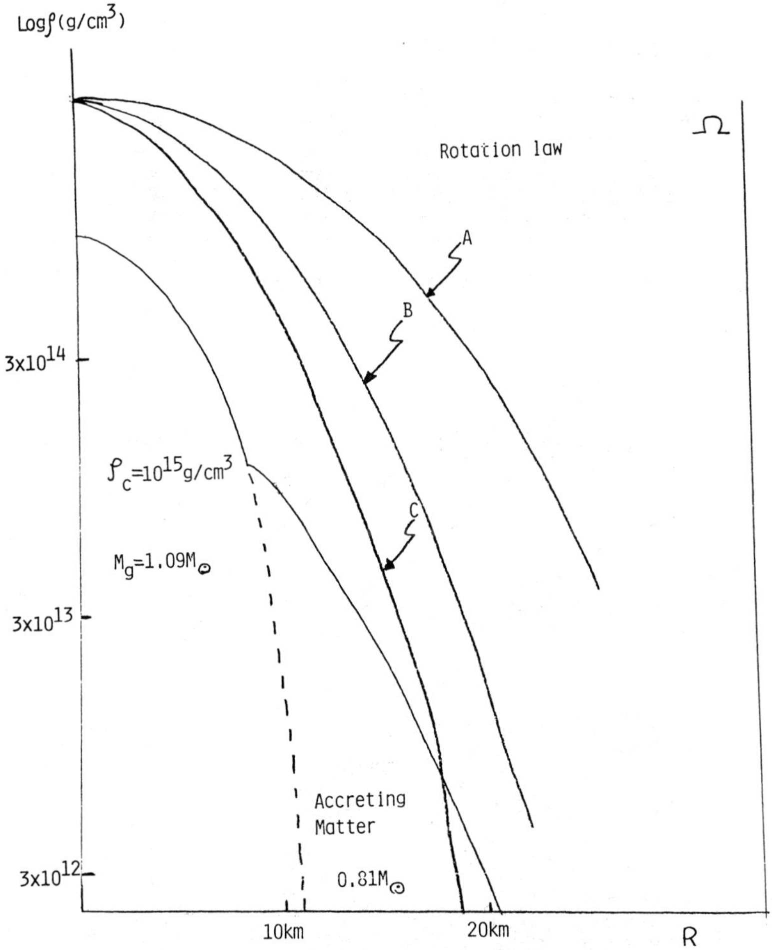

FIGURE 8. The density distribution and the rotation law of accreting neutron stars with rotation. The dashed line shows a neutron star model for $\rho_c = 10^{15}$ g/cm^3. Accreting matter is added to this neutron star. Solid lines with letters A, B, and C show the adopted rotation laws.

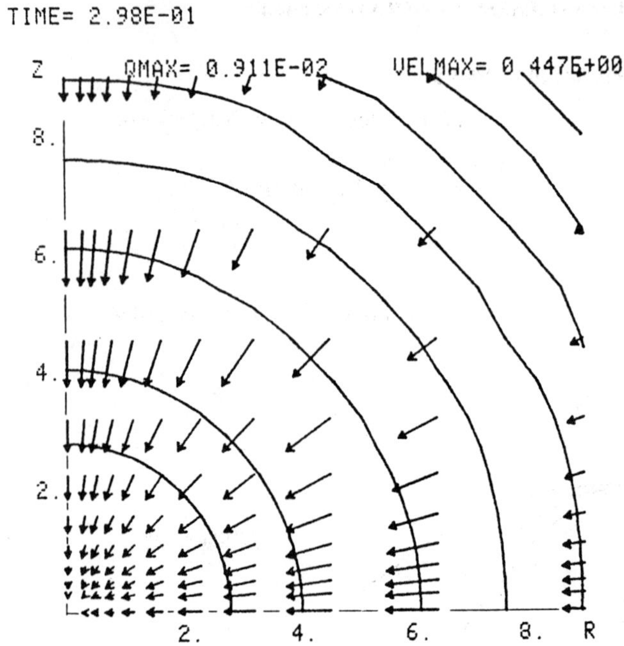

FIGURE 9. The contour lines of Q_b for A07. **FIGURE 9a** (above).

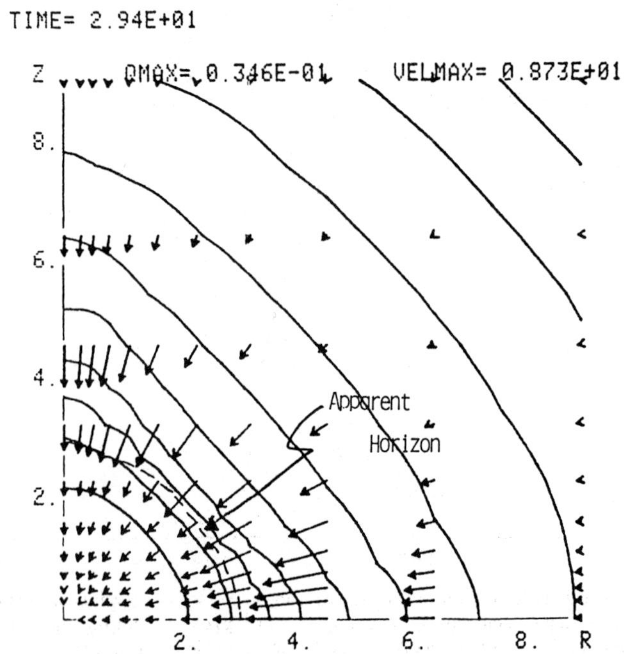

FIGURE 9b. The dashed line shows an apparent horizon.

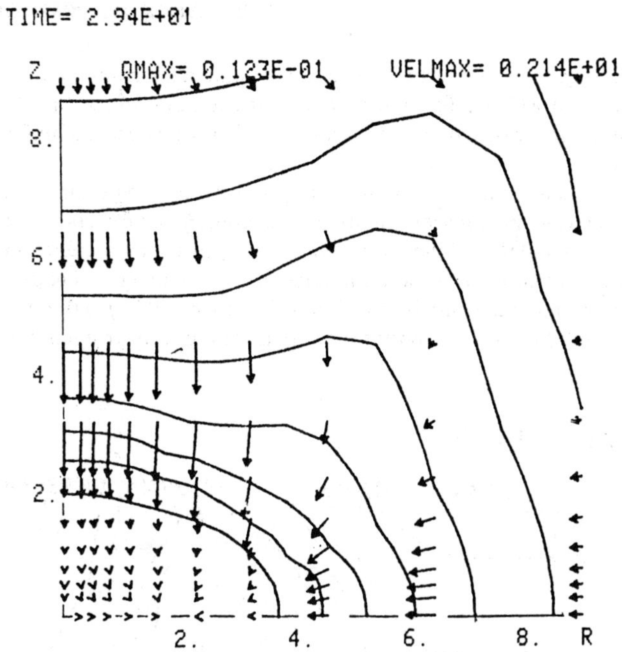

FIGURE 10. The contour lines of Q_b for A97.

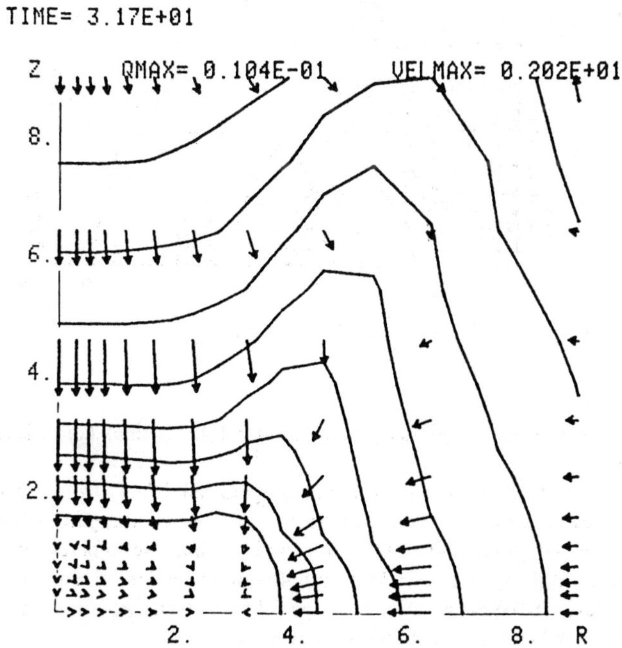

FIGURE 11. The contour lines of Q_b for B97.

For nonrotating cases, in the above equation of state the maximum mass of a neutron star becomes 1.4 $M_\odot$. We first put a nonrotating neutron star with $\rho_c = 10^{15}$ g/cm^3. Then we add an envelope to this neutron star, which is a model of an infalling envelope (see FIGURE 8).

To this system, we add the infall velocity and the rotation using Equations 23 and 24. These determine the initial conditions of accreting neutron stars with rotation. In FIGURE 8, we show three kinds of rotation law used in the present calculations. As before, we use the rotation law and the value of q as the name of a model.

FIGURE 9a shows the almost initial state for model A07 ($q = 0.07$). In this model, the effect of rotation is not important. As total mass of the system is 1.9 $M_\odot$, which is

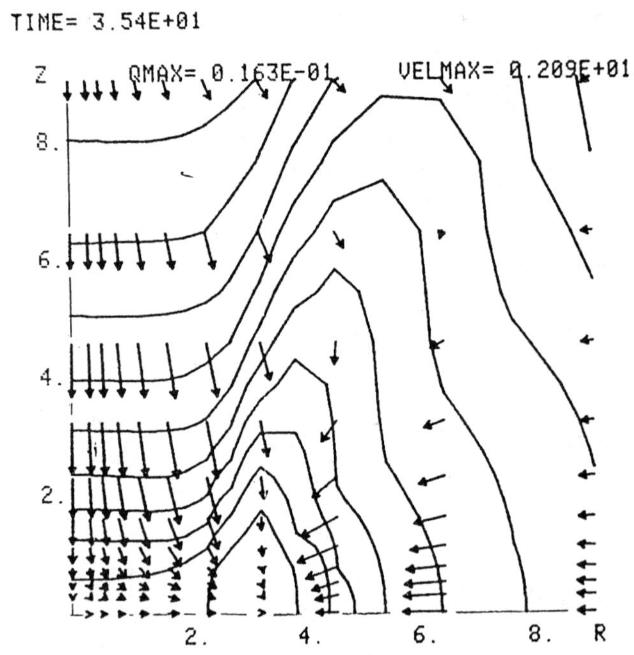

FIGURE 12. The contour lines of Q_b for C93.

greater than the maximum mass of a neutron star 1.4 $M_\odot$, a black hole is formed finally (FIGURE 9b).

In FIGURES 10 to 12 we show the final stage of models A97 ($q = 0.97$), B97, and C93. For these three models, we have tried to identify an apparent horizon, but no horizons have been found. Therefore, they will not collapse to black holes. This is not strange because Wilson showed the rotational enhancement of the maximum mass is 50%.[24] However, the aspect ratio of the core for A97 and B97 is about 2:1, which is close to the Ostriker-Tassoul instability criterion.[25]

The question arises, are rapidly rotating cores (A97), disks (B97), and rings (C93) stable for non–axially symmetric perturbation? Do they fragment into several pieces? To answer these questions, we must construct a code in stage 4.

In conclusion, the present numerical results tell us that q is the determinant of the general relativistic collapse of rotating stars leading to the formation of neutron stars and black holes.

REFERENCES

1. SMARR, L., A. CADEZ, B. DEWITT & K. EPPLEY. 1976. Phys. Rev. **D14:** 2443.
2. PENROSE, R. 1973. Int. Astron. Union Symp. **64:** 82.
3. SMARR, L. 1977. Ann. N.Y. Acad. Sci. **302:** 569.
4. EPPLEY, K. 1979. *In* Sources of Gravitational Radiation. L. Smarr, Ed.: 275. Cambridge University Press. Cambridge, England.
5. MIYAMA, S. 1981. Prog. Theor. Phys. **65:** 894.
6. PIRAN, T. 1982. *In* Gravitational Radiation. N. Deruelle & T. Piran, Eds.: 203. North Holland. Amsterdam, the Netherlands.
7. BARDEEN, J. 1982. *In* Gravitational Radiation. N. Deruelle & T. Piran, Eds.: 433. North Holland. Amsterdam, the Netherlands.
8. DEFELICE, F. & Y. YUNQUIANG. Astrophys. J. (In press.)
9. BACKER, D. C. Ann. N.Y. Acad. Sci. (This volume.)
10. GEROCH, R. 1971. J. Math. Phys. **12:** 98.
11. MAEDA, K., M. SASAKI, T. NAKAMURA & S. MIYAMA. 1980. Prog. Theor. Phys. **63:** 719.
12. PIRAN, T. 1980. J. Comp. Phys. **35:** 254.
13. YORK, J. 1982. *In* Gravitational Radiation. N. Deruelle & T. Piran, Eds.: 175. North Holland. Amsterdam, the Netherlands.
14. NAKAMURA, T. 1981. Prog. Theor. Phys. **65:** 1876.
15. NAKAMURA, T., K. MAEDA, S. MIYAMA & M. SASAKI. 1980. Prog. Theor. Phys. **63:** 1229.
16. NAKAMURA, T., K. MAEDA, S. MIYAMA & M. SASAKI. 1982. *In* Proceedings of the Second Marcell Grossmann Meeting. R. Ruffini, Ed.: 675. Elsevier-North Holland. New York, N.Y.
17. NAKAMURA, T., K. MAEDA, S. MIYAMA & M. SASAKI. 1980. Int. Astron. Union Symp. **93:** 326.
18. NAKAMURA, T. 1981. Prog. Theor. Phys. Suppl. **70:** 202.
19. NAKAMURA, T. & H. SATO. 1981. Phys. Lett. **86A:** 318.
20. NAKAMURA, T. & H. SATO. 1981. Prog. Theor. Phys. **66:** 2038.
21. NAKAMURA, T. & H. SATO. 1982. Prog. Theor. Phys. **67:** 1396.
22. SASAKI, M., K. MAEDA, S. MIYAMA & T. NAKAMURA. 1980. Prog. Theor. Phys. **63:** 1051.
23. HILLEBRANDT, W. Ann. N.Y. Acad. Sci. (This volume.)
24. WILSON, J. R. 1973. Phys. Rev. Lett. **30:** 1082.
25. OSTRIKER, J. P. & J. L. TASSOUL. 1969. Astrophys. J. **155:** 987.

Adiabatic Theories of Galaxy Formation and Pancakes

J. RICHARD BOND

Institute of Astronomy
Cambridge University
Cambridge CB3 0HA, England

Physics Department
Stanford University
Stanford, California 94305

ALEXANDER S. SZALAY

Department of Atomic Physics
Roland Eötvös University
1088 Budapest, Hungary

Astronomy and Physics Departments
University of California
Berkeley, California 94720

INTRODUCTION

The structure of the universe is dominated by distinctive large-scale features: superclusters, with lengths of order 25–100 Mpc, that are often more stringlike than sheetlike.[17,19] Rather than being isolated, there are hints of a network structure to the superclusters, with large voids almost free of galaxies in between. In this paper we review how the presently observed distribution of galaxies may arise if the neutrino or some other collisionless relic of the big bang is endowed with a rest mass of the order 10–10^3 eV.

Gershtein and Zeldovich first suggested that a small neutrino mass may have important cosmological consequences by contributing a larger fraction of the overall density of the universe than do baryons.[18] Marx and Szalay,[20] Cowsik and McClelland,[12] Szalay and Marx,[32] and Schramm and Steigman[28] have refined their arguments: the present limits on the Hubble constant and the deceleration parameter of the universe result in an upper bound to the sum of the neutrino masses: $\Sigma m_{\nu i} < 100$ eV. If particles that interact more weakly than neutrinos—such as the gravitinos and photinos of supersymmetric theories—are massive and long-lived, they must satisfy a similar but higher bound, $m_x \lesssim 1$ keV, though this is dependent upon the particle physics of the very early universe.[23,11] In this paper, we refer to the class of more weakly interacting particles as "gravitinos," or X's.

Calculations of the primordial He and D abundance indicate that baryons can only contribute a small fraction of the critical density. The likely value of the baryon density parameter at the time of primordial nucleosynthesis lies in the range of $0.01 < \Omega_B < 0.1$.[22] This suggests that if baryons dominate the mass density, then the universe is open by a large margin. In such a universe the growth of the initially adiabatic perturbations is highly suppressed: the presence of nonlinear structure now is incompatible with the tight upper bounds on small-scale density perturbations arising from observations of the microwave background temperature fluctuations, and effectively rules out such a case. Below $T \sim 1$ MeV, neutrinos are collisionless particles and are thus not affected by radiation drag as ordinary matter is. This enables much longer fluctuation growth to occur, without strongly influencing the microwave background. On the other hand, neutrinos and gravitinos are subject to phase mixing of their orbits, which results in a characteristic mass scale corresponding to superclusters. For gravitinos, the scale is that of galaxies or of clusters. The collapse of such systems was shown by Zeldovich to lead to highly anisotropic structures, the pancakes.[35] These are not isolated: a cellular structure would form with huge voids in between; and filaments would appear as well. The neutrino mass explains this structure in a simple and elegant way. In a gravitino-dominated universe, many of these features may also arise.

LINEAR PERTURBATIONS IN THE EARLY UNIVERSE

Small fluctuations may emerge naturally near the big bang itself. In the presently popular grand unified theories of particle physics, the most important fluctuations are those that preserve the baryon/photon ratio, namely, the adiabatic ones. These fluctuations are presumed to have a smooth scale-free spectrum existing over a wide range of mass scales. As the expansion proceeds, larger and larger masses come within the horizon. There are two competing effects: gravity attracts the particles toward the highest densities, while the pressure due to thermal motion tries to prevent this. On large scales gravity always wins, matter condenses in some regions and is rarified in others. On small scales pressure is more important. The perturbations behave like acoustic waves; excess density is accompanied by excess pressure, and the local density oscillates.

Until the temperature has dropped to a few thousand degrees, the radiation is still sufficiently energetic to keep the matter ionized. The growth of baryon fluctuations by gravitational instability is inhibited in the ionized phase because the radiation provides a strong source of viscosity. The radiation is scattered mostly by free electrons, so, once the electrons recombine into H-atoms, radiation streams independently of the matter. There is no longer any resistance to fluctuation growth and gravitational instability proceeds, but only at a relatively late stage, $z \gtrsim 1{,}000$. During recombination the smallest scale fluctuations are subject to viscous damping.[29] The compressed radiation tends to diffuse and thereby smooth out all baryon fluctuations below 10^{13}–$10^{15}\ M_{\odot}$.

Even when pressure becomes negligible, the rate of fluctuation growth is small in low-density universes for red shifts $< \Omega^{-1}$, when the curvature dominates the expansion. This can lead to a total growth factor from decoupling to today as small as 15, to be compared with the total growth of 1,000 if $\Omega = 1$. Since the photons were not scattered after decoupling, the present value of the temperature fluctuations of the CBR reflects the amplitude of the baryon density perturbations at this epoch. Upper limits on small-scale fluctuations imply that growth by even this factor of 1,000 will not be enough to give nonlinear structure by now in baryon-dominated universes.

As Bond *et al.*[4] and Doroshkevich *et al.*[15] have emphasized, one of the principal triumphs of a universe dominated by neutrinos or other weakly interacting particles such as gravitinos or photinos is that they can beautifully sidestep this difficulty. This occurs for two reasons. Firstly, Ω can be larger, hence the curvature-dominated era, if it exists at all, will have occurred for a much shorter time. More importantly, once neutrinos become nonrelativistic, which occurs before recombination, their fluctuations become gravitationally unstable on subhorizon scales, as long as these scales are above the instantaneous value of the neutrino Jeans mass. Consequently, the amplitude of these neutrino density fluctuations greatly exceeds those of the baryon perturbations by recombination. Once decoupling occurs, the baryon fluctuations experience an accelerated growth until they equal the neutrino fluctuations. This new feature implies that one can get nonlinearity occurring at $z \sim 5$, and still have the induced temperature fluctuations just smaller than the upper limits on small scales. Though safely lower than the upper bounds set by observers so far, the $\Delta T/T$ structure on all scales larger than $\sim 10'$ is predicted to be rich on the 10^{-5} to 10^{-4} level in a neutrino-dominated universe.

The occurrence of a damping cutoff in the neutrino fluctuation spectrum determines the nature of large-scale structure formation in neutrino-dominated universes. Since momentum suffers the cosmological red shift, once neutrinos go nonrelativistic their velocity slows from the speed of light down to a present rms value of 6 km s^{-1} (30 eV$/m_\nu$). The total comoving displacement that the typical freely streaming neutrino traverses converges to a finite time-independent value given by the mass scale ~ 0.5 $M_{\nu m}$,[6] where the damping scale is the maximum value that the neutrino Jeans mass attains,[4,10,15]

$$M_{\nu m} = 3.6\, m_p^3\, m_\nu^{-2} = 6 \times 10^{15}\, (30\ \mathrm{eV}/m_\nu)^2\, M_\odot$$

Masses are related to lengths by $M = \rho\lambda^3$. Below this scale orbits of neutrinos streaming in different directions and/or with different speeds phase mix, effectively erasing any inhomogeneities initially present. Doroshkevich *et al.* and Wasserman discuss this limiting regime.[15,34] Above this scale, neutrinos cannot move from one lump to another, and the self-gravity of compressions causes them to grow in magnitude via the usual Jeans instability. Bond and Szalay and Peebles have numerically integrated the coupled Einstein-Boltzmann equations to determine the detailed shape of the fluctuation spectra,[5,6,25] which ties together these two regimes. They find that most damping occurs between $z = 10^5\ (m_\nu/30\ \mathrm{eV})$, when the neutrinos are just becoming semirelativistic, and $z = 10^3\ (m_\nu/30\ \mathrm{eV})$, which is near recombination; after this, the spectral shape down to $\ll M_{\nu m}$ is frozen in.

A very similar story holds for gravitinos.[7,24,6,26] The damping mass is now ~ 0.08 $m_p^3/m_x^2 \sim 10^{11}\ (1\ \mathrm{keV}/m_x)^2\ M_\odot$, which suggests galaxies may be the first objects to form if $m_x \sim 1$ keV ($Q = 1$, $H_0 = 100$ km s^{-1} Mpc^{-1}), or clusters if $m_x \lesssim 250$ eV ($\Omega = 1$, $H_0 =$ 50 km s^{-1} Mpc^{-1}, or $\Omega < 1$). It is also possible to have damping scales very much less than that of galaxies, but we do not consider these cases here (see Bond and Szalay).[6]

NONLINEAR STRUCTURE IN THE UNIVERSE

Once the first mass scale in a spectrum with a damping cutoff reaches nonlinearity, neutrino trajectories cease expanding away from each other and begin converging, resulting in the temporary formation of caustics. The density becomes very high and a flat "pancake" is formed, as Zeldovich originally suggested.[35] At first they form at isolated spots where the initial velocity perturbations had the largest gradient. Soon these regions grow, turning into huge surfaces which intersect, forming the walls of a cell structure that is itself gravitationally unstable. The universe may be at this cellular stage today, as detailed numerical calculations indicate.

In the nonlinear phase, mode-mode coupling among Fourier components sends power to short wavelengths, and correlates phases even though the initial fluctuation spectrum may have had random phases. The evolution of structure is then best calculated in real space. Arnold, Zeldovich, and Shandarin have applied the methods of catastrophe theory to analyze structure that develops in potential motion,[1] i.e., with no vorticity, which, if present, should be small by the end of the linear phase anyway. They found that the two-dimensional pancakes are only the lowest order singularities; six other singular topological structures should also appear. Stringlike features are one example, and they can be seen in the N-body simulations of Clypin and Shandarin,[37] Frenk *et al.*,[38] and Melott.[21]

When the intersection of trajectories takes place, gas pressure builds up, the velocity of the collapsing gas exceeds the sound speed, and a shock wave is formed.[31,13] The gas is shock heated up to ~keV temperatures and cools by emitting radiation over a broad spectrum especially in the central layers where the density is higher. Recently, Bond, Centrella, Szalay, and Wilson, hereafter BCSW, have calculated the cooling of collapsing neutrino-baryon pancakes,[8] the details of which are considerably different from those in a pure baryon pancake: the baryon density is lower, but the infall velocities are higher, and thus the cooling rate is much slower.[16,8]

FIGURE 1, which is taken from this work,[8] shows how the shock develops in red shift for the collapse of a $4.4 \times 10^{15}\ M_\odot$ region, occurring at $1 + z_c = 6$, $\Omega_\nu = 0.91$, $\Omega_B = 0.091$. The central regions are highly compressed, and about 10% of the gas has cooled below $T \sim 100$ eV, which is the temperature needed for gas to be bound to galaxies and also coincides with the point at which rapid He recombination cooling takes over from bremsstrahlung. FIGURE 2 illustrates how the cooling time at the instant the shock reaches mass fraction q compares with the Hubble time for collapses of various masses. It is evident that the fraction that can cool significantly is a sensitive function of the mass of the collapsing region.

This cooling is required, since only cool gas is able to form smaller lumps, the seeds of galaxies. Transverse flow in the pancake toward line singularities will increase the local density, which enhances cooling in some cases. Strings may well be the locations of the most effective galaxy formation (see BSCW for details). The UV and soft x-ray emission can photoionize the intergalactic medium, making galaxy formation in regions that have not yet formed pancakes more difficult, which would accentuate the contrast in galaxy density between the strings and pancakes vs. voids, even though the density contrast may be only ~3–10.[36] One of the principal difficulties in this picture is how to drive fragmentation at all, since isolated pancakes have no power on small transverse scales. This question was addressed by Doroshkevich,[14] but has not yet been satisfactorily answered.

Cosmic neutrino pancakes may lead to an attractive explanation of the dark halos of galaxies. In the pancake collapse most of the neutrinos acquire large velocities. However, those neutrinos that were initially close to the midplane of the pancake move

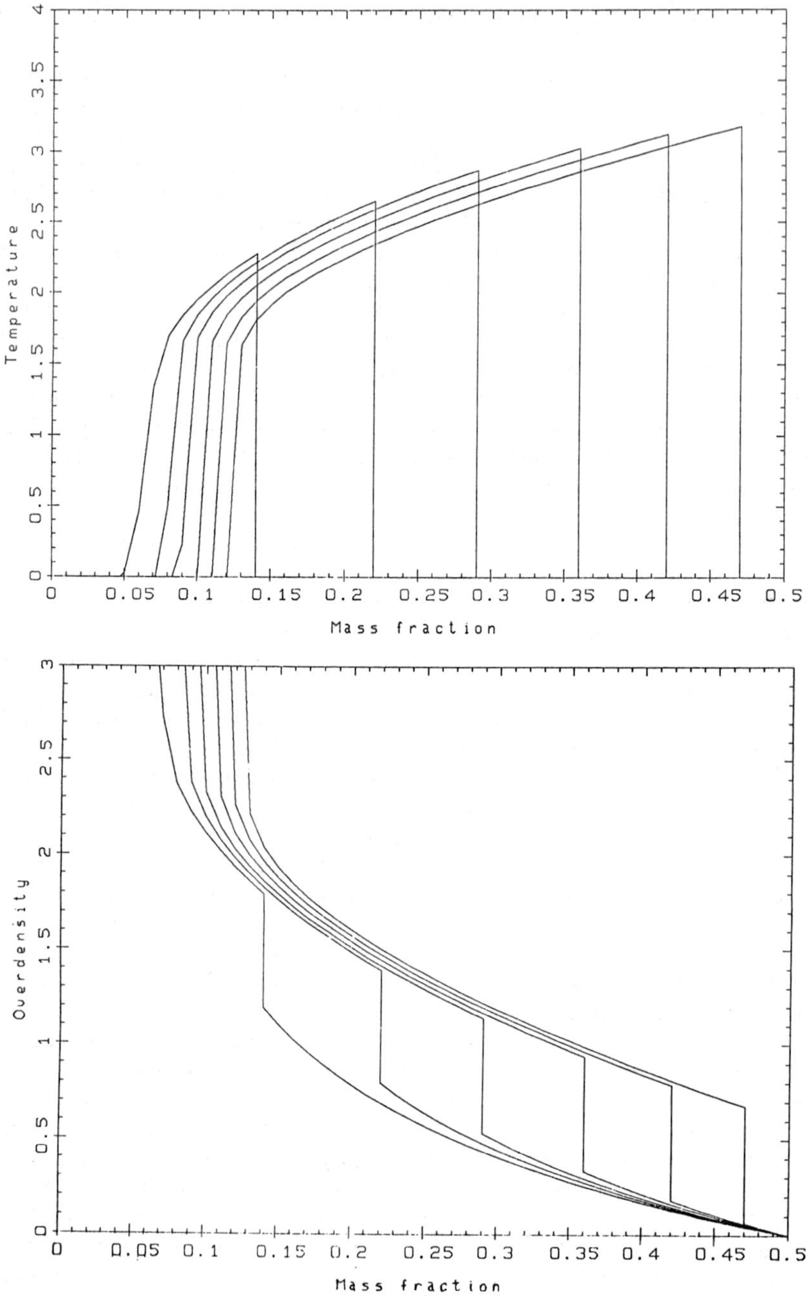

FIGURE 1. A semianalytic model of the evolution of the temperature (top) and overdensity (bottom) relative to the background of the shocked region created by a collapsing pancake of comoving wavelength 25 Mpc in a universe with $\Omega = 1$, $\Omega_B = 0.091$, and $H_0 = 50$ km s^{-1} Mpc^{-1}. The temperature scale is keV and the overdensity is relative to the background; both are logarithmic. The collapse red shift was $z_c = 5$, and the snapshots of the structure are given at $z =$ 4.8, 4.5, 4.2, 3.8, 3.4, and 3.2. This model does not include conduction. It agrees with the equivalent full numerical no-conduction run in Bond, Centrella, Szalay, and Wilson.[8]

only slowly at first. Around this midplane, we assume a thin gas layer condenses and fragments. Bond, Szalay, and White have constructed a simple one-dimensional model, which demonstrates that the slowest moving neutrinos will add on to the baryonic seeds first, followed by progressively faster ones, resulting in a halo which has the total-to-baryonic mass ratio and velocity dispersions needed to describe the halos of spirals.[9] Further, the one-at-a-time addition makes it more likely that a $1/r^2$ halo will form than the $1/r^3$ found in violent relaxation studies, which would be expected if the dark matter and the baryons collapsed together. On the other hand, the slow neutrinos in the neighborhood of the pancake plane are Jeans unstable, especially in the

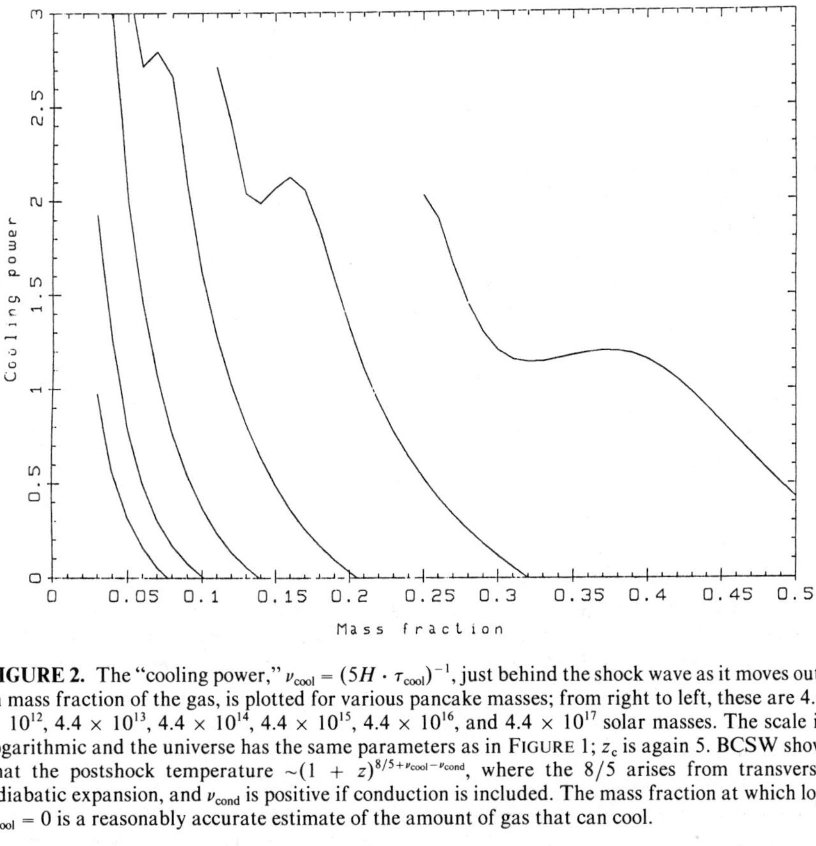

FIGURE 2. The "cooling power," $\nu_{cool} = (5H \cdot \tau_{cool})^{-1}$, just behind the shock wave as it moves out, in mass fraction of the gas, is plotted for various pancake masses; from right to left, these are 4.4 $\times$ 10^{12}, 4.4 $\times$ 10^{13}, 4.4 $\times$ 10^{14}, 4.4 $\times$ 10^{15}, 4.4 $\times$ 10^{16}, and 4.4 $\times$ 10^{17} solar masses. The scale is logarithmic and the universe has the same parameters as in FIGURE 1; z_c is again 5. BCSW show that the postshock temperature $\sim(1 + z)^{8/5+\nu_{cool}-\nu_{cond}}$, where the 8/5 arises from transverse adiabatic expansion, and ν_{cond} is positive if conduction is included. The mass fraction at which log $\nu_{cool} = 0$ is a reasonably accurate estimate of the amount of gas that can cool.

transverse direction. One therefore expects at least some of the neutrinos to condense with the baryons in the development of the seed. However, at fragmentation, the central baryon density is much higher than that of the neutrinos (BCSW), hence the baryon perturbation will presumably run away from the neutrino one; this is especially so if a cooling instability is involved. In any case, our one-at-a-time addition mechanism is itself an expression of the Jeans instability of the neutrino gas. We have found that the fraction of the neutrinos captured and their velocity dispersion rise with the baryonic mass, which is certainly a desirable feature. The problem of halo formation if galaxies arise along strings has not been addressed, though presumably there will again be some low-velocity neutrinos available for capture.

Tremaine and Gunn first emphasized that the conservation of fine-grained phase space density implicit in the Vlasov equation describing neutrino motion imposes a severe constraint upon the viability of neutrino halos.[33] The adiabatic compression of the background neutrinos to an overdenisty $1 + \delta_\nu$ would enforce a three-dimensional velocity dispersion $280\,(30\ \mathrm{eV}/m_\nu)\,[(1 + \delta_\nu)/10^5]^{1/3}\ \mathrm{km\ s^{-1}}$; phase mixing during any violent relaxation that occurs would lower this velocity. However, this becomes a severe restriction only near the central regions of galaxies, where the density of the halo component is poorly determined—even in the Galaxy.[2] Within our simple model of halo formation from pancakes we note: (1) there is little dilution of phase space density for the low-velocity neutrinos; (2) by selecting only slowly moving neutrinos, we are able to get larger overdensities for a given velocity dispersion; (3) the halo component may be less concentrated in the central regions of the galaxies than the usual values quoted for the core radii would suggest, since the more compact baryon component would dominate in their determination.

For heavier gravitinos, one still expects pancake formation though the mass of the collapsing region is smaller in this case. From FIGURE 2, it is evident that the cooling fraction is much higher; indeed, it has been suggested that galaxies are such prominent objects precisely because above galactic masses cooling is inefficient.[3,27,30] Still, the cooled gas is sufficiently abundant for all but the most massive pancakes for galaxy formation from such regions to present attractive models. In the collapse of a gravitino pancake, the dark matter would already be in halos, and, in fact, one must invoke tidal stripping of the outer parts to distribute the dark matter on larger scales. Also, dwarf elliptical galaxies would presumably have to arise from fragmentation of the initial galactic pancake. On the other hand, disk galaxies would be a natural outcome. Superclusters, strings and voids would arise as a consequence of gravitational clustering. There is a shoulder in the gravitino density fluctuation spectrum at this scale which may allow this;[6] N-body experiments with these initial conditions are required to test this possibility.

In conclusion, we hope it has become clear that once dark matter is included, the pancake model seems capable of explaining most features of the large-scale structure. At several points in the development of the theory (CBR fluctuations, gas cooling, the phase space constraints of Tremaine and Gunn,[33] the formation of galaxy halos), the neutrino-dominated universe just works; and gravitino-type theories have no difficulties passing these requirements. Models based upon colder relics of the big bang such as monopoles or axions also have some features attractive for galaxy formation.[6,25] Thus, even if we specify that the perturbations were adiabatic, there are still many possibilities which depend upon the nature of the dark matter that dominates the structure of our universe.

SUMMARY

The adiabatic theory of galaxy formation in universes dominated by massive neutrinos or other more weakly interacting particles such as gravitinos or photinos is reviewed. Collisionless damping leads to a density fluctuation spectrum with a cutoff, the nonlinear evolution of which naturally results in the formation of pancakes, strings, voids, galaxies, and galactic halos.

ACKNOWLEDGMENTS

We thank our collaborators Simon White, Joan Centrella, and Jim Wilson for a stimulating interplay in the development of these ideas.

REFERENCES

1. ARNOLD, V. I., S. F. SHANDARIN & YA. B. ZELDOVICH. 1982. Geophys. Astrophys. Fluid Dyn. **20:** 111.
2. BAHCALL, J. N., M. SCHMIDT & R. SONEIRA. 1982. Astrophys. J. Lett. **258:** 23.
3. BINNEY, J. 1977. Astrophys. J. **215:** 483.
4. BOND, J. R., G. EFSTHATHIOU & J. SILK. 1980. Phys. Rev. Lett. **45:** 1980.
5. BOND, J. R. & A. S. SZALAY. 1981. Proc. Neutrino 81 **1:** 59.
6. BOND, J. R. & A. S. SZALAY. 1983. Astrophys. J. (In press.)
7. BOND, J. R., A. S. SZALAY & M. S. TURNER. 1982. Phys. Rev. Lett. **48:** 1636.
8. BOND, J. R., J. CENTRELLA, A. S. SZALAY & J. R. WILSON. 1983. Mon. Not. R. Astron. Soc. (To be published.)
9. BOND, J. R., A. S. SZALAY & S. D. M. WHITE. 1983. Nature **301:** 584.
10. BISNOVATY-KOGAN, G. S. & I. D. NOVIKOV. 1980. Sov. Astron. **24:** 516.
11. CABBIBO, N., G. R. FARRAR & L. MAIANI. 1981. Phys. Lett. **105B**: 155.
12. COWSIK, R. & J. MCCLELLAND. 1972. Phys. Rev. Lett. **29:** 669.
13. DOROSHKEVICH, A. G., S. F. SHANDARIN & E. SAAR. 1978. Mon. Not. R. Astron. Soc. **184:** 643.
14. DOROSHKEVICH, A. G. 1980. Sov. Astron. **57:** 259.
15. DOROSHKEVICH, A. G., M. YU. KHLOPOV, R. A. SUNYAEV, A. S. SZALAY & YA. B. ZELDOVICH. 1980. Ann. N.Y. Acad. Sci. **375:** 32.
16. DOROSHKEVICH, A. G. & M. YU. KHLOPOV. 1981. Sov. Astron. **25:** 521.
17. EINASTO, J. 1982. Int. Astron. Union Symp. **104**.
18. GERSHTEIN, S. S. & YA. B. ZELDOVICH. 1966. JETP Lett. **4:** 174.
19. KIRSHNER, R. P., A. OEMLER, JR., P. L. SCHECHTER & S. A. SHECTMAN. Ann. N.Y. Acad. Sci. (This volume.)
20. MARX, G. & A. S. SZALAY. 1972. Proc Neutrino '72 **1:** 123.
21. MELOTT, A. 1983. Mon. Not. R. Astron. Soc. **202:** 595.
22. OLIVE, K. A., D. N. SCHRAMM, G. STEIGMAN, M. S. TURNER & J. YANG. 1981. Astrophys. J. **246:** 557.
23. PAGELS, H. & J. PRIMACK. 1982. Phys. Rev. Lett. **48:** 223.
24. BLUMENTHAL, G., H. PAGELS & J. PRIMACK. 1982. Nature **299:** 37.
25. PEEBLES, P. J. E. 1982. Astrophys. J. **258:** 415.
26. PEEBLES, P. J. E. 1982. Astrophys. J. Lett. **263:** 1.
27. REES, M. J. & J. OSTRIKER. 1977. Mon. Not. R. Astron. Soc. **179:** 541.
28. SCHRAMM, D. N. & G. STEIGMAN. 1981. Astrophys. J. **243:** 1.
29. SILK, J. 1968. Astrophys. J. **151:** 459.
30. SILK, J. 1977. Astrophys. J. **211:** 638.
31. SUNYAEV, R. A. & YA. B. ZELDOVICH. 1972. Astron. Astrophys. **20:** 189.
32. SZALAY, A. S. & G. MARX. 1976. Astron. Astrophys. **49:** 437.
33. TREMAINE, S. & J. E. GUNN. 1979. Phys. Rev. Lett. **42:** 407.
34. WASSERMANN, I. 1981. Astrophys. J. **248:** 1.
35. ZELDOVICH, YA. B. 1970. Astron. Astrophys. **5:** 84.
36. ZELDOVICH, YA. B. & S. F. SHANDARIN. 1982. Sov. Astron. Lett. **8:** 259.
37. CLYPIN, A. A. & S. F. SHANDARIN. Mon. Not. R. Astron. Soc. (In press.)
38. FRENK, C., *et al.* 1983. Astrophys. J.

Pancakes and the Formation of Galaxies

Neutrinos and Other 'Inos

PAUL R. SHAPIRO AND CURTIS STRUCK-MARCELL

Department of Astronomy
University of Texas
Austin, Texas 78712

The view that structure in the universe—galaxies, clusters, superclusters, voids—formed when large-scale adiabatic density perturbations grew by gravitational instability in the postrecombination Friedmann universe has been examined in the context of recent suggestions that the mass-density of the universe is dominated by certain finite mass, weakly interacting, elementary particles, or 'inos (e.g., massive neutrinos, gravitinos, photinos), predicted by grand unified theories of particle physics. In order to test the hypothesis that galaxies can form within this so-called pancake scenario, we have performed detailed, numerical hydrodynamical calculations of the one-dimensional, plane-symmetric growth of combined baryon-'ino pancakes in a homogeneous, isotropically expanding universe for a range of pancake sizes and collapse red shifts. The effects of ionization, recombination, radiative and Compton cooling, and thermal conduction are all explicitly included.

We find, for example, that, if $\Omega_{\text{ino}} = 0.9$, $\Omega_{\text{baryon}} = 0.1$, $H_0 = 75$, and the perturbation amplitude is large enough to bring about shock and caustic formation by $z = 6$, then (1) for pancakes of characteristic, comoving scale length 40 Mpc at present (e.g., which corresponds to the case of one massive neutrino type with $m_\nu \simeq 50$ eV and two spin states) and no thermal conduction, ~10% of the baryons form a cold, thin, dense layer capable of fragmenting and forming galaxies (i.e., $\Omega_{\text{galaxies}} \sim 0.01$ in luminous matter), while the rest of the baryons are left as a hot ($10^6 \lesssim T \lesssim 5 \times 10^7$ K), diffuse, intergalactic gas with $\Omega_{\text{IGM}} \sim 0.1$; (2) for the same case *with* thermal conduction, a somewhat smaller fraction (~6%) of the baryons cool; and (3) for galaxy-sized pancakes (e.g., scale lengths ~ 1 Mpc), most of the baryons cool and compress into a thin, dense sheet.

The Texture of the Universe[a]

ROBERT P. KIRSHNER

Department of Astronomy
The University of Michigan
Ann Arbor, Michigan 48109

AUGUSTUS OEMLER, JR.

Department of Astronomy
Yale University
New Haven, Connecticut 06520

PAUL L. SCHECHTER

Kitt Peak National Observatory
Box 26732
Tucson, Arizona 85726

STEPHEN A. SHECTMAN

Mount Wilson and Las Campañas Observatories
of the Carnegie Institution of Washington
Pasadena, California 91101

You ain't seen nothin' yet. (Al Jolson)

Just as the texture of an antique fabric is the result of the vanished loom that made it, the present texture of the universe is a result of physical processes operating in a distant and inaccessible epoch. An adequate justification for detailed study of the three-dimensional distribution of galaxies would be to derive an accurate description of the ways galaxies cluster. Yet the value of this work is greatly enriched by the prospect that studies of today's galaxies may reveal important properties of conditions arising from a hot big bang. No one doubts (except Ostriker and Cowie)[16] that the principal mechanism for forming galaxies and galaxy clusters is gravitational instability. Evolution from very large structures ($10^{15}\ M_\odot$) *down* to galaxy sizes is favored in the picture advocated by Zeldovich and his collaborators,[20] while the assembly of large structures from smaller ones characterizes the clustering picture described by Peebles.[17] Looking for structure, both density enhancements and voids in the galaxy distribution, on large scales (of order 50 h^{-1} Mpc), may provide important clues on the fossilized structure derived from primeval galaxy evolution.

One approach to studying the large-scale structure is to examine in detail the largest nearby structure: the local supercluster. Assembling the nearby structure requires accurate distance measures to individual galaxies. During the past decades, this work has been carried forward by de Vaucouleurs[3,4] and has recently been

[a]This research has been supported in part by the Alfred P. Sloan Foundation and by National Science Foundation Grant AST 82-02930.

described in great detail by Tully, who concludes that the structure of the local supercluster is extremely stringy and inhomogeneous on scales of a few megaparsecs.[19] Evocative, but much less detailed, studies of distant superclusters by Joeveer, Einasto, and Tago and Einasto, Joeveer, and Saar have suggested similar properties on much larger scales.[10,5]

Red shifts provide the evidence for the most convincing investigations of the three-dimensional structure of galaxy distributions on scales of 10 h^{-1} Mpc or more. Ignoring the possible effects of velocity fields (which can be assessed separately), the measurement of a galaxy's red shift to an accuracy of 100 km s^{-1} provides its distance to a precision of 1 h^{-1} Mpc. Surveys of galaxy red shifts by Gregory *et al.* start in the vicinity of well-known density enhancements, such as nearby dense clusters of galaxies, and then trace the extent of the dense region.[7,8] These surveys show that filaments of galaxies stretch over tens of degrees from the dense cluster core, corresponding to 20–30 h^{-1} Mpc. Intervening volumes often show large regions of low galaxy density. Long, stringy density enhancements extending over a radian on the sky at a distance of 40–53 h^{-1} Mpc, but with a width of only 2°, have been identified from a red shift survey by Giovanelli and Haynes.[6]

Large-scale inhomogeneity has also been revealed from the thorough red shift survey carried out at the Harvard-Smithsonian Center for Astrophysics by Davis *et al.*[2] Their survey, which includes 2,400 galaxies with red shifts averaging about 5,000 km s^{-1} (distances of 50 h^{-1} Mpc), reveals a "frothy" structure of dense strings and voids on scales of tens of megaparsecs. Having a large sample, as in the CFA survey, allows a detailed comparison with N-body simulations of galaxy clustering.

In order to study the structure of galaxy clustering on still larger scales, a significantly fainter (hence deeper) sample is required. Narrower fields, but greater depth, in a complete red shift survey provide the data for evaluating the luminosity function for galaxies, and the mean luminosity density. Kirshner, Oemler, and Schechter carried out a red shift survey to an average depth of about 10,000 km s^{-1} (100 h^{-1} Mpc),[11,12] but found that the study of average properties was hampered by fluctuations in the space density of galaxies by a factor of 3 on scales of roughly 50 Mpc. Our attention then shifted to the clustering properties. Fluctuations on scales of 50 Mpc demanded a survey of even greater depth, which we carried out in six fields, three in the north galactic cap and three in the south of 2 square degrees each, to an apparent magnitude of 17.5. Yielding 280 red shifts with an average depth of 200 h^{-1} Mpc, this survey provided data for an evaluation of the luminosity function and luminosity density for a fair sample of the universe.[14] Our attention, however, centered on a peculiar feature in the northern fields of the survey. Unlike the southern portion, the north showed a large enhancement of galaxy density at a red shift of 10,000 km s^{-1} (100 h^{-1} Mpc) and at a red shift near 22,000 km s^{-1} (220 h^{-1} Mpc), with a vast region 6,000 km s^{-1} across centered at 15,000 km s^{-1} which contained only 1 galaxy where about 20 had been expected.[13] Bravely asserting that this large (10^6 Mpc^3 for $H = 50$), low-density region, seen in each of the three northern fields separated by about 35° on the sky, was present in the triangle that they bound, we dubbed this region the Boötes void.

Extension of the survey to the interior of the triangle, as illustrated in FIGURE 1a, has now been carried out.[15] We have constructed a magnitude-limited sample in 282 small fields, each 15′ on a side, placed on a grid that covers the triangle formed by the original three fields. Extending from the center of the proposed void (at $\alpha = 14^h\ 40^m$, $\delta = +50$) we doubled the density of our survey fields. Large gaps in the galaxy distribution at the location of the proposed void are apparent from comparing FIGURES 1b, 1c, and 1d, which use the radial velocities measured for 231 galaxies in the 282 fields. Low-velocity galaxies, in the range 6,000 to 12,000 km s^{-1}, are distributed over

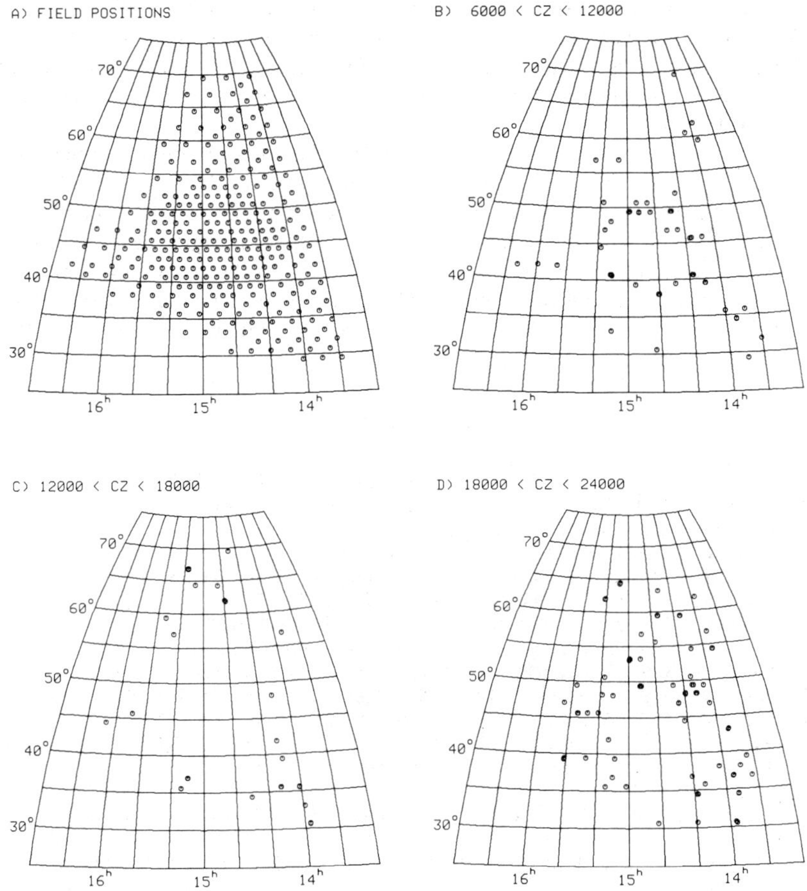

FIGURE 1. (A) Distribution on the sky of 282 sample fields. (B) Galaxies with red shifts between 6,000 and 12,000 km s^{-1}. (C) Galaxies with red shifts between 12,000 and 18,000 km s^{-1}. (D) Galaxies with red shifts between 18,000 and 24,000 km s^{-1}.

all the survey fields as shown in FIGURE 1b. More distant galaxies, in the range 18,000 to 24,000 km s^{-1}, have a similar distribution shown in FIGURE 1d. And in the range 12,000 to 18,000 km s^{-1}, the proposed void, we observe a conspicuous absence of galaxies just where the density of survey fields is highest. Knowledge of the three-dimensional positions shows that a sphere at $\alpha = 14^h\ 48^m$, $\delta = +47$ with a diameter of 6,000 km s^{-1} contains no galaxies from this sample.

Emission-line galaxies in the direction of the void have been investigated by Balzano and Weedman,[1] who found that some Markarian galaxies are within the triangle bounded by the original three fields, but detailed study shows none in the 6,000 km s^{-1} sphere. More emission-line galaxies in the Boötes region were detected in a survey by Sanduleak and Pesch,[18] but red shift measures by Gregory *et al.* show that these galaxies are not in the volume of the void.[9] Of course, surveys of emission-line objects may eventually prove useful for delineating the large-scale distribution of galaxies, because it is so easy to measure their red shifts. However it is important to establish whether galaxies with emission lines provide a reliable tracer of the overall galaxy distribution.

Very crudely, we estimate that there is a 90% chance that the density of the Boötes void is at least four times lower than the average in this survey. It is worth recalling that this survey shows that the region in Boötes is missing bright galaxies comparable to M31 or the Milky Way, but it does not demonstrate that the region has a corresponding mass fluctuation. Evenly distributed matter (such as the many -inos emitted by theoretical physicists) or low-luminosity galaxies are not excluded. Still, this investigation shows that the study of holeography may provide useful constraints on models of galaxy formation.

REFERENCES

1. BALZANO, V. A. & D. W. WEEDMAN. 1982. Astrophys. J. Lett. **255:** L1.
2. DAVIS, M., J. HUCHRA, D. W. LATHAM & J. TONRY. 1982. Astrophys. J. **253:** 423.
3. DE VAUCOULEURS, G. 1953. Astron. J. **58:** 30.
4. DE VAUCOULEURS, G. 1976. Astrophys. J. **203:** 33.
5. EINASTO, J., M. JOEVEER & E. SAAR. 1980. Mon. Not. R. Astron. Soc. **185:** 357.
6. GIOVANELLI, R. & M. P. HAYNES. 1982. Astron. J. **87:** 1355.
7. GREGORY, S. A. & L. A. THOMPSON. 1978. Astrophys. J. **222:** 784.
8. GREGORY, S. A., L. A. THOMPSON & W. G. TIFFT. 1981. Astrophys. J. **243:** 411.
9. GREGORY, S. A., W. G. TIFFT & R. P. KIRSHNER. 1983. Private communication.
10. JOEVEER, M., J. EINASTO & E. TAGO. 1978. Mon. Not. R. Astron. Soc. **185:** 357.
11. KIRSHNER, R. P., A. OEMLER & P. L. SCHECHTER. 1978. Astron. J. **83:** 1549.
12. KIRSHNER, R. P., A. OEMLER & P. L. SCHECHTER. 1979. Astron. J. **84:** 951.
13. KIRSHNER, R. P., A. OEMLER, P. L. SCHECHTER & S. A. SHECTMAN. 1981. Astrophys. J. Lett. **248:** L57.
14. KIRSHNER, R. P., A. OEMLER, P. L. SCHECHTER & S. A. SHECTMAN. 1983. Astron. J. (In press.)
15. KIRSHNER, R. P., A. OEMLER, P. L. SCHECHTER & S. A. SHECTMAN. 1983. IAU Symp. **100.** (In press.)
16. OSTRIKER, J. P. & L. L. COWIE. 1981. Astrophys. J. Lett. **243:** L127.
17. PEEBLES, P. J. E. 1980. Large Scale Structure in the Universe. Princeton University Press. Princeton, N.J.
18. SANDULEAK, N. & P. PESCH. 1982. Astrophys. J. Lett. **258:** L11.
19. TULLY, B. 1982. Astrophys. J. **257:** 389.
20. ZELDOVICH, YA. B., J. EINASTO & S. F. SHANDARIN. 1982. Nature **300:** 407.

The Development of Galaxy Clusters[a]

E. E. SALPETER

Center for Radiophysics and Space Research
Cornell University
Ithaca, New York 14853

INTRODUCTION

Unlike the two previous speakers, I will not discuss the very large structures but concentrate on an individual cluster of galaxies, the matter surrounding it, and the galaxies of which it is composed. A number of different questions are related to each other and to the dynamic development of a cluster: (1) Were individual galaxies formed earlier or later than the first contraction of the cluster in which they now live? How is the Hubble type of a galaxy related to the galaxy's environment? (2) Is much of the "invisible mass" in the universe as a whole attached to individual galaxies or is most of it loose? More specifically, does the "loose" invisible matter associated with a cluster reside mainly at smaller or larger distances from the cluster center than the visible galaxies? (3) What is the physical nature of the invisible matter? As we heard earlier, there are many candidates but I will concentrate on just two possibilities—low-mass stars and massive neutrinos.

I will not give a comprehensive review of clusters, but concentrate on a few subtopics. In particular, I will not discuss intergalactic gas and its x-ray emission. This gas provides a very useful diagnostic tool, but contributes little to the total mass of a cluster. In passing I will also enunciate two "important theorems about observational cosmology." These two theorems (meant only partially in jest) are well known to all astronomers, but theoretical physicists may not be so familiar with them.

MASSIVE HALOS OF INDIVIDUAL GALAXIES

For many spiral galaxies the rotational velocity $V_{rot}(r)$ as a function of radial distance r has been measured quite accurately, but only out to moderate values of r. Optical rotation curves, using spectral lines from stellar emission in the visible, have been known the longest but are difficult to extend far beyond where the optical stellar emission has its peak (typically $r \sim 10$ kpc). Fortunately, neutral hydrogen gas in the disks of spiral galaxies extends out about twice as far as the starlight and rotation curves can be observed farther out. The λ 21-cm line emitted by this gas (and some refinements of the optical work) have shown that most spiral galaxies have "flat rotation curves," i.e., $V_{rot}(r) \approx$ constant at large r where the optical surface brightness is low.[1-4] Since Kepler's Law would have V_{rot} decrease there as $r^{-1/2}$, if the mass to light ratio were constant, we thus have evidence for "invisible mass" in the outer parts of a galaxy. More correctly, we should say that the overall ratio M/L is larger if the outer

[a]This work was supported in part by National Science Foundation Grant AST81-16370.

parts are included than the local ratio in the inner part of the galaxy, which is dominated by "ordinary, visible stars."

If some spherical mass distribution had a density law $\rho(r) \propto r^{-2}$ then the mass contained inside radius r, $M(r) \propto r$, would lead to a flat rotation curve. An inference has sometimes been drawn (from inverting the last sentence) that the massive halo of a spiral galaxy "is spherical, has density varying as r^{-2}, and extends very far out." Some doubt has been thrown on this inference recently (in keeping with stage 2 of the theorem enunciated in the sixth section of this paper), for two reasons: On the observational side it is difficult to obtain unequivocal data beyond about two optical radii—one still finds some hydrogen, but is has a clumpy distribution and warping of the disk makes it difficult to derive true rotation velocities. On the theoretical side it has been shown that one cannot derive a *unique* density distribution in the galactic plane of the invisible matter from a given observational rotation curve;[5] furthermore, the assumption of a spherical distribution for the massive halo is popular but not derived from any direct observations. This caution should not be interpreted as skepticism about massive halos altogether: the virial theorem still works and one still needs a massive halo, but (a) we do not know if the density distribution follows an r^{-2} law or (b) how far out the halo extends. Observational data on velocity differences in binary galaxies[6] are even harder to interpret uniquely, but are more compatible with the presence rather than the absence of very extensive halos.

As discussed by previous speakers, massive neutrinos are good candidates for "closing the universe"; the neutrino contribution Ω_ν to the cosmological density parameter is

$$\Omega_\nu \sim (g_\nu/2)(m_\nu/30\ \text{eV}) \tag{1}$$

if there are g_ν species of neutrinos with mass m_ν. Could massive neutrinos provide the massive halos of individual galaxies? If one could succeed in compressing the neutrinos purely adiabatically during the galaxy formation, they would remain partially degenerate and would settle down to a polytrope with $n = 1.5$. The mass M as a function of radius R differs only by a factor of two from a fully degenerate zero-temperature star,[7]

$$M^{1/2}R^{3/2}g_\nu m_\nu^4 = \text{constant} \tag{2}$$

Expressing mass, radius, and density in units typical of spiral galaxies, we have

$$\frac{M}{10^{11}M_\odot} \gtrsim 0.4\left[\frac{M}{R^3}\frac{(10\ \text{kpc})^3}{10^{11}M_\odot}\right]^{1/2}\frac{2}{g_\nu}\left(\frac{30\ \text{eV}}{m_\nu}\right)^4 \tag{3}$$

The inequality arises because the energy release and the "violent relaxation" during the gravitational collapse[8] could change the adiabat and make the neutrinos less degenerate. Melott argues that the inequality will raise the coefficient in Equation 3 by only a moderate amount,[9] so that the mass-density relation makes neutrinos particularly suitable for ordinary galaxies if the neutrino mass is of order 30 eV. Some dwarf irregular galaxies have a particularly low luminosity but a large luminosity density; if *all* of these "compact" galaxies also contained "invisible mass," neutrinos would be ruled out for that purpose: Equation 3 would require a very large total mass per dwarf galaxy and these galaxies are very numerous. Observational data are accumulating only slowly, both for dwarf irregular and dwarf elliptical galaxies.[10]

Another uncertainty in associating massive neutrinos with individual galaxies concerns the Jeans mass $M_{J,\nu}$, the minimum mass in neutrinos for a region that can separate gravitationally from the cosmological background. With $m_\nu \sim 30$ eV, $g_\nu \sim 2$, the cosmological density as a function of red shift z substituted into Equation 3 gives

$$M_{J,\nu}(z) \sim 10^8 M_\odot (1 + z)^{3/2} \qquad \textbf{(4)}$$

if the neutrinos are nonrelativistic. This mass is less than galaxy masses (and thus presents no problem) now and for low red shifts, but $M_{J,\nu}$ had a maximum value of about 10^{15} $M_\odot$ (much larger than galaxy masses) sometime before the recombination era when the neutrinos were just barely relativistic.[11,12] If one insists on forming individual galaxies straight from the cosmological background, before galaxy groups and clusters formed, massive neutrinos would be excluded. If one believes that protoclusters with $M \gtrsim 10^{15}$ $M_\odot$ formed first and then fragmented into successively smaller chunks, the possibility of neutrinos reemerges.[9]

Another class of candidates for the "invisible mass" is "stellar population III." This involves the postulate that stars with a different initial mass function (IMF) from the present formed efficiently from gas without any heavy elements (beyond helium), *before* ordinary stars formed and thus before the interstellar medium was polluted with heavy elements. One possibility is that the IMF was tilted toward very massive stars which live only a short time and leave compact remnants, neutron stars and black holes, which are optically faint. This possibility has one drawback—one has to make the compact remnants without ordinary supernovae as an intermediary to avoid excessive production of heavy elements. Faint main sequence stars of very low mass (or "Jupiters") would not have this drawback, but it has been argued that interstellar gas clouds without heavy elements have another drawback: they might have so low an opacity that they could not cool radiatively and the large Jeans mass in hot gas would prevent the formation of low-mass stars. More recent work has shown that molecular hydrogen can be produced in abundance and can give lots of radiative cooling.[13] This leaves open the possibility of low-mass stars for stellar population III, but with one cautionary note: hydrogen molecules are easily destroyed in high-velocity shocks and stellar population III may form preferentially in regions of *low velocity*. I return to this point in the fifth section.

DISSIPATIONLESS COLLAPSE AND NEARBY CLUSTERS

I will return to some quantitative details of the dynamic development of a galaxy cluster in the fifth section, but start here with an overview. Like most galaxies, we live in a "local group" of galaxies, dominated by Andromeda and our own galaxy. Of various nearby "rich" clusters of galaxies ("rich" meaning $\gtrsim 50$ galaxies as bright as ours), one—the Virgo cluster—is about five times closer to us than any other. As suggested some time ago by de Vaucouleurs,[14] the Virgo cluster (which is a disk in the sky of about 6° radius) is surrounded by a flattened structure[15,16] about 10 times larger in diameter. Our local group is close to the edge of this structure, which is known as the "local supercluster"[14] (not to be confused with "multicored superclusters," much larger and more complex structures which I will not discuss here). Expressing distances in red shift units (i.e., multiplying distances by the Hubble constant, $H_0 \approx 50$ to 100 km s^{-1}), the Virgo cluster center is about 1,200 km s^{-1} away whereas the next few (Cancer, Coma, Pegasus, Pisces, Perseus, etc.) clusters are about 4,000 to 8,000 km s^{-1} away. It is convenient to think of cells in space, each one containing one cluster, and to approximate each cell by a "Wigner-Seitz sphere," with an internal galaxy density equal to the large-scale cosmological average $\bar{\rho}_\infty$. The average radius of such Wigner-Seitz spheres in our vicinity is about 3,700 km s^{-1}; red shift data on a larger number of somewhat more distant clusters[17] are in reasonable agreement with this number.

We now have a rather systematic sample of galaxy red shifts—the Center for Astrophysics optical survey,[18] which contains over 2,000 galaxies (to apparent magnitude 14.5 m_B) giving good coverage to a distance of at least 6,000 km s^{-1}. There used to be controversy as to whether one should think of the Virgo cluster plus the surrounding local supercluster as a sphere or as a uniform flat pancake. We now have the galaxy distribution accurately and the answer is "neither" (or "both"): the cluster core (defined as a radius of 6° on the sky, or ≈ 120 km s^{-1}) is roughly spherical and contains only about a fifth of the total luminosity L_{VI} of a Virgo-centered region V_{VI} extending to our location; a flat disk of uniform surface density equal to the local value also contains only about a fifth of the total luminosity, leaving three-fifths for an intermediate distribution. The region V_{VI} occupies a volume of $\sim 1/30$ that of the local Wigner-Seitz sphere but contains $\sim 1/10$ of the luminosity, so that the luminosity density contrast is $(\bar{\rho} + \delta\rho)/\bar{\rho} \sim 3$.

The formation of a galaxy cluster (or other bound system) by dissipationless collapse starts with the growth of a small density perturbation at early times (large red shift). Even for an open universe (i.e., overall cosmological density parameter $\Omega < 1$), Ω was very close to unity at early times and a small positive $\delta\rho/\rho$ then can already represent a bound system. The growth of the perturbation becomes nonlinear when $\delta\rho/\rho \gtrsim 1$, the cluster "separates out from the cosmological background," collapses and undergoes "violent relaxation" with a small fraction of the matter ejected and a larger fraction remaining bound. Qualitatively one can say that a cluster keeps its internal density ρ_{int} constant while the cosmological density $\rho_\infty(t)$ decreases, so that ρ_{int} reflects the value of ρ_∞ in the separation era and should give the red shift z_{coll} when the collapse took place. Quantitatively, it is difficult to pin down z_{coll} accurately, even when numerical results for dynamic evolution are available,[19–32] because violent relaxation leads to an appreciable variation of density with radial distance inside a cluster. Nevertheless it is clear that galaxy clusters formed fairly recently,[22,23] with z_{coll} between about 0.5 and 3 (I personally favor $z_{coll} \sim 1$ to 2). It is much less clear when individual galaxies formed—their internal density is large enough so they *could* have formed as early as $z \sim 50$, long before clusters formed, but they could also have formed much later. I discuss this controversy next.

DID GALAXIES FORM EARLY OR LATE?

Theoretically, this controversy is related to the spectrum of primordial density perturbations, discussed in more detail by other speakers. I will describe only two extreme cases and then present arguments against these extremes. (1) In one extreme the spectrum[24] is heavily weighted toward smaller wavelengths, so that an individual galaxy condenses out of the cosmological background before groups of galaxies congregate together, small groups before large clusters, etc. If a galaxy forms as early as $z \sim 50$, the density of a surrounding volume that later will become a cluster core is only a few percent higher than that of the "general field." (2) In the other extreme, perturbations corresponding to individual galaxies are strongly suppressed and a large protocluster separates out (at $z \sim 1$ to 3) still in gaseous form. During the collapse process, the mean internal density increases as do slightly smaller scale perturbations, which can lead to successive fragmentation. In the purest form of this extreme, one postulates further that galaxies do not form until *after* the collapse has led to shocks in the gaseous protocluster; radiative cooling then allows the gas to contract even further under gravity and reach very high densities. This case was first described in terms of the collapse of a flat "pancake,"[25] but for a shape as I described for the local supercluster, such a dissipative collapse would lead also to smaller radial distances from the cluster center (as well as a thinner pancake) than for dissipationless collapse.

Evidence against extreme 1 comes, indirectly, from a strong, uncontroversial observational fact about the different Hubble types: the ratio of numbers of elliptical galaxies to spiral galaxies increases continuously with the present density of the environment, from sparsely populated regions outside superclusters on one extreme to cores of rich clusters on the other. There used to be controversy as to whether this ratio was already built in when the galaxies first formed or whether it evolved later on. On the evolutionary picture all galaxies formed as spirals and those living in a dense environment now were converted into ellipticals (via intermediate Hubble types) via rampressure and tidal stripping. Rampressure stripping by intergalactic gas can indeed shrink the gaseous disk of a spiral, or SO, galaxy somewhat,[26,27] but it cannot convert one Hubble type into another: the inner bulges are very different for different Hubble types,[28,29] whereas rampressure and tidal effects[30] are strong only in the outer regions of a galaxy. Galaxy mergers (or "cannibalism") can produce giant elliptical (or cD) galaxies in a dense cluster core, but *not* moderate-sized ellipticals in a cluster of moderate density (such as the Virgo cluster, which contains quite a few ellipticals). As mentioned, the density enhancement due to a developing protocluster was only a few percent at $z \sim 50$ and it is difficult to see how such a small environmental effect could determine what Hubble type a galaxy should become.

There are also various arguments against the other extreme (2), at least in its purest form: spiral galaxies forming in the aftermath of shock during pancake collapse should have their spin vectors lying preferentially in the plane of the pancake,[31] but there is no observational evidence for such a preference[32] (see FIGURE 1). Furthermore, strong dissipation would lead to greatly reduced velocity dispersion and pancake thickness.[33] Finally, there are some galaxies, but little evidence for distributed gas, in regions outside superclusters where the overall density is too low to have produced shocks yet.

FIGURE 1 demonstrates the absence of the predicted "pancake preference," but it also illustrates the first of the two "important theorems about observational cosmology." This theorem can be stated as, A careful observer will always find *some* "two sigma effect" by serendipity, even if it is not real. This is due to the great complexity of the data, allowing at least 20 questions in addition to the original question and some of

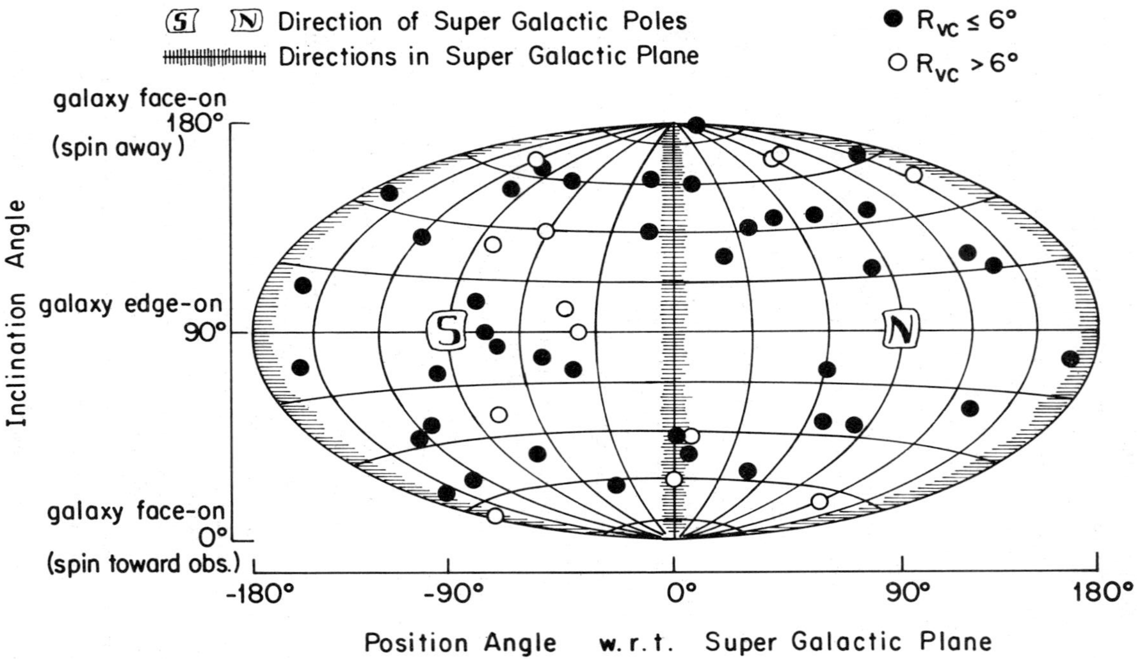

FIGURE 1. An equal area projection of the directions of galaxy spin vectors in the Virgo cluster and its immediate surrounding (the local supercluster). Vector directions in the supergalactic plane are crosshatched; contrary to the predictions of an extreme pancake model, these directions are *not* preferred.

these will have a positive fluctuation. In FIGURE 1 there is a slight preference for galaxy spins to correlate with the south supergalactic polar direction, but one could also have asked for correlation between spin and position in the cluster or between spin and systemic velocity, etc.

Because of the difficulties with the extremes, I favor an intermediate hypothesis where perturbations of all scale sizes, from an individual galaxy (or at least small galaxy group) up to a rich supercluster, coexisted after the recombination era. Assume a perturbation spectrum weighted only slightly toward the larger masses. A gaseous protocluster then goes into the nonlinear contraction phase first, but this large-scale compression also accelerates the growth of a galaxy-size perturbation; individual galaxies are thus formed rapidly in a collapsing protocluster, presumably after only mild dissipation rather than extensive shocks. A galaxy-size perturbation that does not find itself inside a positive cluster-size perturbation could still form an individual galaxy (or small galaxy group), but more slowly. The slower time scale and lower environment density might favor forming galaxy disks,[29] i.e., spiral galaxies, whereas a short time scale and large density might favor elliptical galaxies.

DYNAMIC MODELS AND VARIATIONS IN M/L

For the Virgo cluster and its surrounding, we now have great observational detail on the galaxy number (or luminosity) density distribution and on $\sigma(\theta)$, the dispersion of systemic galaxy velocities as a function of distance θ from the cluster center. We have similar, although less detailed, data for a number of other clusters, especially the Coma cluster.[34] One of the main aims of observational cosmology is to determine the overall cosmological density parameter Ω, and one might think that the cluster observations plus dynamics should be sufficient to furnish Ω fairly reliably. We know the contrast in *luminosity* density between cluster cores, their immediate surroundings, and the general field quite well from the CfA survey.[18] If one *assumes* a constant mass to light ratio M/L everywhere, then one finds a value of $\Omega \sim 0.2$ or 0.3. However, since 1980 my colleagues and I have taken the point of view that one *cannot* assume a constant M/L because of our ignorance about the physical nature of the "invisible mass."[20] Since our warning has mainly fallen on deaf ears so far, I want to spell it out in some detail.

The ordinary stellar populations I and II in a spiral galaxy like ours have a mass to light ratio of $M/L \sim 2$ to 5 (using solar units throughout). Rotation curves for individual spiral galaxies, out to a little beyond the optical radius, give $M/L \sim 10$ to 20 (assuming a Hubble constant $H_0 \sim 75$ km s^{-1} Mpc^{-1}). Since even the cores of galaxy clusters are not in a fully relaxed equilibrium state (the velocity distribution is far from Gaussian, some subgroups are falling in, others are being ejected, etc.), one cannot use the virial theorem to determine masses but must fit observational data to a detailed dynamic model. Using only the data on velocity dispersion $\sigma(\theta)$ and dynamic models, it turns out that one can determine the mass of the *core* of a cluster fairly reliably even without any assumption about M/L: for the Virgo cluster core (radius $\sim 6°$) we find $M/L \approx (400 \pm 150)$;[20,21] Rivolo and Yahil for the same cluster and a different model find $M/L \approx 600$;[23] for the Coma cluster, Kent and Gunn find $M/L \approx 270$.[34] These two clusters are fairly typical of moderately rich clusters,[35] and there is now no doubt of a considerable amount of "invisible mass" residing somewhere other than the visible inner regions of individual galaxies.

It is conceivable that (a) all invisible matter resides in outer halos of individual galaxies with M/L the same in all galaxies everywhere, in which case $\Omega \sim 0.2$ or 0.3. It is also conceivable, and in fact more likely, that much of the invisible mass is "loose." There are at least two such possibilities: (b) Massive neutrinos (or other "inos") are dissipationless whereas baryon matter must have undergone some dissipation inside clusters; moderate dissipation from galaxy-galaxy encounters if galaxies formed before the full protocluster collapse and very drastic dissipation if galaxies formed from shocked gas. For a given cluster the visible galaxies thus lie closer in to the center (deeper potential well) than the massive neutrinos and M/L is larger in the outer parts of the supercluster than in the cluster core, the only region where M/L is known accurately. The density parameter Ω is then larger than in case a; it could for instance be unity but it is in fact quite uncertain because (1) it depends on the extent of the dissipation and (2) we know little about the neutrino perturbation spectrum for scales corresponding to isolated galaxy groups. (c) If most of the invisible mass in a cluster is in the form of loose stellar population III stars, we do not understand their formation process sufficiently to know where they mainly reside. However, we saw in the second section that low velocities might increase the formation probability if stellar population III consists of low-mass stars. If this probability is also increased by large gas density, then the invisible mass could be concentrated to the deep potential well in cluster cores. M/L outside of cores could then be appreciably smaller than inside (individual galaxies only require $M/L \gtrsim 20$) and Ω could be as small as $\sim$0.03.

DEVIATIONS FROM HUBBLE FLOW

We saw that velocity dispersion data plus dynamic models can give masses for cluster cores quite accurately (and have done so already for the Virgo and Coma clusters), but not for the regions surrounding the core. If direct distance indicators for individual galaxies (not involving red shifts) are available, one can also measure a different kind of velocity, ΔV, the deviation from pure Hubble flow due to the gravitational retardation of the Virgo cluster. In principle such a measurement plus dynamic models should give a handle on the mass of the whole Virgo cluster and hence on the cosmological density parameter Ω. In practice, there are complications, especially if one adopts my cautious point of view that M/L need not be a universal constant but may vary with radial distance.

Neglect at the moment deviations from spherical symmetry in the mass density distribution around the center of the Virgo cluster. Pure Hubble flow contains the gravitational retardation effects due to the average cosmological mass density $\bar{\rho}_\infty$. Let R be the distance from the Virgo cluster center to our local group and M_R the total mass in a Virgo-centered sphere of radius R. The deviation ΔV from pure Hubble expansion, measured at the local group position, depends not on the total mass M_R in the sphere but only on the *excess* over what the mass would have been in a homogeneous universe,

$$M_{\text{exc}} = M_R - (4\pi R^3/3)\bar{\rho}_\infty \qquad (5)$$

We know the distribution of luminosity density, and hence the excess luminosity in the sphere of radius R, quite accurately. If one assumes a constant mass to light ratio M/L everywhere, then M_{exc} as well as the cosmological Ω is proportional to this ratio; ΔV is approximately proportional to $M_{exc}^{1/2}$, so Ω is then approximately proportional to ΔV^2. The velocity dispersion in the bound cluster core depends on the total mass density there (not merely the excess), but for constant M/L this also increases with Ω. The situation is radically different if one adopts our point of view[20] of (1) choosing only models that fit the observed velocity dispersion in the core but (2) allows M/L to vary with distance from the center. Because of 1, all models have almost the same central density, but, because of 2, different models have different values of $\Omega \propto \bar{\rho}_\infty$. Because of the negative term in Equation 5 it can happen that, although M_R increases monotonically with Ω, M_{exc} and therefore ΔV actually decrease as Ω increases!

For models that both fit the core velocity dispersion and assume constant M/L the predicted ΔV is close to 250 km s^{-1} and $\Omega \sim 0.2$ or 0.3.[20] For the simplest kinds of models,[21] where M/L is allowed to vary slowly and smoothly, there is indeed a slight decrease of ΔV with increasing Ω. More realistically, there is a great asymmetry between the two extreme cases b and c in the previous section: For case b with $\Omega < 0.1$, most of the loose invisible mass is concentrated in the cluster core and the density outside the core is relatively small and well known. The effect of the flattening of the outer regions is small and can be estimated and the predicted ΔV is well constrained—it must be between 200 and 400 km s^{-1} and point accurately toward the Virgo center. The situation is qualitatively different for case c with $\Omega \sim 1$, where most of the massive neutrinos are in the supercluster outside the core. For the *simplest* spherical models, M_{exc} is very much smaller than M_R and the predicted ΔV would be small, but there is no guarantee that the mass distribution of the neutrino supercluster is simply related to the visible cluster core. The flattening of the supercluster can be important, but not even the sign of the effect is clear (the matter inside and outside of our position having opposite effects) and the predicted ΔV could be "almost anything." In particular,[21] the radial component of ΔV could lie between $\sim(-100$ to $+500)$ km s^{-1} (thus mimicking case b) and it could have an appreciable component in directions other than the Virgocentric one.

The observational situation on ΔV is in a state of flux and I will not quote any individual observation to date, but I will state the second important theorem about observational cosmology: There are three stages in the quest for the measured value of some important quantity (such as ΔV). In the first stage different methods give violently different numbers, each one with a claim for very great accuracy. For instance, the first (unpublished) values for ΔV ranged from about 100 km s^{-1} to about 600 km s^{-1}, each with a small claimed standard error. In the second stage the claimed accuracy becomes more modest and the *increased* standard error actually signifies progress because the observers have now learned what the systematic errors in their method are. This is the present stage for ΔV, with current estimates ranging from $\sim$250 to 500 km s^{-1}. Toward the end of the second stage, the "bandwagon effect" is supposed to push all the measured values close to each other, but that has not happened yet for ΔV. In the third stage the measured values are supposed to approach the real values, but I am too inexperienced in observational cosmology to predict when that will happen!

RELATED TOPICS

One related topic concerns even larger structures than our local supercluster, such as large voids or superclusters that contain two or three cluster cores or even strings of associated superclusters. This topic has been covered by other speakers and I will not discuss it, but I want to mention briefly two other topics.

1. I have argued that it is difficult to obtain mass densities in the outer parts of the galaxy and to estimate the cosmological density parameter Ω. Fortunately, an indirect method is more promising—the measurement of temperature fluctuations in the microwave background radiation on an angular scale of $\sim$10 arcmin.[36] This gives a measure of the density perturbations just after the recombination era, which later evolved into galaxy clusters and superclusters (smaller angular scales are not useful because of smearing due to radiative diffusion). Calculations of the development of clusters (or voids) show what perturbations were required during recombination to give the observed clustering today.[22,37] The required density perturbations, and the predicted microwave background fluctuations, are largest for small Ω and observational upper limits already rule out $\Omega \gtrsim 0.1$ (unless one finds some loophole, such as some process of reionizing the intergalactic medium at red shift $z > 10$).[36] Case b in the fifth section overcomes this difficulty by a combination of three factors: Ω is large, which requires less fluctuation even for the same clustering density contrast today; the contrast today is less because the neutrinos are spread out more than the galaxies; and the baryon and photon fluctuations probably lagged behind the neutrinos at recombination.

2. One should keep at least a slightly open mind for radical departures from standard physics, such as the one advocated by Milgrom.[38] He postulates that there is no "invisible mass" anywhere but instead he modifies the law for the Newtonian gravitational acceleration g_N: with a_0 a universal constant of acceleration, approximately equal to cH_0, he essentially replaces g_N by $(a_0 g_N)^{1/2}$ when $g_N \ll a_0$. For a circular orbit at radius r around a point mass M, Kepler's law for the orbital velocity V is then replaced by

$$V^2 = (a_0 GM)^{1/2} \tag{6}$$

independent of radius r. The same value of a_0 and a mass to light ratio of (2 to 5) $M_\odot/L_\odot$ is then compatible with the observed rotation velocities of individual galaxies and with the velocity dispersions of clusters. Milgrom's postulate makes particularly striking predictions for galaxies of low surface brightness L/R^2. With M, the real mass, proportional to L and with R the radius of the galaxy, the predicted rotation velocity V scales as follows

$$\frac{(G^{-1}V^2R)}{M,} \sim \frac{(a_0R^2)^{1/2}}{GM,} \propto \left(\frac{a_0}{L/R^2}\right)^{1/2} \tag{7}$$

i.e., V should be much larger than it would be with Newtonian mechanics and without invisible mass when the surface brightness is very low. Dwarf spheroidal galaxies should eventually provide a good test.

ACKNOWLEDGMENT

I am indebted to Dr. G. Helou for FIGURE 1.

REFERENCES

1. BOSMA, A. & P. VAN DER KRUIT. 1979. Astron. Astrophys. **79:** 281.
2. FABER, S. & J. GALLAGHER. 1979. Annu. Rev. Astron. Astrophys. **17:** 135.
3. KRUMM, N. & E. SALPETER. 1979. Astron. J. **84:** 1138.
4. RUBIN, V. C. 1979. Comments Astrophys. **8:** 79.
5. BAHCALL, J., M. SCHMIDT & R. SONEIRA. 1982. Astrophys. J. **258:** L23.
6. WHITE, S., J. HUCHRA, D. LATHAM & M. DAVIS. 1983. Mon. Not. R. Astron. Soc. **203:** 701.
7. TREMAINE, S. & J. GUNN. 1979. Phys. Rev. Lett. **42:** 407.
8. LYNDEN-BELL, D. 1967. Mon. Not. R. Astron. Soc. **136:** 101.
9. MELOTT, A. L. 1982. Phys. Rev. Lett. **48:** 894.
10. FABER, S. & D. LIN. 1983. Astrophys. J. **266:** L21.
11. BOND, J., G. EFSTATHIOU & J. SILK. 1980. Phys. Rev. Lett. **45:** 1980.
12. WASSERMAN, I. 1981. Astrophys. J. **248:** 1.
13. PALLA, F., E. SALPETER & S. STAHLER. 1983. Astrophys. J. **271:** 632.
14. DE VAUCOULEURS, G. 1956. Vistas Astron. **2:** 1584.
15. DE VAUCOULEURS, G. 1982. Astrophys. J. **253:** 520.
16. TULLY, R. B. 1982. Astrophys. J. **257:** 389.
17. HOESSEL, J., J. GUNN & T. THUAN. 1980. Astrophys. J. **241:** 486.
18. DAVIS, M., J. HUCHRA, D. LATHAM & J. TONRY. 1982. Astrophys. J. **253:** 423.
19. SILK, J. & M. WILSON. 1979. Astrophys. J. **228:** 641.
20. HOFFMAN, G. L., D. OLSON & E. SALPETER. 1980. Astrophys. J. **242:** 861.
21. HOFFMAN, G. L. & E. SALPETER. 1982. Astrophys. J. **263:** 485.
22. HOFFMAN, G. L., E. SALPETER & I. WASSERMAN. 1983. Astrophys. J. **268:** 527.
23. RIVOLO, A. R. & A. YAHIL. 1983. Astrophys. J. (In press.)
24. PEEBLES, P. J. E. 1980. The Large-Scale Structure of the Universe. Princeton University Press. Princeton, N.J.
25. SUNYAEV, R. & YA. ZELDOVICH. 1972. Astron. Astrophys. **20:** 189.
26. CHAMARAUX, P., C. BALKOWSKY & E. GERARD. 1980. Astron. Astrophys. **83:** 38.
27. GIOVANARDI, C., G. HELOU, E. SALPETER & N. KRUMM. 1983. Astrophys. J. **267:** 35.
28. SANDAGE, A. & N. VISVANATHAN. 1978. Astrophys. J. **225:** 742.
29. DRESSLER, A. 1980. Astrophys. J. **236:** 351.
30. FAROUKI, R. & S. SHAPIRO. 1980. Astrophys. J. **241:** 928.
31. DOROSHKEVICH, A., S. SHANDARIN & E. SAAR. 1978. Mon. Not. R. Astron. Soc. **184:** 643.
32. HELOU, G. 1983. Cornell University Report. Cornell University. Ithaca, N.Y.
33. DEKEL, A. 1983. Astrophys. J. **264:** 373.
34. KENT, S. & J. GUNN. 1982. Astron. J. **87:** 945.
35. BAHCALL, N. A. 1977. Annu. Rev. Astron. Astrophys. **15:** 505.
36. PARTRIDGE, R. B. 1980. Astrophys. J. **235:** 681.
37. PEEBLES, P. J. E. 1982. Astrophys. J. **258:** 415.
38. MILGROM, M. 1983. Astrophys. J. **270:** 365, 371.

Big Bang Nucleosynthesis[a]

Gateway to the Very Early Universe

MICHAEL S. TURNER

Astronomy and Astrophysics Center
Enrico Fermi Institute
The University of Chicago
Chicago, Illinois 60637

INTRODUCTION[b]

The standard, hot big bang model (Friedmann-Robertson-Walker cosmology)[c] is a very successful and economical model of the evolution of the universe. Based upon the concordance of this model with the observational data available, we believe that it provides a reliable description of the evolution of the universe from about 10^{-2} after "the bang" ($kT \simeq 10$ MeV) until the present epoch (10–20 B yr after "the bang"; $kT \simeq 3 \times 10^{-4}$ eV). In the standard model the age and temperature of the universe are related by

$$T \simeq 1.6\, g_*^{-1/4}\, (t/10^{-6}\text{ s})^{-1/2}\ \text{GeV} \tag{1}$$

where $g_*(T)$ is the effective number of very relativistic species at temperature T ($= \sum_{\text{Bose}} g_{\text{Bose}} + 7/8 \sum_{\text{Fermi}} g_{\text{Fermi}}$) and throughout $\hbar = k_B = c = 1$, so that 1 GeV $= 1.16 \times 10^{13}$ K $= (6.58 \times 10^{-25}\text{ s})^{-1} = (1.97 \times 10^{-14}\text{ cm})^{-1}$. Equation 1 is only valid when the energy density of the universe is dominated by relativistic particles ($t \lesssim 10^{10}$ s).

There is nothing in our present understanding of the fundamental laws of physics that would lead us to believe that the standard model cannot be extrapolated back to times as early as 10^{-43} s and temperatures as high as 10^{19} GeV; earlier than this, quantum corrections to general relativity should be significant. The fundamental constituents of matter—quarks and leptons—are believed to be pointlike particles, whose interactions become asymptotically weak at high energies ("asymptotic freedom"), so that the dilute gas approximation [embodied in (**1**)] should remain valid at the very high energies achieved shortly after "the bang." Indeed, this extrapolation to very early times and very high energies is what makes the very early universe so interesting. I hasten to remind the reader that while such an extrapolation is not unreasonable, it may nonetheless prove to be very wrong.

The connection between cosmology and elementary particle physics facilitated by the very early universe is obvious—the early universe is the ultimate high-energy physics laboratory, one without any budgetary constraints! Research done at the interface of cosmology/astrophysics and particle physics has already produced many important and interesting results: a framework for understanding the origin of the

[a]This work was supported in part by the Department of Energy through contract AC02 80ER10773 A003 (at Chicago).

[b]This article is a review of work done in collaboration with D. N. Schramm, G. Steigman, J. Yang, and K. A. Olive, and is described in greater detail in References 1 and 2.

[c]For a discussion of the standard model, see, e.g., Reference 3.

baryon asymmetry (for a review of baryogenesis, see, e.g., Reference 4), the inflationary universe scenario,[5–7] countless candidates for the dark matter, and numerous constraints on the properties (mass, lifetimes, relic abundances, etc.) of a variety of particle species (neutrinos, other -inos, axions, monopoles, etc., see, e.g., Reference 8) which in turn serve to constrain the theories that predict them.

Precisely what evidence do we have that testifies to the validity of the standard model? (1) The universal (Hubble) expansion—direct observations of distant galaxies allow us to see the Universe as it was a few billion years after "the bang." (2) The cosmic microwave background—the surface of last scattering for this radiation is the Universe at a red shift of $0(10^3)$ ($t \simeq 10^5$–10^6 yrs, $T \simeq 1$ eV). The black-body spectrum and spatial uniformity ($\Delta T/T \simeq 10^{-3}$) are direct evidence that the universe was once very hot and very isotropic and homogeneous. (3) The abundances of D, ^{3}He, ^{4}He, and ^{7}Li—these light isotopes are predicted to be synthesized during an epoch of primordial nucleosynthesis ($t \simeq 10^{-2}$–10^2 s, $T \simeq 10$ MeV$-$0.1 MeV). The comparison of the predicted abundances with inferred determinations of their primordial abundances is by far the most stringent test of the standard model available to us at the present time. Being the earliest epoch from which we have relics (the light elements) whose abundances permit a verification of the standard model, primordial nucleosynthesis is the gateway to the very early universe. The importance of continued scrutiny of the predictions of big bang nucleosynthesis is manifest.

In the remainder of this article I will review the status of the concordance of the standard model of primordial nucleosynthesis with current observational data. As both a preview and a brief summary I would say that at present the standard model is consistent with all the observational data provided that the baryon-to-photon $\eta \simeq$ (3–7) $\times 10^{-10}$ and the number of light neutrino species (or the equivalent number of other light species) $N_\nu \leqq 4$; if one accepts the validity of the standard model, these can be taken to be constraints on η and N_ν. Should the primordial mass fraction of ^{4}He be found to be $\leqq 0.22$, the standard model is in serious trouble.

BRIEF REVIEW OF THE EPOCH OF PRIMORDIAL NUCLEOSYNTHESIS

The two fundamental assumptions that underlie big bang nucleosynthesis are the validity of general relativity, and the fact that the universe was once hotter than a few MeV. An additional assumption (which, however, is not necessary) is that the lepton number $[n_L/n_\gamma = (n_{e^-} - n_{e^+})/n_\gamma + (n_\nu - n_{\bar{\nu}})/n_\gamma \simeq \eta + (n_\nu - n_{\bar{\nu}})/n_\gamma]$ like the baryon number ($\simeq\eta$) is small. Having swallowed these assumptions, the rest follows like one-two-three.

Frame 1: $t \simeq 10^{-2}$ s, $T \simeq 10$ MeV. The energy density of the universe is dominated by relativistic species: $\gamma, e^+e^-, \nu_i\bar{\nu}_i$ ($i = e, \mu, \tau, \ldots$); $g_* \simeq 10.75$. Thermal equilibrium is maintained by weak interactions ($e^+ + e^- \leftrightarrow \nu_i + \bar{\nu}_i$, $e^+ + n \leftrightarrow p + \bar{\nu}_e$, $e^- + p \leftrightarrow n + \nu_e$) as well as electromagnetic interactions ($e^+ + e^- \leftrightarrow \gamma + \gamma$, $\gamma + p \leftrightarrow \gamma + p$, etc.), both of which are occurring rapidly compared to the expansion rate $H = \dot{R}/R$ (R = cosmic scale factor). Thermal equilibrium implies that $T_\nu = T_\gamma$ and that $n/p = \exp(-\Delta m/T)$; here n/p is the neutron to proton ratio and $\Delta m = m_n - m_p$. No nucleosynthesis is occurring yet because of the tiny equilibrium abundance of D: $n_D/n_B \simeq \eta \exp(2.2 \text{ MeV}/T) \simeq 10^{-10}$, where n_B, n_D, and n_γ are the baryon, deuterium, and photon number densities, $\eta = n_B/n_\gamma \simeq 0(10^{-10})$, and 2.2 MeV is the binding energy of the deuteron. This is the so-called deuterium bottleneck.

Frame 2: $t \simeq 1$ s, $T \simeq 1$ MeV. At about this temperature the weak interaction rates become slower than the expansion rate and thus weak interactions effectively cease occurring. The neutrinos decouple and thereafter expand adiabatically ($T_\nu \propto R^{-1}$). This epoch is the surface of last scattering for the neutrinos; detection of the cosmic neutrino seas would allow us to directly view the universe as it was 1 s after "the bang." The neutron to proton ratio no longer "tracks" its equilibrium value, but instead "freezes out" a value $\simeq 1/6$, and only slowly decreases, due to occasional free neutron decays. A little bit later ($T \simeq m_e$), the $e^\pm$ pairs annihilate and transfer their entropy to the photons, heating the photons relative to the neutrinos, so that from this point on $T_\nu \simeq (4/11)^{1/3}\ T_\gamma$. The "deuterium bottleneck" continues to operate and prevent nucleosynthesis.

Frame 3: $t \simeq 200$ s, $T \simeq 0.1$ MeV. At about this temperature the "deuterium bottleneck" breaks [$n_D/n_B \simeq \eta \exp(2.2\ \text{MeV}/T) \simeq 1$], and nucleosynthesis begins in earnest. Essentially all the neutrons present ($n/p \simeq 1/7$) are quickly incorporated first into D, and then into the very stable ^{4}He nucleus. Trace amounts of D and ^{3}He remain unburned; substantial nucleosynthesis beyond ^{4}He is prevented by the lack of stable isotopes with $A = 5$ and 8, and by coulomb barriers. A small amount of ^{7}Li is synthesized by $^4\text{He}(t, \gamma)^7\text{Li}$ (for $\eta \lesssim 3 \times 10^{-10}$) and by $^4\text{He}(^3\text{He}, \gamma)^7\text{Be}$ followed by the eventual β-decay of ^{7}Be to ^{7}Li (for $\eta \gtrsim 3 \times 10^{-10}$).

The nucleosynthetic yields depend upon η, N_ν (which parameterizes the number of light ($\lesssim 1$ MeV) species, other than γ and $e^\pm$), and in principle all the nuclear reaction rates that go into the reaction network. In practice, most of the rates are known to sufficient precision that the yields only depend upon a few rates. ^{4}He production depends only upon η, N_ν, and $\tau_{1/2}$, the neutron half-life, which determines the rates for all the weak processes that interconvert neutrons and protons. The mass fraction Y_p of ^{4}He produced increases monotonically with increasing values of η, N_ν, and $\tau_{1/2}$—a fact that is simple to understand. Larger η means that the "deuterium bottleneck breaks" earlier, when the value of n/p is larger. More light species (i.e., larger value of N_ν) increases the expansion rate [since $H \propto (G\rho)^{1/2}$], while a larger value of $\tau_{1/2}$ means slower weak interaction rates ($\propto \tau_{1/2}^{-1}$)—both effects cause the weak interactions to freeze out earlier, when n/p is larger. The yield of ^{4}He is determined by the n/p ratio when nucleosynthesis commences, $Y_p \simeq 2(n/p)/(1 + n/p)$, so that a higher n/p ratio means more ^{4}He is synthesized. At present the value of the neutron half-life is only known to an accuracy of about 2%: $\tau_{1/2} = 10.6$ min $\pm$ 0.2 min.[9–12] Since ν_e and ν_μ are known (from laboratory measurements) to be light, $N_\nu \geqq 2$. Based upon the luminous matter in galaxies, η is known to be $\gtrsim 0.3 \times 10^{-10}$. If all the mass in binary galaxies and small groups of galaxies (inferred by dynamical measurements) is baryonic, then η must be $\gtrsim 2 \times 10^{-10}$.

To an accuracy of several percent, the yields of D and ^{3}He only depend upon η, decreasing rapidly with η. Larger η corresponds to a higher nucleon density and earlier nucleosynthesis, which in turn results in less D and ^{3}He remaining unprocessed. Because of large uncertainties in the rates of some reactions that create and destroy ^{7}Li, the predicted primordial production is only accurate to within about a factor of 2.

In 1946 Gamow suggested the idea of primordial nucleosynthesis.[13,14] In 1953, Alpher, Follin, and Herman all but wrote a code to determine the primordial production of ^{4}He.[15] Peebles (in 1966) and Wagoner, Fowler, and Hoyle (in 1967) wrote codes to calculate the primordial abundances.[16,17] Yahil and Beaudet (in 1976) independently developed a nucleosynthesis code and also extensively explored the effect of large lepton number [$n_\nu - n_{\bar{\nu}} \simeq 0(n_\gamma)$] on primordial nucleosynthesis.[18] Wagoner's 1973 code has become the "standard code" for the standard model.[19] In

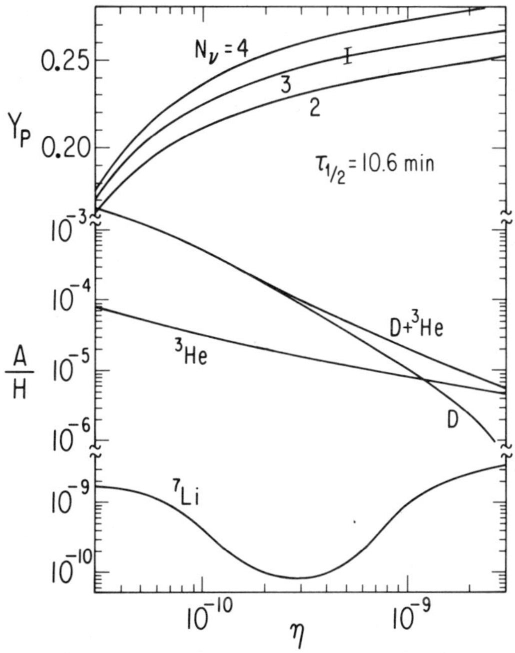

FIGURE 1. The primordial abundances of D, ^{3}He, D + ^{3}He, ^{4}He, and ^{7}Li predicted by big bang nucleosynthesis (based on Wagoner's 1973 code[19] as updated by Yang *et al.*,[2] and modified by Dicus *et al.*[20]). Y_p denotes the mass fraction of ^{4}He, which is related to ^{4}He/H by $Y_p = 4(^4\text{He/H})/[1 + 4(^4\text{He/H})]$; A/H is the abundance (by number) of isotope A relative to H; $\eta \equiv n_B/n_\gamma$ is the present baryon-to-photon ratio, which is related to Ω_B by Equation 2. A neutron half-life of 10.6 min was used for these calculations; the error bar indicates the effect of a ±0.2 min charge in $\tau_{1/2}$. N_ν is the number of light neutrino species.

1981 the reaction rates were updated by Olive *et al.*;[1] the only significant change that resulted was an increase in the predicted ^{7}Li abundance by a factor of 0(3). In 1982 Dicus *et al.* corrected the weak rates in Wagoner's 1973 code for finite temperature effects and radiative/coulomb corrections,[20] which led to a systematic decrease in Y_p of about 0.003 (see also Reference 21). FIGURES 1–4 show the predicted abundances of D, ^{3}He, ^{4}He, and ^{7}Li, as calculated by the most up-to-date version of Wagoner's 1973 code. The numerical accuracy of the predicted abundances is about 1%.

D, ^{3}He, AND ^{7}Li

The abundance of D has been determined in solar system studies and by UV absorption studies of the local interstellar medium (ISM). The solar system determinations are based upon measuring the abundances of deuterated molecules in the atmosphere of Jupiter and inferring the presolar D/H ratio from meteoritic and solar data on the abundance of ^{3}He. These determinations are consistent with $(D/H)_{\text{pre-}\odot} \simeq (2 \pm \frac{1}{2}) \times 10^{-5}$.[22–25] An average value $(D/H)_{ISM} \simeq 2 \times 10^{-5}$ has been derived from UV absorption studies of the local ISM ($\lesssim$ few 100 pc),[26,27] with individual measurements spanning the range $(1–4) \times 10^{-5}$. Note that these measurements are consistent with the solar system determinations of D/H.

The deuteron being very weakly bound is easily destroyed and hard to produce, and to date, it has been difficult to find an astrophysical site where D can be produced in its observed abundance.[28] Thus, it is generally accepted that the presently observed

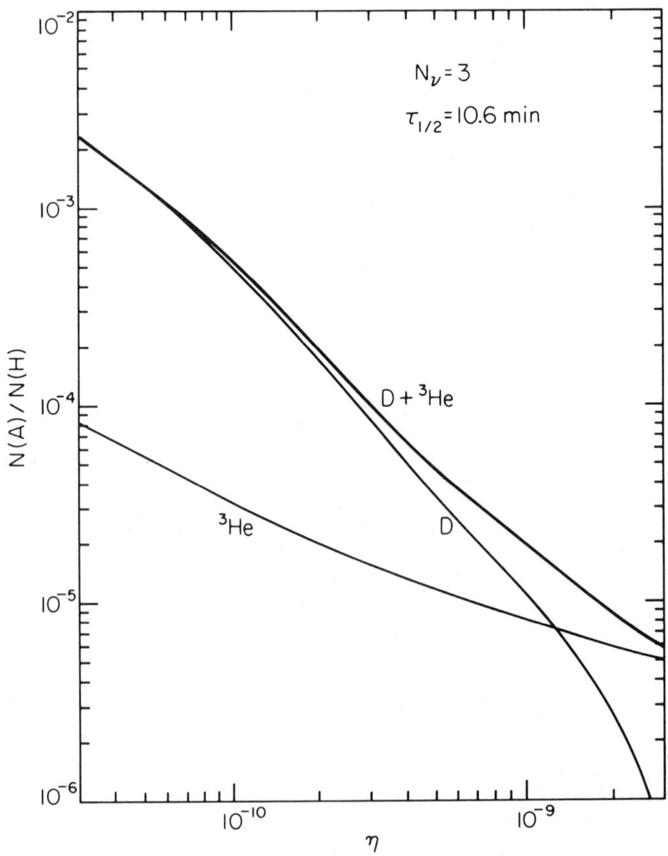

FIGURE 2. The predicted primordial abundances of D, ^{3}He, and D + ^{3}He (for more information, see FIGURE 1).

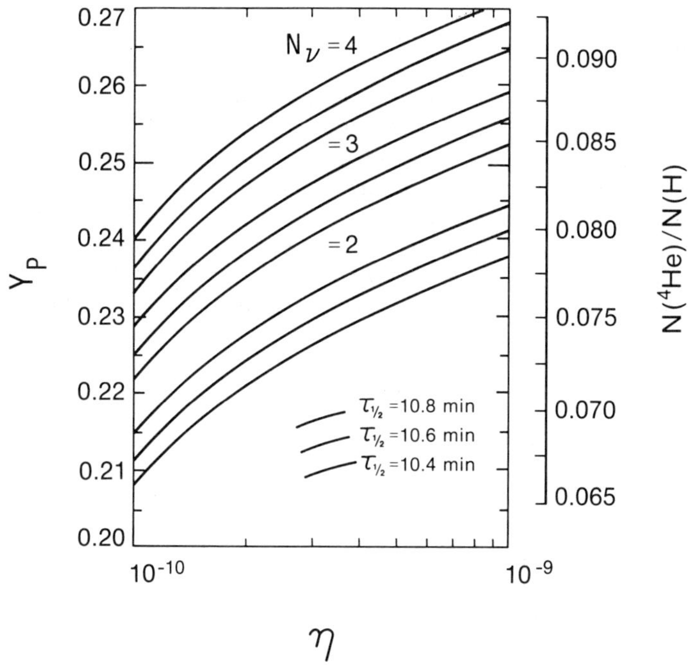

FIGURE 3. The predicted primordial abundance of ^{4}He (for more information, see FIGURE 1).

deuterium abundance provides a *lower* bound to the primordial abundance. Using $(D/H)_p \gtrsim 1 \times 10^{-5}$ it follows that η must be less than about 10^{-9}, in order for the predictions of primordial nucleosynthesis to be concordant with the observed abundance of D. [Note: because of the rapid variation of $(D/H)_p$ with η, this upper bound to η is rather insensitive to the precise lower bound to $(D/H)_p$ used.] The baryon to photon ratio η is related to Ω_B, the fraction of critical density contributed by baryons,

$$\Omega_B = 3.53 \times 10^{-3}(\eta/10^{-10})h^{-2}(T_\gamma/2.7\text{K})^3 \quad \textbf{(2)}$$

where the Hubble parameter $H_0 = 100\ h$ km s^{-1} Mpc^{-1} ($\frac{1}{2} \leqq h \leqq 1$), and T_γ is the present temperature of the microwave background ($2.7\text{K} \leqq T_\gamma \leqq 3.0\text{K}$). From Equation 2 and $\eta \leqq 10^{-9}$ it follows that $\Omega_B \leqq 0.19$. One would like also to exploit the sensitive dependence of $(D/H)_p$ upon η to derive a *lower* bound to η for concordance; this is not possible because D is so easily destroyed. However, as we shall soon see, this end can be accomplished by instead using both D and ^{3}He.

The abundance of ^{3}He has been measured in solar system studies and by observations of the $^3He^+$ hyperfine line in galactic HII regions (analogue of the 21 cm line of H). The abundance of ^{3}He in the solar wind has been determined by analyzing gas-rich meteorites, lunar soil, and the foil placed upon the surface of the moon by the Apollo astronauts. Since D is burned to ^{3}He during the sun's approach to the main sequence, these measurements represent the presolar sum of D and ^{3}He. These determinations[29,30] of D + ^{3}He are all consistent with $[(D + {}^3He)/H]_{pre\text{-}\odot} \simeq$

$(4.0 \pm 0.3) \times 10^{-5}$. Earlier measurements[31] of the $^3He^+$ hyperfine line in galactic HII regions and very recent measurements[32] lead to derived present abundances of ^{3}He: $^3He/H \simeq (3\text{–}20) \times 10^{-5}$. The fact that these values are higher than the presolar abundance is consistent with the idea that the abundance of ^{3}He should increase with time due to the stellar production of ^{3}He by low-mass stars.[33]

^{3}He is much more difficult to destroy than D. It is very hard to efficiently dispose of ^{3}He without also producing heavy elements or large amounts of ^{4}He (environments hot enough to astrate ^{3}He are usually hot enough to burn protons to ^{4}He). Yang *et al.* have argued that in the absence of a Pop III generation of very exotic stars which process essentially all the material in the universe and in so doing destroy most of the ^{3}He without overproducing ^{4}He or heavy elements, ^{3}He can have been astrated by a factor

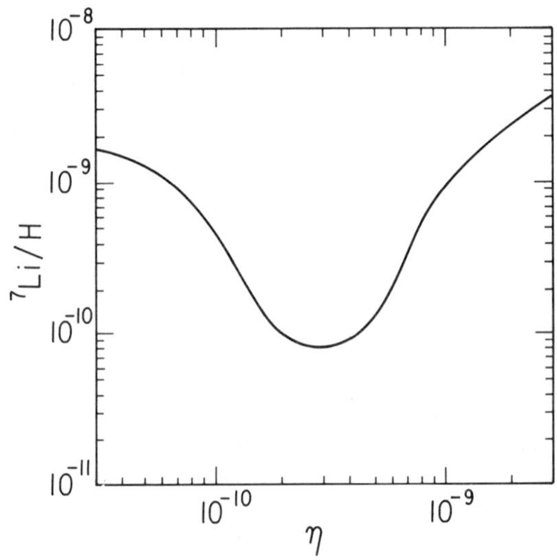

FIGURE 4. The predicted primordial abundance of ^{7}Li (for more information, see FIGURE 1). Due to uncertainties in some of the cross sections for processes that create and destroy ^{7}Li, these predicted abundances are uncertain by about a factor of 2 (for details see References 1 and 2).

of no more than $f_a = 4$.[2] Using this argument and the inequality

$$\left(\frac{D + {}^3He}{H}\right)_p \leq \left(\frac{D}{H}\right)_{pre\text{-}\odot} + f_a\left(\frac{{}^3He}{H}\right)_{pre\text{-}\odot} = f_a\left(\frac{D + {}^3He}{H}\right)_{pre\text{-}\odot} + (1 - f_a)\left(\frac{D}{H}\right)_{pre\text{-}\odot} \quad (3)$$

the presolar abundances of D and D + ^{3}He can be used to derive an upper bound to the primordial abundance of D + ^{3}He: $[(D + {}^3He)/H]_p \lesssim 13 \times 10^{-5}$. [For a reasonable (but less conservative) astration factor, $f_a \simeq 2$, the upper limit becomes 8×10^{-5}.] Using 13×10^{-5} as an upper bound on the primordial D + ^{3}He production implies that for concordance, η must be greater than 3×10^{-10} (for the upper bound of 8×10^{-10}, η must be greater than 3.5×10^{-10}). To summarize, consistency between the predicted big bang abundances of D and ^{3}He and the derived abundances observed today requires η to lie in the range $\simeq 3\text{–}10 \times 10^{-10}$.

Until very recently, our knowledge of the ^{7}Li abundance was limited to observations of meteorites, the local ISM, and Pop I stars, with a derived present abundance of $^7\text{Li/H} \simeq 10^{-9}$ (to within a factor of 2).[34–37] Given that ^{7}Li is produced by cosmic ray spallation and some stellar processes, and is easily destroyed (for $T \gtrsim 2 \times 10^6$ K), there is not the slightest reason to suspect (or even hope!) that this value accurately reflects the primordial abundance. Recently, Spite and Spite have observed ^{7}Li lines in the atmospheres of 13 unevolved halo and old disk stars with very low metal abundances ($Z_\odot/12$–$Z_\odot/250$), whose masses span the range of $\simeq 0.6$–$1.1\ M_\odot$.[38] Stars less massive than about 0.7 $M_\odot$ are expected to astrate [by factors $\gtrsim 0(10)$] their ^{7}Li abundance during their approach to the main sequence (MS), while stars more massive than about 1 $M_\odot$ are not expected to significantly astrate ^{7}Li in their outer layers. Indeed, they see this trend in their data, and deduce a primordial ^{7}Li abundance of $^7\text{Li/H} \simeq (1.12 \pm 0.38) \times 10^{-10}$. Remarkably, this is the predicted big bang production for η in the range 2–5 $\times 10^{-10}$. If we take this to be the primordial ^{7}Li abundance, and allow for a possible factor of 2 uncertainty the predicted abundance of ^{7}Li, then concordance for ^{7}Li restricts η to the range 1–7 $\times 10^{-10}$. Note, of course, that their derived ^{7}Li abundance is the pre–Pop II abundance, and may not necessarily reflect the true primordial abundance (e.g., if a Pop III of stars processed significant amounts of material).

In sum, the concordance of big bang nucleosynthesis predictions with the derived abundances of D and ^{3}He requires $\eta \simeq 3$–10×10^{-10}; moreover, concordance for D, ^{3}He, and ^{7}Li further restricts η: $\eta \simeq 3$–7×10^{-10}.

^{4}He

In the past few years the quality and quantity of ^{4}He observations have increased markedly. In FIGURE 5 all the ^{4}He abundance determinations derived from observations of recombination lines in HII regions (galactic and extragalactic) are shown as a function of metalicity Z (more precisely, 2.2 times the mass fraction of ^{16}O).

Since ^{4}He is also synthesized in stars, some of the observed ^{4}He is *not* primordial. Since stars also produce metals, one would expect some correlation between Y and Z, or at least a trend: lower Y where Z is lower. Such a trend is apparent in FIGURE 5. From FIGURE 5 it is clear that there is a large primordial component to ^{4}He: $Y_p \simeq 0.22$–0.26. Is it possible to pin down the value of Y_p more precisely?

There are many steps in going from the line strengths (what the observer actually measures) to a mass fraction of ^{4}He (e.g., corrections for neutral ^{4}He, reddening, etc.). In galactic HII regions, where abundances can be determined for various positions within a given HII region, variations are seen within a given HII region. Observations of extragalactic HII regions are actually observations of a superposition of *several* HII regions. Although observers have quoted statistical uncertainties of $\Delta Y \simeq \pm 0.01$ (or lower), from the scatter in FIGURE 5 it is clear that the systematic uncertainties must be larger. For example, different observers have derived ^{4}He abundances of between 0.22 and 0.25 for I Zw18, an extremely metal-poor dwarf emission line galaxy.

Perhaps the safest way to estimate Y_p is to concentrate on the ^{4}He determinations for metal-poor objects. From FIGURE 5 $Y_p \simeq 0.23$–0.25 appears to be consistent with all the data (although Y_p as low as 0.22 or high as 0.26 could not be ruled out). Recently Kunth and Sargent have studied 13 metal-poor ($Z \lesssim Z_\odot/5$) blue compact galaxies.[45,46]

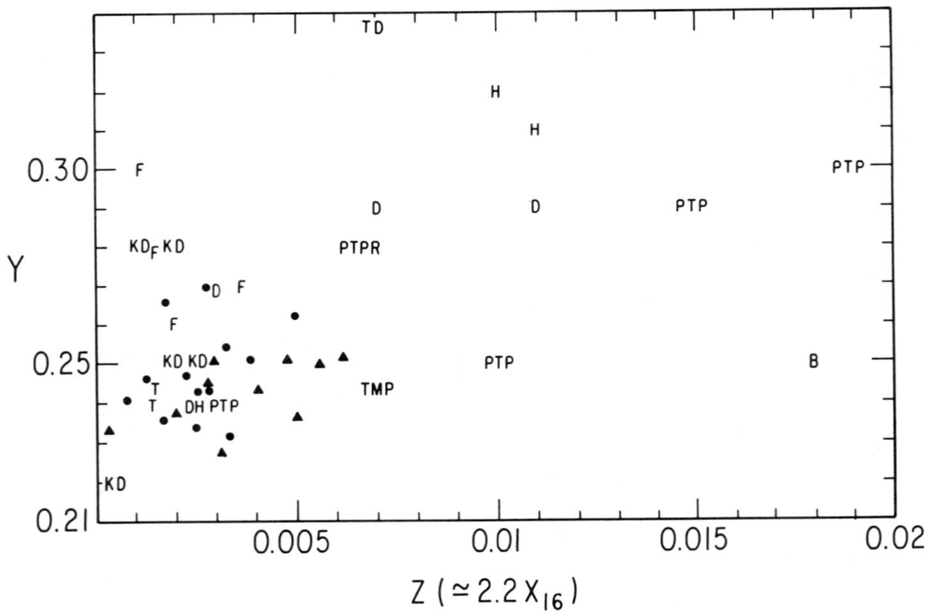

FIGURE 5. Summary of ^{4}He abundance determinations from recombination lines in HII regions vs. fraction of heavy elements Z ($\simeq$2.2 times the mass fraction of ^{16}O). Filled circles are the data of Kunth and Sargent for 13 metal-poor, emission-line galaxies,[45,46] filled triangles are the data of Lequeux *et al.* for 9 metal poor, irregular and compact galaxies;[47] D's denote Dufour;[48] DH's denote Dufour and Harlow;[49] F's denote French;[50] H's denote Hawley;[51] KD's denote Kinman and Davidson;[52] PTP's denote Peimbert and Torres-Peimbert;[53–55] PTPR's denote Peimbert *et al.*;[56] TD's denote Talent and Dufour;[57] T's denote Tully *et al.*[58] Note that authors do not usually quote errors for individual objects.

From a weighted average for their sample they derive a *primordial abundance* $Y_p \simeq 0.245 \pm 0.003$; allowing for a 3σ variation this suggests $0.236 \leqq Y_p \leqq 0.254$.

For the concordance range deduced from D, ^{3}He, and ^{7}Li ($\eta \geqq 3 \times 10^{-10}$) and $\tau_{1/2} \geqq$ 10.4 min, the predicted ^{4}He abundance is

$$Y_p \geqq \begin{cases} 0.226 & N_\nu = 2 \\ 0.240 & = 3 \\ 0.253 & = 4 \end{cases}$$

[Note that $N_\nu = 2$ is permitted only if the τ-neutrino is heavy ($\gtrsim$10 MeV) and unstable; the present experimental upper limit on its mass is 230 MeV.] Thus, since $Y_p \simeq 0.23$–0.25 (0.22–0.26?), there are values of η, N_ν, and $\tau_{1/2}$ for which there is agreement between the abundances predicted by big bang nucleosynthesis and the primordial abundances of D, ^{3}He, ^{4}He, and ^{7}Li derived from observational data.

CONCLUDING REMARKS

The only elements that are predicted to be produced in significant amounts during the epoch of primordial nucleosynthesis are D, ^{3}He, ^{4}He, and ^{7}Li. At present there is concordance between the predicted primordial abundances of all four of these elements and their observed abundances for values of N_ν, $\tau_{1/2}$, and η in the following intervals: $2 \leqq N_\nu \leqq 4$, 10.4 min $\leqq \tau_{1/2} \leqq$ 10.8 min, and $3 \times 10^{-10} \leqq \eta \leqq 7 \times 10^{-10}$ (or 10×10^{-10} if the ^{7}Li abundance is not used). Given that the predicted abundances span some 9 orders of magnitude, this is a truly remarkable achievement, and strong evidence that the standard model is valid back to times as early as 10^{-2} after "the bang."

The standard model will be in serious straits if the primordial mass fraction of ^{4}He should be determined to be less than 0.22. What alternatives exist if $Y_p \leqq 0.22$? If a generation of Pop III stars that efficiently destroyed ^{3}He and ^{7}Li existed, then the lower bound to η based upon D, ^{3}He, (and ^{7}Li) no longer exists. The only solid lower bound to η would then be that based upon the amount of luminous matter in galaxies (i.e., the matter inside the Holmberg radius): $\eta \geqq 0.3 \times 10^{-10}$. In this case the predicted Y_p could be as low as 0.15 or 0.16. Although small amounts of anisotropy increase the primordial production of ^{4}He,[39–41] recent work suggests that larger amounts could *decrease* the primordial production of ^{4}He.[42] Another possibility is neutrino degeneracy; a large lepton number [$n_\nu - n_{\bar{\nu}} \simeq 0(n_\gamma)$] drastically modifies the predictions of big bang nucleosynthesis.[43] Finally, one might have to discard the standard cosmology altogether.

If, based upon its apparent success, we accept the validity of the standard model, we can use primordial nucleosynthesis as a probe of cosmology and particle physics. For example, no more than four light ($\lesssim$1 MeV) neutrino species (or the equivalent number of other light species) can be tolerated. The baryon to photon ratio η is restricted to the range $\eta \simeq 3$–7×10^{-10} which corresponds to $0.011 \leqq \Omega_B \leqq 0.14$, implying that baryons alone cannot close the universe, and that if Ω is greater than about 0.14 (as various dynamical studies suggest), some other component (e.g., a massive neutrino species, axions, monopoles, etc.) must account for a significant (or dominant) portion of the present mass density of the universe. Primordial nucleosynthesis has also been used to constrain the amplitude of isothermal density perturbations[2,44] $\delta\eta/\eta \leqq 0(3)$ and the level of anisotropy during the epoch of nucleosynthesis.[39–41]

Given the important role occupied by big bang nucleosynthesis, it is clear that continued scrutiny is in order. The importance of new observational data cannot be overemphasized: extragalactic D abundance determinations (Is the D abundance universal? What is its value?); more measurements of the ^{3}He abundance (What is its primordial value?); continued improvement in the accuracy of ^{4}He abundances in very metal-poor HII regions (recall that the difference between $Y_p = 0.22$ and $Y_p = 0.23$ is crucial); and further study of the ^{7}Li abundance in very old stellar populations (Has the primordial abundance of ^{7}Li already been measured?). Data from particle physics will prove useful too: a high-precision determination of $\tau_{1/2}$ (i.e., $\Delta\tau_{1/2} \leqq \pm 0.05$ min) will all but eliminate the uncertainty in the predicted ^{4}He primordial abundance; an accurate measurement of the width of the recently found Z° vector boson will determine the total number of neutrino species [with mass $\lesssim 0(40$ GeV$)$] and thereby bound the total number of light neutrino species.[59] All these data will not only make primordial nucleosynthesis a more stringent test of the standard cosmology, but they will also make primordial nucleosynthesis a more powerful probe of the early universe.

[**Note added in proof:** A preliminary determination of the Z° width, $\Gamma_Z < 7$ GeV, has already resulted in the most stringent direct laboratory limit on N_ν: $N_\nu \lesssim 25$.][59,60]

REFERENCES

1. OLIVE, K. A., D. N. SCHRAMM, G. STEIGMAN, M. S. TURNER & J. YANG. 1981. Astrophys. J. **246:** 557.
2. YANG, J., M. S. TURNER, G. STEIGMAN, D. N. SCHRAMM & K. A. OLIVE. Astrophys. J. (Submitted.)
3. WEINBERG, S. 1972. Gravitation and Cosmology: Chapter 15. John Wiley & Sons. New York, N.Y.
4. KOLB, E. W. & M. S. TURNER. 1983. Annu. Rev. Nucl. Part. Sci. **33:** 645.
5. GUTH, A. 1981. Phys. Rev. **D23:** 347.
6. LINDE, A. 1982. Phys. Lett. **108B:** 389.
7. ALBRECHT, A. & P. J. STEINHARDT. 1982. Phys. Rev. Lett. **48:** 1220.
8. DOLGOV, A. D. & YA. B. ZEL'DOVICH. 1981. Rev. Mod. Phys. **53:** 1.
9. CHRISTENSEN, C. J., A. NIELSEN, A. BAHNSEN, W. K. BROWN & B. M. RUSTAD. 1972. Phys. Rev. **D5:** 1628.
10. BONDARENKO, L. N., V. V. KURGUZOV, YU. A. PROKOF'EV, E. V. ROGOV & P. E. SPIVAK. 1978. JETP Lett. **28:** 303.
11. BYRNE, J., J. MORSE, K. F. SMITH, F. SHAIKH, K. GREENE & G. L. GREEN. 1980. Phys. Lett. **92B:** 274.
12. WILKINSON, D. H. 1979. Prog. Part. Nucl. Phys. **6:** 328.
13. GAMOW, G. 1946. Phys. Rev. **70:** 572.
14. ALPHER, R. A., H. A. BETHE & G. GAMOW. 1948. Phys. Rev. **73:** 803.
15. ALPHER, R. A., J. W. FOLLIN & R. C. HERMAN. 1953. Phys. Rev. **92:** 1347.
16. PEEBLES, P. J. E. 1966. Astrophys. J. **146:** 542.
17. WAGONER, R. V., W. A. FOWLER & F. HOYLE. 1967. Astrophys. J. **148:** 3.
18. YAHIL, A. & G. BEAUDET. 1976. Astrophys. J. **206:** 26.
19. WAGONER, R. V. 1973. Astrophys. J. **179:** 343.
20. DICUS, D. A., E. W. KOLB, A. M. GLEESON, E. C. G. SUDARSHAN, V. L. TEPLITZ & M. S. TURNER. 1982. Phys. Rev. **D26:** 2694.
21. CAMBIER, J.-L., J. R. PRIMACK & M. SHER. 1983. Nucl. Phys. **B209:** 372.
22. BEER, R. & F. W. TAYLOR. 1973. Astrophys. J. **179:** 309.
23. TRAUGER, J. T., F. L. ROESLER, N. P. CARLETON & W. A. TRAUB. 1973. Astrophys. J. **184:** L137.

24. BLACK, D. C. 1971. Nature Phys. Sci. **234:** 148.
25. BLACK, D. C. 1972. Geochim. Cosmochim. Acta **36:** 347.
26. YORK, D. G. & J. B. ROGERSON, JR. 1976. Astrophys. J. **203:** 378.
27. VIDAL-MADJAR, A., C. LAURENT, R. M. BONNET & D. G. YORK. 1977. Astrophys. J. **211:** 91.
28. EPSTEIN, R., J. LATTIMER & D. N. SCHRAMM. 1976. Nature **263:** 198.
29. GEISS, J. & H. REEVES. 1972. Astron. Astrophys. **18:** 126.
30. ANDERS, E., D. HEYMANN & E. MAZOR. 1970. Geochim. Cosmochim. **34:** 127.
31. ROOD, R. T., T. L. WILSON & G. STEIGMAN. 1979. Astrophys. J. **227:** L97.
32. WILSON, T. L., R. T. ROOD & T. BANIA. 1983. *In* Proceedings of the ESO Workshop on Primordial Helium. D. Kunth, P. Shavers & K. Kjär, Eds.: 379. ESO. Garching, Federal Republic of Germany.
33. ROOD, R. T., G. STEIGMAN & B. M. TINSLEY. 1976. Astrophys. J. **207:** L57.
34. CAMERON, A. G. W. 1981. *In* Essays in Nuclear Astrophysics. C. Barnes, D. D. Clayton & D. N. Schramm, Eds. Cambridge University Press. Cambridge, England.
35. DUNCAN, D. K. 1981. Astrophys. J. **248:** 651.
36. VANDEN BOUT, P. A., R. L. SNELL, S. S. VOGT & R. G. TULL. 1978. Astrophys. J. **221:** 598.
37. ZAPPALA, R. H. 1972. Astrophys. J. **172:** 57.
38. SPITE, M. & F. SPITE. 1982. Astron. Astrophys. **115:** 357; Nature **297:** 483.
39. HAWKING, S. W. & R. J. TAYLER. 1966. Nature **209:** 1278.
40. THORNE, K. S. 1967. Astrophys. J. **148:** 51.
41. BARROW, J. D. 1976. Mon. Not. R. Astron. Soc. **175:** 359.
42. MATZNER, R. & T. ROTHMAN. 1982. Phys. Rev. Lett. **48:** 1565.
43. DAVID, Y. & H. REEVES. 1980. *In* Physical Cosmology: Les Houches XXXII. R. Balian, J. Audouze & D. N. Schramm, Eds.: 23. North-Holland. Amsterdam, the Netherlands.
44. BARROW, J. D. & J. MORGAN. 1983. Mon. Not. R. Astron. Soc. **202:** 393.
45. KUNTH, D. & W. L. W. SARGENT. 1983. Astrophys. J. (Submitted.)
46. KUNTH, D. 1982. Ph.D. Thesis. University. Paris, France.
47. LEQUEUX, J., M. PEIMBERT, J. F. RAYO, A. SERRANO & S. TORRES-PEIMBERT. 1979. Astron. Astrophys. **80:** 155.
48. DUFOUR, R. J. 1975. Astrophys. J. **195:** 315.
49. DUFOUR, R. J. & W. V. HARLOW. 1977. Astrophys. J. **216:** 706.
50. FRENCH, H. B. 1980. Astrophys. J. **240:** 41.
51. HAWLEY, S. A. 1978. Astrophys. J. **224:** 417.
52. KINMAN, T. D. & K. DAVIDSON. 1981. Astrophys. J. **243:** 127.
53. PEIMBERT, M. & S. TORRES-PEIMBERT. 1974. Astrophys. J. **193:** 327.
54. PEIMBERT, M. & S. TORRES-PEIMBERT. 1976. Astrophys. J. **203:** 581.
55. PEIMBERT, M. & S. TORRES-PEIMBERT. 1977. Mon. Not. R. Astron. Soc. **179:** 217.
56. PEIMBERT, M., S. TORRES-PEIMBERT & J. F. RAYO. 1978. Astrophys. J. **220:** 516.
57. TALENT, D. L. & R. J. DUFOUR. 1979. Astrophys. J. **233:** 888.
58. TULLY, R. B., A. M. BOESGAARD, H. M. DYCK & W. V. SCHEMPP. 1981. Astrophys. J. **246:** 38.
59. ARNISON, G., *et al.* 1983. Phys. Lett. **126B:** 398.
60. RUBBIA, C. 1983. Talk given at the NATO Advanced Study Institute, Munich, Federal Republic of Germany, September.

Nature and Significance of the Mass Distribution in the Universe

P. J. E. PEEBLES

Dominion Astrophysical Observatory
Victoria, British Columbia, Canada V2A 6K3

Measures of the mass distribution on scales greater than a few kpc have considerably improved in the last few years. I review here the status of some of the measures and the interpretations the results suggest to me.

OBSERVATIONS

Large-Scale Anisotropy of the Background Temperature

Inhomogeneities in the mass distribution produce a gravitational potential field that perturbs the microwave background temperature. This effect provides an important constraint on possible large-scale departures from homogeneity under the assumption, which I adopt throughout this paper, that conventional general relativity theory is a useful approximation.

The background temperature anisotropy $\delta T/T$ produced by the potential of a given mass distribution in an Einstein–de Sitter model was derived by Sachs and Wolfe.[1] A convenient measure of the large-scale anisotropy is the quadrupole moment,[2]

$$\langle |a_2^m|^2 \rangle = (a_2)^2 = 4\pi J_3 (H_0/c)^3/105 \tag{1}$$

where a_l^m is the coefficient of the spherical harmonic Y_l^m in the expansion of $\delta T/T$, and J_3 is an integral over the dimensionless mass autocorrelation function $\xi(r)$,

$$J_3(R) = \int_0^R r^2 \mathrm{d}r \xi \tag{2}$$

where R is roughly equal to the Hubble length c/H_0. In an open cosmological model with $\lambda = 0$, a_2 is slightly reduced from the de Sitter value.[3] We must add to a_l^m any other sources of anisotropy, but it is not likely that that could cancel the Sachs-Wolfe effect.

If ξ agrees with the galaxy two-point correlation function at $hr \lesssim 25$ Mpc ($H_0 = 100\,h$ kms^{-1}Mpc^{-1}) and $\xi = 0$ at larger r, then[4]

$$J_3 \sim 800\,h^{-3}\,\mathrm{Mpc}^3,\ a_2 \sim 6 \times 10^{-5} \tag{3}$$

Preliminary results of Boughn *et al.* suggested a_2 is significantly different from zero at about this value,[5] but subsequent measurements indicate that this was due to imperfect galaxy subtraction.[6-8] Following the method in Reference 2 [Equation 19 should read $\chi^2 = \Sigma\, Q_i^2/(Q^2 + m_i^2)$], one finds from the Lubin *et al.* data[6] $a_2 < 1.3 \times 10^{-4}$ at the 90% level. The recent Princeton data yield a similar limit.[7,8] These are consistent with

Equation 3 but we no longer have any evidence that a_2 and J_3 are as large as is indicated in Equation 3: mass fluctuations could be anticorrelated on large scales so $\xi < 0$ at large r and $J_3 \sim 0$.

We do still have a useful limit on any positive large-scale correlation in the mass distribution. For example, if $\xi = 0.1$ at $hr < 100$ kpc and $\xi = 0$ at larger r, then $a_2 \sim 4 \times 10^{-4}$, well in excess of the observational limit. That is, if we have the right gravity theory we must give up the idea that there is appreciable mass correlation as measured by $\xi(r)$ on very large scales. There can be very large clusters, but they must be rare enough or well enough balanced by large-scale holes to make $\xi << 1$.

A second interesting example is based on the galaxy two-point correlation function observed on small scales. It is well approximated by the power law model,

$$\xi_g = (r_0/r)^\gamma, \quad \gamma = 1.77, \quad hr_0 = 5.4 \pm 0.3 \text{ Mpc} \tag{4}$$

The index γ was derived from catalogs of angular distributions of galaxies,[9] but a closely similar result is obtained more directly from the Center for Astrophysics (CFA) red shift sample.[4] The length r_0 is derived from the CFA sample. If we assumed the mass autocorrelation function ξ followed this same power law to $hr = 100$ Mpc, say, with $\xi = 0$ at larger r, we would get $a_2 = 1.4 \times 10^{-4}$ which is just at the observational limit. Thus ξ must break below this power law at separations $hr \lesssim 100$ Mpc, appreciably less than the Hubble length. Analysis of the Lick catalog suggested ξ_g breaks below the power law at $hr \sim 10$ Mpc.[9] The significance of this result is called into question by the discovery that the cross-correlation between the Lick and Zwicky catalogs breaks from a power law at about the same angle as does the angular correlation function for the Lick sample, rather than at the larger angle expected from scaling.[10] On the other hand, the direct estimates of ξ_g at large r in the CFA and Durham-Australian red shift samples both indicate that ξ_g breaks from the power law and is negative at $hr \sim 20$ Mpc.[4,11] Better samples are needed to clarify the shape of ξ_g at $hr \gtrsim 10$ Mpc. If, as the red shift samples suggest, $\xi < 0$ at $hr \sim 20$ Mpc, it could substantially reduce J_3 and so make a_2 much less than the present observational limits.

The Mean Mass Distribution around a Galaxy

We can estimate the mean mass concentration around a galaxy from the rms value $\sigma(r_p)$ of the line-of-sight velocity difference of galaxy pairs as a function of the projected separation r_p. The observational situation for this statistic has greatly improved with the recent completion of several important red shift catalogs.

The open circles of FIGURE 1 are based on Tifft's measurements of red shift differences Δv for Karachentsev's sample of close galaxy pairs.[12,13] Only Tifft's high weight (CL) pairs are used so measuring error should be quite unimportant for our purpose. The central problem in estimating σ is the correction for accidental pairs. That is dealt with by the method of Reference 4 using the fact that, as in the other samples mentioned below, the distribution of Δv approximates an exponential. Therefore $\langle(\Delta v)^2\rangle$ is computed for the distribution cutoff at Δv_{max} and then corrected to σ^2 for an exponential distribution. Results are listed in TABLE 1. The corrected σ shows some tendency to increase with increasing Δv_{max}, which could be due to

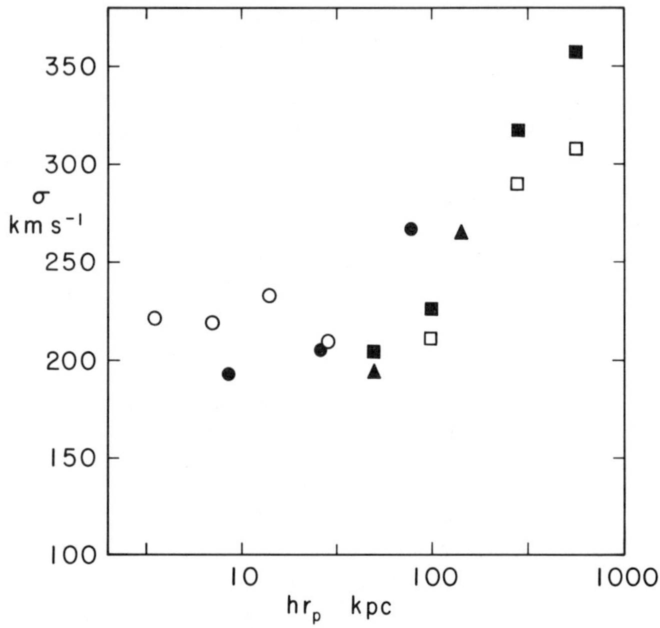

FIGURE 1. The rms line-of-sight velocity difference σ of galaxy pairs as a function of projected separation r_p. The open circles are based on Tifft's measurements of Δv for Karachentsev's close isolated pairs. The filled circles are based on White *et al.*'s measurements of Turner's isolated pairs. The other symbols come from a statistical analysis of the CFA red shift catalogue.

accidentals. However, the effect of accidentals should increase with increasing r_p, and no significant trend is apparent in the table. Thus it is reasonable to adopt as the best value for σ the mean of the three estimates for each r_p. The filled circles are derived by the same method[4] from the measurements by White *et al.* of Turner's isolated binaries.[14,15] The other symbols are based on a statistical analysis of the CFA sample.[4] The two-point correlation function ξ_v in red shift space is measured as a function of separations along and perpendicular to the line of sight, and the elongation of ξ_v along the line of sight is used as a measure of σ.[16] At small r_p the contribution to the width of ξ_v along the line of sight by Hubble flow is negligible so the second moment of ξ_v is a good measure of σ^2. Second moments are shown as the two triangles in the figure. The squares are based on model fits to ξ_v; the filled squares assume the clustering pattern is statistically stable, the open squares assume that clusters on the average expand with the general expansion.

A complication for the Turner and Karachentsev pairs should be mentioned. The galaxy luminosity tends to increase with increasing r_p because broader pairs are more likely to appear isolated when they are farther away.[14] For example, in Tifft's high-weight subset of Karachentsev pairs the rms red shift varies as $\sim r_p^{0.25}$. If mass varied as luminosity and mean rotation curves were flat, we would then expect $\sigma \sim r_p^{0.25}$, which is steeper than the trend of the open circles in the figure. However circular velocities in isolated spirals vary rather slowly with luminosity. For example Burstein *et al.* find that in Sc spirals $L \sim v_c^{5.7}$,[17] which would give $\sigma \sim r_p^{0.1}$. If a similar relation applied in all close pairs, this bias would not greatly increase the slope of σ. It will be noted that the bias is not present in the CFA analysis because all pairs are used here.

The figure shows that σ is about the same for isolated galaxy pairs and the CFA sample of all pairs. Since close pairs of galaxies seldom have bright neighbors very close by, that might have been expected, but it does indicate that the CFA estimates are not strongly influenced by the denser spots where velocities would be higher.

It is striking that σ varies so slowly with r_p, changing by just $\sim$ 50% when r_p varies by a factor $\sim$ 300. The value at small r_p, $\sigma \sim 200$ km s^{-1}, is about what would be expected from typical circular velocities $v_c \sim 200$ in galaxies. The slow variation of σ with r_p is very similar to the nearly flat rotation curves seen in isolated spirals at $hr \lesssim$ 25 kpc (see Rubin *et al.* and earlier references therein).[18] The latter imply the mass density averaged over spherical shells within an isolated spiral varies as $\rho \sim G^{-1}v_c^{2}/r^2 \sim r^{-1.7}$.[17] By a similar argument it has been noted that the roughly similar velocities within galaxies and clusters of galaxies imply that masses scale roughly as lengths, densities as $\rho \sim r^{-2}$.[19,20] More explicitly, the nearly flat shape of σ says the mean value of the mass density found at distance r from a fair sample of galaxies varies as[4,16,21,22]

$$\rho \sim G^{-1}\sigma^2/r^2 \sim r^{-1.8}$$

(This argument is avoided if orbits are radial so pairs are forming and dissolving at the same rate, but that is an artificial arrangement.) With the closer density of points in FIGURE 1 it becomes clear that these results point to a remarkable underlying continuity between galaxies and clusters of galaxies: the mean value of the mass density as a function of the distance r from a bright galaxy varies as a close approximation to a power law $\rho \sim r^{-1.8}$ from the internal structures of galaxies to the structures of clusters of galaxies.

This continuity does not apply to the mean value of the optical luminosity per unit volume, $j(r)$, as a function of the distance r from the center of a galaxy. At $hr \gtrsim 100$ kpc, j varies as the mean abundance of neighbors, $j \sim \xi_g \sim r^{-\gamma}$, $\gamma \sim 1.8$, which is the same shape as $\rho(r)$ within the noise. At smaller r, $j(r)$ would be dominated by the light of the galaxy, and at hr equal to a few kpc, j would be $\sim$ 100 times larger than the

TABLE 1. Relative Velocity Dispersions

	Δv_{max}	n	$\langle \Delta v^2 \rangle^{1/2}$	σ
$2.5 < hr_p < 5$	350	23	129	167
	500	26	192	258
	750	27	217	238
$5 < hr_p < 10$	350	41	140	198
	500	46	182	234
	750	48	207	224
$10 < hr_p < 20$	350	56	142	205
	500	60	177	220
	750	64	238	273
$20 < hr_p < 40$	350	38	134	179
	500	40	167	199
	750	43	225	250

extrapolation of the power law. (This is based on $M/L_B \sim 500\ h$ for the large-scale average and $M/L_B \sim 5\ h$ within galaxies). This break from a power law shape of j is consistent with the fact that galaxies are seen as discrete objects. The evidence from $\sigma(r)$ and rotation curves is that if we saw mass instead of light the picture would be much more fuzzy and, to judge from the statistic σ, we might be hard pressed to decide where galaxies end and clusters commence.

INTERPRETATIONS

The underlying continuity between galaxies and clusters could be arranged whether galaxies formed first or protoclusters formed and then fragmented, so long as the process introduces no characteristic lengths at 3 kpc $\lesssim hr \lesssim$ 10 Mpc. However if we accept the conventional expanding relativistic world model we must conclude that the clustering pattern is still growing by gravitational interactions at the upper end of this range and that this has been going on for some time (since red shift $z > 3.5$ if galaxies formed before quasars) because galaxies by and large are old. If the smaller scale end of the clustering hierarchy formed by a different process the observed continuity underlying the mass distribution would have to be only a temporary coincidence, which seems unlikely. This argument applies to the protocluster picture mentioned above and to the Ostriker explosive intergalactic medium picture,[23] where galaxies and perhaps groups are produced by nongravitational forces. The simpler, and perhaps more likely, interpretation is that galaxies, clusters, and superclusters all formed by the same gravitational instability process.

A related argument might be mentioned: it is difficult to see how the galaxies in the local group could be fragments of a protocluster.[24] If these galaxies were ejected from the Virgo cluster our peculiar velocity ought to be away from the Virgo cluster, contrary to the evidence from the anisotropies of the microwave background and the Hubble flow.[25] If these galaxies formed by collapse of the local supercluster we would have to explain why the local supercluster appears to be only now in the process of collapse, while the galaxies clearly are old. The same argument applies to the assumption that the local group is the protocluster. One could imagine that some galaxies formed as individuals while other are fragments of protoclusters, but the basic similarities of cluster and field galaxies—sizes, luminosities, masses—argue against it.[23]

If we have the sequence of creation right it places some interesting constraints on the nature of the mass clustering at high red shift. I will discuss here only initially adiabatic departures from homogeneity (where there is a fixed value for the local entropy per baryon). As is described by Guth at this conference, the adiabatic constraint is attractive because it agrees with modern microscopic physics.

If the mass of the universe now were dominated by baryons, dissipation during decoupling of matter and radiation would produce a minimum value of the coherence length that is comparable to the size of a cluster of galaxies. That was one motivation for the Zel'dovich pancake picture, where protoclusters fragment to produce galaxies.[26] It seems to me that the continuity result argues against this picture. Another perhaps more direct argument against the model is the isotropy of the microwave background.[27–29] Since clusters are not all that dense, protoclusters would have to form at red shifts well under 10, and therefore the universe would be transparent to the microwave background at $z \lesssim 1{,}000$. Thus we have a limit $\delta\rho/\rho \lesssim 3\delta T/T$ on the adiabatic density fluctuations at decoupling. For example, Uson and Wilkinson find $\delta T/T \lesssim 1.1 \times 10^{-4}$ at $\theta \sim 4.5'$.[30] This angle subtends the comoving length $8(\Omega_0 h)^{-1}$ Mpc at high red shift, which exceeds the present characteristic galaxy clustering length $r_0 \sim 5\ h^{-1}$ Mpc (Equation 4). As the growth factor from decoupling to $z \sim 3.5$, when we would want the first quasars to form, is $\lesssim 200$, the wanted amplitude strongly violates the isotropy measurements. We avoid this problem if we assume the universe is dominated by neutrinos at thermal abundance and with mass a few tens of eV, because $\delta\rho/\rho$ can start to grow well before decoupling.[31] However, the coherence length prior to decoupling still is comparable to clusters of galaxies rather than galaxies so the continuity argument still applies.[31,32] (If the primeval spectrum drops rapidly with increasing wavelength, the coherence length, at which the autocorrelation function of the mass distribution is $\xi(r) = \xi(0)/2$, can be reduced to $r_c \sim 4(\Omega_0 h^2)^{-1}$ Mpc, which perhaps is tolerably small if $\Omega_0 \gtrsim 1$, but then space curvature fluctuations rapidly approach the relativistic limit at $r < r_c$ unless the spectrum flattens out again at $r < r_c$.) These conclusions all are based on linear perturbation calculations, and one can ask whether nonlinear processes at $z >> 1{,}000$ might change the results. However nonlinear density fluctuations on scales larger than the horizon tend to collapse to black holes rather than galaxies or clusters of galaxies, and while low-mass black holes might even be beneficial we would not want a large abundance at masses $\gtrsim$ galaxy mass because we do not often see galaxies being disrupted by unseen companions.

Davis, Lecar, Pryor, and Witten were the first to note that the coherence length problem is relieved if we assume the universe is dominated by weakly interacting particles with masses M_x appreciably greater than 30 eV.[33] If M_x is very large then a coincidence involving M_x and the abundance of these particles is needed to make the red shift z_{eq} at which the mass densities of nonrelativistic matter and radiation agree assume the observed value $z_{eq} \sim 10^4$. But given that, and assuming $M_x \gtrsim 1$ keV, we derive a pleasant coincidence, that the coherence length of the mass distribution prior to the first generation is independent of M_x (now the coherence length is fixed by radiation pressure rather than particle streaming) and about agrees with the comoving size of a galaxy.[34] This would seem to offer a natural explanation of why galaxies appear at about the sizes they do. In this scenario the first generation would be protogalaxies. Baryons and electrons would tend to settle dissipatively into the dense spots, in the manner described by White and Rees and Silk and Norman,[35,36] making galaxies appear as isolated islands. Since protogalaxies would be closer together when they formed there would be considerable interaction between neighbors; and as the initial distribution of protogalaxies need not be smooth, one could imagine that in dense spots such parameters as the abundance ratio of elliptical to lenticular galaxies were systematically different from the mean and that such differences were preserved as dense patches coalesced to form clusters.

Let us return now to mass density fluctuations on scales comparable to the Hubble length cH_0^{-1}. As far as I am aware, no one has found a way to perturb this part of the mass fluctuation spectrum by nongravitational forces: we seem to have a clear shot at the primeval spectrum. It was noted above that we have a measure of the spectrum through the large-scale anisotropy of the microwave background by the Sachs-Wolfe effect. The observations do not yet exclude the assumption that the mass autocorrelation function ξ agrees with the galaxy two-point correlation function on small scales, and vanishes at $hr \gtrsim 30$ Mpc. This situation, where the primeval power spectrum of mass fluctuations is flat, might be considered attractive because it means that whatever perturbed the mass distribution away from exact homogeneity did so statistically independently (so far as the mass is concerned) in each volume element, consistent with the fact that at high red shift the horizon would have been exceedingly small. On the other hand, if the initial conditions were set up at the Planck time, when the horizon contained only a few photons, the initial amplitude would have to be chosen so that the space curvature perturbation amounts to ~ 1 part in 10^5 on the scale of $\sim 10^{88}$ photons (the number in the present horizon), which seems to be an unlikely arrangement. For that reason many people like a primeval power spectrum that varies as $|\delta_k|^2 \sim k$, where the amplitude of the space curvature perturbation per octave of wavelength is independent of wavelength.[37–39] In that case the theory must only generate a single number, the space curvature perturbation, and as that number is not all that small, $\sim 10^{-4}$ to 10^{-5}, it might naturally arise in some fundamental theory. If the power spectrum does vanish at $k \rightarrow 0$, which is at large wavelength, then the mass autocorrelation function ξ must have a negative part so the integral $J_3(\infty)$ of ξ, which is the spectrum at $k = 0$, can vanish. It was noted above that there is an interesting hint in the galaxy distribution that $\xi_g < 0$ at 20 to 40 h^{-1} Mpc, but that will have to be tested with better samples. By pushing down J_3 we suppress the Sachs-Wolfe effect (Equation 2). For example, if the universe were dominated by very massive weakly interacting particles with a primeval spectrum $\propto k$ then the quadrupole moment would be only $a_2 \sim 3 \times 10^{-6}$, safely out of reach of present experiments.[34]

CONCLUSIONS

People looking for an *ab initio* theory of the big bang and the origin of galaxies and clusters of galaxies should be cautious in assuming we know the answer for which they are seeking a theory. Even the sequence of creation is controversial, though to my mind the recent measures of the mean mass distribution around galaxies provide strong evidence that galaxies formed before clusters and by the same gravitational instability process. If this is so, it means that the mass fluctuation spectrum at decoupling had to have been arranged so that the first generation was no larger than a typical bright galaxy. That situation could evolve out of primeval adiabatic perturbations if the universe were dominated by very massive weakly interacting particles. To avoid a conflict with the measurements of the large-scale anisotropy of the microwave background we would have to suppose the primeval spectrum was flat or decreased with increasing wavelength; the spectrum $|\delta_k|^2 \propto k$ would do, and is attractive because it is characterized by a single scale-independent amplitude of curvature fluctuations. If it proves difficult to outfit the universe with very massive weakly interacting particles, we might fall back on neutrinos at $M_x \sim 20$ to 100 eV, but here the mass coherence length prior to the first generation is uncomfortably large unless we make special arrangements: M_x must be close to the upper limit (so $\Omega_0 \gtrsim 1$) and the primeval spectrum of mass fluctuations must be nearly flat at $hr \lesssim 1$ Mpc (to avoid diverging curvature fluctuations) but drop rapidly at longer wavelengths. If that cannot be arranged we will have to start thinking more seriously about scenarios in which the entropy per baryon is not fixed, but rather the primeval mass distribution is homogeneous and the baryons or other particles are placed in clumps that act as seeds.

REFERENCES

1. SACHS, R. K. & A. M. WOLFE. 1967. Astrophys. J. **147:** 73.
2. PEEBLES, P. J. E. 1981, Astrophys. J. Lett. **243:** L119.
3. PEEBLES, P. J. E. 1982. Astrophys. J. **259:** 442.
4. DAVIS, M. & P. J. E. PEEBLES. 1983. Astrophys. J. **267:** 465.
5. BOUGHN, S. P., E. S. CHENG & D. T. WILKINSON. 1981. Astrophys. J. Lett. **243:** L113.
6. LUBIN, P. M., G. L. EPSTEIN & G. F. SMOOT. 1983. Phys. Rev. Lett. **50:** 616.
7. FIXSEN, D. J. 1982. Doctoral Dissertation. Princeton University. Princeton, N.J.
8. WILKINSON, D. T. 1982. *In* The Book of the Universe: 17th Recontre de Moriond. J. Audouze & J. Taan Thanh Van, Eds.: 159. Frontiéres. Paris, France.
9. GROTH, E. J. & P. J. E. PEEBLES. 1977. Astrophys. J. **217:** 385.
10. SELDNER, M. & J. M. USON. 1982. Astrophys. J. Lett. **261:** L65.
11. SHANKS, T. 1982. *In* Progress in Cosmology. A. W. Wolfendale, Ed.: 335. D. Reidel Publishing Co. Dordrecht, the Netherlands.
12. TIFFT, W. G. 1982. Astrophys. J. Suppl. **50:** 319.
13. KARACHENTSEV, I. D. 1972. Commun. Special Astrophys. Obs. USSR **7:** 3.
14. WHITE, S. D. M., J. HUCHRA, D. LATHAM & M. DAVIS. 1983. Mon. Not. R. Astron. Soc. **203:** 701.
15. TURNER, E. L. 1976. Astrophys. J. **208:** 20.
16. PEEBLES, P. J. E. 1980. The Large-Scale Structure of the Universe: Section 76. Princeton University Press. Princeton, N.J.
17. BURSTEIN, D., V. C. RUBIN, N. THONNARD & W. K. FORD. 1982. Astrophys. J. **253:** 70.
18. RUBIN, V. C., W. K. FORD, N. THONNARD & D. BURSTEIN. 1982. Astrophys. J. **261:** 439.
19. OSTRIKER, J. P., P. J. E. PEEBLES & A. YAHIL. 1974. Astrophys. J. Lett. **193:** L1.
20. DAVIS, M., J. TONRY, J. HUCHRA & D. LATHAM. Astrophys. J. Lett. **238:** L113.
21. PEEBLES, P. J. E. 1976. Astrophys. Space Sci. **45:** 3.
22. WHITE, S. D. M. 1981. Mon. Not. R. Astron. Soc. **195:** 1037.
23. OSTRIKER, J. P. 1982. Astrophysical Cosmology. Scripta Varia **48:** 473.
24. PEEBLES, P. J. E. 1983. Astrophys. J. **274.** (In press.)
25. DAVIS, M. & P. J. E. PEEBLES. 1983. Annu. Rev. Astron. Astrophys. **21:** 109.
26. ZEL'DOVICH, YA. B., M. VAN DYKE & J. V. WEHAUSEN. 1977. Annu. Rev. Fluid Mech **9:** 215.
27. PRESS, W. H. & E. T. VISHNIAC. 1980. Astrophys. J. **236:** 323.
28. WILSON, M. L. & J. SILK. 1981. Astrophys. J. **243:** 14.
29. ROZGACHEVA, I. K. & R. A. SUNYAEV. 1981. Pis'ma Astron. Zh. **7:** 323.
30. USON, J. M. & D. T. WILKINSON. 1982. Phys. Rev. Lett. **49:** 1463.
31. DOROSHKEVICH, A. G., M. YU. KHLOPOV, R. A. SUNYAEV, A. S. SZALAY & YA. B. ZEL'DOVICH. 1981. Ann. N.Y. Acad. Sci. **375:** 32.
32. PEEBLES, P. J. E. 1982. Astrophys. J. **258:** 415.
33. DAVIS, M., M. LECAR, C. PRYOR & E. WITTEN. 1981. Astrophys. J. **250:** 423.
34. PEEBLES, P. J. E. 1982. Astrophys. J. Lett. **263:** L1.
35. WHITE, S. D. M. & M. J. REES. 1978. Mon. Not. R. Astron. Soc. **183:** 341.
36. SILK, J. & C. NORMAN. 1981. Astrophys. J. **247:** 59.
37. HARRISON, E. R. 1970. Phys. Rev. **D1:** 2728.
38. PEEBLES, P. J. E. & J. T. YU. 1970. Astrophys. J. **162:** 815.
39. ZEL'DOVICH, YA. B. 1972. Mon. Not. R. Astron. Soc. **160:** 1P.

The 300-Inch New Texas Telescope

HARLAN J. SMITH

McDonald Observatory
University of Texas at Austin
Austin, Texas 78712

WHY NEW BIG TELESCOPES AT A TEXAS SYMPOSIUM

There has been some question as to whether telescopes form an appropriate subject for a Texas Symposium on Relativistic Astrophysics. The topic is certainly unconventional, but perhaps can be tolerated once every 20 years as a brief change of pace in the meetings. More seriously, large telescopes are one of if not even the principal source of astrophysical data, and a case can be made that every decade or two theorists should get a glimpse of the future possibilities in the field. Most importantly, it is a very long, hard, and expensive task to bring any of these instruments into existence. Those of us involved in such projects need your encouragement and help with any and all relevant authorities, if *any* of the new-generation telescopes are to come to pass. Because so many of these instruments are now under active consideration, it seemed timely to the Organizing Committee this year to devote one morning session to the topic.

THE NEED FOR NEW GROUND-BASED TELESCOPES

The opinion is sometimes expressed, especially in popular literature, that space astronomy—in particular space telescope—will render ground-based work obsolete. There are a number of reasons why this is not the case, and why there will be a greater need than ever over the next several decades for new big ground-based telescopes. These reasons include:

- The past 20 years have seen a veritable explosion in the number of astronomers, without a proportional increase in the number of large instruments able to work effectively on some of the most important problems. Over the same period there has been a comparable explosion in the number of new phenomena and kinds of objects to study. The effects have led to a fierce oversubscription pressure on present large telescopes.
- The gratifying improvement in photon detectors is beginning to approach its limits, so that increase in size of telescopes once again represents the principal way of making substantial additional gains. Simultaneously, the improvements in detectors have brought enormously more objects of any given kind within range of telescopes to observe, thus in a sense increasing the pressure on the telescopes along with the improvement in detectors.
- Thanks to improved design concepts the new generation of telescopes can be built for little more than a tenth of the cost of the older, classical designs, yet perform better.

- Even when it flies, probably in 1986, there will be only one space telescope (ST) which can look essentially at only one object at a time. Many of its exposures will be very long. Thus in a given year, or even a decade, ST—although it may well prove the most valuable telescope ever built—will still only be able to work on some of the problems and a minute fraction of the objects of interest or importance to astronomers. Furthermore, it will be a long time before other STs are routinely available, because of the enormous price tag, approaching $1 billion, for such instruments. Photons collected in space cost more than 100 times as much as those on the ground, at the current state of technology.
- The new generation of ground-based telescopes can be very large, collecting from 10 to 50 times the number of photons as ST per unit time, this being of particular importance for spectroscopy. The ground-based telescopes can have much larger fields than ST, imaging 100 times or more the area of sky at any given time. The new ground-based telescopes, by design and location, will have better-quality imagery than the older ones, in extreme cases almost competitive with the ST. Finally, in some specialized applications such as speckle interferometry, even their resolution will substantially exceed that of ST.

The above tabulation makes clear the growing worldwide pressure for a new generation of very large ground-based telescopes. In a later talk at this session, Sandra Faber will discuss some of the observational possibilities for this new generation of telescopes.

WORLDWIDE PLANS FOR NEW LARGE TELESCOPES (GREATER THAN 6 METERS)

At least seven very large new telescopes are in various stages of active planning or discussion:

- University of Texas—7.6 meters
- University of California—10 meters
- Japan—10 meters
- United States National NNTT—15 meters
- European Southern Observatory—16 meters (?)
- United Kingdom—16 meters (?)
- Union of Soviet Socialist Republics—25 meters (?)

It is perhaps appropriate that the University of Texas telescope leads the list, not only because Texas is hosting this symposium, but because there is a substantial probability that the 7.6-meter may be the only one completed during this decade of the 1980s. (Indeed several on the list are unlikely to be finished even in the nineties.) Although our goals included frugality, speed of design and construction, and a willingness to utilize good ideas from anyone, the telescope nevertheless inevitably includes several relatively radical new approaches. The remainder of this paper outlines the present status of the 7.6-meter Texas project.

THE UNIVERSITY OF TEXAS APPROACH

Background of the Project

McDonald Observatory, with its 2.1- and 2.7-meter telescopes plus several smaller instruments, is relatively well equipped but nevertheless is increasingly less able to compete with the more-than-a-dozen 3- to 6-meter-class instruments around the world. Our goal is to gain about a factor of 10 in light-gathering power over our present two big telescopes. Along with other improvements in image quality, pointing accuracy, power of instrumentation, etc., this will increase by about 100-fold the number of accessible objects of any arbitrary type, and open up a long generation of frontier research possibilities.

Our philosophy, best articulated by Ed Nather, has been to push technology to the limits, then back off to where things appear relatively comfortable; also to invent only where truly necessary, otherwise to use relatively proven approaches. In this way we hope to achieve the telescope as rapidly and securely as possible, while staying within the limits imposed by our being fundamentally an observatory of a single state.

McDonald Observatory is a subcampus of the University of Texas at Austin. It serves other astronomers from Texas and, by special request, a certain percentage each year of non-Texas and even non–United States astronomers. The resources available to build the telescope must come primarily from Texas, which, contrary to popular belief, is far from being the wealthiest American state.

In view of these factors, our initial decision was for a monolithic mirror telescope, in the belief that this would probably give us the quickest and least-expensive construction, and the greatest certainty of good performance. We briefly considered various sizes of monolithic mirror up to 10 meters, but quickly settled on an approximately 7-meter size as the "comfortable" limit for Texas in terms of cost and technological risks. A number of factors conduced to the approximately 7-meter size including the availability of facilities at Corning to fabricate a mirror of about that size, the existence of a facility at Arizona's Optical Sciences Institute able to figure such a mirror, and the ability of a 7-meter still to be transported over the United States interstate highway network. Very early in the project, however, it became clear that America is still far from being on the metric system! Scarcely any of the people with whom we were discussing the project had any intuitive understanding of just what a 7-meter-class telescope really meant. When we explained that it was essentially a 300-inch, however, the matter immediately became clear. Indeed, there is a certain symmetry in the successive steps from the 100-inch to the 200-inch and now the 300-inch as being each in their time the world's largest monolithic telescopes. Finally, preliminary costing indicated that the 300-inch telescope and dome could be completed for around $30 million 1980 dollars (although to be sure the entire project would cost more), and this was approaching the upper limit of Texas funding that might be expected—to strive for 10 meters would double the cost and put the project out of reach.

The Telescope Project

Around 1970 I first began discussing with university authorities and prospective donors the idea of a $50 million class telescope at McDonald Observatory. It became possible to take this seriously around the beginning of 1980, with a new and sympathetic university administration. Our initial thoughts for the telescope were outlined in the winter of 1979–80 and a detailed conceptual design was worked out by Aden and Marjorie Meinel in collaboration with our scientists and engineers in the spring of 1980. Engineers of the Western Development Labs of Ford Aerospace carried this conceptual design through full-fledged preliminary design and costing during the summer and autumn of 1980. The total project cost, including all peripherals such as aluminizing tank, observing instruments, site development, administration, etc., came to $44 million 1981 dollars, to be spent over a five- or six-year period. In December 1980 the university regents accepted the project. Unfortunately at this time they did not initiate funding, but specified that the project should begin with outside funding. This made it necessary for the observatory to set up a fund-raising operation; indeed, since that time a very large fraction of my time has been spent in seeking support for the project. Results have been moderately successful; through the end of 1982 nearly a million dollars in outside funds were raised, in addition to a $250,000 special grant from the regents. These funds have allowed final design studies and decisions to progress. We believe that most of this work will be completed in 1983, and hope to begin some aspects of the construction no later than 1984. [During the 1983 session of the Texas legislature, the project was budgeted for a $5 million appropriation to initiate construction; unfortunately the expectation of declining revenues resulting from a worldwide drop in oil prices caused the legislature in April 1983 to delete the telescope appropriation.]

The Telescope Steering Committee consists of H. J. Smith (Chairman), Robert Tull (Project Scientist), Eric Silverberg (Project Manager 1980/82), Gerald Carr (Project Manager beginning 1983), Thomas Barnes (Assistant Director, McDonald Observatory), Ed Nather (computers and general wisdom), David Dittmar (electrical engineering and control systems), Woon-Yin Wong (structural engineering), Frank Ray (finite element analysis of mirrors and structures), and Aden and Marjorie Meinel (principal consultants).

The Site—McDonald Observatory

Another early decision of the project was to locate the 300-inch at or very close to McDonald Observatory. This stems partly from the realization that the costs of construction and operation go up by about 20% to 30% at one of the extreme remote sites such as Mauna Kea or Chile, and our budget is already near the upper limit for what is feasible. Also McDonald Observatory is fully developed with six operating telescopes and very substantial support facilities so that it is not necessary to duplicate these at a totally new site. McDonald is relatively easy of access from Austin, close to major roads and railroads, hence economical for both construction and operation. Its southerly latitude of 30° is the lowest of any major observatory in the continental United States, giving better views of the southern skies. Its altitude of 2,100 meters (6,800 feet) is quite favorable, and there are adjoining higher peaks going all the way up to 2,600 meters (8,400 feet) which may be accessible and available. The climate is semidesert, with long-term averages of around 65% good observing weather of which perhaps 35% to 40% is photometric and the remainder of spectroscopic or photographic quality. (While occasionally very good, the seeing is ordinarily not exceptional. This is partly due to the design and construction of the older domes, also probably due in part to the relatively low altitude of Mt. Locke compared to its immediately surrounding terrain and to the considerably higher surrounding mountains.)

In addition to the above generally favorable aspects, the skies at McDonald are among the darkest of any developed site in the world. The population in the Big Bend country of west Texas is exceedingly sparse, and there is simply no problem with light pollution now or in the forseeable future.

Finally, as a general principle, it is probably good for not all of the world's great telescopes to be concentrated at just two or three sites.

HOW TO REDUCE TELESCOPE COSTS

For the Texas project to succeed the costs must be kept within the limits stated above. Yet to build a classical telescope of 300-inch size the cost would probably exceed $300 million. There are two principal methods for drastically reducing telescope costs. First

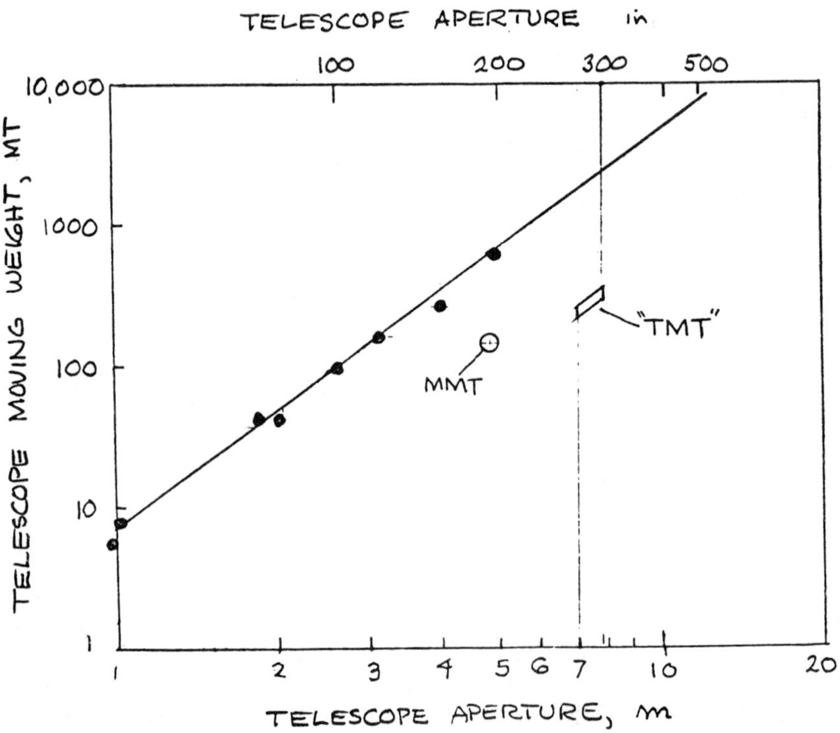

FIGURE 1. Scaling law for moving weight of existing telescopes showing that the calculated weight for the 300-inch Texas monolithic telescope (TMT) would be well over 1,000 tons if classical designs were to be used; the target weight is an order of magnitude reduction.

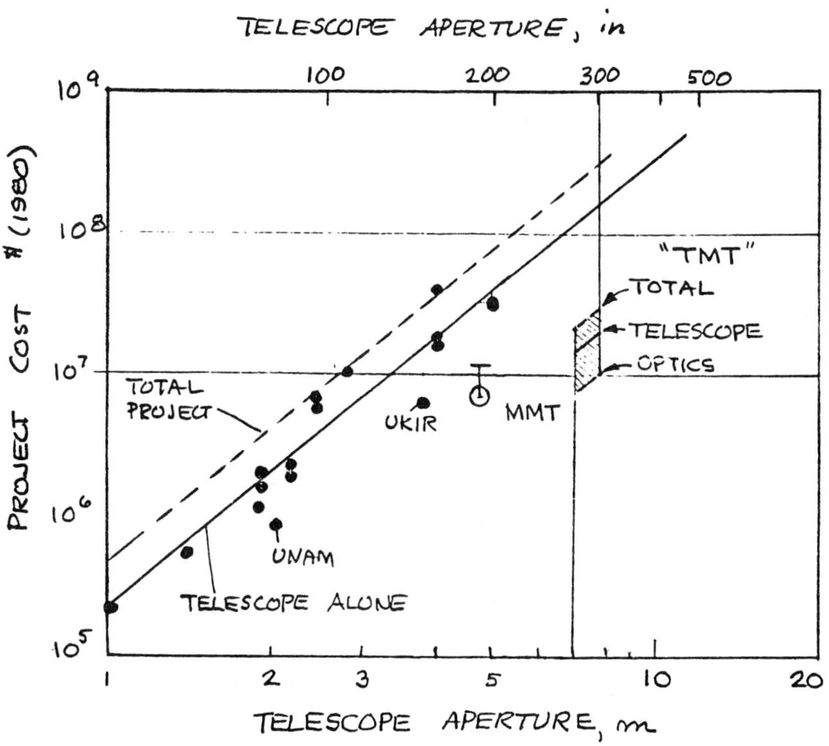

FIGURE 2. Cost scaling law for existing telescope costs and for total project costs, extrapolating to a projected cost of over $300 million (1980) for the 300-inch if classical designs were chosen.

is to realize that a telescope is simply a device for pointing a few grams of accurately shaped aluminum at some celestial object. Why then should a telescope weigh hundreds of, or by classical design for the 300-inch, even well over a thousand tons? FIGURE 1 is a graph taken from the work of Aden Meinel showing that the weights of existing more or less classical telescopes are proportional to their aperature raised to the approximately 2.7 power. The cost, in FIGURE 2, follows a similar progression.

Of course the few grams of aluminum must be coated on a glass mirror for stability. Older telescopes were so heavy because, for the mirror to hold its figure, the rule of thumb was that the aspect ratio (ratio of diameter to thickness) should be about six, in order for the mirror to be more or less self-supporting against gravitational and wind flexures which would distort its figure. By this rule, the 300-inch mirror would weigh more than 100 tons, leading to an enormously massive tube and a stupendously massive mounting, with correspondingly astronomical cost. The fundamental solution is to go to a very thin and lightweight mirror, also to use very lightweight construction throughout the telescope.

Secondly, older telescopes had relatively slow focal ratios in the range of f/3.3 to f/5. This was to ease the task of the optician, but had the undesirable consequence of making the tube very long and in particular the dome immensely large, expensive, and conducive to poor seeing conditions. Thus, for example, if one were to reduce a telescope from f/4 to f/2, the tube would be only half as long, and the dome in turn only half as long, wide, and high, or only about 1/8 the cost otherwise required. This is a

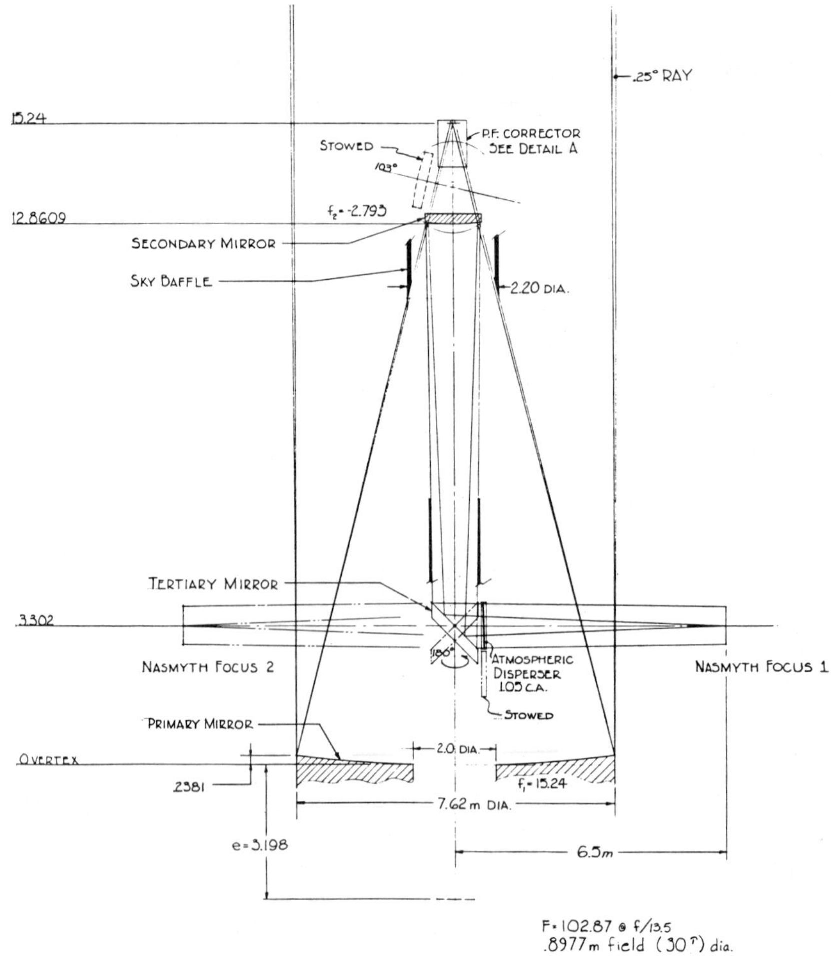

FIGURE 3. Optical schematic of the 300-inch telescope.

vital point, since the cost of the dome for a big telescope is roughly equal to that of the telescope itself.

The above principles, to be sure applied in a variety of ways, offer a new generation of telescopes better than any that have ever existed, at only about 1/10 of the cost that would be required using the older and more classical approaches.

THE 300-INCH TELESCOPE

Optical Design

Our primary mirror will be f/2.0, with a modified Ritchey-Chrétien figure (see FIGURE 3). The secondary will probably work at f/13.5 (the McDonald standard), using a tertiary flat to feed a pair of Nasmyth platforms through large holes in the elevation bearings.

A prime focus is planned (if it can be afforded) with at least one Richardson f/2.2 three-element corrector, having a 30+ arc-minute field with images everywhere better than an arc second and useful color range from 3,000 to 10,000 angstroms. The replaceable third element can be incorporated with a grism or special filters. It is possible that for special purposes a small wobbling IR secondary may be used.

One Nasmyth focus (FIGURE 4) will contain a MacFarlane camera working at f/3, having at least a 10 arc-minute unvignetted field; this camera will have ultrahigh resolution as well as correction for atmospheric dispersion. Also the Nasmyth foci will include a large high-resolution spectrograph, one or more low-resolution spectrographs, and appropriate photometers, polarimeters, and IR and speckle equipment.

We do not plan to utilize the Cassegrain focus. An important part of the design is to allow extremely rapid change from any instrument to any other, to accommodate changing programs and changing conditions through the night.

Mechanical Design

The telescope will feature an ultralight space-frame construction rotating on wheels on a precision track (see FIGURE 5). A central pintle bearing will keep the rotation in true alignment.

The basic mounting is alt-azimuth. This of course has the defect of field rotation, requiring all instruments that deal with more than a single on-axis image to have built-in field rotators.

The telescope tube will also be of a lightweight space-frame construction. The Nasmyth platforms will be extremely large, of the order of 6 by 15 meters, to accommodate the various instruments.

Dome Design

The dome design (see FIGURE 5) is not yet as detailed as that of the telescope, but provisionally it will be of the same style as the MMT telescope in Arizona. In any event its rotation will be synchronized with the telescope. Whatever the final configuration, it will be wide open for good ventilation and improved seeing, following the principle of the telescope "living out on a porch." Also, consistent with MMT experience, the telescope slot will be fully and highly insulated to improve seeing. The dome will have only minimum human habitation, and those areas will be extremely well insulated to prevent any significant heating effect on the telescope.

Because of the difficulty in handling and moving the enormous mirror safely,

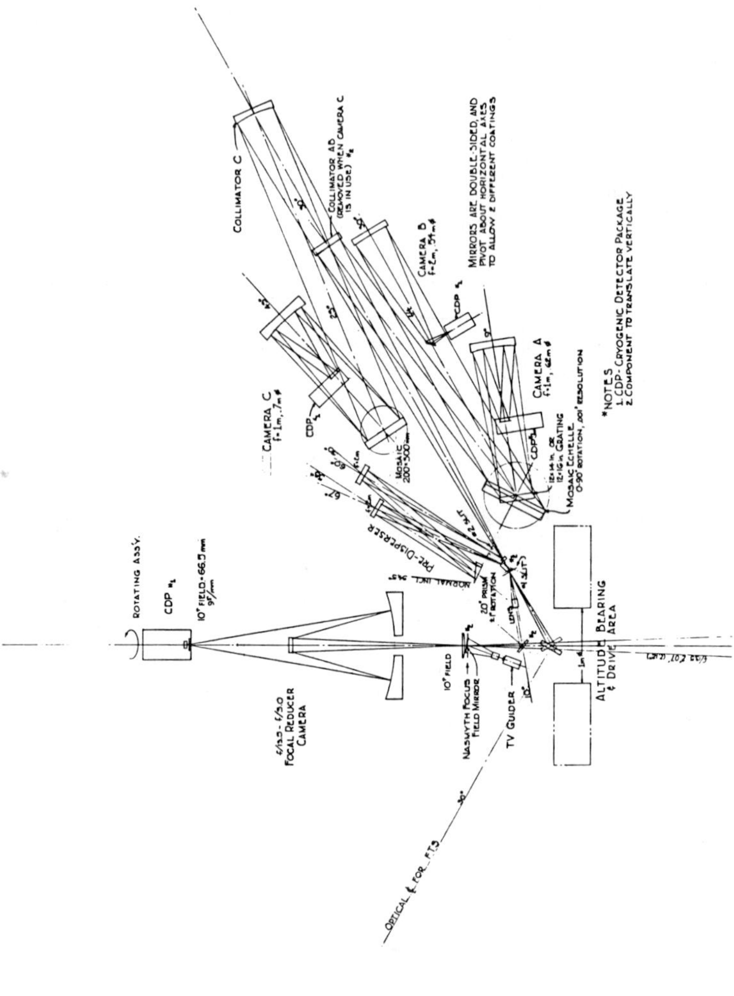

FIGURE 4. Schematic layout of several Nasmyth-platform instruments.

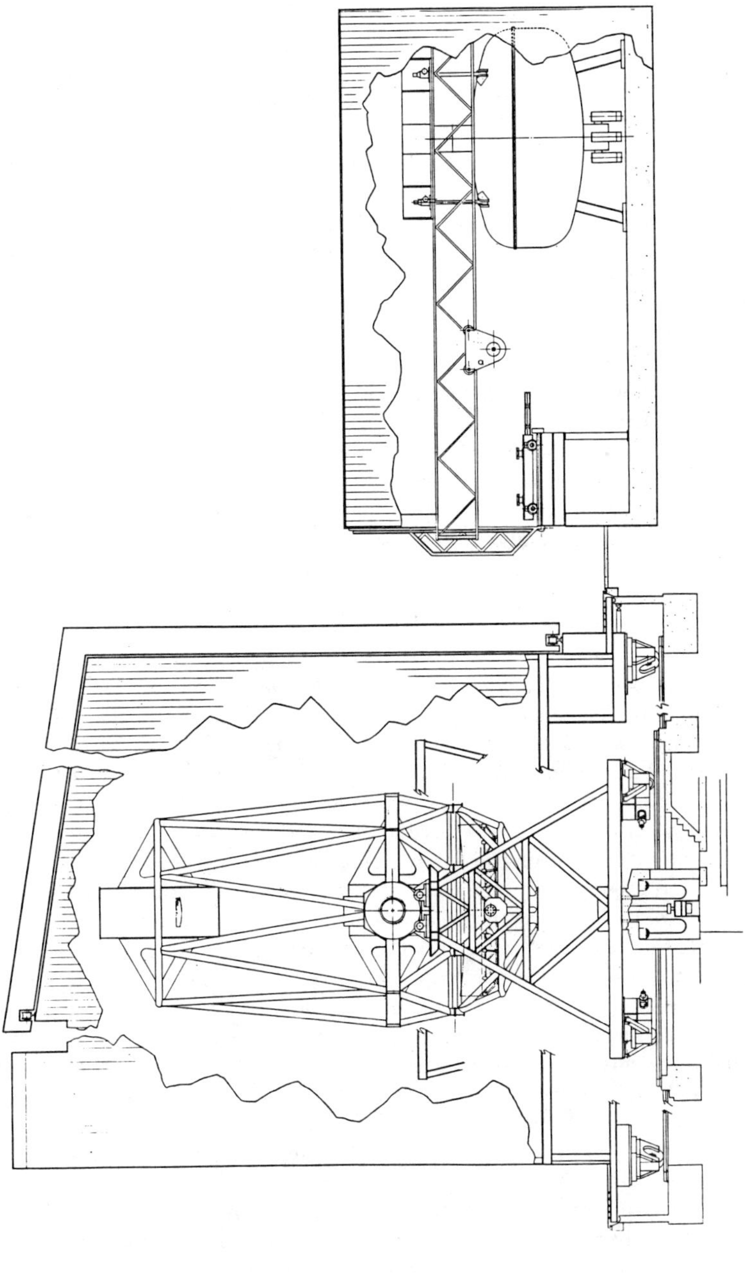

FIGURE 5. General mechanical layout of telescope, dome, and aluminizing area.

unless some extremely careful provision is made in advance, aluminizing will probably be done in a separate building reached by a drawbridge and railroad track (FIGURE 5).

THE 300-INCH MIRROR

The Thin Meniscus

The heart of this telescope is of course the large, thin monolithic mirror. Our original plan, and still the most likely alternative, is a 12-centimeter thick meniscus of Corning ULE (ultra-low-expansion) glass (see FIGURE 6). This choice has a number of advantages. These include guaranteed prompt delivery of the raw blank in less than two years, at reasonable cost (of the order of $3.5 million). The ULE will give no significant thermal distortion of the mirror figure. The ULE process is also so well controlled that it will be of highly uniform quality for figuring. Finally, the very thinness of the mirror, which of course requires active supports, also offers the major potential advantage of allowing the support system to correct for some errors in figure. Thus, with the aid of image analyzers and appropriate actuators, we have the opportunity to insure that the mirror in principle generates an extremely good image at all times. Under certain conditions it may even be possible to correct for some of the very slow components of atmospheric seeing, although this is not a primary plan at present.

To be sure there are some disadvantages with the thin monolith. Perhaps the most serious is the fact that the surface temperature can respond only relatively slowly to changing air temperature. The thermal time constant of the mirror, even if energetically ventilated, will be about two hours. The temperature of the mirror surface will thus not always be nearly identical to that of the ambient air. Should those temperatures differ by even a degree or two, convection currents tend to be set up over the surface of the mirror, which cause the seeing to degrade. We do propose to ventilate the mirror very actively in order to speed up its reaction to temperature. Blowing air over the surface of the mirror also helps to break up the convection cells before they have time to form significantly. (Further experimentation will be necessary to see how effective this process can be.) A second disadvantage of the meniscus is that the mirror, instead of needing only a few dozen relatively cheap and passive supports, will now require against gravity and wind at least 120 passive flotation points and some 40 to 60 active "tweaker" points (which may or may not be at the same places as flotation points). Effective operation will clearly require mirror and image sensing, interpretive algorithms, and control and actuator systems of a sophistication far higher than has ever been used with astronomical telescopes before, at least outside of the classified area. However, with the technology of the 1980s these challenges are readily solvable, and as noted above in the end may even prove to be an advantage for this telescope.

The most probable support mechanisms for the thin meniscus will consist of astatic levers with flex-pivot joints, to take essentially all the weight of the mirror, plus active stiff support points to provide up to about a percent of additive or subtractive force. The astatic levers will thus support the weight of the mirror. The stiff actuator points will prevent bending under wind forces, and will respond to appropriate sensors by applying

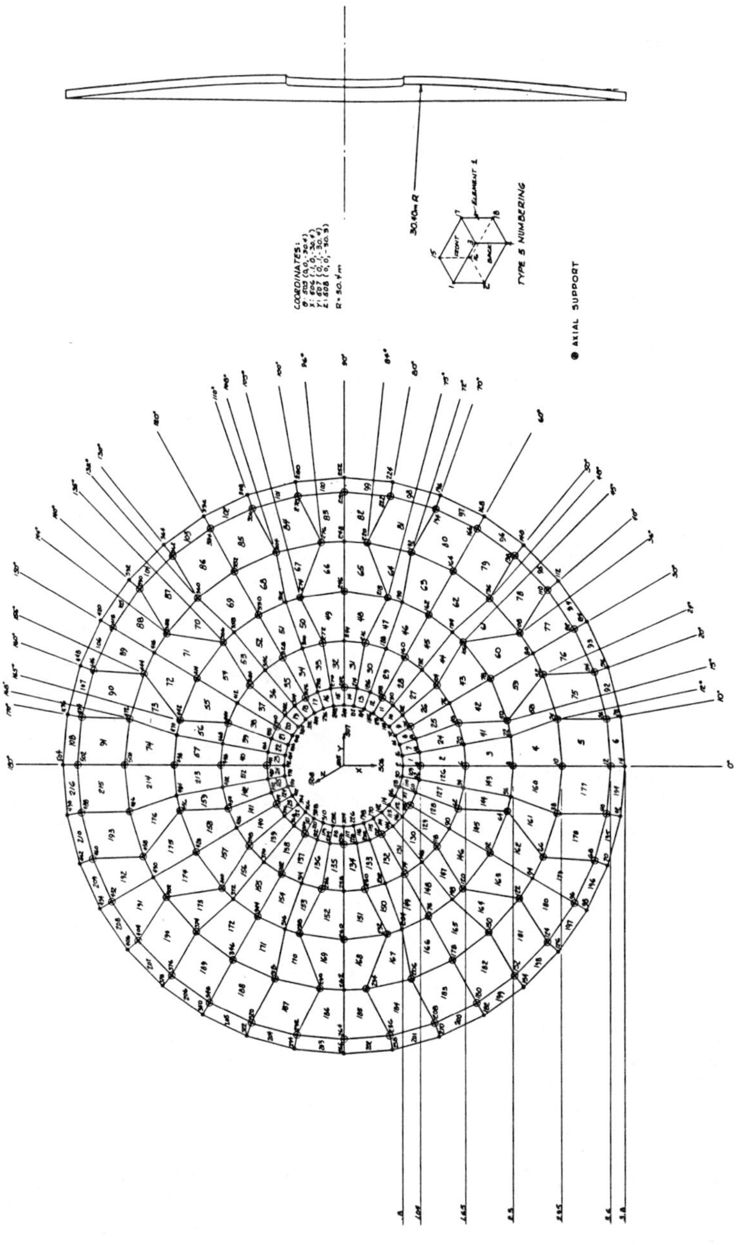

FIGURE 6. One of many finite-element analysis programs testing the effect of different sets of support points on the support of a thin-meniscus 300-inch primary mirror.

slight forces to improve the figure of the mirror. Three levels of command are contemplated for these actuator points. Tull has designed a laser back-plane reference for the mirror which, once calibrated, should be able to hold the figure of the mirror under all conditions to a prearranged shape. Calibration, and further correction for other difficulties farther along the optical train or for very slow atmospheric effects, will be provided through an image sensor able to work on relatively bright stars to monitor the figure of each part of the primary mirror. This is likely to be a sensor similar to ones recently developed by Roland Shack of the Optical Sciences Laboratory in Tucson. Finally, a computer will be learning the performance of the telescope under various conditions. Even in the event of failure of the other two devices, it should be able to instruct the actuators to do a relatively good job based upon previous knowledge of the performance of the mirror at various elevation angles and of the telescope under various ambient conditions. As still one final level of basic operation, the entire actuator system can be backed off the mirror in which case the astatic flotation, once properly adjusted, should do a first-order job of producing an image of at least spectroscopic quality when the wind is not strong.

Pyrex Honeycomb

Roger Angel of Tucson has proposed that our mirror be derivative of the 200-inch design, consisting of an approximately 60-cm-deep Pyrex honeycomb structure with face plate and ribs of the order of 5-cm thick. This approach has several potential advantages. The surface temperature will tend to follow atmospheric changes with a time constant of only about 30 minutes. A relatively simple set of supports, comparable to the 1930s' technology with the 200-inch, should be adequate. The mirror will be relatively stiff against wind, and—once the technology is adequately developed by Angel—the cost of such a mirror might be as low as about $1 million per blank.

However, there are certain disadvantages to this approach. The worst of these is the uncertain delivery date, likely to be at least three years, possibly longer. The cost is also quite uncertain, including development of the large furnace and further debugging of the technique. The finished mirror will also have appreciable thermal distortions because of the expansion coefficient of Pyrex. Finally, figure-control possibilities are much reduced, although still present.

TARGETS

We believe the technical problems with the telescope are relatively well in hand and offer interesting challenges but no insuperable difficulties. Funding will be the primary problem, with total project cost expected to be at least $45 million (1982) dollars. We hope to get about 40% of this cost (namely, that of the telescope) largely from private sources over the next few years, and that the remainder may be funded largely if not entirely through university resources. We also hope that the final commitment will be made for the telescope within the next year. If so the original goal to "dedicate in 88" could still be achieved. More likely, we will do well to finish the telescope by the end of this decade. We believe that construction of the 300-inch new Texas telescope will demonstrate what can be done even by a single astronomical institute, thanks to thin-mirror technology. It may well remain one of the—if not *the*—largest monolithic mirror telescopes, as various types of multiple-mirror designs are introduced in order to make telescopes effectively larger yet. Finally, we are sure that it will play a major role in astronomy for many decades to come.

The University of California Ten-Meter Telescope Project

JERRY NELSON

Lawrence Berkeley Laboratory
and Astronomy Department
University of California
Berkeley, California 94720

INTRODUCTION

In recent years it has become increasingly obvious that the primary telescope for University of California (UC) astronomers, the Mt. Hamilton 120-inch, is losing its competitive edge due to light pollution from neighboring San Jose. In an attempt to overcome this problem, a group of astronomers within the university began discussing the possibility of a much larger telescope at a dark, high-quality site. In 1977 these discussions matured to the point that we began to develop the ideas for a 10-meter optical and infrared telescope.

After about two years of research effort and much discussion, UC astronomers decided upon the basic design for this telescope. It was clear that achievement of the desired aperture of 10 meters would require substantial innovation and improvements upon existing designs of telescopes. By far the most radical step taken is the decision that the primary mirror should be composed of mirror segments, rather than the single piece of glass that is traditional for optical telescopes. Because of the extreme accuracy desired in the optical surface, the decision to make a segmented primary carries with it the acceptance of the need for an actively controlled primary, another unusual concept for this telescope.

In designing such a large instrument, one must be extraordinarily conscious of costs, and do the utmost to find ways of greatly reducing the costs without compromising the desired performance. Since in many situations costs are driven by size and weight, two guiding principles have influenced almost every aspect of the design: make it compact and make it light. The segmented primary exemplifies this idea well where we have substituted "electronic" stiffness for mechanical (and massive) stiffness. The rest of the telescope and the telescope enclosure also exemplify these principles.

BASIC DESIGN

Historically, the surrounding building and dome for a telescope have cost about as much as the telescope itself. In order to minimize the cost of the enclosure for the 10-meter telescope (TMT) we have shortened the telescope as much as possible, since the telescope length determines the basic dimension of the dome. We have adopted an f/1.75 primary, thus giving the TMT a 17.5-meter focal length, just slightly longer than that of the f/3.3 200-in. telescope. The telescope structure itself is a carefully designed space frame, designed to give the maximum stiffness at the minimum weight.[1] The basic telescope structure can be seen in FIGURE 1. The lowest resonant frequency in the telescope-yoke system is about 5 Hz, as high as or higher than any other large telescopes. This stiffness is particularly important because of the anticipated large wind loads on the telescope, and because of the active control system required for the primary mirror. The excellent performance of the structure is the

FIGURE 1. A model of the TMT showing the space frame structure of the telescope and yoke. The mounting is altitude-azimuth. The Nasmyth platforms are not shown.

result of three factors: an altitude-azimuth mount; a short, squat telescope; and extensive computer modeling of the structure, allowing careful optimization. The success of the design can also be seen in FIGURE 2, where the weight of the telescope is shown in comparison with the weights of other large optical telescopes.

The surrounding building and dome were designed to minimize the size of the

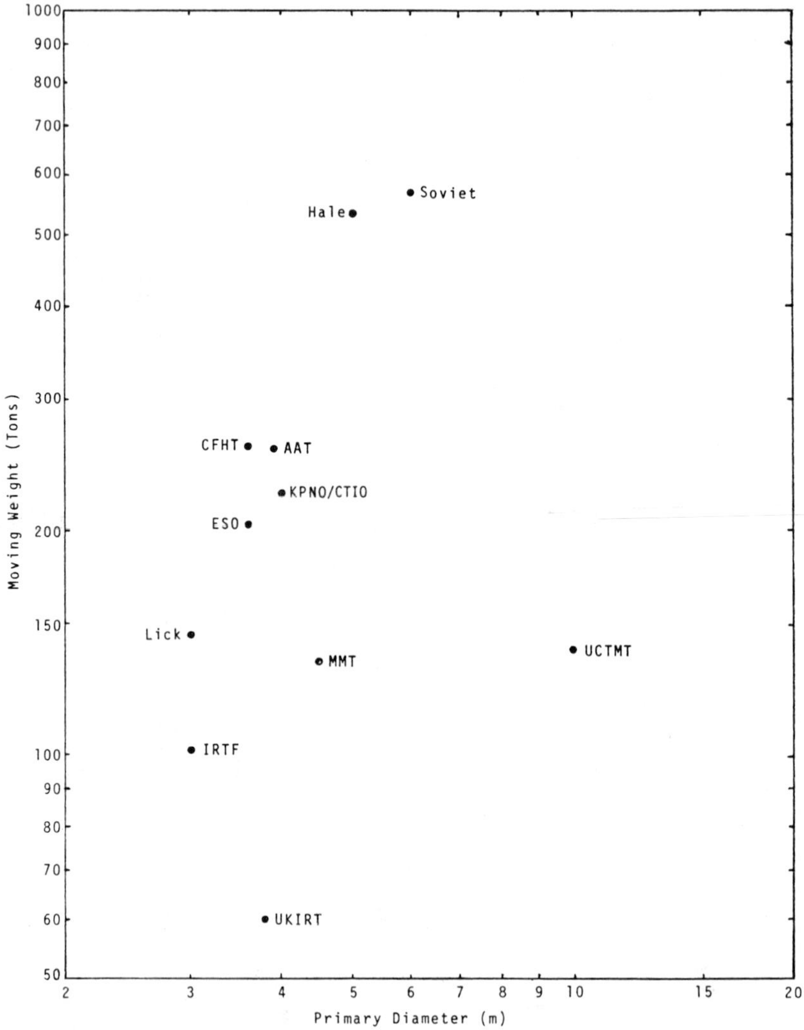

FIGURE 2. The total moving telescope weight plotted against telescope aperture for several large telescopes, including the currently anticipated 10-meter telescope weight.

enclosure, consistent with providing the necessary functions. The short focal length, the altitude-azimuth mount, and compact designs for instrument-changing devices have all contributed to the relatively small size of the enclosure. The basic configuration of the building and dome is seen in FIGURE 3. Details of the design are described by Rose and Hoggan.[2] The success of the design in terms of size is exemplified by the comparison to other domes as shown in FIGURE 4. Note that the TMT is appreciably smaller than the 200-in. building and dome.

The primary mirror figure must be maintained to give an image quality of 0.3

arcseconds. To achieve this, the surface of the mirror must be specified to a fraction of the wavelength of light. Maintaining this shape in the face of gravitational, thermal, and wind-induced forces becomes very rapidly more difficult for larger mirrors. Gravitationally induced deflections grow as the fourth power of the mirror diameter, so mirror flexibility grows extremely rapidly. Because of this, segments, being much smaller, can be quite stiff and appreciably thinner as well. The basic geometry of the primary mirror is seen in FIGURE 5. The mirror is composed of 36 hexagonal segments, each 1.8 m in diameter and 7.5 cm thick. Since these segments are relatively small, we will be able to treat them as rigid mirrors. The segments are to be composed of a low thermal expansion coefficient ceramic material. The total glass weight is about 14 tons, slightly less than the weight of the 200-in. mirror.

The basic advantages of the segmented design include several aspects. The weight

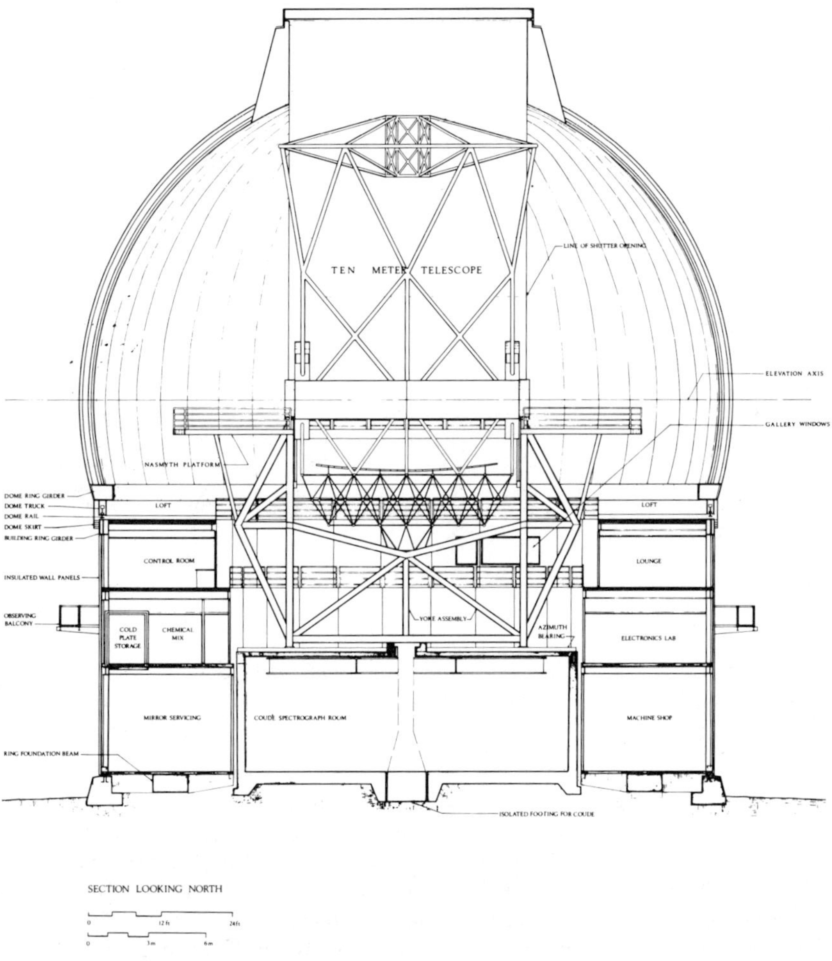

FIGURE 3. Elevation view of the building and dome surrounding the telescope.

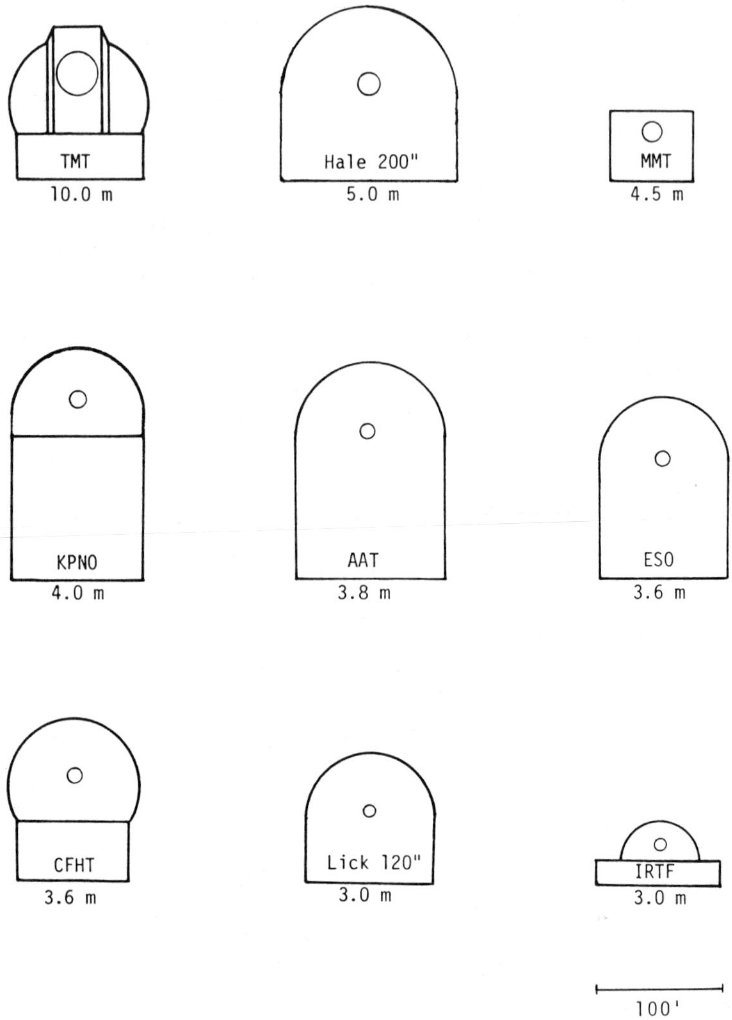

FIGURE 4. Relative sizes of telescope buildings, domes, and apertures. A comparison of features of existing telescopes with the 10-meter telescope. All are drawn to the same scale. The telescope aperture is given beneath each building.

of the mirror can be greatly reduced of course, leading to a general reduction in the weight and cost of the telescope. Smaller mirror blanks are also much more readily available, and at a substantially lower unit cost. The handling of segments during polishing and periodic aluminizing is much easier, and also the consequences of breakage are very much less (we will have extra segments available). With a single mirror primary, the consequences of single point failure are obviously extreme.

PASSIVE SEGMENT SUPPORT

Although we wish to assume that the individual segments are rigid plates, the extreme thinness of the segments makes them naturally rather flexible. Because of this, they must be supported very carefully in both the axial and radial directions to prevent them from being objectionably warped under the influence of gravity. The axial support system consists of a 36-point support system which adequately approximates a uniform support. Since we wish to control the piston (axial position) and tilt of each segment, these 36 points must be reduced to 3 points of support or attachment to the mirror cell. At these 3 points we locate the position actuators which are a fundamental part of the control system. These are shown schematically in FIGURE 6. The system for distributing the forces from 3 points to 36 points is a whiffletree system as shown in FIGURE 7. This represents an adaptation of a design by Kitt Peak National Observatory and, by employing flex pivots, provides a kinematic mounting system for the mirror without any moving parts.

The radial support of the segments differs from the commonly used edge-support systems because the segment edges are not easily accessible due to the presence of

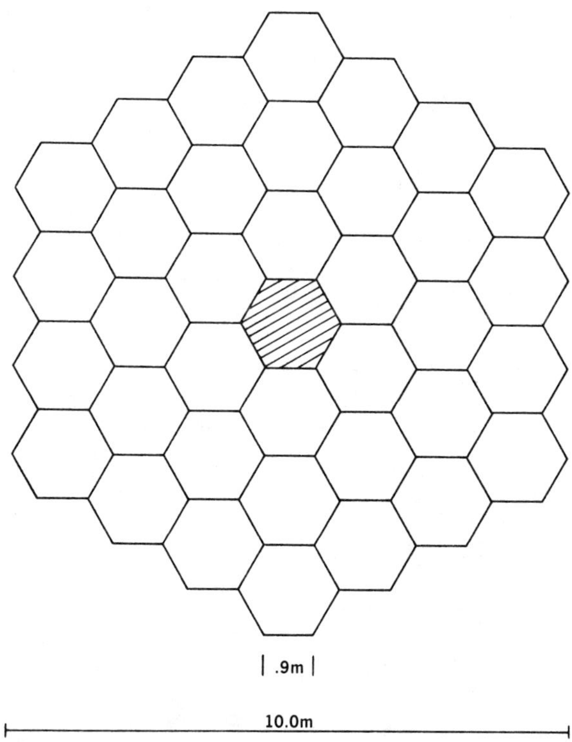

FIGURE 5. The 36-segment mirror geometry. The primary mirror of the TMT showing the 36 segments that make up its surface. The central hole allows for the Cassegrain focus.

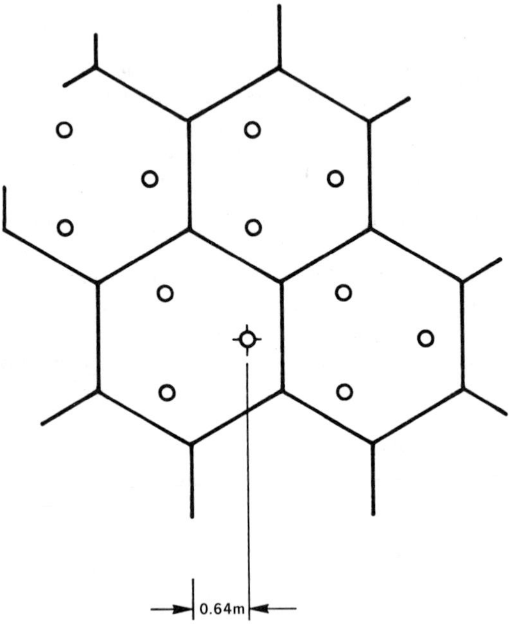

FIGURE 6. Mirror actuator locations. A detail of the segments, showing the locations of the position actuators that control the position of the mirror segments.

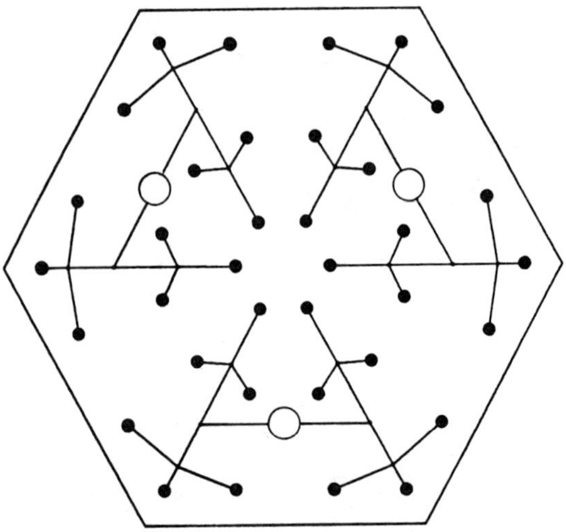

FIGURE 7. Schematic showing the 36 points of axial support for each segment and the three whiffletrees used to connect those points to the three actuators.

neighboring segments. We have developed a support that uses a single post attached to the center of gravity of the segment. By using a flexible diaphragm for the actual attachment, we can maintain robust radial stiffness while generating virtually no axial forces at all.

The combination of these two supports provides an adequate support system so that each segment behaves as though it were perfectly rigid, at least to the necessary optical

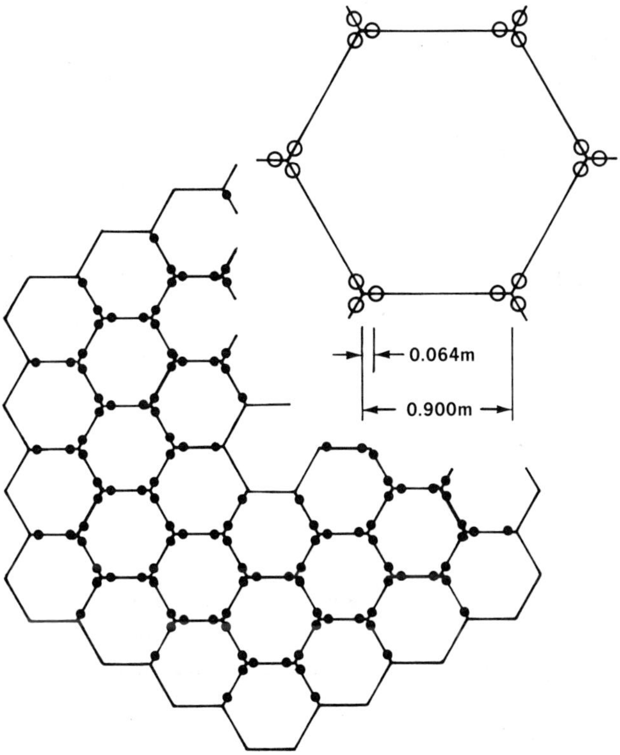

FIGURE 8. Displacement sensor locations. Schematic showing the locations of the capacitive sensors which measure the relative piston and tilts of the mirror segments. The 168 sensors in the system are mounted on the backs of the segments. With 108 actuators, the control system is highly redundant.

tolerances. Passive support system deflections of the segment are expected to be under 10 nm.

ACTIVE SEGMENT CONTROL SYSTEM

Because of the large size of the primary mirror, gravitational distortions will surely exceed the necessary optical tolerances. Thus, an active alignment system to keep the

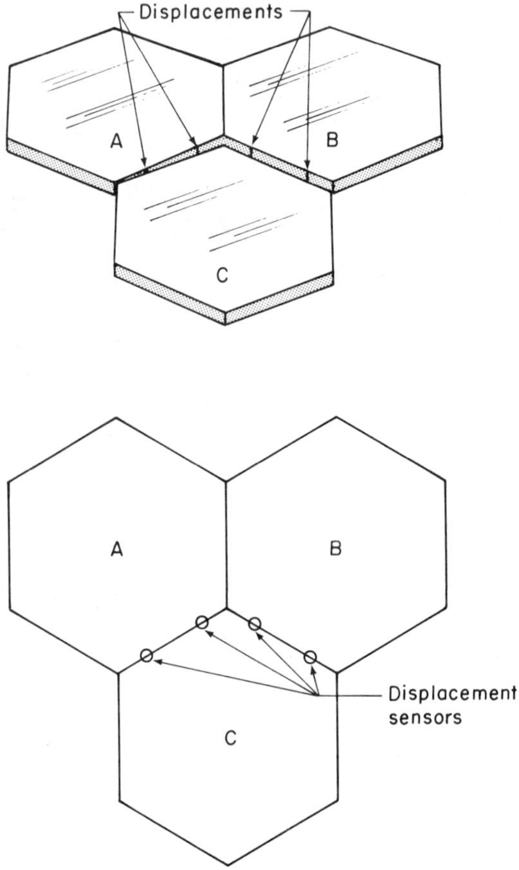

FIGURE 9. The locations of the sensors are shown, to give some idea of how the interlocking segment geometry allows control over the segment positions.

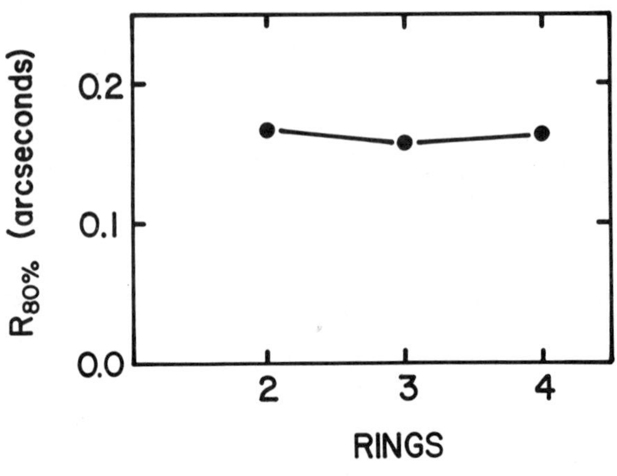

FIGURE 10. The image radius in arcseconds containing 80% of the energy from a point source for arrays with 2, 3, and 4 rings of segments. An offset $L = 3$ cm and a sensor noise of 50 nm are assumed.

segments in their desired positions is needed. A basic assumption for the control system is that the individual segments can be treated as rigid bodies, thus piston and tilt will suffice to restore the segments to their desired positions.

We have developed a sensing system to measure the positions of the segments relative to each other. It avoids any direct use of the front surface of the mirror, since this may interfere with astronomical observations. Rather, we will sense the segment positions by sensing the locations of the back surface of the segments. This is effective since the front surface and the back surface are related by the robust and thermally stable glass ceramic mirror material. We will use capacitive displacement sensors to measure the displacements of one segment relative to its neighbor at the segment edges. The locations of the sensors are indicated in FIGURE 8. By design, these sensors are sensitive to only the component of displacement that is normal to the mirror surface, and completely insensitive to any other motion at the sensor. By placing two

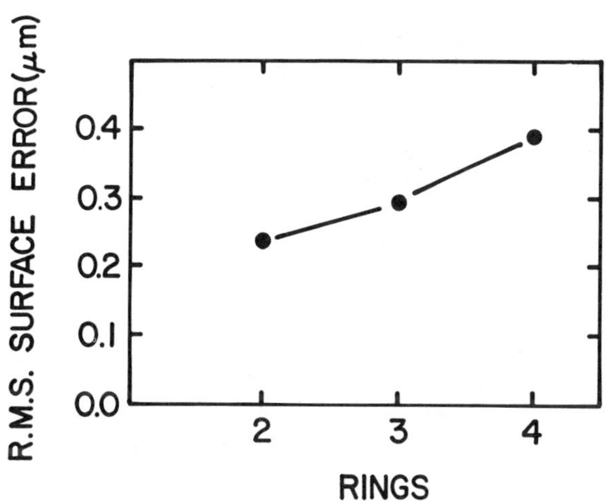

FIGURE 11. The rms surface error in microns for arrays with 2, 3, and 4 rings of segments. An offset $L = 3$ cm and a sensor noise of 50 nm are assumed.

sensors along every intersegment edge, a total of 168 sensors will be used. Combined, these give all necessary information to determine the segment positions. One can see this intuitively by use of FIGURE 9. Imagine one segment interlocked into two neighboring segments that are assumed to be held perfectly rigidly. Any rigid body motion (tilt or piston) by segment C will lead to changes in the edge displacements, and hence in the sensor readings. This interlocking character is present for all the segments. It is easy to show that the 168 sensor readings are linearly related to the 108 actuator positions. These equations can be solved (given the sensor readings, find the actuator values) in the least-squares sense to provide a unique algorithm for directing the actuator motions from the sensor information.

Detailed calculations and Monte Carlo computations have been used to study the control system behavior and to establish the necessary tolerances for the sensors and

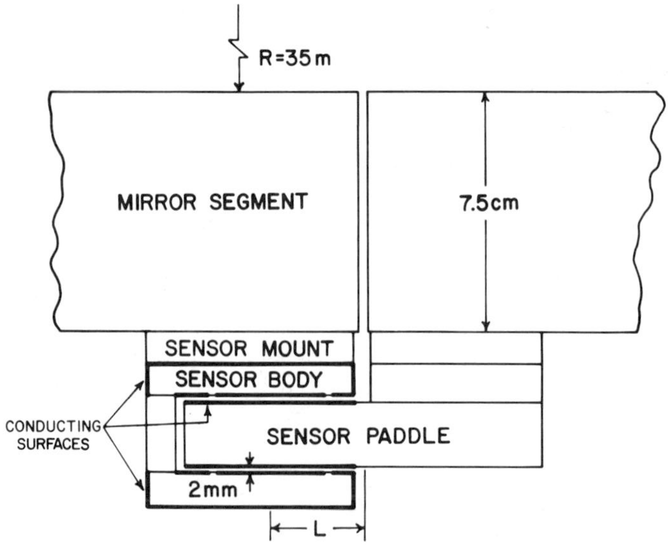

FIGURE 12. Schematic of a capacitive sensor. It is mounted on the backs of the segments and spans the gap between two adjacent segments to measure the relative normal displacements of the two segments.

actuators. Details of this analysis are described by Mast and Nelson.[3,4] The results indicate that both the geometric image quality and the root-mean-square (rms) surface error are almost independent of the number of rings of segments that are used. This can be seen in FIGURES 10 and 11. This means that this control technique is applicable equally well to mirrors with small numbers of segments and to mirrors with large numbers of segments. The control system studies show that sensor accuracy at the level of 50 nm is needed, and actuator smoothness at the same level is desired.

FIGURE 13. Roller screw (cutaway).

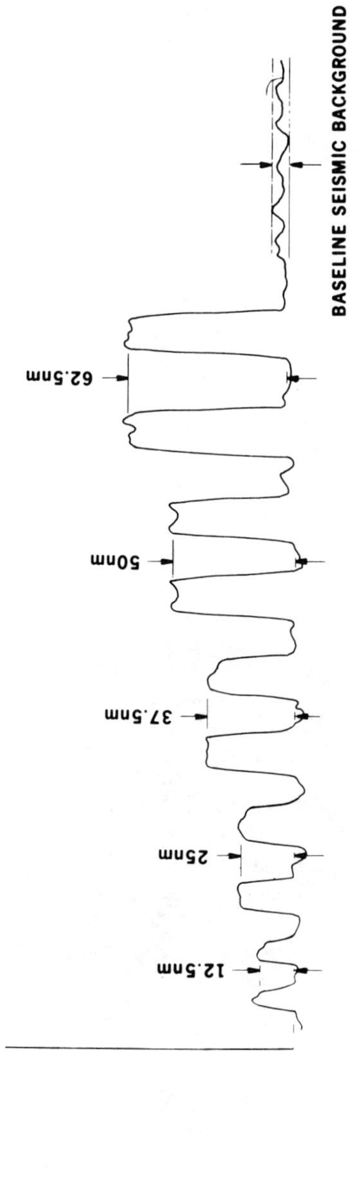

FIGURE 14. Roller screw displacement response for incremental step input.

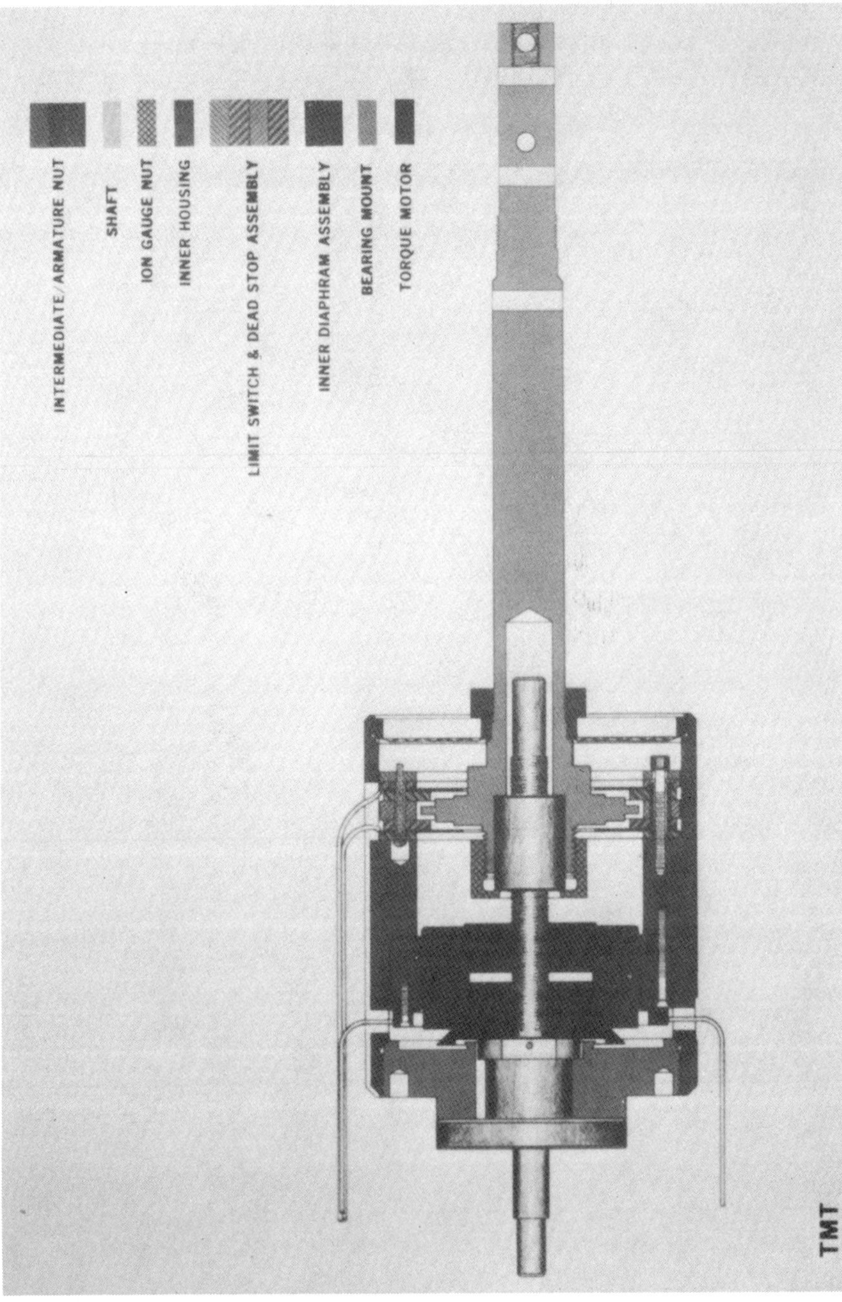

FIGURE 15. Actuator inner housing assembly.

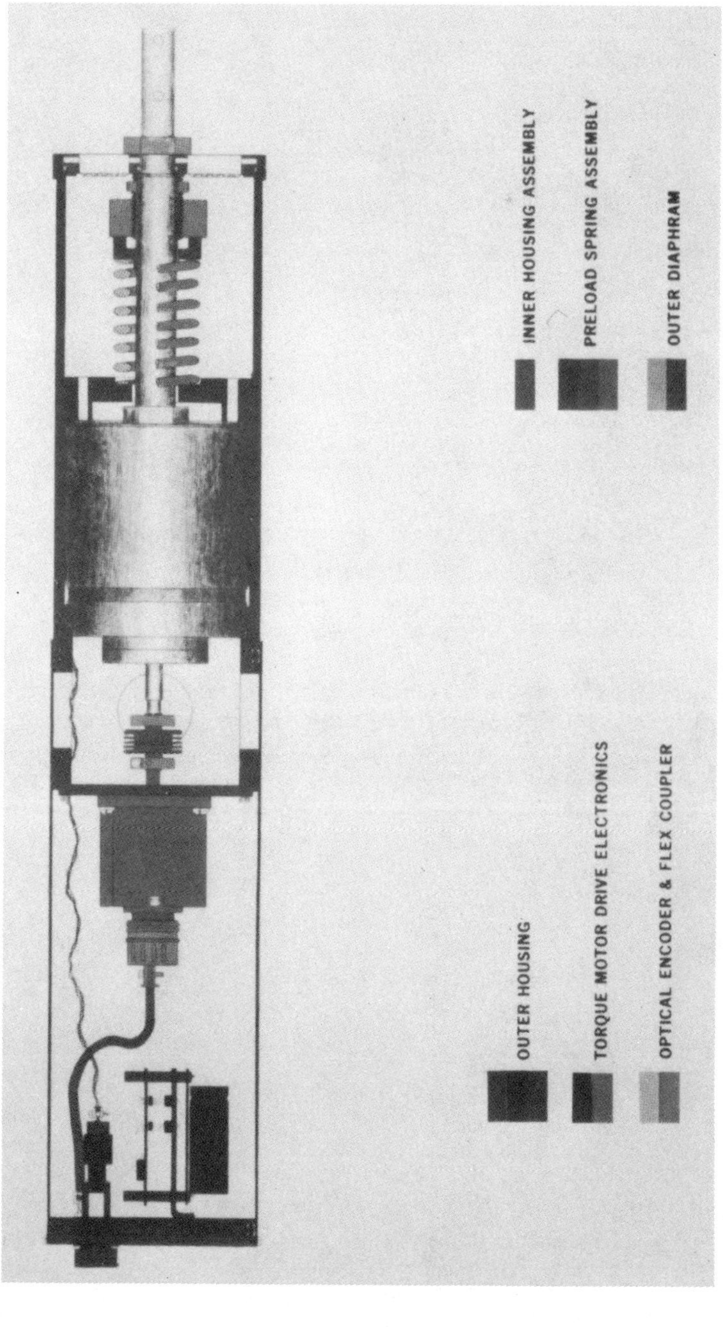

FIGURE 16. Ten-meter telescope precision servo actuator, outer housing assembly.

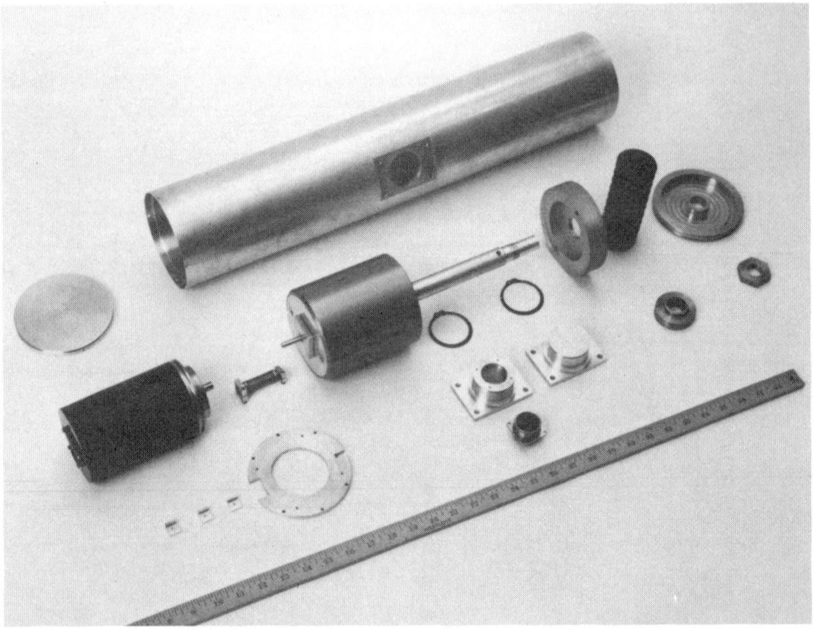

FIGURE 17. Partially assembled actuator.

The ratiometric capacitive sensors and their basic configuration are seen in FIGURE 12. Prototype sensors have been built and tested. These have measured noise levels below 1 nm at a 200 Hz bandwidth, and time and temperature stability better than 10 nm. Thus sensors capable of providing the needed accuracy have been developed and are practical. Details of the sensor design are available in a series of reports by Gabor.[5–7]

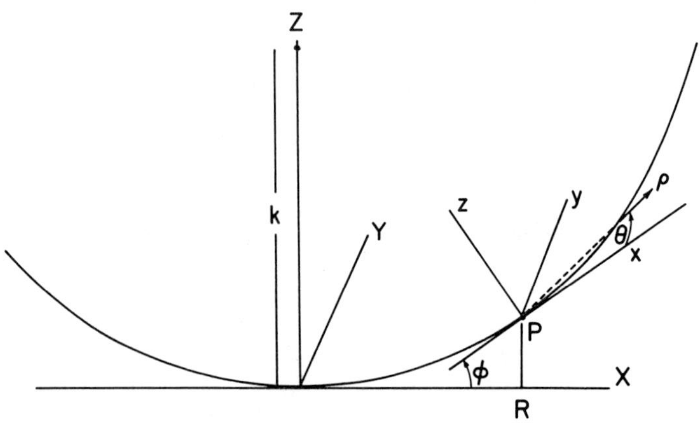

FIGURE 18. Diagram defining global (X, Y, Z) and local coordinates (x, y, z = r, θ, z) of mirror segment on paraboloid.

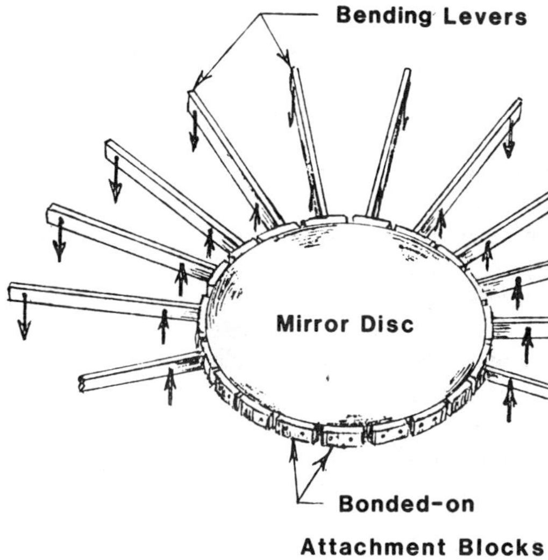

FIGURE 19. Stressed mirror polishing mechanical lever arrangement.

The actuators represent another technically challenging aspect of the control system. The actuators must carry the full weight of the mirror, move over a range of about 1 mm, and behave smoothly at a scale of 50 nm. After studying a variety of candidate devices, we have developed a motor-driven actuator using a roller screw as the basic rotational to linear device. A roller screw is a commercially available device made by La Technique Integral in France. A cutaway of the screw is shown in FIGURE 13. This screw operates by use of rolling friction only, with no sliding friction between the shaft and the nut, and this is the basis for its extremely smooth motion. Tests of this screw showed it was very smooth, and FIGURE 14 indicates one of the results of the test.

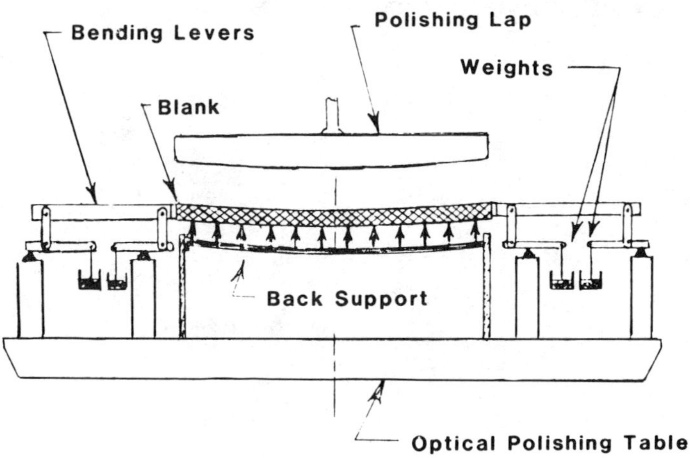

FIGURE 20. Stressed mirror polishing mechanical setup.

Having found that the screw smoothness was acceptable, the screw was incorporated as part of a full actuator. The inner assembly of the actuator showing the screw and the torque motor is shown in FIGURE 15. The full assembly of the actuator is shown in FIGURE 16. A photograph of one of the first prototype actuators is shown in partial assembly in FIGURE 17. Details of the design, construction, and performance of the actuator are described by Gabor.[5–7]

In summary, the control system works by sampling the sensors periodically and then by direct calculation determining what actuator displacements are needed to restore the sensor readings to zero, and hence restore the segments to their desired locations. The command is then sent out to all the actuators, and the motion is executed. This cycle of sensing, computing, and moving is repeated periodically at a frequency sufficient to maintain the segments in their desired positions to optical tolerances. We expect that correcting the segment positions several times per second will be sufficient. The mathematical algorithm has been thoroughly tested on the computer, and sensors and actuators have been successfully prototyped and tested.

SEGMENT FABRICATION

The second major challenge in making a segmented primary has to do with the polishing of the optical figure on the segments. Since the overall primary is a paraboloid of revolution, it is axisymmetric about the optical axis. However, the individual segments are not axisymmetric about their own centers. This is the heart of the technical fabrication problem since opticians very much prefer to polish axisym-

FIGURE 21. Stressing jig used to make a 36-cm diameter off-axis section of a paraboloid.

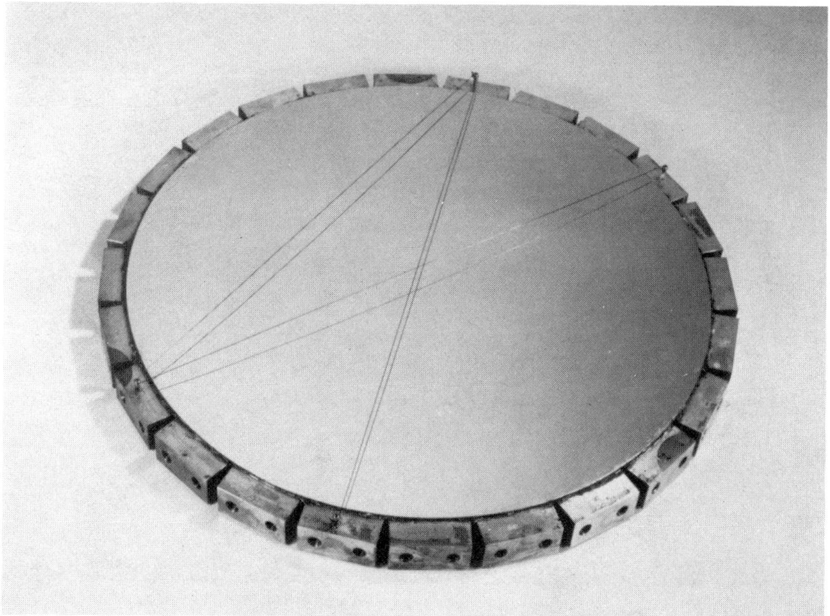

FIGURE 22. The 36-cm diameter off-axis section of a paraboloid, fabricated by the stressed mirror polishing technique. The blocks on the edges are used to attach the levers for applying the stresses.

metric mirrors. Easiest of all for the optician is the polishing of spheres, since large tools can be used.

We have developed a method for polishing these segments that is a generalization of the method first used by Schmidt in making corrector plates. The basic idea is that we will apply forces to the mirror blank to warp it elastically from the desired off-axis paraboloidal shape to a sphere. In this stressed state, a sphere will be ground and polished into the blank. Once this is done, the stresses can be removed and the mirror should elastically relax to the unstressed state and the spherical surface will be transformed into the desired off-axis shape.

The details of the problem can be explored by first establishing the form of the optical surface in the coordinate system of the individual segment. Such a coordinate system is indicated in FIGURE 18. Making the appropriate mathematical transformations reveals that the deviation of the segment surface is in fact small, and its form is dominated by astigmatism and coma. This very simple form is most fortuitous, since the elastic deformations of a plate when in this form are readily produced by the appropriate application of forces and moments to just the edge of the plate. The details of the elastic properties of plates and their controlled deformations are described by Lubliner and Nelson.[8]

A practical method of applying the edge forces and moments involves the use of weighted levers attached to the blank edge. This is indicated schematically in FIGURE 19. Another view showing schematically the mirror back support and the rest of the

FIGURE 23. The KPNO mirror in the stressing fixture.

FIGURE 24. A photograph of the existing telescope domes on Mauna Kea, taken from the proposed TMT site.

stressing jig is shown in FIGURE 20. To test that this method does in fact work as predicted, we performed a test on a 36-cm diameter mirror. The actual stressing jig used for this test is shown in FIGURE 21, and the mirror with its attachment blocks on the edge is shown in FIGURE 22. The test mirror was an off-axis paraboloid, and it matched the desired figure to within $\lambda/20$ rms, a completely acceptable figure for an optical mirror. Details of this test are described by Nelson *et al.*[9] The success of this test, and the simple physical and mathematical basis for the procedure have encouraged us that this is the preferred technique for making the segments. A project at Kitt

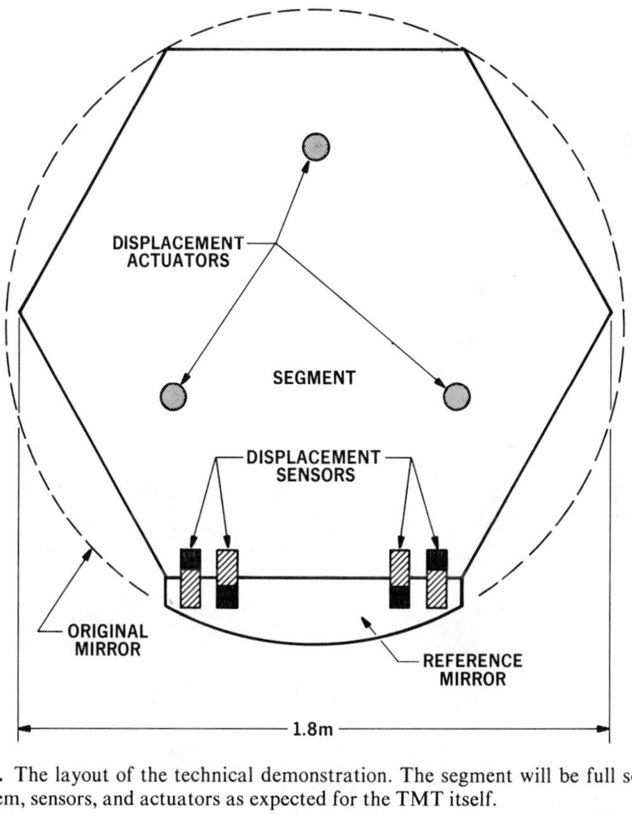

FIGURE 25. The layout of the technical demonstration. The segment will be full scale with a support system, sensors, and actuators as expected for the TMT itself.

Peak National Observatory (KPNO) is now under way to make two identical off-axis paraboloids with this technique. These will be 2 m in diameter, somewhat larger than those needed for the TMT; hence their work will serve as an excellent full-scale test of the method. The KPNO work is part of the Technology Development Program for the National New Technology Telescope, a program for a 15-m telescope. Details of the Kitt Peak work on off-axis mirrors are described by Golden *et al.*[10] The KPNO stressing jig is shown in FIGURE 23, and bears a great similarity to the one we used in the initial tests, except for the change in scale.

SITE

Since the TMT is to be used extensively at both optical and infrared wavelengths, a site that gives both superlative visible seeing and excellent IR transmissivity is desired. Since water vapor is a major source of IR opacity, high cold sites are preferred. After a study of a number of presently developed and other potential sites, we have tentatively selected an area near the summit of Mauna Kea, Hawaii, as our site. This site is at 4,000 m elevation and is well known to have excellent seeing and good weather conditions. It is presently the site of the University of Hawaii 88-inch, the National Aeronautics and Space Administration infrared telescope facility, the Canada-France-Hawaii 3.6-m telescope, and the United Kingdom infrared telescope. The site is shown in FIGURE 24, a picture that was taken from our proposed site. Negotiations with the University of Hawaii are currently under way to secure the site.

WORK IN PROGRESS

Although the basic ideas have now been developed for the TMT, and prototypes of key components have been developed and tested, a more complete engineering demonstration is desired. We are now in the process of fabricating a full-size segment and a reference mirror on which to test the control system. The basic layout of the mirror for this test is shown in FIGURE 25. We will use four sensors and three actuators for the control system, and the segment will be supported by the axial and radial support

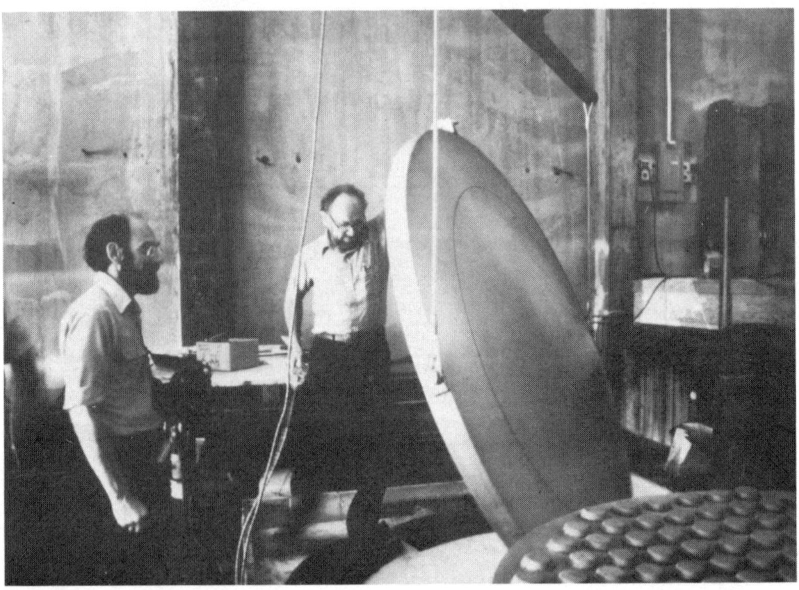

FIGURE 26. The Zerodur blank that is being ground and polished by Tinsley Laboratories for the technical demonstration mirror.

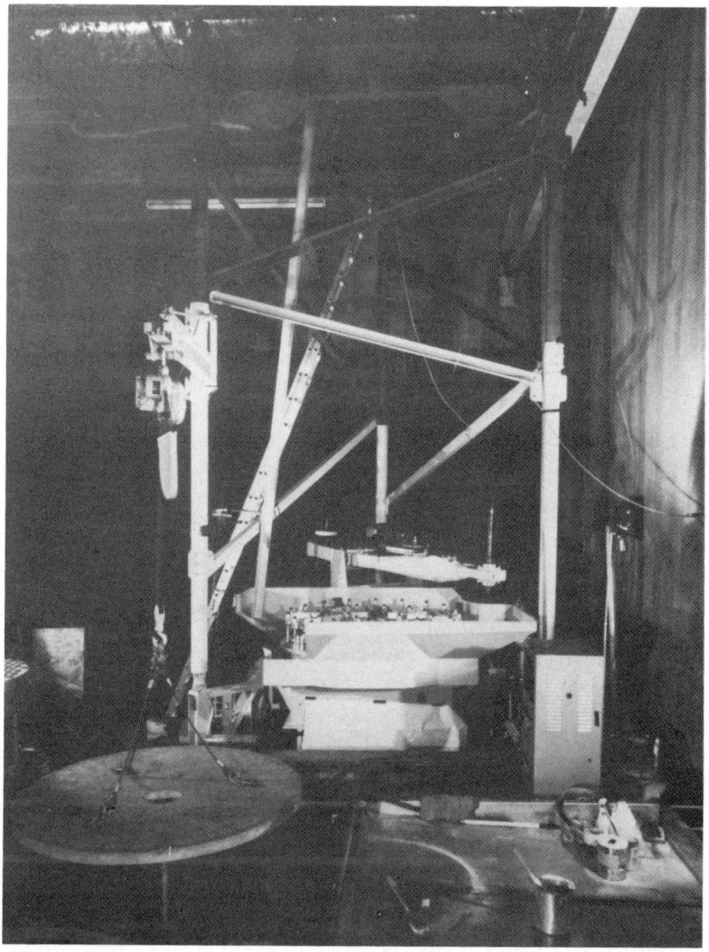

FIGURE 27. The polishing table and test tower at Tinsley Laboratories. A concrete blank of the same size as the Zerodur blank is in the foreground.

system planned for the TMT. The test mirror will be mounted into a special telescope so the effects of gravitational, thermal, and other perturbing forces can be studied. The basic objective is to show that the two mirror surfaces can be held in place as though they were a single optical surface, even in the face of outside perturbations. With the completion of this test in 1983, the basic ideas and hardware of the control system and the mirror support system will have been tested full scale. FIGURES 26 and 27 show some of the work on the optical fabrication.

The other key issue is the fabrication of the off-axis segments. As mentioned before, KPNO is presently testing the proposed technique of stressed mirror polishing full scale, so the completion of their test in 1983 will conclusively show the practicality of this method.

After the successful completion of these engineering tests we hope to begin the detailed design and construction of the TMT. If the necessary funding is made available, we expect completion of the telescope in 1988.

ACKNOWLEDGMENTS

This project is only possible through the unflagging efforts of many astronomers and engineers throughout the University of California system (UC San Diego, UCLA, UC Santa Cruz, and UC Berkeley). The generous financial support of the TMT by the UC President, David Saxon, has made this work possible. An index listing the available TMT reports can be obtained from the author.

REFERENCES

1. MEDWADOWSKI, S. 1981. Conceptual Design of the Structure of the U.C. Ten Meter Telescope. TMT Report No. 59. University of California. Berkeley, Calif.
2. ROSE, R. & H. R. HOGGAN. 1981. Building and Dome Schematic Design. TMT Report No. 85. University of California. Berkeley, Calif.
3. MAST, T. & J. NELSON. 1982. Appl. Opt. **21:** 2631.
4. MAST, T. 1982. Figure Control for a Segmented Mirror: Curvature Control without Tilt Sensors. TMT Technical Note No. 46. University of California. Berkeley, Calif.
5. GABOR, G. 1979. SPIE J. **172:** 39.
6. GABOR, G. 1980. Optical and Infrared Telescopes for the 1990's: 587. Kitt Peak National Observatory. Tucson, Ariz.
7. GABOR, G. 1982. SPIE J. **332:** 382.
8. LUBLINER, J. & J. NELSON. 1980. Appl. Opt. **19:** 2332.
9. NELSON, J., G. GABOR, L. HUNT, J. LUBLINER & T. MAST. 1980. Appl. Opt. **19:** 2341.
10. GOLDEN, L., P. GILLETT, R. RADAU, J. RICHARDSON & G. POCZULP. 1982. SPIE J. **332:** 357.

Steps Toward Eight-Meter Honeycomb Mirror Blanks[a]

I. Rationale and Approach

J. R. P. ANGEL AND NEVILLE J. WOOLF

Steward Observatory
University of Arizona
Tucson, Arizona 85721

RATIONALE

There is now a widely felt need for new optical/infrared telescopes of considerably larger light grasp than afforded by those currently available. There is also a growing realization that careful choice of site and thermal control of the telescope and its enclosure can give distinctly better images than are now generally achieved. The question of how to realize a new generation of large, high-quality telescopes is attracting much attention. Simply scaling up the present generation of 4-m class instruments having equatorial mounts and thick heavy mirrors would be inordinately expensive and probably not even give very good results, because of high thermal inertia. A variety of different approaches have been put forward, generally making use of alt-azimuth mounts to control telescope size and flexure, and some degree of servo control to maintain high optical quality in the presence of thermal, gravitational, and wind-driven disturbances.

The tried and tested principle for large telescope mirrors, universally used in ground- and space-based telescopes, is to take a rather rigid piece of glass, preferably lightweight and of low thermal expansion, figure it as a paraboloid or hyperboloid of revolution, testing and then using it in a support that resembles as closely as possible neutral bouyancy flotation. Of course, in space the last step comes for free! Even on the ground though, pneumatic, hydrostatic, and astatic lever systems have been devised that are quite adequate to balance gravity.

What is the practical size limit for this approach? However well we float ever bigger mirrors against gravitational deflection, ultimately wind will be our downfall. If wind pressure can bend the glass beyond tolerable limits, or gusting can excite low-frequency resonance, then the simplicity of flotation is gone—the mirror can no longer hold its shape by virtue of its own stiffness. Fortunately the sharp peak at 4 m at the right-hand side of the existing mirror diameter versus number diagram does not reflect the above limit. While much bigger mirrors with good rigidity would become impractically heavy and unmanageable if they were solid, given a good lightweight structure a doubling of the popular 4-m size looks very practical, as we shall discuss below.

For some of the most important problems requiring optical and infrared observa-

[a]This work is supported by the National Science Foundation and the National Aeronautics and Space Administration.

tion even an 8-m diameter is not adequate. The recent Astronomy Survey Committee calls for a telescope of twice this aperture, and similar goals have been set by European, British, and Russian astronomers. Single glass mirrors of 15-m diameter do go beyond the limits of passive support, and would be difficult to handle, transport, and aluminize. Fortunately, this does not force us into a radical change of mirror technology, because 8-m mirrors serve well as building blocks for larger telescopes. Several can be operated together in a single pointed mount, as in the multiple mirror telescope (MMT), or they can be separately mounted and used together when necessary. For most observations the combined data from a number of separately instrumented telescopes will yield the same performance as a single telescope of the same total area. This is true as long as each gathers enough light to overcome any detector noise, and each is large enough so that its own diffraction limit is smaller than the limit to resolution already imposed by the atmosphere. These conditions are satisfied by 8-m mirrors. Modern detectors already allow telescopes much smaller than 8-m to operate generally in a photon noise or sky background noise domain. The diffraction-limited resolution of an 8-m telescope is 0.3 arcseconds at 10 microns, 0.6 at 20 microns.

Of course, the spatial resolution that can be achieved with an MMT or array of 8-m telescopes is not limited to that of each one. Techniques for interferometry in the optical and infrared are now being vigorously explored. Speckle, Michelson, heterodyne, and intensity interferometry techniques have all shown recovery of information at the limit of the longest baseline. This would be 20–40 meters for an MMT configuration and hundreds of meters for separately mounted telescopes. At visible wavelengths the resolution is 5 milliarcseconds for 20 m baseline, 0.5 for 200 m. Speckle and Michelson methods are not limited to bright stars, but have already been used to observe quasars, jets from galactic nuclei, and so on.

A practical consideration in choosing the optimum unit size for a telescope is the ease and efficiency of coupling into instrument and detectors. This is not the place for a critical discussion of issues such as detector resolution and format, image and pupil reshaping. It is our conclusion though that good instruments for an 8-m aperture will not need to be inordinately large or heavy. Similarly, the size of secondaries for reasonably fast wide-field Cassegrain or Ritchey-Chrétien foci, in the 2–3 m range, is not unreasonable. The plate scales they give, around 5 arcsec/mm, are not inconveniently large, given the need to properly sample images under the best seeing conditions with detectors having pixels of around 25 μ.

MIRROR STRUCTURE

As a starting point, we consider that an 8-m mirror must be relatively light in weight. To a large extent the mass (and cost) of the telescope structure is driven by the mirror weight. There is also a premium on keeping the thermal time constant short, as will be discussed below. Our goal has been to keep the surface density at no more than 20–30 g/cm^2. The resulting mass of 10–15 tons for an 8-m mirror is about the same as that of existing 4-m blanks, and should result in single-mirror 8-m telescopes also of the same mass as the present generation. It is worth noting that there is no advantage in making ultralightweight mirrors for ground-based telescopes, because wind loading of the mirror and structure demand a certain minimum stiffness.

Given the weight, we have a choice of structure. A simple sheet of glass of 25 g/cm^2 is about 10-cm thick. Higher stiffness is achieved by structuring the glass like the Palomar blank, as a face sheet with ribs behind (waffle plate). Best of all is the form of a honeycomb sandwich, front and back sheet with ribs between. The space telescope and MMT mirrors are examples of this type. Rigidity in a honeycomb sandwich depends on the structural details, but analytical studies, for example by Barnes,[1] show that optimized designs deflect under their own weight somewhat less than a solid blank of the same thickness. Since the self-stiffness of a uniform blank is proportional to (thickness)^2, we see that the same 25 g/cm^2 can be made >36 times stiffer if it is in the form of a honeycomb sandwich 60-cm thick.

For mirrors of 8-m diameter, the improved stiffness of honeycomb structure is not simply a useful bonus, it is fundamentally required if the mirror is not to distort under wind forces. A support can be designed to balance the steady force of gravity to very high precision. The wind though is a maverick, whose gusts and pressure over the mirror face are unpredictable. If the telescope were in a dome much larger than the mirror diameter, as is now usual, wind might not be too serious a problem. Reasons of economy and thermal balance though will drive enclosures to be much more like that of the MMT, whose compactness and large opening provide little shelter during observations. Allowing only for a typical mountain top wind speed of 5 m/sec, and assuming this is halved by the shelter, typical wind force on the mirror surface will be 0.03 g/cm,^{2} or about 10^{-3} of the mirror weight. Such a force, blowing the mirror against three optimally located hard points, will degrade the image of a 10-cm-thick mirror to about 1 arcsec. This is not acceptable for a telescope whose overall image quality should be 0.25 arcsec or better. Little more than 0.1 arcsec of the error budget can be allowed to wind deformation. The additional stiffness of the 60-cm-thick honeycomb allows us to meet this strict criterion even for external wind speeds up to ~10 m/sec.

Independent of the mechanical advantage of honeycomb, a very significant thermal advantage is to be had for a telescope whose goal is excellent image quality. There is usually a seeing contribution to the image size caused by the mirror being out of thermal equilibrium with the air. In a typical telescope, this is caused by the excess heat in the mirror needing to be dissipated across the mirror boundary layer. The seeing effects for a given temperature depend on the inclination of the mirror, as does the cooling rate. When pointing at zenith, the mirror seeing effect is about [T(°C)/2] arsec FWHM. Typical values at other elevation angles are about half this. If we allow a further 0.1 arcsec to the image error budget from this cause, the mirror should be no more than 0.2–0.4°C above the ambient air temperature. Given 0.5°C/hr as a typical nighttime temperature gradient, it follows that the mirror should have a thermal relaxation time of no more than an hour.

The time constant for cooling of solid mirrors is set by the rate of convection at the mirror face(s). From telescope tests at the MMT the rate of cooling from the front face corresponds to about 1/4 hour per gram of material below a surface. Bowen reports a consistent cooling rate from convection measured at the 200-inch telescope.[2] A sheet of glass of 25 g/cm^2 would thus have a cooling time of 6 hours, or 3 hours if air at ambient temperature is blown across the back of the mirror to cool at the same rate as at the front. By contrast, honeycomb structure has a large ratio of surface area to mass. By blowing air at ambient temperature through the honeycomb as well as across the back, time constants of under 1 hour can be achieved. Wong and Wang have demonstrated that acceptable flow rates, corresponding to 6,000 cfm for a 7.6-m mirror, can in practice realize a cooling time constant of 1 hour in a honeycomb blank.[3]

We conclude that, considering both the structural and thermal characteristics, there is an overwhelming advantage to using honeycomb, or possibly waffle plate, structure for 8-m mirrors.

MATERIAL

It is all very well to conclude that honeycomb is great stuff, but how do you get it? The largest honeycomb mirrors produced are 2.4-m diameter (space telescope). Steps to make similar 4-m mirrors are under way, but 8 m (by current commercial technology) looks rather remote and very expensive. We have to reexamine the material and fabrication methods afresh to find viable alternatives. Even metal or composite mirrors may provide a solution, and experiments with pure aluminum are being undertaken by Franza and Wilson.[4] Our belief is that some form of glass is still likely to be best, because of its very high chemical and mechanical stability and uniformity.

Glasses used to make mirror substrates fall into three distinct families: common glasses of relatively low melting temperature; glass ceramics; and refractory glasses, which are pure or nearly pure fused silica. Nearly all glass for domestic and industrial use falls into the first family, which is inexpensive and easy to produce and fabricate. Low-expansion borosilicate glass is of this class, and is currently produced for such things as laboratory ware, ovenware, and so on. The Palomar 200-inch, Lick 120-inch, and Kitt Peak National Observatory 84-inch mirrors were made of borosilicate. Its composition is mostly silica, but the addition of 20% of other oxides, mostly B_2O_3, dramatically reduces the melting point. Manufacture is by melting sand, borax, and other ingredients in huge tanks at $\sim$1,500°C. The liquid glass is drawn or rolled into sheets, pressed into molds, and so on. The cost of sheets or blocks of glass is in the range of \$2–\$10/lb., depending on the quality and volume of production.

Chemically stable and meltable glasses like borosilicate and aluminosilicate have not been made with expansion coefficients less than about 2.5×10^{-6}/°C. Much lower expansion can be achieved in the second and third families, but for both there is a price to be paid. Glass ceramic achieves its low expansion through controlled devitrification, crystals of negative expansion in a vitreous glass matrix so as to give a net coefficient close to zero. Cervit and Zerodur are examples of this glass, from which most 4-m class mirrors are made. The trouble is that controlled crystal growth requires that the glass be free of any nucleating impurities, and also that all parts of a mirror have virtually the same thermal history. Casting into complex molds is ruled out because of inevitable surface impurities, and fusion of a honeycomb piece by piece is very difficult. Each bond must be made, fused, and cooled in a matter of seconds to avoid significant changes in the nucleation.[5] At this time no large honeycomb mirrors of glass ceramic have been made, except by the expensive and laborious process of hogging out solid blanks. The material does not hold much promise for 8-m honeycomb blanks at affordable cost.

Members of the third glass family are fused silica, with expansion coefficients 0.55×10^{-6}, and Corning's ULE, silica with titanium dioxide tailored for near zero coefficient at room temperature. Both materials are too refractory for the common forming methods of rolling, casting, etc. Corning manufactures the glasses by flame deposition, burning $SiCl_4$ in oxygen to build up boules of ~1.5-m diameter. These are then fitted together as hexagons and fused to make larger solid blanks, or cut up into small pieces which are fused together to make honeycomb core for lightweight mirrors. A description of the manufacturing steps for making the space telescope mirror, and the prospects for making future mirrors by frit bonding, is given by Lewis and Shirkey.[6] The frit for these materials is powdered raw glass of the Cervit type in an organic medium, used like glue. On firing and heat treating it makes a bonding material with matching very low expansion.

We think it is fair to say that while the technology developed by Corning for the refractory glasses could probably be further developed for 8-m honeycomb mirrors, the process is elaborate and costs would remain high unless large volume allowed more efficient automated production. As a benchmark, Corning has quoted $3.5 million for a solid meniscus of ULE 7.6-m in diameter by 12-cm thick. This would be simply boules fused together. A true honeycomb structure of 8-m diameter by the present technology would have to cost probably in excess of $10 million.

Given the uncertainties and high cost of making 8-m honeycomb blanks from the difficult silica and glass ceramic families of very low expansion, we must consider carefully if we really need this property. It is our conclusion that strict tolerances of image quality can be met with borosilicate or aluminosilicate glasses *in honeycomb form*. The problem that can arise in mirrors with nonzero expansion is figure distortion when thermal gradients are present. The expansion of borosilicate is large enough that if temperature differences of a few degrees within a mirror are set up by diurnal or other temperature changes, the resulting distortion of the reflected wave front would be an arcsecond or more. Heavy solid mirrors, like those in the current 4-m telescopes, do develop such gradients, and they persist for many hours because of the long internal thermal time constant, and low rate of heat transfer by convection at the surface.

The solution is either to avoid borosilicate and similar glasses, or, as Ritchey strongly advocated in the 1920s,[7] to use a ventilated honeycomb mirror for rapid thermal equilibration. We have earlier identified a need for ventilation to bring the mirror into thermal equilibrium with the ambient air. This feature will also control figure distortion.

A quantitative estimate of the component of distortion that is not simply change of curvature and focus is difficult. Detailed structural and thermal modeling is still required. The best present estimate comes from extrapolation from the Palomar mirror. Although this is not a lightweight honeycomb, its ribbed construction is a step toward lower thermal inertia, and in fact thermal figure distortion is not usually detected.[8] The effect of sudden changes in ambient temperature, caused by a cold front going through, was investigated by Bowen.[2] A sudden 5° drop of temperature cooled first the edge band of the mirror, pulling down the edges by 0.3 arcsec as measured by a Hartmann test, enough to give a 1.2 arcsec image from spherical aberration. This effect was reduced by insulation, but no details are given. If we assume that gradients of a few degrees were present in the glass during Bowen's test, then we must rely on ventilation and thermal control to hold thermal gradients to a few tenths of a degree to keep figure distortion at the 0.1 arcsec level. This is not trivial, but not an unrealistic goal.

FABRICATION OF BOROSILICATE GLASS

With the realization that low-expansion borosilicate or aluminosilicate glasses in honeycomb form would make excellent mirror blanks, we undertook a research program to develop manufacturing techniques that would take full advantage of the favorable properties of these glasses. We have explored both casting and fusion-bonding techniques. Since detailed current accounts of both are in the literature,[9,10] we will give here only a general description and status report.

The method for casting honeycomb sandwich that we have found most efficient is as follows. A tub the size of the mirror is made from refractory tiles like staves of a barrel. These are compressed against the hydrostatic force of liquid glass by refractory nickel alloy wires, tensioned by weights and pulleys outside the furnace. The base of the mold is made of tiles, held under compression by the outer ring. Above each tile is a block of ceramic fiber which will leave one of the voids in the honeycomb. Nuts and bolts of silicon carbide hold each block down to the base tile, against the bouyancy force of the liquid glass. A collar of ceramic fiber spaces each block from the base, so glass runs under the blocks to form a back face with holes. The mirror is made by piling chunks of cold glass in the mold and heating to 1,200°C, when the liquid glass fills the tub, forming a free surface above the blocks and running into all the gaps and spaces. The ceramic blocks that formed the ribs are removed by waterblasting after the mirror is annealed and cooled. A good quality 1.8-m mirror has now been successfully cast by this method, and we are currently developing the facility for making 8-m mirrors. The process will be essentially the same, except the furnace will be on a rotating base so the curved mirror face is cast in directly. This avoids the cost of buying and subsequently grinding away ten tons of glass, in the case of an 8-m f/2 curve.

While we believe casting will be the most efficient way to make 8-m honeycomb blanks, with a price tag of ~$1,000,000 each, we are also looking at new processes for fusion bonding. The most promising uses round glass tubing as a starting point for making hexagonal honeycomb. The tubes are cut up, hexagonal close packed, and filled with zircon beach sand. On firing the glass softens, and the heavy sand acts like a fluid to push the tubes together into honeycomb. This is a very quick way to make honeycomb but several steps would be involved in making a finished sandwich blank. Machining the honeycomb and fitting and fusing on a thin curved face sheet is not easy, and several firings are needed from start to finish. By comparison, casting makes the complete honeycomb sandwich in a single firing.

CONCLUSION

Our reevaluation of the best form for 8-m mirrors has taken us back 60 years to Ritchey's concept for ventilated glass honeycomb, and to methods of casting borosilicate glass similar to those used for the Palomar 5-m blank in 1934. Our goals for image quality are high, but we have the benefit of 50 years of technological development to help us. The concept we have developed is, if you will, the next step beyond the 200-inch blank, advancing not only in size, but in structural and thermal design. Modern refractories allow us to make a complex mold to form thin ribs and faces, and to make a full honeycomb sandwich rather than the simpler waffle plate construction. The advantages are large, for example a 75% lightweighted honeycomb sandwich with the same diameter and thickness (60 cm) as the Palomar disk would have four times the self-stiffness but only half the weight. Its thermal time constant of 1 hour when ventilated would be several times reduced. These conclusions follow from Rule's data for the 200-inch mirror,[11] and the structural and thermal analysis of Barnes and Wong already discussed. We are currently studying the optimum geometry, support, and thermal stabilization of 8-m blanks, and will present results at the London SPIE meeting in September 1983.

SUMMARY

The present generation of 4-m telescopes does not reach the practical aperture limit for passively supported monolithic glass mirrors. Mirrors up to 8-m diameter made with honeycomb sandwich structure can meet the most exacting demand goals for subarc-second image quality now being made by both optical and infrared astronomers. They are light, have good stiffness against gravitational and wind loading, and have low thermal inertia. The latter allows them to be operated at the ambient air temperature to avoid local seeing problems. Large honeycomb blanks of either ceramic or silica glasses, the very low expansion materials currently needed to make large solid blanks, would be very expensive and difficult to make. However, a ventilated honeycomb does not need such low expansion, because internal thermal gradients are sharply reduced. Borosilicate and aluminosilicate glasses, which are much less expensive and easier to fabricate, are quite adequate. We have developed casting and fusing techniques specially adapted to borosilicate and have already cast a good 1.8-m honeycomb sandwich blank. Assuming the development to 8-m is successful, we anticipate a new generation of relatively inexpensive 8-m telescopes using these mirrors, and also larger instruments using the mirrors mounted together like the MMT, or in arrays for high-resolution interferometry.

REFERENCES

1. Barnes, W. P. 1969. Appl. Opt. **8:** 1191.
2. Bowen, I. S. 1965. Int. Astron. Union Symp. **29:** 1313.
3. Wong, W. & S. Wang. 1983. Report to the Engineering Committee of the University of Texas 300-Inch Project. University of Texas. Austin, Texas.
4. Franza, F. & R. Wilson. 1982. Proc. SPIE **332:** 90.
5. Gardopee, G. J. & D. W. Shen. 1982. Final Technical Report RADC-TR-82-260. Rome Air Development Center. Rome, N.Y.
6. Lewis, W. C. & W. D. Shirkey. 1982. Proc. SPIE **332:** 307.

7. Ritchey, G. W. 1926. L'Astronomie **40:** 57.
8. Oke, J. B. Private communication.
9. Hill, J. M. & J. R. P. Angel. 1983. Proc. SPIE **380.** (In press.)
10. Angel, J. R. P. 1983. Proc. SPIE **383.** (In press.)
11. Rule, B. 1968. *In* Support and Testing of Large Astronomical Mirrors. D. Crawford, A. Meinel & A. Stockton, Eds.: 84. University of Arizona. Tucson, Ariz.

Large Optical Telescopes

New Views into Space and Time

S. M. FABER

Lick Observatory
and
Board of Studies in Astronomy and Astrophysics
University of California
Santa Cruz, California 95064

In the wake of recent intensive debate among optical and infrared astronomers—debate that was occasioned and sponsored by the activities of the Astronomy Survey Committee—there has emerged a new consensus to press for the construction of one or more optical-infrared telescopes much larger than any now in existence. This new resolve was not forged without controversy. Specifically, there were some who believed that several 4-m-class telescopes would better serve the scientific needs of optical-IR astronomy than would one or at most a few very large instruments. These individuals quite rightly pointed to the enormous popularity of existing 4-m telescopes and the excellent 4-m-class astronomy that is not now being carried out owing to lack of adequate telescope time.

While recognizing this weighty argument, the Astronomy Survey Committee nevertheless found the counterarguments in favor of very large telescopes more compelling. The committee noted with favor the development of new techniques that will allow us to build telescopes larger, lighter, and cheaper than ever before. These technical breakthroughs are the legacy of 50 years of industrial and scientific progress since the design of the Palomar 200-inch was conceived in the 1930s. Over decades, the major advances in observational astronomy have come principally from the inexorable, onward push to ever-larger collecting areas. To continue this progress it is incumbent on each generation of astronomers to seize the technical opportunities of the present moment and, in so doing, to prepare the way for further aperture increases in the future.

There is also the undeniable excitement and allure of very large projects. Giant telescopes capture public imagination, are easier to publicize, and are easier to fund. We astronomers must ask ourselves whether we can attract the same number of dollars into astronomy for the construction of many intermediate-class telescopes as we can for the construction of one or two very large instruments. Realistically, I think the answer is no. Neither public nor private funding agencies are likely to be excited by proposals to simply "do more of the same." Friends of astronomy in Washington who are astute observers of the political scene assure us that this conjecture is indeed correct.

More important than any arguments based on political strategy or technology, however, are the unique scientific opportunities that only very large optical telescopes can offer. These are indeed exciting and aim, among other things, at the heart of many of the fundamental cosmological questions that have generated so much interest and discussion at this symposium. If we quantitatively compare the capabilities of a national 15-m telescope with existing 4-m telescopes, for example, we find that the 15-m offers a speed gain of 14 for most problems and gains as large as 200 for certain important problems in the infrared (see below). Limiting angular resolution will also increase by a factor of 3½ to 5. Gains of such a magnitude enable studies that are not possible on existing instruments and thus open the door to qualitatively new science. This is ultimately the decisive argument—the opportunity not merely to increase the rate at which the frontier is explored but, more than that, to open up virgin territory that would otherwise remain beyond our reach.

It is my task today to acquaint you with some of the exciting new opportunities afforded by large optical-IR telescopes. Before mentioning specific projects, however, let us first take a look at the quantitative gains afforded by large aperture. These are summarized in TABLE 1. For definiteness, I have focused on a proposed national 15-m telescope and compared it to the existing national 4-m instruments. Limiting magnitudes and speed gains for several representative types of observations are indicated. It is seen that, for most observations, the expected speed gain is in proportion to the area of the primary mirror but that for certain projects, the gain goes as area squared. These larger gains occur whenever observations are detector-noise limited, as in the 1.2–2.5 μ region, and at wavelengths where existing telescopes are diffraction limited rather than seeing limited, that is, at 5 μ and beyond. Because of these enhanced IR gains, large telescopes are uniquely valuable to infrared astronomy. It is therefore vitally important that the special demands of the IR—coherent optics and low thermal emissivity—be met.

TABLE 1. Theoretical Limits of a Fifteen-Meter Telescope[a]

	Band	S/N	R^b	Seeing (arcsec)	Limiting Magnitude	Gain over 4-m Magnitude	Gain over 4-m Time
Direct:							
Stars	V	10	0.2	1	27.3	1.4	14
	V	10	0.2	0.5	28.1	1.4	14
	K (2.2 μ)	10	0.1	1	18.4	1.4	14
	K (2.2 μ)	10	0.1	0.5	19.1	1.4	14
	K (10 μ)	10	0.1	0.5 (10μ)	9.4	2.9	196
First-ranked cluster galaxy,	V	27	0.2	1	—	1.4	14
$z = 1$	K	15	0.1	1	—	1.4	14
Spectroscopy:							
Low resolution	V	30	1,000	1	22.1	2.0	14
High resolution	V	30	30,000	1	19.1	2.9	14

[a] "Reasonable" reflectivities, detector efficiencies (CCD), and IR detector noise are assumed everywhere. All exposure times are 3600 sec, $H_0 = 50$ km s^{-1}, $q_0 = 0$.

[b] R = wavelength resolution. $R = \Delta\lambda/\lambda$ for filter photometry, where $\Delta\lambda$ = bandwidth. $R = \lambda/\Delta\lambda$ for spectroscopy, where $\Delta\lambda$ = Å/pixel.

Lines 1 and 2 of TABLE 1 illustrate how important good seeing is to ultimate performance. For observations of faint objects that are limited by the sky background, a factor of two improvement in seeing is equivalent to a factor of two increase in primary-mirror diameter. Excellent seeing is therefore equivalent to the expenditure of many extra millions of dollars in construction funds. In choice of site, therefore, the fraction of time given to superb seeing should be given greater weight than the average value of the seeing (e.g., Reference 26). Concomitantly, both the design and scheduling of large telescopes should be flexible enough to allow quick changeovers to programs that can benefit most from good seeing. Adherence to this goal will, I believe, necessitate substantial changes in the operating philosophy in use at most observatories today.

I have selected four basic areas to talk about here that illustrate the new scientific opportunities afforded by large optical-IR telescopes. These were chosen from lengthy lists of projects that have been studied by several groups, including the University of California,[9] Kitt Peak National Observatory,[21] and the UVOIR Panel of the Astronomy Survey Committee.[24] Many additional projects not mentioned here are discussed in these reports. Brief reviews of the science to be done with large telescopes were also given by several speakers at the Tucson Conference on Large Telescopes for the 1990s (e.g., References 10, 22, and 23).

The first area of interest today is *lookback studies of distant galaxies and clusters of galaxies*. The study of the evolution of galaxies by the lookback effect at great distances has just begun. Counts and colors versus magnitude out to a z of 1 show small but distinct signs of evolution, in itself a crucial cosmological datum in favor of the big bang. The first spectra of galaxies below the first-ranked galaxy in clusters are just now becoming available, and they show surprising peculiarities—an abundance of Seyfert-type spectra and anomalously blue objects with featureless continua.[6] Important questions to be answered include the following:

1. *When did galaxies form*? In some currently popular scenarios, galaxy formation is imagined to be triggered by the collapse of larger structures (e.g., the "pancake" theory).[7] Since the low density of such structures tells us that they must have formed close to the present time, galaxy formation is a relatively recent event in such pictures, with an upper limit on formation red shift of roughly $z = 3$. As TABLE 1 shows, this epoch is comparable to or only slightly larger than the ability of a 15-m to penetrate back in time. If galaxies formed as recently as $z = 3$, faint galaxies should show strong signs of evolution in color and luminosity. Other, more conventional views of galaxy formation (e.g., hierarchical clustering)[25] place the formation epoch of galaxies at earlier times, say $z \gtrsim 5$, and the predicted evolutionary effects are correspondingly smaller.

   Counts and color studies of distant galaxies with existing 4-m-class telescopes appear to show measurable but not large evolutionary changes down to 23 B mag. However, the interpretation of these data remains cloudy owing to the fact that there are no red shifts and hence the distance of each individual galaxy is uncertain. Hundreds of red shifts are required to rectify this situation, all at B magnitudes 22 or fainter. Such a task is photon limited and is beyond the limits of 4-m-class telescopes.
2. *How did galaxy clustering evolve*? The two pictures for clustering and collapse of galaxies mentioned above predict drastically different clustering behavior at red shifts of one or two. In the hierarchical picture, small clusters should already be in place by that time, with large structures just beginning to form. In the pancake picture, the reverse is true. Magnitude-limited surveys of the positions

and red shifts of galaxies at B = 23 mag should be able to discriminate between these two pictures quite decisively.

3. *How did Hubble types evolve*? A still unanswered question is the relative importance of "heredity" versus "environment" in the determination of galactic Hubble types. We realize now more clearly than a few years ago that this issue is intimately bound up with questions 1 and 2 above, as the clustering process may have profoundly influenced the development of Hubble types if it overlapped strongly in time with galaxy formation. The red-shift-magnitude data discussed above are therefore relevant here, as well. These data are also complementary to the high-resolution images of distant galaxies to be provided by space telescope (ST), which will show us how the actual appearance and Hubble types of galaxies behaved with time and environment.

It should be emphasized that the prime requirement for a larger telescope in the area of lookback studies arises from the need for red shifts and spectral-energy distributions. The broad-band counts and colors can be obtained with smaller telescopes and are already coming in; ST will provide high-resolution images. Spectroscopic studies for red shifts and stellar population syntheses require many more photons, however. Even with an aperture of 15 m, it will be necessary to employ red shift multiplexing detectors operating over a wide field of view.

The second broad category of studies before us today concerns the *density distribution of nonluminous matter in the universe.* The major question here is whether the two types of matter, nonluminous and baryonic, cluster with similar efficiencies on all mass scales larger than galaxies. Indeed, do galaxies themselves all contain the same ratio of nonluminous to baryonic matter? These questions are important because the two pictures of galaxy formation—pancake and hierarchical—make very different predictions about the relative density distributions of the two types of matter on large and small scales. Moreover, the presence of significant amounts of nonluminous matter in small structures like dwarf galaxies would suffice to rule out certain elementary particles such as neutrinos as the major constituent of nonluminous matter.[17]

The spectroscopic capabilities of a 15-m telescope will help to answer these fundamental observational questions in at least two ways. First, the velocity dispersions of distant groups and clusters can be studied in a manner analogous to that used for nearby structures. We can therefore verify that the "missing mass" problem is ubiquitous in space and in time. Second, we can use the motions of individual objects to study the mass distributions within typical nearby galaxies of various sizes and Hubble types. Such studies have up to now relied mostly on 21-cm rotation curves. These are useful but are restricted mainly to fairly large spirals with gas-rich disks in rotational equilibrium. By studying the motions of globular clusters and individual stars, we will be able to probe the outer regions of spheroidal components in elliptical and spiral galaxies and the entire mass distribution in dwarf galaxies. These measurements are now strongly limited by aperture—a handful of the brightest such objects can be studied but not enough to provide adequate statistical samples. Stellar and globular-cluster dynamics in external galaxies is clearly an area where a 15-m telescope will make a major impact.

A third important category of observations concerns quasi-stellar objects (QSOs). There are a number of basic questions about the nature of QSOs themselves that require large apertures for study. I am thinking here of such issues as how QSOs associate with galaxies and in what kind of galaxies they are found. To answer such questions we need more red shift studies of QSO-galaxy associations (e.g., Reference 20) and spectroscopic studies of the "fuzz" around QSOs (e.g., Reference 27).

Potentially just as interesting, however, is the use of QSOs as a means of studying the intergalactic medium (IGM) at high red shifts. Intervening clouds of gas along the line-of-sight produce absorption lines in QSOs, and these absorption-line spectra have proven to be extremely rich probes of the IGM in the early universe, yielding cloud sizes, density distributions versus red shift, and heavy-element abundances.[18,19] These data are sensitive to the efficiency of gas infall into galaxies and the early rate of nucleosynthetic enrichment of heavy elements versus time. The distribution of clouds along the line of sight also tells us about the evolution of clustering at red shifts up to $z \approx 3.5$. Absorption lines thus provide indirect evidence on the state of the universe at epochs much earlier than we can hope to reach using directly emitted radiation.

These particular QSO studies are strongly photon limited. The "fuzz" and clustering observations involve low-resolution spectroscopy at very low light levels. The absorption-line work is now limited by inadequate spectral resolution on even the brightest QSOs. The 1.4 to 2.9 magnitude gain of a 15-m telescope would transform the status of all of this work from marginally possible to clearly feasible.

The fourth and final area of interest today is the important category of *infrared-red observations at 5 μ and beyond.* Throughout this region, the main source of noise is not photon statistics, as at shorter wavelengths, but thermal background emission from the atmosphere and telescope. This background, B, scales as $B \alpha D^2 \Omega^2$, where D is primary diameter and Ω is the solid angle subtended by the source. Furthermore, at wavelengths longer than 5 μ, existing telescopes are diffraction limited, that is, the diffraction image is larger than the seeing image. In such a situation, since $\Omega \alpha D^{-2}$, we find that the background is constant, independent of telescope aperture. The signal-to-noise ratio on point sources therefore scales as D^2, and the time required to reach a given S/N scales as D^4. This is the origin of the enormous IR speed gain of ~200 mentioned in the introduction.

Not only is a 15-m telescope much faster in detecting point sources beyond 5 μ, but it also offers a substantial increase in angular resolution of a factor between 3.5 and 5, depending on how the primary mirror is configured and with which existing telescopes the comparison is made. One way of quantifying such a comparison is to consider the speed gains that are possible using speckle interferometry, a short-exposure technique that "freezes" the atmospheric modulation-transfer function and reveals high-frequency structural detail that is washed out in long exposures. Infrared speckle interferometry is still in its infancy, and the speckle image is presently being scanned in a one-dimensional manner behind a rapidly moving slit. Greater efficiency is in principle possible with two-dimensional sensing, however, and infrared arrays that will allow such sensing are now under active investigation. With such a two-dimensional detector, FIGURE 1 shows the speed gain in measuring Fourier amplitudes on various angular scales that would result by using a 15-m telescope compared to a 4-m. In addition, I have assumed that the seeing (θ) at visible wavelengths is 1 arcsec for the 4-m but is 0.5 arcsec for the 15-m. Excellent seeing of 0.5 arcsec or better is reported a substantial fraction of the time at Mauna Kea,[26] a site that on account of its high altitude and low water vapor is an all-around excellent IR site.

At low spatial frequencies, the speed gain is constant with angular scale and goes as

$D^2 \theta^{-2}$. Of the total speed gain, then, a factor of 14 is due to the larger aperture and 4 comes from the assumed better seeing—one more graphic example of how important good seeing is to telescope performance. At higher frequencies, the speed gain becomes infinite, and information is sensed by the higher diffraction limit of the 15-m that is not available at all to the smaller telescope. The useful angular frequency limit $(\sim\lambda/D)$[15] is 0.07 arcsec at 5 μ for the 15-m compared to 0.26 arcsec for the 4-m. Over the entire range of 0.1 to 1 arcsec, the 15-m is faster than the 4-m by a factor of $\gtrsim 80$. This gain comes in a particularly useful frequency range and is expected to open up many exciting new avenues of investigation in infrared astronomy, as discussed below.

Because of the primitive detection methods in use up to now, the full power of IR speckle is still several years away, and present work is concerned largely with sources

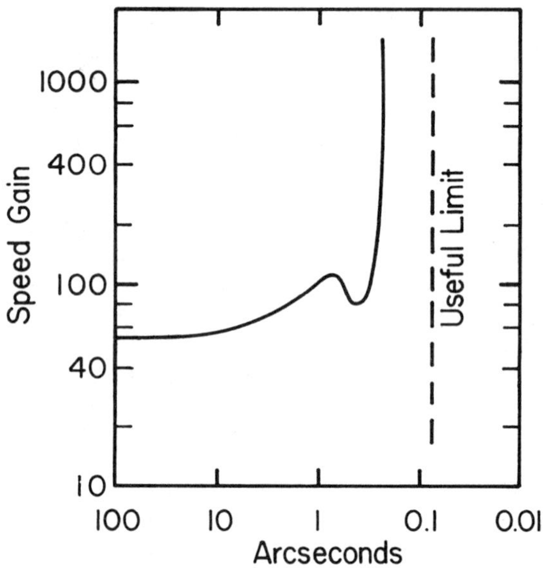

FIGURE 1. The speed gain in measuring Fourier amplitudes at various spatial frequencies is shown at 5 μ for a 15-m telescope compared to a 4-m telescope. The seeing at 5,000 Å is assumed to be 1 arcsec at the 4-m telescope and 0.5 arcsec at the 15-m.

where the geometry may be assumed to be simple. A recent important breakthrough is the discovery that T Tau is an IR double star[8] with a separation of only 140 AU. This duplicity has important implications for the formation of the small IR companion in the strong tidal field of T Tau. The usual isolated stellar collapse models no longer apply, and one chain of argument suggests that the mass of the companion must be less than 0.02 $M_{\odot}$.[11] If this first example is any guide, the discovery of more low-luminosity doubles among T Tauri stars could be important in telling us why binary stars form in some cases instead of planets.

Another exciting area in the infrared is the investigation of material distributed anisotropically around protostellar objects. Shrouded in dense cocoons of dust and gas, protostars radiate strongly at infrared wavelengths. Recent developments have shown that many such protostars are shedding mass in the form of a rapidly flowing, bipolar stellar wind (see Reference 28 for a comprehensive review). It is widely speculated that the bipolar nature of these winds is caused by a dense torus, or disk, of dust and gas, which girdles the equatorial regions of the protostar. This disk somehow channels the outflow into two symmetric polar cones, although exactly how this happens is still not yet fully agreed upon (e.g., References 12 and 14).

Indirect evidence for such equatorial disks comes also from "cometary nebulae," many of which are associated with protostars.[3] Cometary nebulae are fan-shaped reflection nebulae in which rapid mass loss is occurring from an embedded protostellar object near the apex of the fan (e.g., R Mon).[13] Cohen[5] and others have suggested that cometary nebulae are actually symmetric bipolar nebulae in which the far-side outflow is obscured by an optically thick, equatorial dust girdle. CO observations provide direct evidence for such a dense disk around R Mon and the bipolar nebula S106.[4,1] However, the angular resolution of these CO observations is low, and the structures seen are much too far from the star to constrain the central flow near the star itself.

The possibility of disks around protostars raises many intriguing theoretical questions. Disks form naturally in models of rotating, collapsing clouds and are probably crucial in solving the angular momentum problem for stellar collapse (e.g., Reference 2). The presence or absence of a disk may ultimately govern whether a star becomes single or multiple. There is also the obvious fact that our own planetary system is thought to have originated from such a flattened cloud surrounding the youthful sun. Thus, in observing other disks we may be witnessing other solar systems in the process of formation.

Theoretical models of such nebulae indicate temperatures of a few hundred degrees and sizes from a few to a thousand AU.[16] Typical infrared angular diameters are predicted to be 0.1 arcsec to 1 arcsec for nearby sources. Similar estimates for protostellar diameters are obtained under the simple assumption that the observed flux from bright IR sources is black-body radiation.[28] As we have seen, this is just the angular regime where a 15-m telescope offers large gains in speed and resolving power over existing instruments. At the same time, the higher photon flux of the big telescope will allow a happy melding of speckle interferometry and spectroscopy, bringing dynamical analysis of the intimate protostellar environment also within reach.

To summarize, very general considerations indicate that we are now on the threshold of resolving many protostellar sources and that just a factor of a few increase in resolving power will bring many within our grasp. Furthermore, there is every expectation that higher resolution will reveal many asymmetrical disklike structures that relate in an essential way to stellar rotation, binary and planetary formation. The infrared view through a 15-m telescope should be fascinating indeed.

In the brief space available here, I have mentioned only a few of the many areas where large ground-based optical telescopes will have enormous scientific impact. I have purposely concentrated on issues of interest to participants at this meeting—hence the emphasis on cosmology, the early universe, and galaxy formation. These few topics serve to demonstrate the versatility of large optical-IR telescopes. Like their smaller cousins of today, these telescopes will be applied to problems spanning the whole gamut of modern astronomy. Their large photon-gathering capacity will allow us to bring to bear the tremendous power of spectroscopic analysis—at low resolution, applied to ever more distant objects at the limits of the visible universe and, at high resolution, to brighter objects of fascinating spectral complexity in our own and neighboring galaxies. In short, large optical-IR telescopes offer not only the flexibility to study a large array of fundamental problems which we can now clearly foresee but also, more importantly, the versatility to reveal those many serendipitous discoveries through which astronomy has traditionally made its most spectacular advances.

SUMMARY

The scientific opportunities of very large optical-IR telescopes are illustrated via comparison of a proposed national 15-m telescope with existing 4-m-class telescopes. In the optical region, the 15-m gives a factor of 14 gain in speed and a gain of 1.4 to 2.9 in limiting magnitude, depending on the type of observation. In the diffraction-limited infrared, the sensitivity limit for point sources is increased by 2.9 mag, and the speed gain is nearly 200. The increase in angular resolving power is a factor of 3.5 to 5, depending on the geometrical configuration of the primary mirror.

Four sample areas of research are chosen to illustrate what these gains mean in terms of scientific productivity. The topics selected include lookback studies of galaxies and galaxy clustering, the distribution of nonluminous matter in the universe, spectroscopic studies of QSOs, and the structure of infrared protostars. These are all areas in which fundamental information is just out of the reach of existing telescopes and for which the expected gains in speed and angular resolving power would have great impact. Large optical-IR ground based telescopes appear to constitute versatile and cost-effective additions to the array of existing astronomical instrumentation.

REFERENCES

1. Bally, J. & N. Z. Scoville. 1982. Astrophys. J. **255:** 497.
2. Bodenheimer, P. H. & D. C. Black. 1979. Numerical calculations of protostellar hydrodynamic collapse. *In* Protostars and Planets. T. Gehrels, Ed.: 288. University of Arizona Press. Tucson, Ariz.
3. Calvet, N. & M. Cohen. 1978. Mon. Not. R. Astron. Soc. **182:** 687.
4. Canto, J., L. F. Rodriguez, J. F. Barrall & P. Carral. 1981. Astrophys. J. **244:** 102.
5. Cohen, M. 1974. Pub. Astron. Soc. Pac. **86:** 813.
6. Dressler, A. & J. E. Gunn. 1982. Astrophys. J. **263:** 533.
7. Doroshkevich, A. G., M. Yu. Khlopov, R. A. Sunyaev, A. S. Szalay & Ya. B. Zeldovich. 1981. Ann. N.Y. Acad. Sci. **375:** 32.
8. Dyck, H. M., T. Simon & B. Zuckerman. 1982. Astrophys. J. (Lett.) **255:** L103.
9. Faber, S. M. 1979. The Scientific Case for a 10-Meter Telescope. (Unpublished report to the director of Lick Observatory.)
10. Faber, S. M. 1980. The scientific case for a 10-meter telescope. *In* Optical and Infrared Telescopes of the 1990's: 304. Kitt Peak National Observatory. Tucson, Ariz.
11. Hanson, R. B., B. F. Jones & D. N. C. Lin. 1983. Astrophys. J. Lett. **270:** L27–L30.
12. Icke, V. 1981. Astrophys. J. **247:** 152.
13. Jones, B. & G. H. Herbig. 1982. Astron. J. **87:** 1223.
14. Königl, A. 1982. Astrophys. J. **261:** 115.
15. Korff, D. 1973. J. Opt. Soc. Am. **63:** 971.
16. Lin, D. N. C. The nebular origin of the solar system. *In* The Solar System. Rubey Colloquium **3.** M. Kivelson, Ed. (In press.)
17. Lin, D. N. C. & S. M. Faber. 1983. Astrophys. J. Lett. **266:** L21–L25.

18. Peterson, B. A. 1982. Private communication.
19. Sargent, W. L. W., P. J. Young, A. Boksenberg & D. Tytler. 1980. Astrophys. J. Suppl. **42:** 41.
20. Stockton, A. 1978. Astrophys. J. **223:** 747.
21. Strittmatter, P., Ed. 1980. The Scientific Case for a Very Large Aperture Ground-Based Telescope. Report to the director of Kitt Peak National Observatory. Kitt Peak, Arizona.
22. Strittmatter, P. 1980. Optical/IR observations of compact extragalactic sources using very large telescopes. *In* Optical and Infrared Telescopes of the 1990's: 329. Kitt Peak National Observatory. Tucson, Ariz.
23. Wallerstein, G. 1980. Stellar astronomy in the 1990's. *In* Optical and Infrared Telescopes of the 1990's: 342. Kitt Peak National Observatory. Tucson, Ariz.
24. Wampler, E. J. 1982. Phys. Today **35:** 44.
25. White, S. M. & M. J. Rees. 1978. Mon. Not. R. Astron. Soc. **183:** 341.
26. Woolf, N. J. 1982. Annu. Rev. Astron. Astrophys. **20:** 367.
27. Wyckoff, S., P. A. Wehinger & T. Gehren. 1981. Astrophys. J. **247:** 750.
28. Wynn-Williams, C. G. 1982. Annu. Rev. Astron. Astrophys. **20:** 587.

The 1.5-Millisecond Pulsar

D. C. BACKER

Radioastronomy Laboratory
University of California
Berkeley, California 94720

A radio pulsar spinning 20 times faster than the Crab nebula pulsar, 642 Hz, has been found in the constellation Vulpecula, not far from the location of the original pulsar 1919 + 21.[1] This pulsar discovery, which took place at the Arecibo Observatory in the fall of 1982, culminated several years of investigation of the peculiar source catalogued by Cambridge radioastronomers as 4C21.53. Along with the Crab nebula pulsar, the new pulsar, designated 1937 + 21, is a rare case of a catalogued radio source becoming a pulsar.

The radiation from 1937 + 21 arrives in two sharp peaks separated by 173° of rotational longitude, the main pulse and the interpulse. These two peaks have counterparts in the Crab pulsar. The pulse widths are roughly 10° at half-power level. These pulse widths are typical of *all* pulsars despite the ratio of maximum period to minimum period of ~2,500. In naive models the radio emission comes from the cone of open field lines whose width would decrease with the square root of the period. This scaling is not operating in going from the conventional 1-Hz pulsar to the Crab pulsar and now to the millisecond pulsar. One resolution of this inconsistency is that the main pulse-interpulse emission in the Crab and the millisecond pulsars is produced by a spin-independent mechanism located, perhaps, at the light cylinder.

Accurate timing of pulse arrivals began shortly after the discovery in November. The initial measurements indicated stability of the period down to 10^{-7}. The next phase of measurements rapidly showed stability at the level of 10^{-12} and indicated an apparent slowdown of 1×10^{-19} sec/sec.[2] Thus despite the dramatic spin rate of $\Omega \sim$ 642 Hz, the indicated spin down implied a change of only 1.5 Hz in 10^6 years! Since rotational energy losses by electromagnetic torques and by gravitational radiation scale with Ω^4 and Ω^6, respectively, this pulsar, necessarily, has extremely small dipole moments for its magnetic field and quadrupole moment of its mass distribution. The indicated spin down has considerable uncertainty owing to the 0.2 arcsec uncertainty in its astrometric position and the 0.2 arcsec uncertainty in the relationship between the planetary ephemeris and the FKIV star reference frame.

If the stability of this pulsar's spin is maintained for periods of a decade, then it will provide a data base for the best determination of ephemeris time. Orbits of the Earth and other solar system planets could be refined by monitoring this "one-way" radar signal. Realization of this program will require long-term correction of Earth-based proper, or atomic, time to coordinate time in the solar system barycenter. This correction will provide a new test of the equivalence principle. Finally, the accuracy with which the new pulsar can be timed will allow a new search for the effects of gravitational waves on a wide range of time scales.

Just north of the pulsar is an H II region which confused the early investigations of 4C21.53. Recent detection of extensive absorption of interstellar neutral hydrogen allows us to place this object on the other side of the galaxy far from the pulsar whose distance is estimated from its dispersion measure to be a few kpc only.

The rotational energy of the new pulsar is $\sim 10^{52}$ ergs, comparable to the entire mechanical energy release of a supernova event. Yet the low spin-down rate and the absence of debris in the pulsar vicinity (from radio to x-ray wavelengths) indicate that the object was not formed recently. The rapid spin places this pulsar at the brink of instability in conventional neutron star models. However, since its spin down is so slow, there is no difficulty with its initial spin; with an age of 10^7 years its initial spin would only be 657 Hz. The location of 1937 + 21 near the galactic plane suggests an age not much greater than 10^6 years assuming a pulsarlike proper motion of 100 km s^{-1}.

Models for the origin of the 1.5-millisecond pulsar fall into two classes: those that assume all pulsars have 10^{12} gauss surface dipole fields at birth, and those that allow neutron stars to form with low magnetic fields. In the latter case, the pulsar spins up in the initial core collapse, has small electromagnetic torque for slowdown, and lives peacefully as a pulsar for 10^8 years or so until the particle acceleration giving rise to radio emission drops below threshold. In the former case, a binary evolution scenario is required. During the first stage a rapidly rotating, high-field neutron star forms, its surface field decays to 10^9 gauss over 10^7 years, and its spin decays to ~ 1 Hz level. Mass transfer from the now evolved secondary begins, creating an accreting x-ray source and spinning the neutron star back to kHz values. The secondary passes into supernova stage, disrupting the binary and setting the original neutron star free to be observed as a low-field, high-spin pulsar. The binary radio pulsars lend support to this scenario since they also have small values of $\dot{P}$ when compared to the average pulsar.

Searches for new members of the millisecond pulsar "class" are under way using a variety of techniques. The sensitivity of previous all-sky pulsar searches was reduced below periods of ~ 100 msec; searches will be extended down to 1 msec. The trail to 1937 + 21 was blazed by observations of interplanetary scintillation, linearly polarized flux, and interstellar scintillation. The sky can be surveyed for radio sources with similar properties.

REFERENCES

1. Backer, D. C., S. R. Kulkarni, C. Heiles, M. M. Davis & W. M. Goss. 1982. Nature **300:** 615.
2. Backer, D. C., S. R. Kulkarni & J. H. Taylor. 1983. Nature **301:** 314.

The Millisecond Pulsar[a]

ANDREW F. CHENG[b]

Department of Physics and Astronomy
Rutgers University
Piscataway, New Jersey 08854

INTRODUCTION

Backer *et al.* have reported the exciting discovery of a 1.56-ms radio pulsar,[1] whose period is 21 times smaller than that of the Crab pulsar, hitherto the fastest known. The neutron star in this object, now designated PSR 1937+214, is rotating at nearly the maximum possible rate, with an equatorial rotation speed more than one-tenth the speed of light. Despite its extremely rapid rotation, the millisecond pulsar is not detected as either an x-ray source or a γ-ray source.[1] The distance to the pulsar, inferred from its dispersion measure, is $\sim$2.3 kpc, which is comparable to that of the Crab pulsar. Thus the x-ray and γ-ray luminosities of 1937+214 must be far less than those of the Crab. In addition, a star of red magnitude $\sim$20 has been detected at the pulsar position,[2] but no optical pulsations have been detected.[3] As the pulsar is near the galactic plane, the chance that a normal star would be randomly superposed on the pulsar position is not negligible. A radio nebula, identified as an H II region, is also observed near the pulsar position and may or may not be physically associated.[1,4] Finally, the millisecond pulsar is not associated with a supernova remnant.[1]

These observations require PSR 1937+214 to have a magnetic field much smaller than in typical radio pulsars. In that case its spin-down time would not be preposterously short, and it would not be expected to be a strong source of pulsed x-rays and γ-rays. With a magnetic field like that of the Crab, its spin-down time would be $\sim$5 yr.

Moreover, PSR 1937+214 cannot be a very young pulsar, despite its extremely short period.[5] If 1937+214 were formed in a supernova, then the supernova must have occurred at least 10^4 years ago, because the pulsar is not now associated with a supernova remnant and it is not now observable as a thermal x-ray source. A purely statistical argument also shows that 1937+214 cannot be very young. If it were, then our observation of it would be very improbable unless the birthrate were very high. If millisecond pulsars rapidly slow down to become Crab-like pulsars, it becomes very puzzling why only one pulsar like the Crab is now known.[5]

FORMATION OF MILLISECOND PULSARS

Alpar *et al.* recently proposed that millisecond pulsars can be formed in binary systems with accretion disks,[5] in which the progenitor was spun up by accretion torques.[6,7] This

[a]This work was supported by the U.S. Department of Energy.
[b]Present affiliation: Applied Physics Laboratory, Johns Hopkins University, Laurel, Maryland 20707.

model is attractive because (1) it allows 1937+214 to be billions of years old, and (2) it explains why 1937+214 has an abnormally low magnetic field.

A magnetic field much less than 10^{12} G is required for spin up of a neutron star to millisecond periods in an accretion disk. This is because the equilibrium period at which the accretion torque vanishes is roughly the Kepler period at the Alfvén radius. Alpar *et al.* found that the equilibrium period can be as short as a millisecond,[5] if the neutron star surface magnetic field is $\sim 10^8$ G. Sufficient angular momentum is transferred for spin up of a neutron star to millisecond periods, if accretion near the Eddington limit can be maintained for $\sim 10^8$ yr.

Thus Alpar *et al.* were able to predict the magnetic field, and hence the $\dot{P}$, of a millisecond pulsar spun up by an accretion disk:[5]

$$\dot{P} \approx 3 \times 10^{-20} I_{45}^{-1} (P_{\rm ms})^{4/3} M^{5/3} \dot{m}_{17} \quad \textbf{(1)}$$

where I_{45} is the moment of inertia in 10^{45} g cm^2, $P_{\rm ms}$ is the period in ms, M is the mass in solar masses, and $\dot{m}_{17}$ is the accretion rate in 10^{17} g s^{-1}. This predicted $\dot{P}$ agrees well with the value subsequently observed for PSR 1937+214, which is 1.3×10^{-19} ss^{-1}.[8] Indeed, Backus *et al.* have already noted that the observed P and $\dot{P}$ of all three known binary radio pulsars are also consistent with **(1)**.[9] Thus it appears that 1937+214 and the three known binary radio pulsars are members of a new class spun up by accretion disks. The vast majority of known radio pulsars cannot be members of this new class, because their magnetic fields are so great that they could not have been spun up even to their present periods by accretion disks.

FIGURE 1 gives the $\dot{P}$,P distribution for a large sample of radio pulsars.[5] The vast majority of radio pulsars occupy an "inverted triangle" with $P \sim 1$ s and $\dot{P} \sim 10^{-15}$ ss^{-1}. The value of $\dot{P}$ given by **(1)** is plotted in FIGURE 1. This $\dot{P}$ is the *initial* value for a radio pulsar spun up by an accretion disk, since it applies immediately after the accretion rate drops to a sufficiently low value to allow radio pulsar action to begin. Subsequently, the pulsar is slowed by electromagnetic braking torques (at least), so it evolves downward and to the right in FIGURE 1. Thus accretion spin-up pulsars are expected to occupy the crosshatched region in FIGURE 1. FIGURE 1 also gives the theoretical P, $\dot{P}$ relations for accretion spin-up pulsars after various elapsed times assuming a constant braking index $n = 3$. FIGURE 1 shows that the three binary radio pulsars and the millisecond pulsar lie near or within the crosshatched region predicted for accretion spin-up pulsars, whereas the vast majority of radio pulsars do not. Therefore, most radio pulsars cannot have been spun up by accretion disks.

Available polar cap pulsar models[10,11] appear to be able to accommodate the millisecond pulsar despite its extremely small P and $\dot{P}$. These models typically predict cessation of radio pulsar action when (roughly) $B_{12}P^{-2} \lesssim 0.1$, where B_{12} is the surface magnetic field in 10^{12} G. For the millisecond pulsar, the observed P and $\dot{P}$ yield $B_{12}P^{-2} \sim 200$, only a factor of two smaller than for the Vela pulsar. Indeed, if the pulsed γ-ray luminosity of 1937+214 were less than or comparable to that of the Vela pulsar, it would not be detectable at its $\sim$2.3 kpc distance. However, difficulties may be encountered in models for coherent radio emission that invoke two stream instabilities in electron-positron pair plasma, as the relativistic plasma frequency may not fall below the radio frequency within the light cylinder.[5] This difficulty may not be encountered if the radio emission occurs from interpenetrating ion streams.[12]

Finally Alpar *et al.* suggested that the progenitor system for the millisecond pulsar

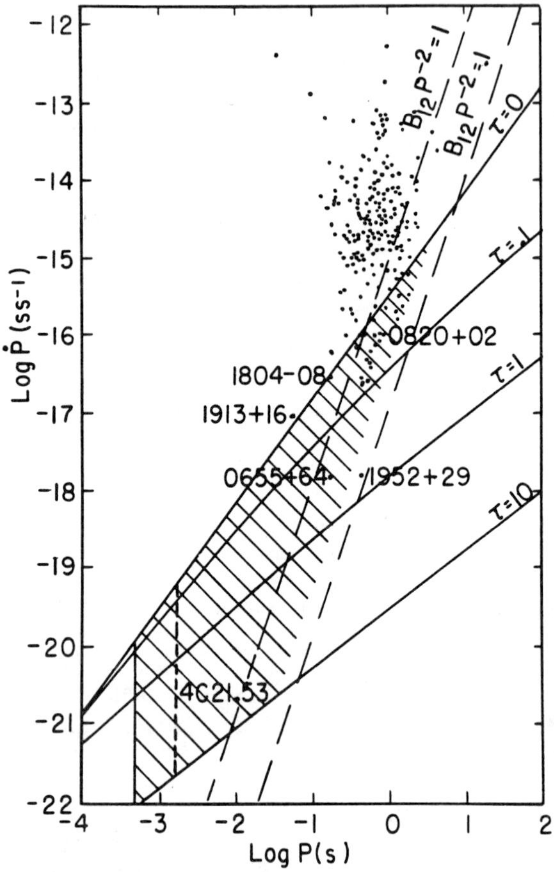

FIGURE 1. The $\dot{P}$-P diagram for pulsars. Constant age curves for the new class of accretion spun-up pulsars are shown with $\dot{P}$ values given in units of $I_{45}^{-1}\,(M/M_{\odot})^{5/3}\dot{m}_{17}\mathrm{ss}^{-1}$. Their pulsar age τ is given in units of $(10^9\,\mathrm{yr})I_{45}\,(M/M_{\odot})^{-5/3}\dot{m}_{17}^{-1}$. On hitting the appropriate $B_{12}P^{-2} = \eta$ line ($\dot{P}$ is in units of $R_6^6 I_{45}^{-1}\mathrm{ss}^{-1}$ on these lines), it will turn off as a pulsar. Thus, all pulsars of the new class should lie in a region similar to the shaded region of the diagram, bounded by the $\tau = 0$ line, the pulsar turn-off line, the $\tau \sim 10$ line, and the line of rotational breakup of neutron stars. These pulsars are confined to that region even when the prebirth spin up did not reach the $\tau = 0$ curve. (Reproduced from Reference 5, by permission from *Nature*. Copyright 1982 Macmillan Journals Limited.)

was a low-mass x-ray binary.[5] In such systems, a disk accretion phase with accretion rates not far below the Eddington limit may last approximately 10^9 yr, allowing spin up of a neutron star to millisecond periods. If such a system evolves to become a neutron star with a degenerate companion, gravitational radiation would cause the orbit to decay, leading ultimately to tidal disruption of the companion. If millisecond pulsars can indeed be made in low-mass x-ray binaries, it may be possible to find millisecond x-ray pulsars among the galactic bulge sources.[5,13] Finally, such scenarios predict that the number of millisecond radio pulsars currently observable is of the order of unity.[5,13]

Alternative evolutionary scenarios may be viable for formation of radio pulsars spun up by accretion disks. In the intermediate-mass x-ray binaries, where the component masses are $\sim$7 to 8 $M_\odot$, a first supernova may form a pulsar which accretes from a disk, and a second supernova may or may not disrupt the binary. In this scenario the disk accretion phase may not last much longer than 2×10^7 yr, so it may be difficult to make millisecond pulsars, but binary radio pulsars with $P \gtrsim 100$ ms might not pose a problem. Another possibility is that the progenitor was a white dwarf binary. If a "nonmagnetic" white dwarf with $B \sim 10^4$ G is spun up in an accretion disk to $P \sim 100$ s, and if it collapsed to a neutron star, it could become an object like the millisecond pulsar.

SUMMARY

Alpar *et al.* have suggested that the millisecond pulsar, like the three known binary radio pulsars, was spun up by an accretion disk in a binary system.[5] The galactic bulge x-ray sources may contain millisecond x-ray pulsars.

REFERENCES

1. Backer, D. H. C., S. Kulkarni, C. Heiles, M. Davis & W. Goss. 1982. Nature **300:** 615.
2. Djorgovski, S. 1982. Nature **300:** 618.
3. Backer, D. C. Ann. N.Y. Acad. Sci. (This volume.)
4. Arons, J. Ann. N.Y. Acad. Sci. (This volume.)
5. Alpar, A., A. F. Cheng, M. A. Ruderman & J. Shaham. 1982. Nature **300:** 728.
6. Davidson, K. & J. Ostriker. 1973. Astrophys. J. **179:** 585.
7. Ghosh, P. & F. K. Lamb. 1979. Astrophys. J. **234:** 296.
8. Taylor, J. H. Private communication.
9. Backus, P., J. Taylor & M. Damashek. 1982. Astrophys. J. **255:** L63.
10. Cheng, A. F. & M. A. Ruderman. 1980. Astrophys. J. **235:** 576.
11. Arons, J. 1981. Astrophys. J. **248:** 1099.
12. Ruderman, M. A. 1981. Int. Astron. Union Symp. **95:** 87.
13. Fabian, A., J. Pringle, F. Verbunt & R. Wade. 1983. Nature **301:** 222.

Models of Type I Supernovae and Observations of SN 1981b[a]

DAVID BRANCH

Department of Physics and Astronomy
University of Oklahoma
Norman, Oklahoma 73019

INTRODUCTION

Type II supernovae usually occur in the arms of spiral galaxies, and their optical spectra contain strong hydrogen lines. They are believed to be the explosions of massive stars ($\gtrsim 8\ M_\odot$) having extended ($\gtrsim 10^{14}$ cm) hydrogen-rich envelopes, i.e., red supergiants. Gravitational energy is released by core collapse, and a shock wave (somehow) ejects the envelope. As the heated, extended envelope expands and cools, diffusive release of the shock energy accounts for the type II light curve.

Observations of type I supernovae (SN I) do not point so unambiguously to the identities of the progenitors. SN I appear in elliptical and irregular galaxies as well as in the disk population of spirals, and their spectra indicate that hydrogen is deficient if not absent completely. If SN I are the explosions of single stars, their initial masses probably are in the range 5–8 $M_\odot$, because more massive stars become type II supernovae and less massive ones eject their envelopes as planetary nebulae and settle down as stable white dwarfs. But there are observational[1] and theoretical[2] reasons to believe that 5–8 $M_\odot$ stars also become planetary nebulae and white dwarfs, there are theoretical reasons to doubt that single stars can lose their hydrogen and still evolve to an explosive configuration, and there are strong observational reasons to doubt that elliptical galaxies contain many, if any, 5–8 $M_\odot$ stars. Increasing attention is being given, therefore, to the idea that an SN I occurs when mass transfer from a close companion in a binary system provokes a white dwarf to explode.

The SN I light curve shows a 15-day rise to maximum light, an initial decline of about 3 magnitudes in 30 days, and then the famous exponential tail corresponding to a half-life of about 60 days. Lasher showed that the early peak of the SN I light curve could be explained, as for type IIs, by the diffusive release of shock energy deposited in an extended envelope.[3] For SN I, an envelope containing about 2 $M_\odot$ of helium and approaching 10^{14} cm in extent would be required. But the tail of the SN I light curve can not be explained this way, so some delayed input of energy is required. If the progenitor is a compact object such as a white dwarf, adiabatic cooling of the expanding ejecta will prevent the shock energy from contributing even to the peak of the light curve. Recently, theoretical light curves have been calculated on the assumption that the energy source is from the radioactive decay of Ni^{56} and its unstable daughter, Co^{56}, to stable Fe^{56}.[4–6] Within the uncertainties involved in comparing the theoretical and observational light curves, such as the opacity of the gas and the

[a]Supported by National Science Foundation Grant AST 7808672.

transformations from observed to bolometric magnitudes, radioactivity is able to account for the entire light curve. The peak is powered by trapped and thermalized gamma rays from the 6-day Ni^{56} decay, and during the tail the progressive escape of the Co^{56} gammas from the expanding shell transforms the 77-day Co^{56} decay to an effective 60-day half-life for the release of optical radiation. But, within the uncertainties mentioned above, it is possible that shock energy from an extended envelope also makes a contribution to the light curve peak. Theoretical light curves including both energy sources have been computed by Sutherland and Wheeler.[7]

Further crucial support for the Ni^{56} radioactivity hypothesis comes from Axelrod's interpretation of the late-time (>200 days) spectra of SN I.[8,9] Axelrod assumes that the ejected shell is initially composed of pure Ni^{56}. For late times, when the shell has become optically thin to its own continuum, he calculates the heating of the gas by the radioactivity decay products, the non-LTE (local thermodynamic equilibrium) level populations, and theoretical optical spectra which consist primarily of collisionally excited forbidden emission lines of iron ions. The theoretical spectra nicely match many of the features observed in the late-time spectra of SN 1972e in NGC 5253.[10] One of the matching features is identified with an emission line of [Co III], decaying at the predicted rate.

The recent modeling of the light curves has involved initial Ni^{56} masses in the range 0.5–1.4 $M_\odot$, motivated primarily by the tendency for evolutionary models of SN I to completely disrupt and synthesize such large amounts of Ni^{56}. Also, the failure of the Einstein Observatory to detect thermal x-ray point sources at the centers of galactic remnants of SN I implies that SN I do completely disrupt, barring anomalously rapid cooling of newborn neutron stars.[11] Given complete disruption, all of the kinetic energy must come from the release of thermonuclear energy. Studies of the early-time spectra show that near the light curve peak the photosphere is embedded in material moving about 11,000 km/sec and that some line formation occurs in material moving as fast as 20,000 km/sec. To accelerate enough material to these velocities, as well as to provide an estimated total kinetic energy of 10^{51} ergs, at least 0.7 $M_\odot$ of Ni^{56} must be synthesized.[12]

The ejection of such large amounts of Ni^{56} would have several implications. One of them is a relatively low value of the Hubble constant. From observations, the peak luminosity of an SN I beyond the local supercluster can be expressed in terms of H_0, and from theory the peak luminosity can be expressed to first order in terms of the Ni^{56} mass and the light-curve rise time, so the Hubble constant can be expressed approximately in terms of the Ni^{56} mass $H \approx 40\,(M_{Ni}/M_\odot)^{-1/2}$ km/sec per Mpc. If $M_{Ni} \gtrsim 0.7\, M_\odot$, then $H_0 \lesssim 50$ km/sec per Mpc. For $H_0 = 100$ km/sec per Mpc, $M_{NI} \approx 0.16\, M_\odot$. This would accelerate a negligible amount of material to 11,000 km/sec,[12] and would lead to a total kinetic energy of less than 10^{50} ergs.[13] A Hubble constant of 100 km/sec per Mpc apparently is incompatible with completely disrupting white dwarf models; a neutron star remnant would be required to provide the kinetic energy. Another implication would be that current estimates of the maximum amount of iron ejected per SN I, based on chemical evolution models of the Galaxy, are in error by a factor of several. For example, Twarog and Wheeler find an upper limit of 0.3 $M_\odot$ of iron,[14] even if SN I have synthesized all of the iron in the local solar neighborhood. Finally, the x-ray spectra of galactic remnants of SN I indicate that only a small fraction of the ejected matter that has been heated to x-ray emitting temperatures is in the form of iron.[15,16] This is not a strong constraint on the total mass of iron produced; Wu *et al.* interpret broad interstellar Fe II absorptions in the ultraviolet spectrum of a star behind the remnant of SN 1006 as evidence for the presence of at least 0.03 $M_\odot$ of ejected iron that has not yet been heated to x-ray temperatures.[17] The total amount of unheated iron could be much greater.

MODELS OF SN I

A review of all recent models for type I supernovae has been given by Wheeler.[7] Only the three that appear to be natural, well-motivated models, as far as stellar evolution is concerned, are discussed here. Two of these arise from the accreting white dwarf scenario, and the third could be the result of single-star evolution. All three of these models disrupt completely, synthesize on the order of 1 $M_\odot$ of Ni^{56}, and provide on the order of 10^{51} ergs of kinetic energy.

The evolution of accreting carbon-oxygen white dwarfs, for ranges of accretion rates and initial masses, has been studied by Fujimoto and Taam.[18] Their results are summarized in FIGURE 1. For very rapid accretion rates, $\sim 10^{-6}$ $M_\odot$/year, the white dwarf cannot accommodate the accreted matter at its surface, it begins to develop an extended envelope, and the outcome is uncertain. For less extreme rates the evolution can be followed to its outcome.

For rapid accretion rates of $\sim 10^{-7}$ $M_\odot$/year, the accreted hydrogen burns to helium near the surface of the white dwarf, and the helium burns to carbon and oxygen. The white dwarf grows in mass until it approaches 1.4 $M_\odot$, when the ignition of degenerate carbon at its center drives a (subsonic) deflagration wave which completely disrupts the star. The details of the outcome depend on the velocity of the deflagration wave, which is difficult to calculate self-consistently and has therefore been treated as a parameter. For $V_{def} = 0.2\ V_{sound}$, a velocity that leads to a kinetic energy of 1.3×10^{51} ergs, the inner 1.0 $M_\odot$ is burned to Ni^{56}, ~ 0.3 $M_\odot$ is burned to intermediate mass elements (oxygen to calcium), and the outer 0.1 $M_\odot$ is unburned carbon and oxygen.[19] Nomoto predicts that some mixing between the three composition regions will occur,[13] and Chevalier has shown that the light curve predicted on the basis of this model is in satisfactory agreement with observation.[6]

For more moderate accretion rates of $\sim 10^{-8}$ $M_\odot$/year, the accreted hydrogen burns to helium but the helium collects in a thick shell. Eventually, when degenerate helium ignites at the base of its shell, (supersonic) detonation waves are driven inward through the carbon and oxygen and outward through the helium. Again the star disrupts, but owing to the high specific energy release of helium fusion, practically all of the star is incinerated to Ni^{56}.[20,21] The light curve for this model also is consistent with observation.[5]

For slower accretion rates of $\sim 10^{-9}$ $M_\odot$/year, Fujimoto and Taam find that the accreted hydrogen burns in shell flashes which are strong enough to eject most of the matter accreted between the flashes, presumably making classical novae in the process. The net accretion rate is therefore very low, and supernovae are not expected to occur, because the time required to provoke the white dwarf to explode exceeds 10^{10} years and/or because the total amount of matter required from the companion exceeds a solar mass.

A model that could arise from single-star evolution has been proposed by

Nomoto.[22] If a star initially in the 6–8 $M_\odot$ range manages to lose its hydrogen during the core helium burning phase, it will become a helium star of about 2 $M_\odot$. Helium stars of this mass develop carbon-oxygen cores which eventually deflagrate inside extended helium envelopes. Thus the light curves may have a contribution from shock energy in the envelope, in addition to radioactivity.

These three models are rather similar in terms of the amount of Ni^{56} and kinetic

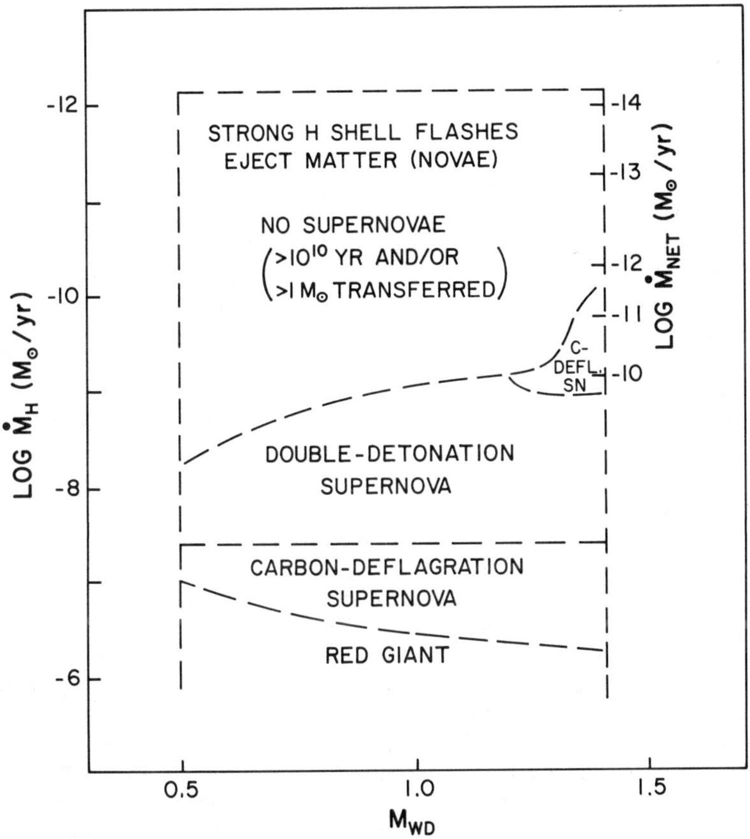

FIGURE 1. Adapted from Figure 1 of Fujimoto and Taam.[18] The predicted outcome of accreting carbon-oxygen white dwarfs for various initial white dwarf masses and hydrogen accretion rates is shown. The net accretion rate for white dwarfs that eject some of the accreted matter by hydrogen shell flashes is shown at the upper right.

energy produced, but their exterior compositions are very different. The white dwarf carbon-deflagration model has mainly intermediate-mass elements mixed with some unburned carbon and oxygen and some mass-56 isotopes (Ni, Co, and Fe). The white dwarf double-detonation model is almost pure mass-56. The helium star should have a helium-rich exterior with cosmic abundances of the heavy elements except for carbon and oxygen, which would have been converted to nitrogen by prior processing by the CNO cycle. The spectra of these three models should appear very different.

SN I comprise a well-defined class observationally. Although the question of whether real differences, rather than just observational errors, have been detected remains controversial, SN I spectra are sufficiently homogeneous that only one (or none) of the models discussed here can contribute to the class. The only loophole is that if the exterior of the helium-star model is optically thin as early as maximum light, we would see into the deflagrated core and the helium star might be hard to distinguish spectroscopically from the carbon-deflagration white dwarf.

OBSERVATIONS OF SN 1981b IN NGC 4536

Observations of SN 1981b provide the most detailed information yet available for the first hundred days of evolution of a type I supernova. Comparison with the well-observed SN 1972e in NGC 5253 shows the two supernovae to have been practically identical photometrically and spectroscopically. Optical[23] and infrared[24] photometry of SN 1981b have been combined and calibrated to make flux distributions covering the first 80 days after maximum light.[25] At maximum light the data from the B-band out to the K-band at 2.2 microns fit a 20,000 K blackbody curve. Satellite observations show that there is an ultraviolet deficiency which increases towards shorter wavelengths; only a small fraction of the flux is radiated in the UV. At later times the data remain generally consistent with ultraviolet-deficient blackbodies. The color temperature decreased to a minimum of 6500 K by 30 days after maximum light, and then slowly increased. The evolution of the angular radius of the photosphere can be derived from the color temperatures and the light curve. The photosphere expanded until 30 days after maximum light, and then slowly receded. This behavior is easily understood. Initially the photosphere is embedded in the steep density gradient at the outer edge of the ejected shell, and it almost keeps up with the expanding matter. Eventually, due to geometrical dilution of the expanding shell, the photosphere recedes into the shallower density gradient inside the shell, and it then shrinks as matter flows through it. Assuming that the optical emissivity at maximum light is that of a 20,000 K blackbody, the distance to the supernova can be estimated by matching the photometric and spectroscopic angular radii at that time. The spectroscopic radius is given by the velocity at the photosphere as determined by spectral synthesis (12,000 km/sec) times the light curve rise time (15 days) divided by the distance. The distance required is 28 megaparsecs, which is a large distance for this galaxy. NGC 4536 is apparently part of the Virgo cluster complex although not in the cluster itself. The corresponding peak absolute magnitude is $M_B = -20.7$, which in turn corresponds to $H = 32$ km/sec per Mpc via calibration of the Hubble diagram for SN I beyond the supercluster. This result certainly is consistent with the synthesis of a solar mass or more of Ni^{56}, but it is dependent on the assumption that the emissivity is that of a blackbody at the color temperature. In any case, the analysis of the photometry is not sufficiently precise to distinguish between the three models discussed above.

The maximum-light spectrum of SN 1981b has been discussed by Branch *et al.*[26] The spectrum was well matched by a synthetic spectrum consisting of blended resonant scattering features of Ca II, Si II, S II, Mg II, and O I on a thermal continuum.

According to LTE estimates these elements appeared to be present in roughly solar proportions. The velocity at the photosphere was found to be 12,000 km/sec. The presence of these elements at this velocity rules out the double-detonation model. Branch *et al.* also found that the introduction of significant amounts of Co II detracted from the spectral fit, but with a new, exact treatment of P Cygni line blending following the work of Olson for blended doublets,[27] Co II can be included without destroying the fit. In fact, Co II lines near 3500 Å provide a fit to a feature for which no other identification had been made (FIGURE 2). The maximum-light spectrum now appears to be entirely consistent with the carbon-deflagration model, and there is some

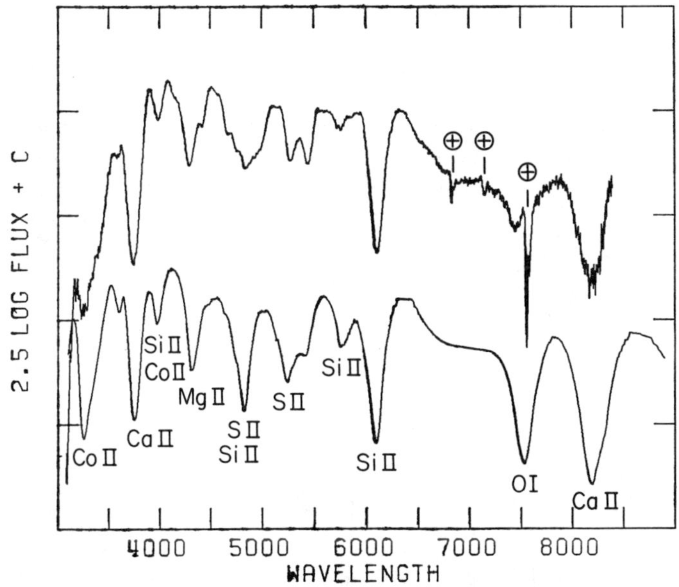

FIGURE 2. The maximum-light spectrum of SN 1981b in NGC 4536 (upper spectrum) is compared to a resonant-scattering synthetic spectrum. The relative abundances of intermediate-mass elements (O–Ca) are assumed to be approximately solar. The relative abundance of iron is also assumed to be solar, and the cobalt-to-iron ratio is taken to be 11, as expected if the cobalt and iron had been synthesized as Ni^{56} at the time of outburst. The density law is assumed to be r^{-7}, and the velocity is fit to be 12,000 km/sec.

weak evidence for the presence of Co II lines. There is little evidence for the presence of helium, and the apparent presence of oxygen is an argument against the helium-star model, unless the envelope is optically thin.

During the first 30 days after maximum light, while the photosphere continued to expand, the red part of the spectrum hardly changed but features appeared in the blue which can be matched by blended Fe II lines.[26] LTE calculations suggest that iron was overabundant relative to the other elements represented at this phase, which would indicate that freshly synthesized iron had been exposed as the photosphere receded into material moving slower than 12,000 km/sec. With the exact blending procedure, Co II lines can be included at their expected strengths relative to iron without destroying the spectrum fit. Spectra obtained at later phases, up to 116 days after maximum light, continued to show similar features in the blue, while the red spectrum changed slowly. If the earlier spectra were dominated in the blue by blended Fe II lines the later spectra must have been also. FIGURES 3 and 4 show that the spectrum around 100 days after maximum light can be remarkably well matched by resonant scattering Fe II lines, along with Ca II and Na I. For these calculations the density law was taken from the hydrodynamical calculations of Sutherland and Wheeler for a completely disrupted white dwarf. Although LTE calculations of abundances would be unjustified for these phases, when the line formation occurred in an extended envelope around a small receded photosphere, the iron lines could barely have been produced at their observed strengths by an amount of iron equal to that contained in the sun, even if all of the iron was singly ionized. It seems likely that the spectra at this phase require that synthesized iron was being seen, but no reliable estimate of the amount of iron is available.

A spectrum obtained 270 days after maximum light is compared in FIGURE 5 to the spectra observed near 100 days after maximum. The 270-day spectrum is well matched by Axelrod's synthetic spectra, but it also shows obvious resemblances to the earlier spectra which have been interpreted here as resonant-scattering spectra. One feature

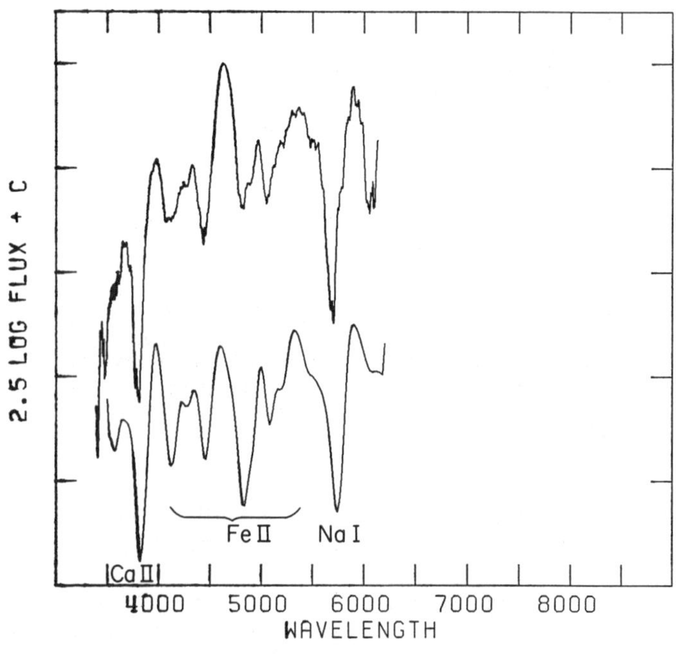

FIGURE 3. The spectrum of SN 1981b in NGC 4536 obtained 95 days after maximum light (upper spectrum) is compared to a resonant scattering spectrum containing only lines of Fe II, Ca II, and Na I. The density law is taken to be that of a disrupted white dwarf following Sutherland and Wheeler.[12]

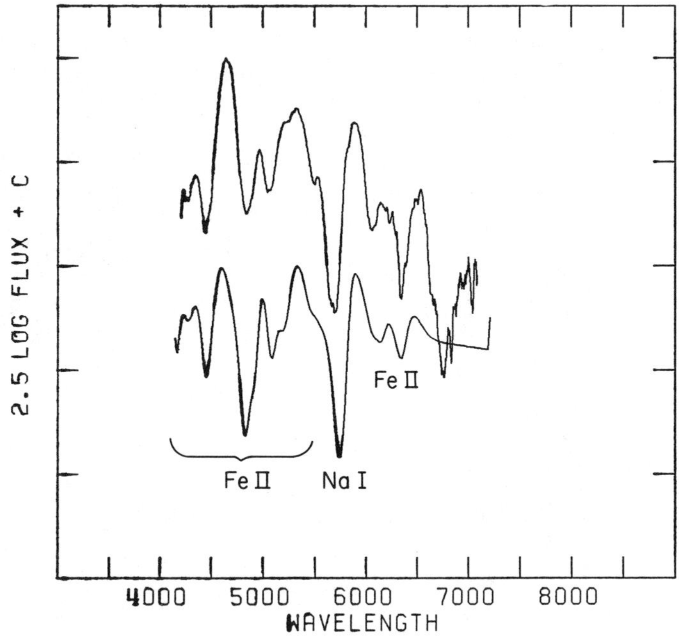

FIGURE 4. Like FIGURE 3, but for 116 days after maximum light.

that certainly seems to be an absorption feature, rather than merely a dip between two emission lines, is the one with sharp red and blue edges, centered on 3,700 Å. This feature appears to be a strengthened and broadened Ca II line. If so, the red and blue edges correspond to 8,000 and 23,000 km/sec. The sharp red edge is particularly interesting because it suggests that the Ca II abundance changes abruptly at this velocity, as expected if fairly little mixing between intermediate-mass elements and mass-56 elements had occurred.

SUMMARY

The analysis of the photometry and spectroscopy of SN 1981b is generally consistent with the carbon deflagration of an accreting white dwarf. The assumption of blackbody emissivity at maximum light leads to a high luminosity, as expected for any model that synthesizes on the order of a solar mass of Ni^{56}. The spectra establish that intermediate mass elements were present with roughly solar relative abundances in the outer ($\gtrsim$8,000 km/sec) parts of the ejecta, and suggest that freshly synthesized iron was present in the inner parts.

In addition to the three issues mentioned in the Introduction, which apply to any model that synthesizes a large amount of Ni^{56}, there are several other questions that need to be answered before the carbon-deflagration model can be accepted. First, no supernova spectrum that could correspond to the double-detonation model has ever been recorded. If the carbon deflagration of a white dwarf occurs regularly, producing SN I, why does the closely related double detonation so rarely, if ever, occur? Second, since the carbon deflagration happens as the accreting white dwarf approaches the Chandrasekhar mass, the white dwarf structure is almost uniquely defined, and SN I

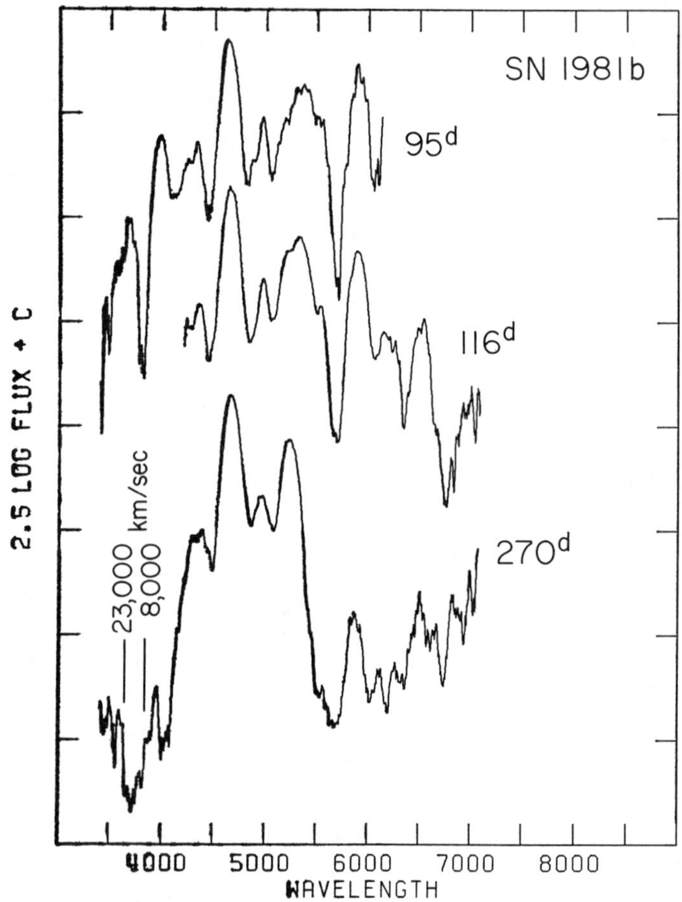

FIGURE 5. The spectrum of SN 1981b in NGC 4536 obtained by Beverley Wills 270 days after maximum light is compared to spectra obtained 95 and 116 days after maximum light. The dip near 3,700 angstroms is interpreted as an absorption feature due to the H and K lines of Ca II, extending from 8,000 to 23,000 km/sec.

might be expected to be excellent standard candles, as some workers find them to be.[28] Other recent studies have found apparent correlations between the observational properties of SN I that suggest that real differences, as well as observational errors, are present in the data.[29–32] Further precise observations are needed to establish that the variations really occur. Further theoretical studies of nonparametrized carbon deflagrations, as begun by Muller and Arnett,[33] are also needed to determine the true outcome of the deflagration, and how much variety can be produced.

ACKNOWLEDGMENTS

I am grateful to Peter Sutherland, Craig Wheeler, and many others for conversations on the origin of SN I.

REFERENCES

1. Reimers, D. & D. Koester. 1982. Astron. Astrophys. **116:** 341.
2. Tuckman, Y., N. Sack & Z. Barkat. 1978. Astrophys. J Lett. **225:** L137.
3. Lasher, G. 1975. Astrophys. J. **201:** 194.
4. Colgate, S. A., A. G. Petschek & J. T. Kriese. 1980. Astrophys. J. Lett. **237:** L81.
5. Weaver, T. A., T. S. Axelrod & S. E. Woosley. 1980. *In* Proceedings of the Texas Workshop on Type I Supernovae. J. C. Wheeler, Ed.: 113. University of Texas. Austin, Texas.
6. Chevalier, R. A. 1981. Astrophys. J. **246:** 267.
7. Wheeler, J. C. 1982. *In* Supernovae, a Survey of Current Research. M. J. Rees & R. J. Stoneham, Eds.: 167. Reidel. Dordrecht, the Netherlands.
8. Axelrod, T. S. 1980. Ph.D. Thesis. University of California. Santa Cruz, Calif.
9. Axelrod, T. S. 1980. *In* Proceedings of the Texas Workshop on Type I Supernovae. J. C. Wheeler, Ed.: 80. University of Texas. Austin, Texas.
10. Kirshner, R. P. & J. B. Oke. 1975. Astrophys. J. **200:** 574.
11. Nomoto, K. & S. Tsuruta. 1981. Astrophys. J. Lett. **250:** L19.
12. Sutherland, P. & J. C. Wheeler. 1984. Astrophys. J. (In press.)
13. Nomoto, K. 1981. *In* Fundamental Problems in the Theory of Stellar Evolution. D. Sugimoto, D. Q. Lamb & D. N. Schramm, Eds.: 295. Reidel. Dordrecht, the Netherlands.
14. Twarog, B. A. & J. C. Wheeler. 1982. Astrophys. J. **261:** 636.
15. Shull, J. M. 1982. Astrophys. J. **262:** 308.
16. Sarazin, C. L., A. J. S. Hamilton & R. A. Chevalier. 1983. *In* Supernovae Remnants and Their X-Ray Emission. I. J. Danziger & P. Gorenstein, Eds. Reidel. Dordrecht, the Netherlands. (In press.)
17. Wu, C., M. Leventhal, C. L. Sarazin & T. R. Gull. 1983. Astrophys. J. Lett. **269:** L5.
18. Fujimoto, M. Y. & R. E. Taam. 1982. Astrophys. J. **260:** 249.
19. Nomoto, K., D. Sugimoto & S. Neo. 1976. Astrophys. Space Sci. **39:** L37.
20. Woosley, S. E., T. A. Weaver & R. E. Taam. 1980. *In* Proceedings of the Texas Workshop on Type I Supernovae. J. C. Wheeler, Ed.: 90. University of Texas. Austin, Texas.
21. Nomoto, K. 1982. Astrophys. J. **257:** 780.
22. Nomoto, K. 1982. *In* Supernovae: a Survey of Current Research. M. J. Rees & R. J. Stoneham, Eds.: 205. Reidel. Dordrecht, the Netherlands.
23. Buta, R. & A. Turner. 1983. Publ. Astron. Soc. Pacific **95:** 72.
24. Elias, J. H., J. A. Frogel, J. A. Hackwell & J. E. Persson. 1982. Astrophys. J. Lett. **251:** L13.
25. Branch, D., C. H. Lacy, M. L. McCall, P. G. Sutherland, A. Uomoto, J. C. Wheeler & B. J. Wills. 1983. Astrophys. J. **270:** 123.
26. Branch, D., R. Buta, S. W. Falk, M. L. McCall, P. G. Sutherland, A. Uomoto, J. C. Wheeler & B. J. Wills. 1982. Astrophys. J. Lett. **252:** L61.
27. Olson, G. L. 1982. Astrophys. J. **255:** 267.

28. TAMMANN, G. A. 1982. *In* Supernovae: a Survey of Recent Research. M. J. Rees & R. J. Stoneham, Eds.: 371. Reidel. Dordrecht, the Netherlands.
29. BARBON, R., F. CIATTI & L. ROSINO. 1973. Astron. Astrophys. **25:** 248.
30. RUST, B. W. 1974. Ph.D. Thesis. University of Illinois. Urbana, Ill.
31. PSKOVSKII, Y. P. 1977. Sov. Astron. AJ **21:** 675.
32. BRANCH, D. 1981. Astrophys. J. **248:** 1076.
33. MULLER, E. & W. D. ARNETT. 1982. Astrophys. J. Lett. **261:** L109.

Core-Collapse Models of Type II Supernova Explosions and Neutron Star Formation

WOLFGANG HILLEBRANDT

Max Planck Institute for Physics and Astrophysics
Institute for Astrophysics
8046 Garching/Munich, Federal Republic of Germany

A FEW OBSERVATIONAL FACTS

It is now generally agreed that the total energies of type II supernovae are typically of the order of a few times 10^{50} to 10^{51} erg.[9] Expansion velocities up to 10^4 km s^{-1}, maximum temperatures around 15,000 K, radii of about 10^{14} cm, and lower limits for the ejected mass of around 5 to 6 $M_\odot$ are found (see, e.g., Reference 13 for SN 1979e). The ejecta are certainly hydrogen rich, indicating red giant envelopes of the progenitor stars,[13,31] whereas the heavy element abundances, as far as we have information, seem to be close to solar values.

There are a few exceptions from this general rule which deserve special mention. The fast-moving knots in the supernova remnant Cas A (for which we do not know the type) are rich in oxygen, sulfur, argon, and calcium, the typical fingerprints of an oxygen-burning zone of a massive star,[18,30] and similar results have been found for the supernova remnants G 292.0 + 18 and N 132 D.[24,37] It is interesting to note that in these latter cases no hydrogen, helium, and nitrogen lines are observed, again in concordance with the oxygen-burning interpretation.

Another exception is the Crab nebula, for which we have mass estimates of around 1.2 to 3.0 $M_\odot$, helium overabundance ($1.6 < x_{He}/x_H < 8$), and some oxygen underabundance ($x_0 \simeq 0.003$, compared to the solar value of 0.007).[19,25] In addition, carbon to oxygen ratios of 0.4 to 1.1 are observed,[19] indicating that the carbon abundance is close to solar. We will come back to an interpretation of these abundances later and will only mention here that they impose strong constraints on the Crab's progenitor mass.

The frequency of galactic type II supernovae has recently been discussed by Tammann,[62] so we will only briefly repeat his conclusions. Assuming a ratio of occurrence of type I to type II supernovae of 1/0.82, he finds a frequency of type II events of 0.023 yr^{-1} and a number of 2.3×10^{-11} per pc^2 per year, the uncertainties being typically of the order of a factor of 2. From these numbers Tammann derives a lower limit for the masses of type II progenitor stars and obtains $M \gtrsim 6\ M_\odot$ for the initial mass function (IMF) of Lequeux[38] or $M \gtrsim 10\ M_\odot$ for the IMF of Miller and Scalo.[48] An important conclusion follows, namely, that even if very massive stars ($M \gtrsim 20\ M_\odot$) should explode as type II supernovae, they will account only for a very small fraction of all events. The majority of progenitor stars should have masses around 8 to 10 $M_\odot$, a mass range that has not yet been explored systematically.

In concluding this section we briefly discuss the relation between the neutron stars and type II supernovae, which may also impose observational constraints upon theoretical models. Unfortunately the observational situation is not clear. Only two of the historical supernovae *may* have been of type II, the Crab and Cas A, and only in the case of the Crab has a neutron star definitely been born. For the Crab the progenitor mass was at least 3 $M_{\odot}$ and if the observations of Murdin and Clark turn out to be correct,[53] this mass will go up to 8 to 10 $M_{\odot}$. The progenitor of Cas A was very likely more massive ($M \gtrsim 15\ M_{\odot}$) and has left no pulsar and presumably no neutron star. So neutron star remnants may not always be the outcome of type II explosions.

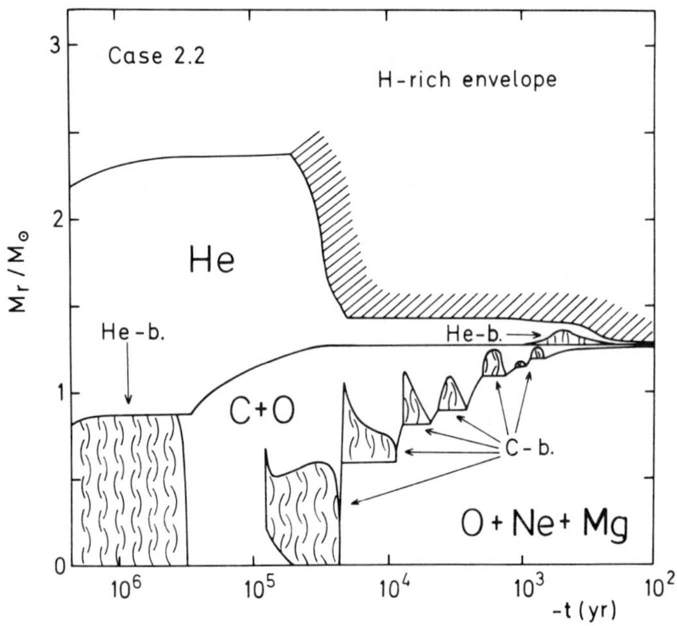

FIGURE 1. Chemical evolution of a 9 $M_{\odot}$ star (from Reference 29). Time is measured in years from the end of the computation. A degenerate O-Ne-Mg core is formed, the core mass being slightly below the critical mass (1.37 $M_{\odot}$) for Ne ignition. Convective regions (curls) as well as the surface convection zone (shaded) are also shown.

The main conclusion from the observational data is therefore that very likely the progenitor stars of type II supernovae have typically around (10 ± 4) $M_{\odot}$, and that there are no good direct observational arguments as to why more massive stars should explode. On the contrary, it has been argued that from a galactic nucleosynthesis viewpoint one encounters difficulties if all stars more massive than about 12 $M_{\odot}$ would explode.[54] On the other hand, most theoretical models have been for rather massive stars ($M \gtrsim 15\ M_{\odot}$), which with realistic microphysics input data for nonrotating initial stellar models did not yield explosions. From the above discussion this failure may not be an indication that all those model computations were wrong, but may be that they gave correct answers after all.

Recently stellar evolution calculations have been performed for the mass range 8 to 10 $M_\odot$, which as we will show do indeed give both supernova explosions and neutron star formation. Therefore, in the following sections we will concentrate mainly on their properties. In addition, we will present arguments why more massive stars face severe problems if they are to eject their envelopes in supernova explosions.

PRESUPERNOVA STELLAR MODELS

From observations we know that the total energy output in supernova explosions is around 10^{51} erg. The origin of this energy can either be thermonuclear burning, which requires that about 1 $M_\odot$ of He, C, or O is burned into iron group elements, or gravitational binding energy, if we recall that the binding energy of a neutron star of 1 $M_\odot$ is about 2×10^{53} erg. Consequently, thermonuclear models require that the explosion mechanism is very efficient, close to 100%, which in turn means that almost all the energy available must be given to the ejecta. In the case of stellar collapse, on the other hand, an efficiency of 1% or even less is needed. In fact what we are supposed to do is to search for a process of such a low efficiency with numerical models. Arnett has called this situation a "theorist's nightmare," and it explains why over the years the final outcome of computer simulations has been so controversial.

Explosive nuclear burning can disrupt white dwarfs, today the fashionable model for type I supernovae, or degenerate carbon-oxygen cores of stars of main sequence masses in the range 6 to 8 $M_\odot$ (see for a recent simulation Reference 52), which then may be of type II without a compact remnant. Here we are interested in those models only that do collapse and therefore the further discussion will be restricted to main sequence masses above 8 $M_\odot$.

Stars in the Range 8 to 10 M$_\odot$

Stars in this mass range, corresponding to He-core masses of 2.2 to 2.7 $M_\odot$, differ from more massive stars because they produce either degenerate cores of O, Ne, and Mg ($M_{He} \simeq 2.2$–$2.6\ M_\odot$)[49,54,55] or Ne and Si[73,55] (see also FIGURES 1 and 2). The mass of the degenerate cores is always slightly above the Chandrasekhar mass ($M_{core} = 1.38\ M_\odot$), so consequently they should contract; the question remains whether or not nuclear burning can for some time prevent the core from collapsing. A feature all those models have in common is an enormous drop in density (by about 10 orders of magnitude) at the core-envelope boundary. This steep density gradient will have important implications for the propagation of the shock fronts as we will discuss later. All models have rather low entropy, about 0.7 k_B/nucleon, and the collapse is initiated by electron captures on Ne, Mg, and Na. Degenerate nuclear burning leads to a subsonic burning front ("deflagration wave") without mass loss ($M < 9.5\ M_\odot$)[49,54,55] or to Ne flashes with mass loss ($M \simeq 10\ M_\odot$).[73] In this latter case the star may lose its envelope even prior to collapse. In both kinds of models nuclear burning goes directly into nuclear statistical equilibrium (NSE) and both should therefore collapse to neutron star densities.

Stars in the Range 12 (or 15) $M_{\odot}$ *to 100* $M_{\odot}$

Stars in this mass range, corresponding to He-core masses of 4 to 30 $M_{\odot}$, burn all their nuclear fuel nondegenerately. After core-Si burning they set up an Si-burning shell and produce Fe-Ni cores of 1.5 to 1.7 $M_{\odot}$, the entropy at the onset of collapse being typically 1 k_B/nucleon.[1,68] The collapse is then initiated by nuclear photodisinte-

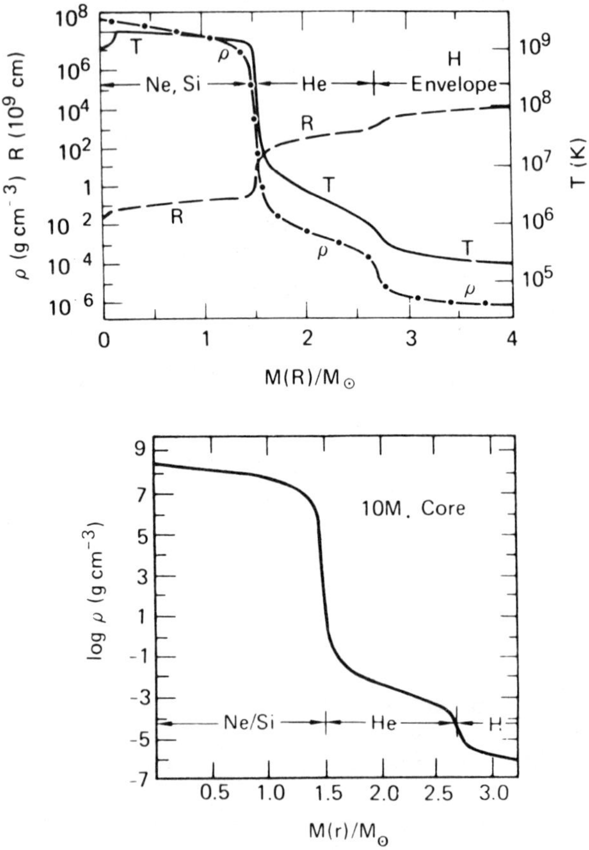

FIGURE 2. Structure of a 10 $M_{\odot}$ star at the time when core neon flashes are occurring (from Reference 73). At the boundary of the Ne-Si core the density drops by about 10 orders of magnitude. In this model computation, the He-H envelope was lost from the star prior to collapse.

grations rather than e^- captures, which are less important. Besides larger core masses, these stellar models do not show the steep density gradient as seen in the models of less massive stars. Here as well the collapse should proceed up to neutron star densities because the entropy is much too low to allow for complete nuclear breakup prior to nuclear matter density.

THE INPUT PHYSICS

The Equation of State

The most important and most crucial input into any numerical study of core-collapse supernovae is the equation of state. This is because the deviation of the actual adiabatic index from the value 4/3 of a relativistic gas of leptons determines the size of the homologously collapsing part of the core. The homologous core, on the other hand, determines where inside the star a shock front forms and how much energy is put into the shock. The adiabatic index, however, is the second derivative of the energy density with respect to the density at constant entropy, and consequently the energy density has to be computed very accurately. We stress this point because recently analytic approximations to realistic equations of state have been presented that do not fulfill this requirement.[8] Considerable progress has been made over the last several years in computing realistic equations of state. At low densities ($\rho \gtrsim 10^{12}$ g cm^{-3}) a Boltzmann gas approach is essentially the exact solution to the nuclear part of the problem, the only unknown quantities being ground state binding energies and partition functions of nuclei.[20,44] Fortunately the composition in NSE is dominated by nuclei not too far away from experimentally known ones and therefore no large extrapolations from experimental nuclear physics are needed. At higher densities ($\rho \gtrsim 10^{12}$ g cm^{-3}) a Boltzmann gas treatment of nuclei becomes gradually less reliable, because the spacing between nuclei is no longer very large compared to their size, and more sophisticated methods are required. Consequently, computations of the equation of state get more complicated and thus more controversial.

A simple approach is the so-called finite temperature liquid drop model of the nuclei, a method that has been used by the Illinois group[34] (see also Reference 36 for a review). Although this model will reproduce correctly the gross properties of the equation of state at high densities, it may not be sufficiently accurate for our present purpose. More sophisticated microscopic models start from an effective two-body interaction (like, e.g., a Skyrme force) and minimize the Helmholtz free energy of a system of neutrons and protons, which may or may not cluster to form nuclei, in either Thomas-Fermi[36,40,57] or Hartree-Fock approximation.[11,72] Fully microscopic calculations, though more reliable, are of course much too expensive for direct use in hydrodynamics, but extended tabulations have recently become available,[72] so we will discuss the Hartree-Fock approach and its results in a little more detail.

First of all one has to adopt a model Hamiltonian consisting of strong interaction terms (the so-called Ska-force[32] has been used by Wolff[72]), terms describing electron and positron contributions, and terms for proton-proton and proton-lepton coulomb interactions (lattice energy). The grand-canonical potential is then minimized on a Wigner-Seitz sphere with respect to single-particle wave functions and energies. Some results are shown in FIGURES 3–5. At zero temperature the Hartree-Fock calculations are in good agreement with measured energies. Toward the neutron and proton drip lines, nuclear-level densities extracted from the computed single-particle levels are significantly lower than those predicted from a back-shifted Fermi-gas model, but are consistent with experimental data to within better than a factor of two whenever such data are available. Consequently one expects a relatively large fraction of dripped neutrons (and protons) even at moderate entropies (in contrast to References 7 and 14), and consequently a stiff equation of state.

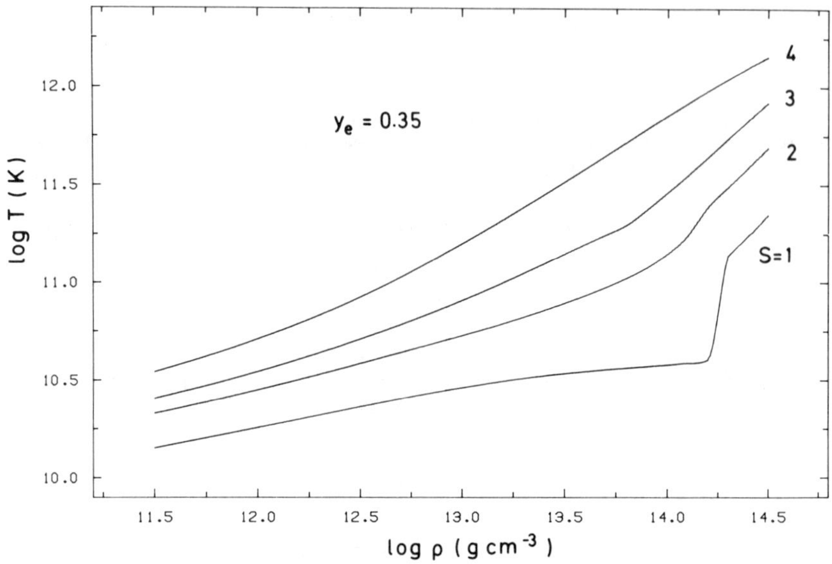

FIGURE 3. Isentropes for fixed $y_e = 0.35$ (from a Hartree-Fock computation of Wolff).[72] The entropy is given in units of Boltzmann's constant per nucleon. The critical density at which nuclei dissolve into a homogeneous nuclear fluid is indicated by a sudden increase in temperature as the entropy stored in nuclear excited states below the critical density (for fixed total entropy) increases the kinetic temperature of the gas after dissolution.

At finite temperatures one finds that for certain values of baryon density and temperature, three types of solutions may be obtained: homogeneous solutions ("homogeneous nuclear matter"), solutions with a maximum density at the center of the cell ("nuclei embedded in a free nucleon gas") and solutions with maximum density at the cell boundary ("bubbles"). According to Wolff the existence of a bubble phase, however, seems to be unlikely, since in those cases where these configurations gave the lowest Helmholtz free energy, a nonzero slope of the density at the cell boundary makes it difficult to establish periodicity. On the other hand it seems that phenomenological nuclear forces with larger compressibility coefficients like the one used by Ravenhall *et al.*[58] favor bubblelike configurations, which is mainly due to a different structure of the single-particle spectrum and thus a change in the nuclear partition function. Numerically the deviations between the different realistic approaches to the equation of state around nuclear matter density are not large, giving us the confidence that we know the most important set of input data quite well. In addition it seems that improved calculations yield stiffer equations of state below and near nuclear matter density, an effect that favors supernova explosions.

Weak Interaction Rates

In the stellar collapse problem most of the nonequilibrium weak interaction processes take place at densities from a few times 10^{10} up to 10^{12} g cm^{-3}, because at lower densities the electron chemical potential is too small for e^- captures to occur and at higher densities due to neutrino trapping a β-equilibrium is established very quickly. At such densities and moderate entropies ($S \simeq (1\text{–}2)k_B$/nucleon) the most abundant nuclei in NSE are those around closed neutron shells (like ^{48}Ca) for electron concentrations $y_e \gtrsim 0.4$.[20] It has been demonstrated that due to shell-blocking effects the e^--capture rates of those nuclei are very small.[21–23] Proton concentrations, on the other hand, are sufficiently large ($y_p \simeq 10^{-4}\text{–}10^{-3}$) so that e^- captures on free protons are the dominant contribution to the total capture rate, again in contrast to statements found in the literature.[7,14] For elementary processes such as $e^- + p \rightarrow n + \nu_e$, $\nu_e + n \rightarrow p + e^-$, etc., the rates can be computed numerically exactly (see e.g. Reference 61) and therefore very few uncertainties do arise.

Neutrino Transport

All recent investigations agree on the fact that neutrino diffusion times become shorter than the collapse time scale once the density of the core exceeds a few times 10^{11} g cm^{-3}, this effect being caused by coherent ν-scattering off nuclei.[2,41,42,59] Since the transition region between complete trapping and free streaming of neutrinos is very narrow in a collapsing stellar core, a flux-limited diffusion approximation to the full transport seems to be sufficiently accurate, at least during the collapse phase.[69] On the other hand, in most of the diffusion region, neutrinos are not in thermal equilibrium

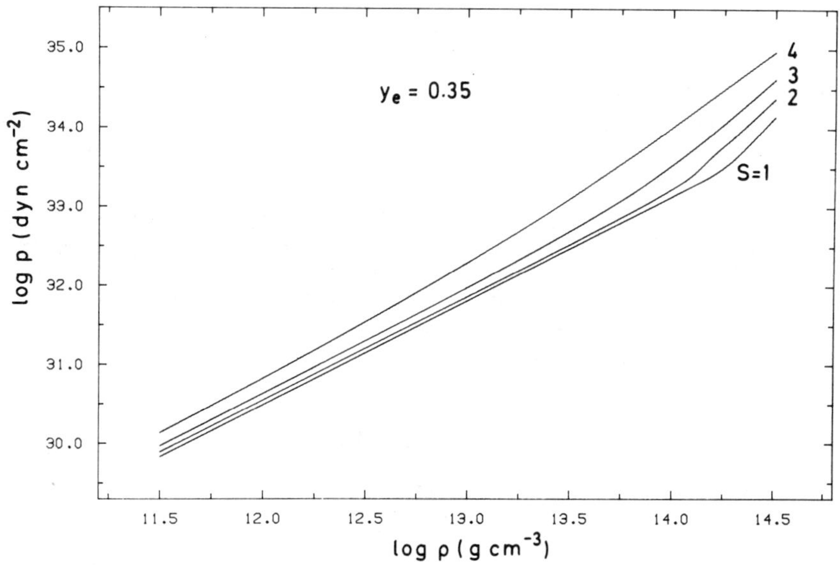

FIGURE 4. Pressure versus density for the isentropes of FIGURE 3. The steep increase in pressure above nuclear matter density reflects the rather large incompressibility of nuclear matter. Note that the equation of state is still temperature dependent above nuclear matter density. So $S = 0$ approximations should be considered with some care. It can also be seen that the transition to homogeneous nuclear matter is smooth, showing no indication for a first-order phase transition.

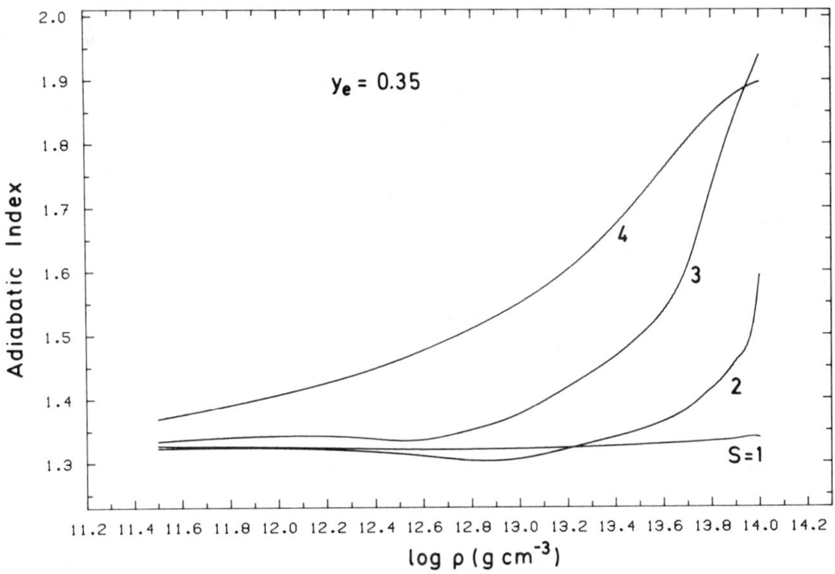

FIGURE 5. Adiabatic index versus density for the isentropes of FIGURE 3. For low entropies (S = 1) γ is only slightly less than 4/3 up to almost 10^{14} g cm^{-3}, so one expects to obtain rather large homologous cores in collapse computations. For large entropies and moderate densities, on the other hand, the equation of state is rather stiff (large values of γ), a property of the equation of state that favors shock propagation.

with matter and therefore a multi-energy-group approach seems to be appropriate,[2,4,12,69,70] but even more simplified transport schemes give similar results as far as the lepton concentration at neutrino trapping and the global dynamics are concerned.[45,46,65] Recently, based on an idea of Arnett,[3] Hillebrandt has used an approximation to the multigroup method by assuming the neutrino distribution function to be a Fermi function, but allowing for ν-temperatures different from the matter temperature.[28] This method, which will be described in more detail elsewhere, combines the simplicity of an equilibrium diffusion approach with the requirement of nonthermal equilibrium locally in the transport region. Some of the results discussed in the next section have been obtained with the help of this method.

NUMERICAL STUDIES

8–10 M$_\odot$ *Stars*

A Model of a 10 M$_\odot$ *Star*

Recently Hillebrandt,[26,27] Bowers and Wilson,[12] and Wilson[71] have performed collapse computations starting from a model of a 10 $M_\odot$ star that has been evolved from the main sequence to the onset of core collapse by Weaver and Woosley[66] (see also Reference 73 for more details). This model consisted of a 1.38 $M_\odot$ "Fe" core in nuclear statistical equilibrium surrounded by an Ne-O shell of 0.13 $M_\odot$ and an He-H envelope expanding with 100–300 km s^{-1} out to 10^{16} cm when the core began to collapse. Further properties of the model were: It had a central density, temperature, and electron concentration of 5×10^9 g cm^{-3}, 7.6×10^3 K, and 0.43, respectively. As the core mass of this model was larger than the Chandrasekhar mass ($\simeq$1.1 $M_\odot$) it was gravitationally unstable. The entropy per nucleon was 0.8 k_B, which is a somewhat lower value than for models of more massive stars. At the edge of the core the density dropped by more than 10 orders of magnitude over a tenth of a solar mass.

In Hillebrandt's computation the collapse phase of this model turned out to be quite similar to those that had been computed previously for more massive stars (for a review see Reference 27). Due to the somewhat lower entropy of the initial model the lepton concentration at neutrino trapping was $y_e = 0.4$, a value slightly larger than usual. This lepton concentration corresponds to a Chandrasekhar mass of about 0.9 $M_\odot$ which is approximately equal to the homologously collapsing part of the core. As a result the shock formed at 0.8 $M_\odot$ when the core bounced at a central density of about 4×10^{14} g cm^{-3}, slightly above nuclear matter density (see FIGURES 6–9). As in most computations using realistic equations of state, we did not observe large-amplitude oscillations of this inner core. However, in following the analysis of Brown *et al.*,[15] we could estimate the energy initially given to the shock and found roughly 8×10^{51} erg. This energy is sufficient to dissociate about 0.5 $M_\odot$ of heavy elements into neutrons and

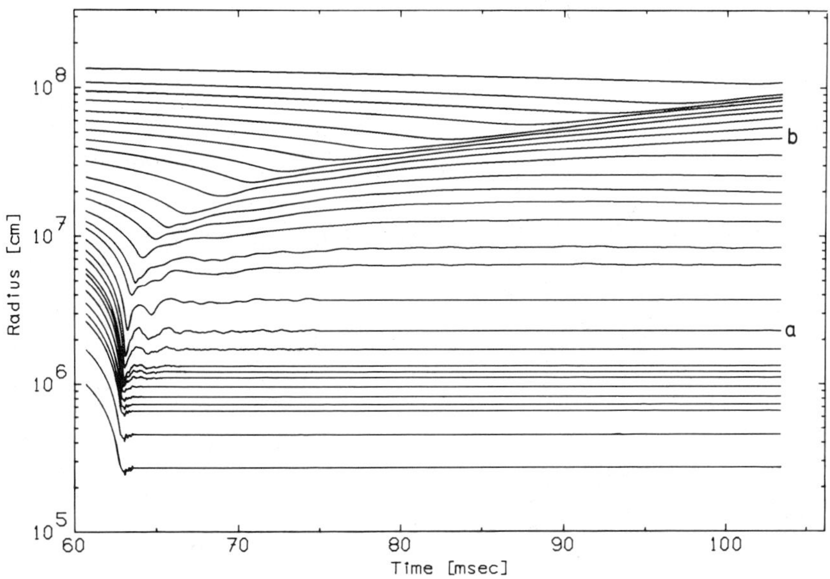

FIGURE 6. Radius versus time for selected mass zones of a 10 $M_\odot$ star (from Reference 27). Only the inner 1.5 $M_\odot$ are plotted. The time is measured in ms from the beginning of the computation when the central density was 2.7×10^{10} g cm^{-3}. The curve labeled **a** gives the boundary of the unshocked inner core of 0.75 $M_\odot$, the curve labeled **b** the boundary of the original Fe-core ($M \simeq 1.36$ $M_\odot$).

protons, and one could therefore expect that the shock would at least move out to the edge of the iron core. The numerical study showed that these considerations were indeed correct. The shock was significantly damped on its way out, and by the time it had progressed to about 1.3 $M_{\odot}$, the outward velocities had dropped to about 10^9 cm s^{-1}. But then, due to the steep density gradient there, the shock speeded up again and matter reached escape velocity approximately 30 ms after core bounce. In this model, 0.06 $M_{\odot}$ from the mass of the core was ejected with velocities of about 25,000 km s^{-1}, leaving behind a neutron star of 1.44 $M_{\odot}$. The total energy in the explosion was found to be 5×10^{50} erg, a little low for a typical type II supernova. It is interesting to note that these numerical results are consistent with analytic considerations by Mazurek,[43] Yahil and Lattimer,[74] and Burrows and Lattimer.[16] Wilson, on the other hand, found an even weaker explosion for the same stellar model;[71] in his computation the explosion mechanism was the ν-flux from the hot neutron star rather than the original hydrodynamical shock wave. As the shock in his computations was much weaker from the beginning, we believe that the final differences were caused by the fact that his equation of state is significantly softer than ours at densities below nuclear matter density.

It is worthwhile to mention that models like the one we have discussed show several properties that are quite similar to those of SN 1054, the Crab. First of all the explosion of a star with such a very extended envelope should give rise to a very bright

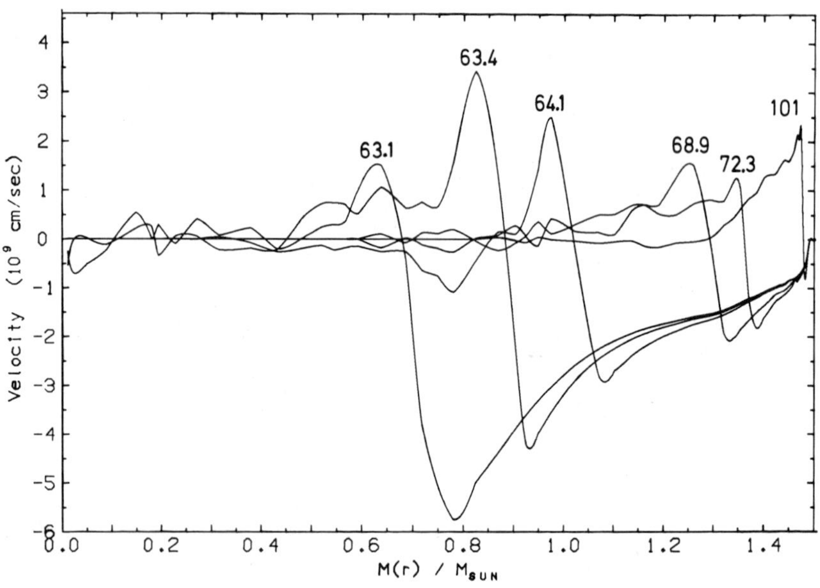

FIGURE 7. Snapshots of velocity profiles after core bounce (from Reference 27). Inward velocities are negative. The curves are labeled with time in ms as in FIGURE 6. One sees the damping of the shock due to nuclear dissociations and the speeding up near the edge of the core. The very outer layers of the original core reach escape velocity approximately 30 ms after core bounce.

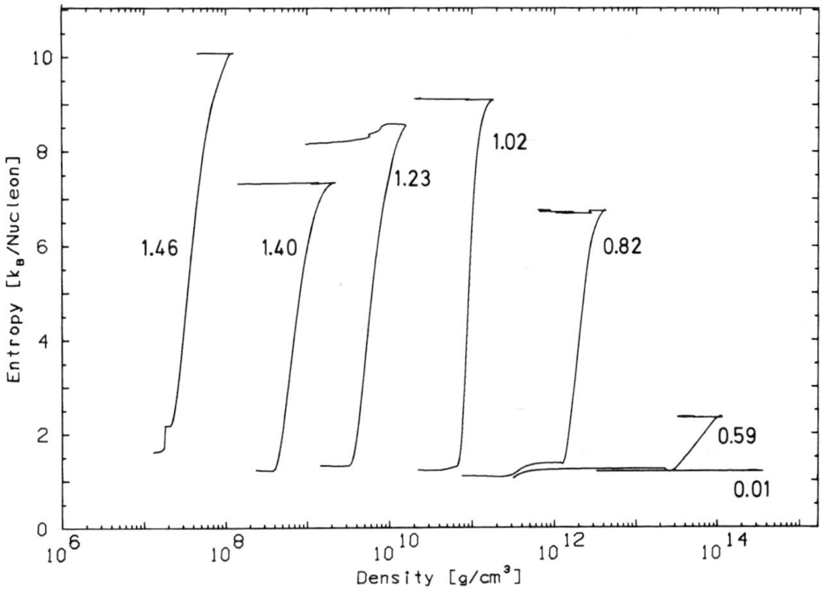

FIGURE 8. Entropy versus density for selected mass zones of the 10 $M_\odot$ model (from Reference 27). Curves are labeled with the mass $M(r)$ in units of $M_\odot$. The steep increase in entropy reflects the passage of the shock front. The time evolution of the mass zones begins at the lower left end of each curve 60 ms after the start of the computations. The decrease in maximum entropy with increasing mass shows the effect of nuclear dissociations. The decrease of entropy after maximum at 1.23 $M_\odot$ is caused by neutrino losses from this zone.

event, as was the case for SN 1054, despite the fact that the total energy output is not very large. Moreover a neutron star is produced, no overproduction of heavy elements can be expected, and the helium enrichment of the Crab nebula can be explained. The agreement would be excellent, if the observed hydrogen halo around the Crab nebula of about (8 ± 2) $M_\odot$ should be confirmed.[53] There is a minor problem with the rather low oxygen abundance of the Crab nebula,[19] which might indicate a slightly lower mass of the progenitor star.[54,56]

A Model of 9 $M_\odot$

Recently a model computation has been performed in which the core mass was reduced even further.[54,55,29] The model had a helium-core mass of 2.2 $M_\odot$ only, corresponding to a main-sequence mass of about 9 $M_\odot$. At the onset of core collapse the model contained an O-Ne-Mg core of 1.36 $M_\odot$, surrounded by a thin shell burning C and an He-H envelope. The core first contracted due to e^- captures on ^{24}Mg and ^{20}Ne. When the central density exceeded 2.5×10^{10} g cm^{-3}, oxygen burning was ignited, a subsonic burning front moved outward leaving behind an inner core of matter in NSE, which was continuously underoing e^- captures. By the time the central density had reached 5×10^{11} g cm^{-3}, the electron concentration had decreased to 0.35 and the

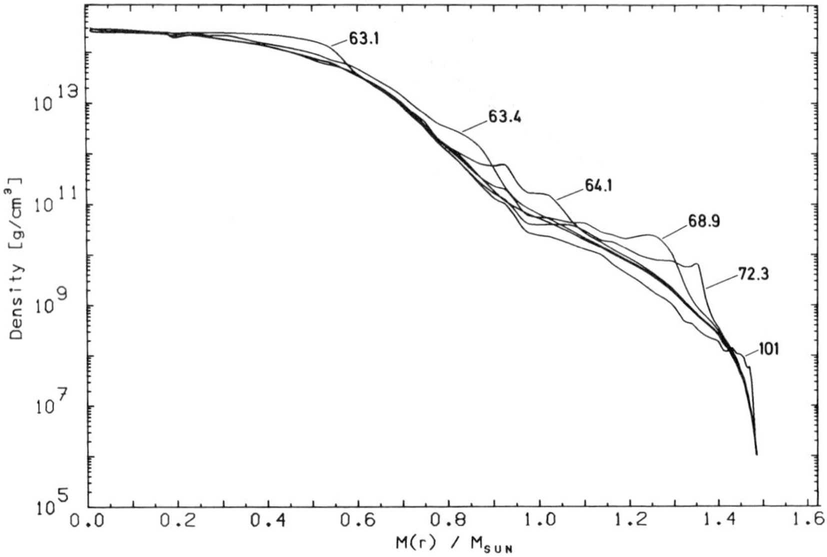

FIGURE 9. Density versus mass for different times (10 $M_{\odot}$ star, from Reference 27). The snapshots are taken at the same time steps as those of FIGURE 7. One can see the steepening of the shock in the region of large density decrease.

oxygen deflagration front had incinerated 0.7 $M_{\odot}$ of the core. It was only then that the further evolution became really dynamical because the NSE part of the core had grown above the Chandrasekhar mass, to approximately 0.7 $M_{\odot}$.

Several aspects of this model computation, which has been performed with both the nonequilibrium neutrino-transport scheme and the Hartree-Fock equation of state, were quite similar to, but others were completely different from, previous models. The central values of density, temperature, and entropy at bounce were almost identical to those found for the 10 $M_{\odot}$ star: $\rho_c \simeq 3 \times 10^{14}$ g cm^{-3}, $T_c \simeq 2 \times 10^{11}$ K, $S_c \simeq 1.0$ k_B/nucleon. The maximum infall velocity at the time when the inner 0.6 $M_{\odot}$ had essentially come to rest was 5.3×10^9 cm s^{-1} (see FIGURE 10), again in good agreement with the other model. Due to the somewhat lower lepton concentration the mass of the unshocked part of the core was 0.1 $M_{\odot}$ smaller and also the energy of the shock was reduced by about 15%.

Nevertheless the shock front gave rise to a very energetic explosion, the reason being twofold (see FIGURES 10–13). First of all, at the time of rebound the oxygen-burning front had incinerated only 1.1 $M_{\odot}$. Therefore the shock had to dissociate a rather small amount of heavy nuclei. Secondly, due to nuclear burning, matter was significantly decelerated when passing through the burning front. As a result both the infall velocity and the density of the matter the shock had to pass through were reduced, thus lowering the ram pressure on the shock and its damping.[74]

We want to close this section with a few final remarks. It has been demonstrated that the hydrodynamical shock wave from a rebounding protoneutron star can indeed lead to mass ejection, but it seems that even in the very narrow mass range that has been investigated (9 and 10 $M_\odot$) the outcome is very different, due to the different structures of both stars prior to collapse. This result calls for very detailed and accurate stellar evolution models. Part of the rather high explosion energy ($\simeq 2.5 \times 10^{51}$ erg) of the 9 $M_\odot$ model as compared to the 10 $M_\odot$ was caused by the excellent resolution of the shock we had in the 9 $M_\odot$ case: a total of 185 mass zones for about 1.4 $M_\odot$ were used, of which 130 zones described the region between 0.5 and 1.3 $M_\odot$. Therefore there was very little numerical damping. Neutrino transport, on the other hand, turned out to be helpful in the 9 $M_\odot$ model though not decisive. Thermal emission of μ- and τ-neutrinos has been neglected in our computations and therefore we will have overestimated the total explosion energy somewhat. It is interesting to note that despite the significantly different explosion energies, the remaining neutron stars had almost the same mass: 1.25 $M_\odot$ and 1.44 $M_\odot$ respectively. This does not come as a surprise since it is caused by the mass of the dengerate core of the initial stellar model. From the point of view of nucleosynthesis, the 9 $M_\odot$ model is also a very interesting case. We found that some of the innermost ejected shells had neutron-to-proton ratios of almost 4 by the time we stopped the computations. This large neutronization was caused by the rather high entropies ($S \simeq 15\ k_B$/nucleon) and low densities ($\rho \lesssim 5 \times 10^9$ g cm^{-3}) of those zones, where neutrinos were freely streaming outward but electrons were still captured. These shells might turn out to be a good site for the rapid neutron capture (r-) process; particularly since they contain very little mass ($\lesssim 0.03\ M_\odot$) and therefore overproduction of r-nuclei would be avoided.

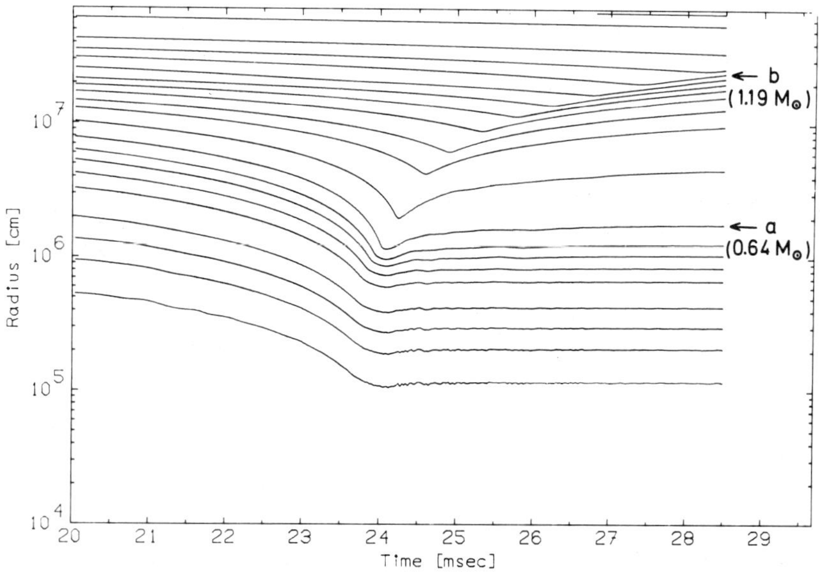

FIGURE 10. Same as FIGURE 6 but for a 9 $M_\odot$ stellar model. Only the inner 1.35 $M_\odot$ are plotted. The time is measured from the beginning of the computation at $\rho_c = 2.7 \times 10^{11}$ g cm^{-3}. The curve labeled **a** gives the boundary of the unshocked inner core of 0.6 $M_\odot$, the curve labeled **b** is the innermost zone that has a velocity larger than escape velocity ($\simeq 1.2\ M_\odot$).

Stars with More Than 12 $M_{\odot}$

The status of core-collapse models based on more massive stars has been reviewed recently by Hillebrandt,[27] so we need not go into details here. There are, however, a few more new investigations which essentially confirm the earlier results.[4,12,64] The general agreement by now is that models with Fe-Ni cores more massive than about 1.5 $M_{\odot}$ and the best input physics available do not explode. Such massive cores are expected to form at the end of thermonuclear burning of stars with main-sequence masses above 12 to 15 $M_{\odot}$. Three things are found to be responsible for this result. First of all, due to the large stellar core, the shock has to dissociate too many heavy nuclei. The dissociation of nuclei causes so much damping that the shock finally cannot overcome the ram pressure of the infalling matter and changes into a standing accretion shock. Second, in those stars the structure does not change at the core-envelope boundary. The presence of such a change, however, made the 10 $M_{\odot}$ model discussed earlier successful. On the other hand it had been argued that neutrino diffusion out of the unshocked inner core, the neutron star, should supply a sufficient amount of energy to the matter behind the shock front to balance nuclear photodissociation losses.[5,6,15] In a recent paper, Burrows and Mazurek have shown that this is not the case.[17] They could demonstrate that based on realistic collapse models, energy-diffusion times of neutrinos are too long compared to shock-propagation times so that neutrino transport cannot strengthen or energize weak shocks.

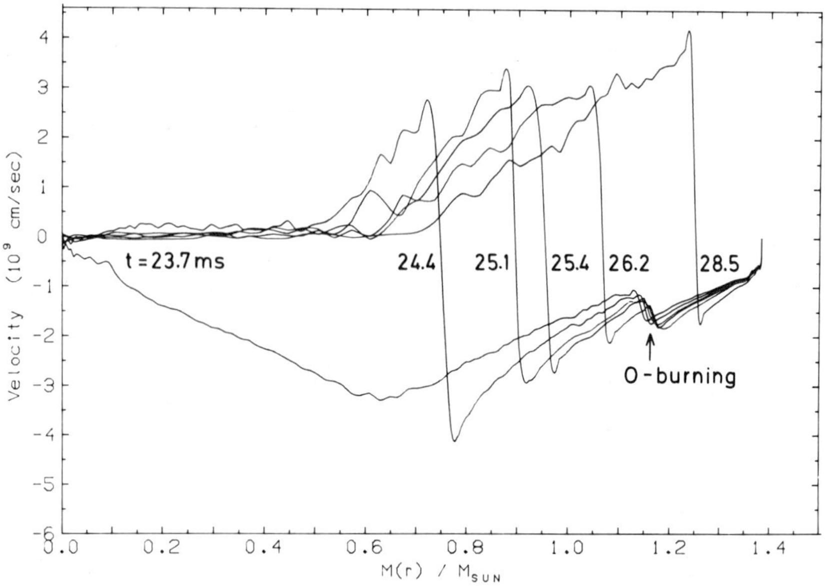

FIGURE 11. Same as FIGURE 7 but for a 9 $M_{\odot}$ star. In contrast to the 10 $M_{\odot}$ model one sees that there is much less damping of the shock, because the ram pressure on the shock is significantly lower. The secondary minimum in the velocity curves around 1.1 to 1.2 $M_{\odot}$ is caused by the oxygen-burning front which reduces both the infall velocity and the density of the infalling matter. After the shock front has passed the oxygen shell, it speeds up and the mass zones reach escape velocity.

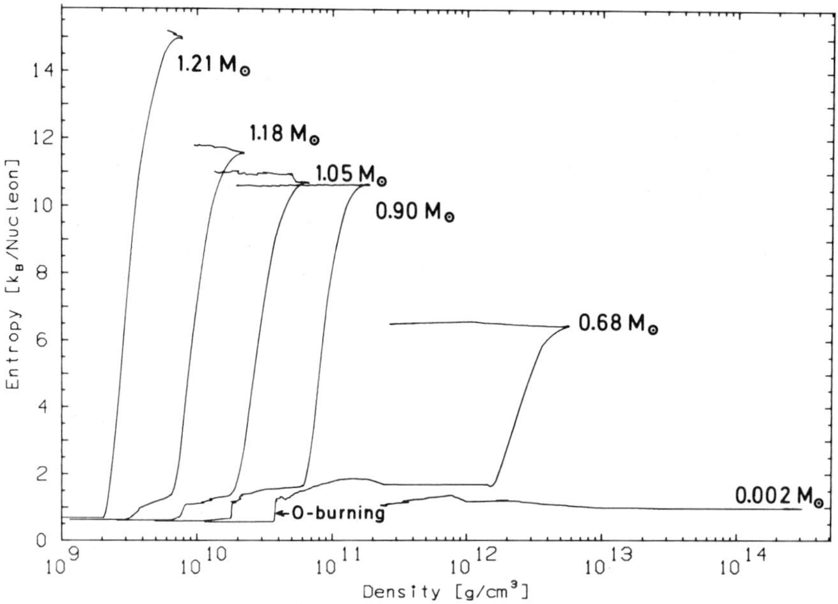

FIGURE 12. Same as FIGURE 8 but for a 9 $M_\odot$ star. The general behavior of the entropy curves is similar to the 10 $M_\odot$ model, the main difference being the lack of much damping as compared to FIGURE 8. Note that after the shock has passed the O-burning shell, maximum entropies around 15 k_B/nucleon are obtained corresponding to temperatures of around 5×10^{10} K. These zones might be the site for the production of heavy and neutron-rich isotopes.

The question then is whether we really need to think that at least some of the very massive stars explode. The only good argument in favor of this requirement at present is the absence of other explanations of the heavy element yields in the galaxy. If we accept this point of view, we have to look for modifications of the standard picture that may alter the conclusions drawn before.

A very straightforward explanation would be that stellar evolutionary models are not correct or, to be more precise, that the predicted core masses and (or) entropies are too large. Reducing either of these two will help to produce explosions, and in fact it has recently been argued that a better treatment of silicon burning leads in this direction.[4,67]

One might also wish to abandon the assumption of spherical symmetry. Realistic stellar models including rotation are much more difficult to construct, but it has been shown that they open several interesting possibilities. Müller and Hillebrandt found that rotation produces entropy gradients which in turn drive large-scale circulations.[51] Those large-scale motions will transport neutrinos much more efficiently to the shock front than does diffusion. Bodenheimer and Woosley have looked at explosive oxygen burning in collapsing rotating massive stars and found in some cases weak thermonuclear explosions leaving behind a black hole rather than a neutron star.[10] Finally, magnetohydrodynamical effects may or may not be negligible in rotating cores,[39,47,50] although large effects can only be obtained from unrealistically high values of the initial magnetic field strength.

Even more speculative possibilities have also been discussed, in most cases motivated by our ignorance of elementary particle physics. One interesting suggestion has been made by Kolb *et al.*[33] who considered that neutrinos might annihilate into a neutral boson field, the majoron. Consequently in β-equilibrium the neutrino chemical potential would have to be zero (as for an equilibrium gas of e^+e^--pairs). This then means that many more electrons can be captured raising the entropy above the value ($\simeq 3\ k_B$/nucleon) required for nuclear dissociations. A large-amplitude thermal bounce of the core at low densities might result but also the mass of the homologous core would be small. The final outcome of such a scenario has not yet been computed but might yield interesting results. Another suggestion postulates the existence of light, neutral,

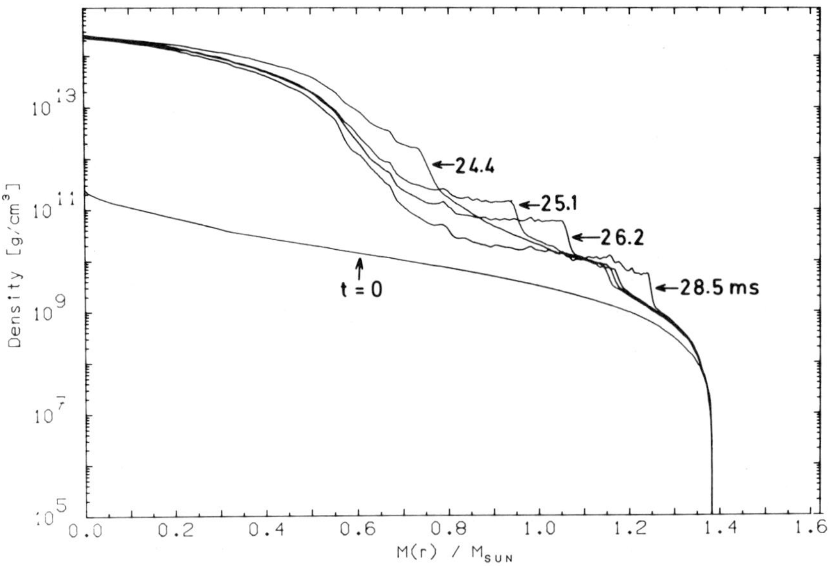

FIGURE 13. Same as FIGURE 9, but for a 9 $M_\odot$ star. Also the density profile of the initial model is shown (labeled $t = 0$). Essentially all the mass outside the O-burning shell (at about 1.2 $M_\odot$) is ejected in this computation.

semiweakly interacting particles ("axions") which should decay into two photons with lifetimes typically around 10^{-3} s.[60] They can, in principle, diffuse out of the core and deposit their energy in the stellar envelope, therefore again removing the difficulties arising from the rather long neutrino-diffusion time scales mentioned earlier.

SUMMARY AND CONCLUSIONS

The theory of core-collapse models of type II supernovae has had many ups and downs over the past 15 years and at present it seems that we are again approaching an up phase. We have learned that even apparently small differences in stellar evolution models can have dramatic effects on the predictions of the final outcome, namely, whether a neutron star forms in a type II supernova explosion or the collapse continues to a black hole. This sensitivity is mainly caused by the rather inefficient way in which the gravitational binding energy of the proto–neutron star is converted into outward momentum by a hydrodynamical shock wave, and by the rather long neutrino diffusion times, which do not allow efficient use of the thermal energy stored in leptons. Whether or not the shock is sufficiently strong for mass ejection is mainly determined by the presupernova stellar structure, by the core mass and the entropy, which are not uncorrelated but depend sensitively on for example convection during the late stages of the evolution. Here one would hope that future work will also increase somewhat the mass range of successful models by reducing both the core mass and the entropy of more massive stars. First promising attempts in this direction have indeed been made and they would help to restore our ideas about galactic nucleosynthesis.

REFERENCES

1. ARNETT, W. D. 1977. Astrophys. J. Suppl. **35:** 145.
2. ARNETT, W. D. 1977. Astrophys. J. **218:** 815.
3. ARNETT, W. D. 1981. Private communication.
4. ARNETT, W. D. 1982. Astrophys. J. Lett. **263:** L55.
5. BETHE, H. A. 1982. *In* Supernovae: a Survey of Current Research M. J. Rees & R. J. Stoneham, Eds.: 35. Reidel. Dordrecht, the Netherlands.
6. BETHE, H. A., J. APPLEGATE & G. E. BROWN. 1980. Astrophys. J. **241:** 343.
7. BETHE, H. A., G. E. BROWN, J. APPLEGATE & J. M. LATTIMER. 1979. Nucl. Phys. **A324:** 487.
8. BETHE, H. A., G. E. BROWN, J. COPPERSTEIN & J. R. WILSON. 1983. Nucl. Phys. **A403:** 625.
9. BLAIR, W. P., R. P. KIRSHNER & R. A. CHEVALIER. 1981. Astrophys. J. **247:** 879.
10. BODENHEIMER, P. & S. E. WOOSLEY. 1980. Bull. Am. Astron. Soc. **12:** 833.
11. BONCHE, P. & D. VAUTHERIN. 1980. Nucl. Phys. **A372:** 496.
12. BOWERS, R. & J. R. WILSON. 1982. Astrophys. J. **263:** 366.
13. BRANCH, D., S. W. FALK, M. L. MCCALL, P. RYBSKI, A. K. NOMOTO & B. J. WILLS. 1981. Astrophys. J. **244:** 780.
14. BROWN, G. E. 1982. *In* Supernovae: A Survey of Current Research: 13. (See Reference 5.)
15. BROWN, G. E., H. A. BETHE, & G. BAYM. 1982. Nucl. Phys. **A375:** 481.
16. BURROWS, A. & J. M. LATTIMER. 1983. Astrophys. J. **270:** 735.
17. BURROWS, A. & T. J. MAZUREK. 1982. Astrophys. J. **259:** 330.
18. CHEVALIER, R. A. & R. P. KIRSHNER. 1978. Astrophys. J. **219:** 931.
19. DAVIDSON, K., T. R. GULL, S. P. MARAN, T. P. STECHER, R. A. FESEN, R. A. PARISE, C. A. HARVEL, M. KAFATOS, & V. L. TRIMBLE. 1982. Astrophys. J. **253:** 696.
20. EL EID, M. F. & W. HILLEBRANDT. 1980. Astron. Astrophys. Suppl. **42:** 215.
21. FULLER, G. M. 1982. Astrophys. J. **252:** 741.
22. FULLER, G. M., W. A. FOWLER & M. J. NEWMAN. 1982. Astrophys. J. **252:** 715.
23. FULLER, G. M., W. A. FOWLER & M. J. NEWMAN. 1982. Astrophys. J. Suppl. **48:** 279.
24. GOSS, W. M., P. A. SHAVER, W. J. ZEALEY, P. MURDIN & D. H. CLARK. 1979. Mon. Not. R. Astron. Soc. **188:** 357.
25. HENRY, R. C. & G. M. MAC ALPINE. 1982. Astrophys. J. **258:** 11.
26. HILLEBRANDT, W. 1982. Astron. Astrophys. **110:** L3.
27. HILLEBRANDT, W. 1982. *In* Supernovae: a Survey of Current Research: 123. (See Reference 5.)
28. HILLEBRANDT, W. (In preparation.)
29. HILLEBRANDT, W., K. NOMOTO & R. WOLFF. Astron. Astrophys. (Submitted.)
30. KIRSHNER, R. P. & R. A. CHEVALIER. 1977. Astrophys. J. **218:** 142.
31. KIRSHNER, R. P. & J. KWAN. 1974. Astrophys. J. **183:** 27.

32. KOHLER, H. S. 1976. Nucl. Phys. **A258:** 301.
33. KOLB, E. W., D. L. TUBBS & D. A. DICUS. 1982. Astrophys. J. Lett. **255:** L57.
34. LAMB, D. Q., J. M. LATTIMER, C. J. PETHICK & D. G. RAVENHALL. 1978. Phys. Rev. Lett. **41:** 1623.
35. LAMB, D. Q., J. M. LATTIMER, C. J. PETHICK & D. G. RAVENHALL. 1981. Nucl. Phys. **A360:** 459.
36. LATTIMER, J. M. 1981. Annu. Rev. Nucl. Part. Sci. **31:** 337.
37. LASKER, B. 1980. Astrophys. J. **237:** 765.
38. LEQUEUX, J. 1979. Astron. Astrophys. **80:** 35.
39. LE BLANC, J. M. & J. R. WILSON. 1970. Astrophys. J. **161:** 541.
40. MARCOS, S., M. BARRANCO & J. R. BUCHLER. 1982. Nucl. Phys. **A381:** 507.
41. MAZUREK, T. J. 1975. Astrophys. Space Sci. **35:** 117.
42. MAZUREK, T. J. 1976. Astrophys. J. Lett. **207:** L87.
43. MAZUREK, T. J. 1982. Astrophys. J. **258:** L13.
44. MAZUREK, T. J., J. M. LATTIMER & G. E. BROWN. 1979. Astrophys. J. **238:** 139.
45. MAZUREK, T. J., J. COOPERSTEIN & S. KAHANA. 1980. *In* Proceedings of the Dumand Workshop. v. J. Steger, Ed.: 142. University of Hawaii. Honolulu, Hawaii.
46. MAZUREK, J. T., J. COOPERSTEIN & S. KAHANA. 1982. *In* Supernovae: a Survey of Current Research: 71. (See Reference 5.)
47. MEIER, D. L., R. I. EPSTEIN, W. D. ARNETT & D. N. SCHRAMM. 1976. Astrophys. J. **204:** 869.
48. MILLER, G. E. & J. M. SCALO. 1979. Astrophys. J. Suppl. **41:** 513.
49. MIYAJI, S., K. NOMOTO, K. YOKOI & D. SUGIMOTO. 1980. Pub. Astron. Soc. Jpn. **32:** 303.
50. MÜLLER, E. & W. HILLEBRANDT. 1979. Astron. Astrophys. **80:** 147.
51. MÜLLER, E. & W. HILLEBRANDT. 1981. Astron. Astrophys. **103:** 358.
52. MÜLLER, E. & W. D. ARNETT. 1982. Astrophys. J. Lett. **261:** L109.
53. MURDIN, P. & D. H. CLARK. 1981. Nature **294:** 543.
54. NOMOTO, K. 1982. *In* Supernovae: a Survey of Current Research: 205. (See Reference 5.)
55. NOMOTO, K. 1982. Proc. IAU Symp. **101:** 139.
56. NOMOTO, K., W. M. SPARKS, R. A. FESEN, T. R. GULL, S. MIYAJI & G. SUGIMOTO. 1982. Nature **299:** 803.
57. OGASAWARA, R. & K. SATO. 1982. Prog. Theor. Phys. **68:** 222.
58. RAVENHALL, D. G., C. D. BENNET & C. J. PETHICK. 1972. Phys. Rev. Lett. **28:** 978.
59. SATO, K. 1975. Prog. Theor. Phys. **54:** 1352.
60. SCHRAMM, D. N. & J. R. WILSON. 1982. Astrophys. J. **260:** 868.
61. TAKAHASHI, K., M. F. EL EID & W. HILLEBRANDT. 1978. Astron. Astrophys. **67:** 185.
62. TAMMANN, G. A. 1982. *In* Supernovae: a Survey of Current Research: 371. (See Reference 5.)
63. TWAROG, B. A. & J. C. WHEELER. 1982. Astrophys. J. **261:** 636.
64. VAN RIPER, K. A. 1982. Astrophys. J. **257:** 793.
65. VAN RIPER, K. A. & J. M. LATTIMER. 1981. Astrophys. J. **249:** 270.
66. WEAVER, T. A. & S. E. WOOSLEY. 1980. Bull. Am. Astron. Soc. **4:** 724.
67. WEAVER, T. A., S. E. WOOSLEY & G. M. FULLER. 1982. Bull. Am. Astron. Soc. **14:** 957.
68. WEAVER, T. A., G. B. ZIMMERMAN & S. E. WOOSLEY. 1978. Astrophys. J. **225:** 1021.
69. WILSON, J. R. 1978. Proc. Int. Sch. Phys. Enrico Fermi **65:** 644.
70. WILSON, J. R. 1980. Ann. N.Y. Acad. Sci. **336:** 358.
71. WILSON, J. R. 1982. Livermore Report. (Unpublished.)
72. WOLFF, R. 1983. Doctoral Thesis. Technical University of Munich. Federal Republic of Germany.
73. WOOSLEY, S. E., T. A. WEAVER & R. E. TAAM. 1980. *In* Type I Supernovae. J. C. Wheeler, Ed.: 164. University of Texas. Austin, Texas.
74. YAHIL, A. & J. M. LATTIMER. 1982. *In* Supernovae: a Survey of Current Research: 53. (See Reference 5.)

The Interaction of Supernovae with a Circumstellar Medium[a]

ROGER A. CHEVALIER

Department of Astronomy
University of Virginia
Charlottesville, Virginia 22903

INTRODUCTION

Models for the optical light curves of type II supernovae have had considerable success in reproducing observations of these events.[1-3] The models are consistent with the progenitor stars being massive red supergiant stars in their final phases of evolution. When red supergiants are observed in our galaxy, they always show evidence for mass loss in a slow ($v_w \approx 10$ km s^{-1}) wind. The mass loss rates, $\dot{M}$, are generally not well determined and the mass loss mechanism is not well understood. The mass loss formula of Reimers[4] implies $\dot{M} \approx 10^{-6}\ M_\odot$ yr^{-1}. However, recent observations have indicated mass loss rates up to about $10^{-4}\ M_\odot$ yr^{-1}. With the higher mass loss rates, the interaction of the supernova with circumstellar matter can lead to observable effects.

There have been recent observations of type II supernovae over the whole wavelength range from radio to x-rays that are not accounted for by the standard light curve models. These observations point to the importance of interaction with a circumstellar medium and are discussed here. There are indications that the interaction can be observed for an extended period at radio wavelengths and that there are a number of extragalactic radio sources that can be identified with the explosions of massive stars. At present there is no direct evidence that interaction with a circumstellar medium plays an important role in the evolution of type I supernovae.

HYDRODYNAMIC THEORY OF THE INTERACTION

When the supernova shock front first emerges from the stellar envelope, the radiation can accelerate the surrounding gas.[1,6,7] The velocity field set up by radiation pressure in the optically thin region is

$$\mathrm{v} = \frac{\kappa E_{\mathrm{rad}}}{4\pi r^2 c} \tag{1}$$

where κ is the gas opacity and E_{rad} is the total radiation energy. This is a converging velocity field, and the flow is expected to steepen to a shock. Epstein showed that the radiative acceleration is sufficiently strong to prevent the formation of a viscous shock wave when the effects of the supernova explosion first reach the photosphere.[6]

[a]My research on supernovae is partially supported by National Science Foundation Grant AST 80-19569.

Fransson suggested that the viscous shock develops at a radius about twice the initial stellar radius,[7] but detailed calculations of this process have not yet been performed.

The expansion of the stellar material is accompanied by the conversion of internal energy to kinetic energy. The internal energy drops as R^{-1}, where R is the supernova radius, because radiation pressure dominates gas pressure. The freely expanding matter tends toward a velocity profile $v = r/t$.

After a few doubling times of the initial radius (typically a few days), the viscous shock in the circumstellar matter heats gas to a high temperature and the cool supernova matter freely expands. The outer shock wave cannot be freely expanding, so the supernova matter interacts with the hot gas and a second, inner shock front develops. By this time, the effects of radiative acceleration should be small and the red supergiant wind velocity is much less than the shock velocity so that the circumstellar medium can be regarded as stationary. It has a density profile $\rho = A_1 r^{-2}$. If the freely expanding supernova gas has a decreasing power law distribution, it has a density profile $\rho = A_2 t^{n-3} r^{-3}$ where A_2 is a constant. Chevalier and Nadyozhin showed that the resulting interaction region can be described by a self-similar solution if $n > 5$.[8,9] The radius of the contact discontinuity between the shocked circumstellar gas and the shocked supernova gas is

TABLE 1. Properties of Self-Similar Solutions

n	R_1/R_c	R_2/R_c	B	ρ_2/ρ_1	p_2/p_1	u_2/u_1	M_2/M_1
6	1.377	0.958	0.62	3.9	0.21	1.006	0.44
7	1.299	0.970	0.27	7.8	0.27	1.058	0.82
9	1.250	0.981	0.096	19	0.33	1.090	1.6
12	1.226	0.987	0.038	46	0.37	1.104	2.7
14	1.218	0.990	0.025	70	0.38	1.108	3.4

$$R_c = \left(B\frac{A_2}{A_1}\right)^{(1/(n-2))} t^{((n-3)/(n-2))}$$

where B is a number that is given in TABLE 1 for various values of n. TABLE 1 also gives values for some ratios, where the subscript 1 refers to the outer shock wave and 2 to the inner shock wave, except for the mass ratio which is the mass of shocked supernova gas divided by the mass of shocked circumstellar gas. It can be seen that larger values of n lead to larger density contrasts between the inner and outer regions. At the contact discontinuity, the density goes to infinity and the temperature goes to zero.

The value of n for a type II supernova is not well determined. Numerical hydrodynamic calculations by Jones, Smith, and Straka gave $n \approx 12$.[10] More recent computations of the same problem with higher resolution gave $n \approx 9$.[11] However, the initial stellar models did not include mass loss and this could alter the outer density distribution. Also, these models tend to produce too little high-velocity matter when compared to a supernova like SN 1979c.[12]

The observations of supernovae give some information on density structure when synthetic spectral fits are carried out. Branch *et al.* were able to obtain good fits to spectra of SN 1979c with $n = 7$,[13] but the results are probably not very sensitive to n and the presence of a fairly steep profile ($n > 4$) is the only firm conclusion. However, the model fits of Branch *et al.* do allow a determination of A_2 if the gas opacity is estimated. Near maximum light, $\kappa = 0.3\ \mathrm{cm^2\ g^{-1}}$ is a good estimate and the spectral fits yield $\rho = 5.30 \times 10^{67}\ t^4\ r^{-7}$ in cgs units. The density in the external medium can be expressed as $\rho = 5.05 \times 10^{14}\ r^{-2}\ (\dot{M}_{-4}/v_{wl})\mathrm{g\ cm^{-3}}$ where $\dot{M}_{-4}$ is the presupernova mass loss rate in units of $10^{-4}\ M_\odot\ \mathrm{yr^{-1}}$ and v_{wl} is the wind velocity in units of $10\ \mathrm{km\ s^{-1}}$. With these parameters, the radius of the contact discontinuity is $R_c = 2.75 \times 10^{14}$ $(\dot{M}_{-4}/v_{wl})^{1/5}\ (t/1\ \mathrm{day})^{0.8}$ cm. Other properties of the interaction region can be obtained from TABLE 1 and from Chevalier.[8]

The self-similar solutions assume adiabatic flow. This is not the case if the cooling time scale is shorter than the hydrodynamic time. The time scale for optically thin radiative cooling in the shocked circumstellar gas is long, but radiative cooling of the shocked supernova gas is marginally important and is important near maximum light for values of n larger than 7. Inverse Compton cooling by photospheric photons is marginally important near maximum light for the SN 1979c parameters. It was probably able to effectively cool the interaction region before maximum light. For the SN 1979c model with $\dot{M}_{-4} \leqq 1$, cooling is not important at and after maximum light.

One property of the solutions is that the density of the inner shocked region is larger than that of the outer shocked region. For a decelerating shell, this situation is subject to the Rayleigh-Taylor instability. The instability is expected to grow into the nonlinear regime for length scales considerably less than the thickness of the interaction region. Some mixing of the two shocked regions is likely, but the detailed results of the instability are not known.

Deviations from adiabatic flow can result if heat conduction is not impeded by magnetic fields. In a decelerating flow, the pressure is expected to increase with radius. If conduction is sufficiently strong to make the gas isothermal, the density profile will follow the pressure. Thus, the inner density is no longer larger than the outer density and the flow is not subject to the Rayleigh-Taylor instability. An additional effect occurs if the heat conduction penetrates the shock waves, giving thermal waves outside the shock waves. Liang has performed numerical calculations showing how the internal energy carried out by the conduction lowers the temperature and increases the density in the interaction region.[14]

The theory described here assumes spherical symmetry. There is some evidence for an asymmetric distribution of circumstellar mass (most of the mass in a plane) around some late type stars. This is particularly true for the extreme carbon stars.[15] However, the circumstellar matter around α Orionis does appear to be fairly symmetric.[16]

OBSERVATIONS OF TYPE II SUPERNOVAE AND THEIR INTERPRETATION

The most complete observations of type II supernovae are of SN 1979c in M100 and SN 1980k in NGC 6946. There are optical data on SN 1969l and SN 1970g, but data at other wavelengths are generally lacking. SN 1970g was detected as a radio source,[17] but the source was confused with an HII region so that the radio properties were not well determined. The discussion here will concentrate on SN 1979c and SN 1980k.

Radio

The discovery of radio emission from SN 1979c is described by Weiler *et al.*,[18] and Weiler *et al.* summarize the data on SN 1979c and SN 1980k.[19,20] Radio light curves at 6 and 20 cm are available through April 1982, and observations have continued past that date. When the large radio luminosity of the supernova is combined with size of the supernova, a brightness temperature $\gg 10^4$ K is indicated, which implies that the radiation is nonthermal in origin. Synchrotron radiation is a plausible emission mechanism. The generation of the required magnetic field and relativistic particles is unlikely in the cool, uniformly expanding supernova gas. While it has been suggested that a central pulsar plays a role,[21-23] these models have problems with the shielding of the pulsar by the supernova matter and radiation field.

Chevalier suggested that the circumstellar interaction region is the site of the radio emission.[12,24] The magnetic field may be built up by random motions generated by the Rayleigh-Taylor instability in the interaction region (cf. References 25 and 26). The relativistic electrons may also be generated by the random motions or may be accelerated in the shock waves. If the efficiency of the generation of magnetic fields and relativistic electrons is comparable to that known to occur in the galactic supernova remnants, the observed radio luminosity can be reproduced.

The observations show that there is a low-frequency turnover to the radio spectrum at early times. In the circumstellar model, the turnover is due to free-free absorption in the circumstellar medium external to the interaction shell. This mechanism is expected to be more important than other possible mechanisms, and it can explain the sharp turn on of the radio emission at one frequency. The time it takes for the interaction region to move out to a point where the gas is optically thin gives a measure of the amount of circumstellar matter. For SN 1979c, Chevalier estimated $\dot{M}_{-4}/v_{wl} = 0.5$ and for SN 1980k, $\dot{M}_{-4}/v_{wl} = 0.1$.[12] These results have an accuracy of a factor of about 2 or 3.

On the assumption that the ratios of magnetic energy density and relativistic electron energy density to thermal energy density remain constant in the interaction shell, the model predicts that the radio flux at a particular frequency varies as

$$F_\nu = \text{const.} \left(\frac{\lambda}{20\ \text{cm}}\right)^{(\gamma-1)/2} \left(\frac{t}{t_{20}}\right)^{-((\gamma+5-6m)/2)} \exp\left[-\left(\frac{\lambda}{20\ \text{cm}}\right)^2 \left(\frac{t}{t_{20}}\right)^{-3m}\right] \quad \textbf{(2)}$$

where λ is the wavelength of observation, γ is the power law index of the electron energy spectrum, t_{20} is the time at which the circumstellar medium exterior to the emitting shell has a free-free absorption optical depth of unity at 20 cm, and $m = (n - 3)/(n - 2)$. In this model γ and the temperature of the absorbing medium are assumed to be constant in time. The circumstellar temperature is generally expected to be close to 10^4 K, but at early times temperatures of the order of 10^5 K are possible because of x-ray and ultraviolet heating close to the shock front.[7,27] Radio light curves based on Equation 2 are shown in FIGURE 1 for $\gamma = 2.0$ and 2.8 and $m = 0.8$ ($n = 7$), 0.9 ($n = 12$), and 1.0 ($n = \infty$). It can be seen that the relative peak fluxes at different wavelengths provide a good measure of m. The value of γ can be determined from the late time (optically thin) spectrum; it plays a role in rate of decline from the peak flux.

A comparison of the models with the observations of Weiler *et al.*[20] shows that while all models have the correct general behavior, there are specific models that

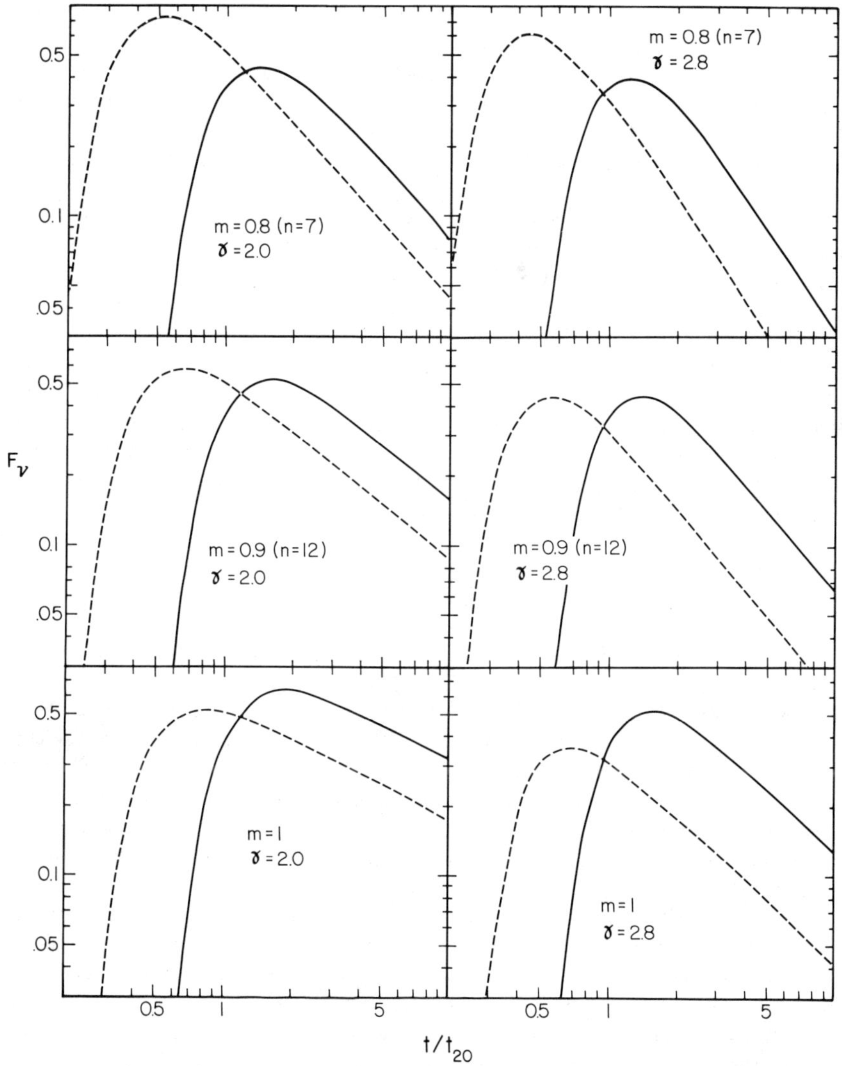

FIGURE 1. Model predictions of the radio flux as a function of time (see Equation 2). t_{20} is the time at which the free-free optical depth of the circumstellar is unity at 20 cm, F_ν is the radio flux normalized to $F_\nu(20\text{ cm}) = e^{-1}$ at $t = t_{20}$, γ is the electron energy spectral index, and m is given by $R_s \propto t^m$ where R_s is the outer shock radius. For a circumstellar medium with $\rho \propto r^{-2}$, $m = (n - 3)/(n - 2)$. The solid line is the 20-cm flux and the dashed line is the 6-cm flux.

provide a best fit. For SN 1979c, the best fit is $m = 1.0$, $\gamma = 2.8$ with $t_{20} = 950$ days and for SN 1980k it is $m = 0.9$, $\gamma = 2.0$ with $t_{20} = 190$ days. The fits are shown in FIGURE 2, where it has been assumed that the explosion of SN 1979c occurred on April 1, 1979, and that of SN 1980k on October 15, 1980. It is clear that both supernovae show fluctuations in their radio flux superposed on the general trends given by the model. The origin of the fluctuations is not known, but they indicate that the emitting region is inhomogeneous. The fits allow predictions of the future evolution of the radio sources. SN 1979c should show clear signs of a decreasing flux within the next few years. If the model assumptions remain valid, SN 1979c should be observable for several tens of years.

The circumstellar model predicts the power law dependence of F_ν on time until the interaction region reaches a place in the wind where $\dot{M}/v_w$ decreases. Then a more rapid decrease in F_ν is expected. Stellar evolution and atmosphere theory do not predict how long high values of $\dot{M}/v_w$ last before the star explodes, but wind extents of close to a parsec are possible if one value of $\dot{M}/v_w$ persists throughout the red supergiant evolution. There is some evidence that the radio flux from SN 1970g decreased suddenly at an age of about four years,[21] but the data are of poor quality. If this result is correct, a truncated wind is indicated. There are also many upper limits on the radio emission from supernovae[28-30] (see TABLE 2). Some of the upper limits are incompatible with the predicted evolution of SN 1979c, but they are compatible with the predicted evolution of SN 1980k (FIGURE 2).

X-Rays

While searches for x-rays from supernovae had previously been carried out, the Einstein Observatory observations of SN 1980k provided the first detection of an extragalactic supernova.[31] At a distance of 10 Mpc, the x-ray luminosity about 40 days after discovery of the supernova was 2×10^{39} ergs s^{-1}. This luminosity could be thermal emission from the shocked supernova gas in the interaction region.[12] The value of $\dot{M}/v_w$ derived from the radio observations predicts approximately the correct flux. The observed decline in the x-ray flux can also be reproduced if the evolution of the interaction shell was adiabatic. During the early cooling phase, the x-ray flux is expected to be constant.[27]

Another possible source of x-rays is the interaction of the radio-emitting relativistic electrons with photospheric photons.[32] In the circumstellar model, the thermal flux is likely to dominate the inverse Compton flux, but the inverse Compton mechanism cannot be ruled out. A possible problem with inverse Compton emission is that inverse Compton losses would be an important energy loss for the radio-emitting electrons around the time of the first x-ray observations. Yet the radio emission did not show a strong rise as the optical photographic light faded.

Attempts were made to detect x-rays from SN 1979c, but they were not detected.[33] The circumstellar model indicates that it should have had a flux close to the detection limit.

X-ray observations at early times are of special importance because it may be possible to measure the circumstellar absorption. A successful measurement would lead to a value for $\int n\, dr$ external to the interaction shell, while the free-free radio

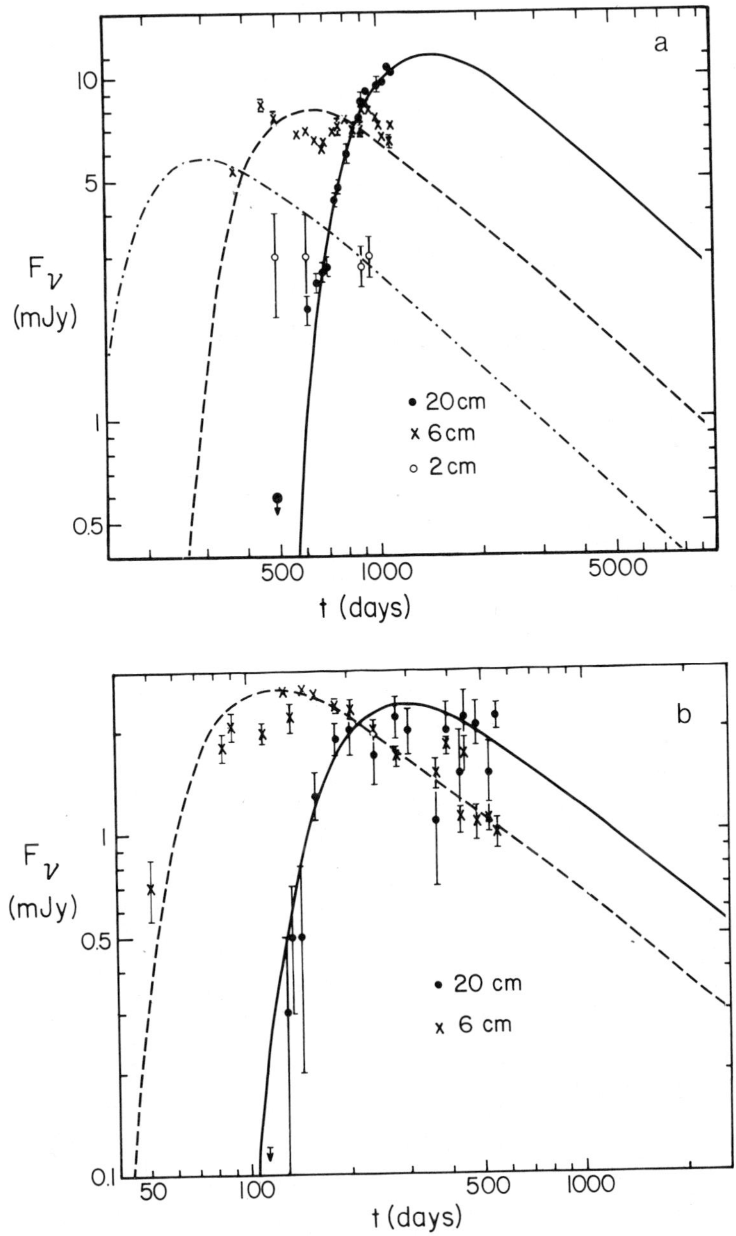

FIGURE 2. Model fits to the radio data on (a) SN 1979c and (b) SN 1980k. The data have been presented by Weiler *et al.*[20] The model curves are at wavelengths of 20 cm (solid line), 6 cm (dashed line), and 2 cm (dash-dot line). The model fit to SN 1979c has $m = 1$, $\gamma = 2.8$; and that fit to SN 1980k has $m = 0.9$, $\gamma = 2.0$.

TABLE 2. Radio Sources

Object	Supernova Type	Age (years)	Galaxy	Distance (Mpc)	τ_{decay} (years)	α ($F_\nu \sim \nu^{-\alpha}$)	Peak 6-cm Flux (mJy)	$\frac{L_{6cm}}{L_{Cas\,A}}$	Reference
Cas A	—	320	Galaxy	0.0028	60	0.77	9×10^5	1	
SN 1981b	I	1	NGC 4536	20			≦0.06	≦3	20
SN 1980k	II	0.5	NGC 6946	10	1	≥0.5	3	44	20
SN 1979c	II	1.3	NGC 4321	23	—	0.9	8	550	20
SN 1970g	II	1	M101	7	~1		5	35	17, 20, 21
SN 1968l	II	13	NGC 5236	4			<0.3 (20 cm)	<0.3	30
SN 1968d	II	11	NGC 6946	10			≦0.6	≦9	29
SN 1939c	II	40	NGC 6946	10			≦0.6	≦9	29
41.9 + 58		15–25?	M82	3.2	10	0.9	200	290	56, 57
—		50–200	NGC 4449	5	10	0.7	11	39	61, 62
—			NGC 4656	5.2			20	77	62
~30 Sources			M82	3.2			0.5–100	0.7–150	57
—			NGC 6052 (Mrk 297)	63			12	6750	64

absorption leads to a value for $\int n^2 dr$ external to the interaction shell. These measurements would give information on clumpiness in the circumstellar matter, although the analysis would be complicated by the fact that the integrals are unlikely to refer to the same shell position.

Ultraviolet

The launch of the International Ultraviolet Explorer (IUE) has made possible ultraviolet spectroscopy of nearby extragalactic supernovae. Panagia *et al.* present ultraviolet spectra of SN 1979c near the time of discovery and for a month thereafter.[34] Most of the radiation is continuum radiation; it is the extension of the optical continuum, although there is evidence for excess emission (above a blackbody spectrum) at wavelengths below 1,600 Å. There are many absorption features that can be attributed to absorption in the interstellar medium of our galaxy and in that of M100. The lines attributed to the supernova are more controversial. Emission features of N V (λ1240), Si IV (λ1400), and C IV (λ1550) were clearly present. Panagia *et al.* noted some other emission features and claimed that all the supernova features were approximately symmetrical emission lines with velocities $\leqq$ 4,000 km s^{-1}.[34] The relatively low velocity and relatively low temperature implied that the emitting gas had not yet passed through the main supernova shock wave. The lack of occultation effects in the line profiles implied that the emitting region was at least about two photospheric radii from the center of the star. This was taken as evidence for large amounts ($\sim 10^{-2}$ $M_\odot$) of circumstellar matter.

The fact that the emitting gas had not yet passed through the supernova shock but was moving at several thousand km s^{-1} suggested radiative acceleration, and two models were developed. Chevalier suggested that the gas was radiatively accelerated in ultraviolet lines,[24] a process similar to that thought to occur in hot star winds. The line photons are absorbed in a small region ahead of the main shock wave so that there is a radiatively accelerated shock precursor. However, the model only succeeded in reproducing the observed velocities for the low circumstellar densities and required $\dot{M}_{-4} v_{w1}^{-1} \leqq 0.01$. The radio observations discussed above imply a much higher value of $\dot{M}_{-4} v_{w1}^{-1}$.

The second model was that of Fransson, who suggested that radiative acceleration through electron scattering by the total supernova radiated energy could give the required velocities[7] (see Equation 1). The main problems with this suggestion are that the observed value of E_{rad} for SN 1979c fell short of that required by at least an order of magnitude and that the emission lines did not clearly become narrower with time, as would be expected for the velocity profile given by Equation 1. Fransson also performed detailed calculations of the effect of the inverse Compton scattering in the hot interaction shell on the photospheric photons and calculated the ionization structure of the circumstellar gas. He assumed $\dot{M}_{-4} v_{w1}^{-1} \approx 1$. The scattering of photospheric photons in the interaction shell did give rise to excess ultraviolet, and Fransson identified the observed excess with this effect.

The ultraviolet spectra of SN 1979c have been reduced with improved resolution and reanalyzed by Fransson *et al.*,[35] who obtain quite different conclusions from Panagia *et al.* Fransson *et al.* find that when interstellar absorptions are carefully taken into account, the emission features of highly ionized ions have velocities up to about 8,400 km s^{-1}, which is close to the velocity of the optical supernova photosphere. They conclude that the lines are formed in a narrow region close to the photosphere where the density is in the range 2×10^9 to 10^{10} cm^{-3}. This density is in agreement with the density of the expanding matter in the model discussed here in the second section. The gas is probably ionized by radiation from the hot interaction region, as shown by Fransson.[27,36] The gas can be ionized by either bremsstrahlung emission from the hot gas or Comptonized photospheric emission. Fransson argues for the latter situation in SN 1979c based on the luminosity of the ultraviolet lines and the time evolution of the line flux.

Ultraviolet spectra have also been obtained for SN 1980k.[37] There is no evidence for strong emission lines from the supernova.

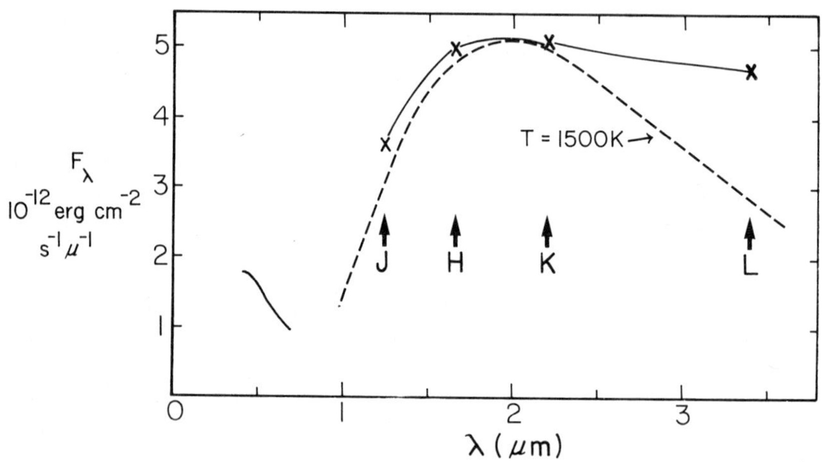

FIGURE 3. Optical and infrared[38] spectral fluxes from SN 1979c on January 3, 1980. The infrared spectrum is based on observations in four infrared photometric bands. A blackbody spectrum with $T = 1{,}500$ K is shown for comparison.

Infrared

Near maximum light, the thermal continuum emission from a supernova photosphere extends into the near infrared. However, after maximum light the infrared radiation does not decline in the same way as the optical emission. For both SN 1979c and SN 1980k, the infrared luminosity exceeded the optical luminosity starting about half a year after maximum optical light.[38,39] The infrared fluxes from SN 1979c in the standard infrared bands are shown in FIGURE 3 for January 3, 1980. The observed optical continuum from about the same time is also shown.[68] The optical continuum is actually an upper limit because it did not change with time and may not be from the supernova. The figure also shows that the spectral distribution is broader than that of a black body of a single temperature and that the maximum temperature of the emitting material was about 1,500 K if it radiated as a black body. At a distance of 23 Mpc, the infrared luminosity of SN 1979c was about 10^{42} erg s^{-1}.

Two possible emission sources for the infrared radiation are (1) emission from dust that formed in the expanding supernova matter and (2) reradiation of the supernova light by dust in the circumstellar region. Chevalier argued that the emission was associated with the supernova envelope partially on the basis of the fact that SN 1979c showed a broad Hα line that was much stronger to the blue than to the red at the time of the infrared.[24] The asymmetry could be due to dust absorption. There are two arguments against the dust-formation hypothesis. The first is that a powerful energy source is required for the infrared emission. In order to produce the 10^{42} erg s^{-1} in SN 1979c by radioactive decay of ^{56}Co, several $M_\odot$ of ^{56}Ni must have been synthesized in the explosion. This is unlikely. Pulsar energy input is another possibility, but the rotational energy of the pulsar would have to be greater than is thought to occur for newly born pulsars. For SN 1980k, optical continuum radiation was present at late times, which implies that slow-moving gas near the center absorbed the energy input from either radioactivity or a pulsar. The lack of strong reddening of the continuum then argues against dust absorption of most of the continuum.

The second argument against dust formation in SN 1979c in the envelope is stronger. If several $M_\odot$ of gas have an energy input $\gtrsim 10^{42}$ erg s^{-1} at the densities expected in the supernova envelope, the gas cannot cool to below 2,000 K, which is the temperature required for dust formation.

The alternative model of circumstellar infrared emission was first suggested by Bode and Evans.[40] Chevalier[24] discussed the circumstellar model in the context of a constant-velocity, slow presupernova wind[24] (see also References 41 and 42).

There is a critical radius, r_{cr}, inside of which the dust is evaporated by the supernova light. It is

$$r_{cr} \approx 1 \times 10^{18} \left(\frac{L_{SN}}{10^{45} \text{ ergs s}^{-1}} \right)^{1/2} \left(\frac{T_{cr}}{1{,}500 \text{ K}} \right)^{-5/2} \text{cm} \tag{3}$$

where L_{SN} is the maximum supernova luminosity and T_{cr} is the temperature at which dust grains evaporate. The luminosity 10^{45} erg s^{-1} is higher than has been observed from a type II supernova but it is expected to occur when the shock wave emerges from the supernova photosphere. The dust optical depth to the radiated energy is

$$\tau_d = 0.15 \left(\frac{\kappa_d}{300 \text{ cm}^2 \text{ g}^{-1}} \right) \left(\frac{r_{cr}}{10^{18} \text{ cm}} \right)^{-1} \dot{M}_{-4} v_{w1}^{-1} \tag{4}$$

where κ_d is the dust opacity. For SN 1979c, the total radiated infrared energy was about $E_{rad}/4$, where E_{rad} is the observed total radiated energy. Considering..hat there may have been a substantial amount of radiation emitted before discovery of the supernova, the circumstellar model is energetically feasible for $\dot{M}_{-4} v_{w1}^{-1} = 0.5$.

In the model, the dust emission flux and effective temperature remain constant until time

$$t_{cr} = \frac{2 r_{cr}}{c} = 2.1 \left(\frac{r_{cr}}{10^{18} \text{ cm}} \right) \text{yr} \tag{5}$$

After this time, the flux drops as t^{-2} and the effective temperature drops as $t^{-2/5}$. The expected spectral distribution is broader than that of a single black body because the emission is from a range of radii and because at any one radius, there is a range of dust temperatures due to the time variation of the supernova luminosity.

SN 1979c was discovered in April 1979. Between January and July 1980, the infrared flux dropped by a factor of 8 while the effective temperature remained approximately contant.[38] The constant temperature is compatible with evolution during the time $t < t_{cr}$. However, the flux decrease is not compatible with the simple model. It may indicate an asymmetry in the distribution of circumstellar dust.

SN 1980 was discovered in October 1980. Between May and October 1981, the infrared flux dropped by a factor of 3.4 while the dust temperature dropped from 1,100 K to 840 K.[43] This evolution is approximately consistent with the predicted $t > t_{cr}$ and implies that $t_{cr} \leqq 0.6$ yr. For $T_{cr} = 1{,}500$ K, this implies a peak supernova luminosity $\lesssim 8 \times 10^{43}$ erg s^{-1}. The fact that the dust temperature was lower than that implied for SN 1979c at a comparable time also indicates a lower peak luminosity. The observed infrared radiated energy was only about 1% of the optical radiated energy.[43] For $\dot{M}_{-4} v_{w1}^{-1} \approx 0.1$, $r_{cr} = 2.5 \times 10^{17}$ cm, and similar dust properties to those in the SN 1979c circumstellar medium, the infrared energy would be several times greater than that observed. The observations imply that either there is a decrease in the circumstellar density steeper than r^{-2} or the grain properties are different from those in the SN 1979c medium. In the present theory, the circumstellar mass around SN 1979c is at least about 2 $M_\odot$. In his analysis of the infrared emission, Dwek estimated circumstellar masses of 1–5 $M_\odot$ and 0.1–0.4 $M_\odot$ for SN 1979c and SN 1980k respectively.[44]

If the circumstellar emission theory is correct, the observations imply the presence of extended regions of circumstellar matter. The light pulse from the supernova travels at about 30 times the supernova shock velocity so that the interaction of the shock wave with circumstellar matter could continue for about 30 years. Another implication is that it is possible for a substantial fraction of the supernova light to be absorbed and reradiated in the infrared. If the circumstellar matter is somewhat denser than that around SN 1979c, the supernova could be an infrared event rather than an optical one. Morris and Jura have suggested that NML Cygnus will give rise to such an "infrared supernova."[45]

The presence of dense, dusty circumstellar matter implies that the supernova progenitor stars of SN 1979c and SN 1980k were dust enshrouded if the dust formed in the stellar wind close to the stellar surface. The progenitors may have been strong infrared sources, but emitted little visual light. Under these circumstances, plates taken in visual light could not set useful limits on the luminosity of the progenitor stars, and Thompson's upper limit of 18 $M_\odot$ on the mass of the SN 1980k progenitor may not be applicable.[46]

Optical

It has been known for some time that type II supernovae show strong Hα emission near maximum light and that this emission persists and becomes more pronounced at late times. The emission presents a problem because the photospheric temperature is too low to significantly ionize hydrogen out of its ground state. Kirshner and Kwan suggested that the density in the envelope is sufficiently high that there is a substantial population of hydrogen atoms with electrons in the $n = 2$ level.[47] These atoms can be ionized and excited by Balmer continuum photons, and there is the possibility of emitting a large Hα flux. However, this mechanism fails to produce the observed Hα fluxes at late times. Gordon called attention to this and suggested that the source of ionization is energetic protons created by a central pulsar.[48]

The Hα problem appeared again in observations of SN 1979c. Branch *et al.* computed synthetic spectra to compare with observed spectra near maximum light.[13] The computed spectra, which took the radiation field corresponding to the continuum temperature, failed to reproduce the Hα emission. Branch *et al.* suggested that shock phenomena were responsible for the emission. The Hα emission becomes more pronounced at late times; Branch *et al.* show a spectrum from seven months after maximum, and Kirshner and Chevalier obtained a spectrum one year after maximum.[68] There was blue continuum emission present, but it did not change with time which suggests that it was not associated with the supernova. The Hα luminosity declined from about 1.3×10^{40} ergs s^{-1} in November 1979 to about 5×10^{39} ergs s^{-1} in April 1980.

The late observations of SN 1979c require an excitation source other than photospheric emission. Radioactivity or pulsar energy input are possibilities, as discussed for the infrared emission. The energy requirements are more modest for the Hα emission, which makes these mechanisms more plausible. However, for SN 1979c near maximum light and for other supernovae even at late times, there is photospheric emission as well as strong Hα emission. The presence of photospheric emission implies that there is a large column density of gas that is likely to absorb the energy input from a central source (radioactivity or a pulsar). The energy input is thus thermalized and appears as photospheric emission.

This argument suggests that the energy source is external to the photosphere and that the circumstellar interaction region is a plausible source. The energy transfer from the interaction region can be by radiation or particles. Radiation from the interaction region in SN 1979c may not be the source because there are strong limits on the x-ray flux.[33] A very high efficiency of conversion of hard radiation to Hα would be required. Particles are another possibility, but the details of the energy transfer process are poorly understood. The particles can be either thermal with energy transfer by electron heat conduction or relativistic. The radio observations imply that relativistic particles are present. However, the transfer of charged particles in the collisionless limit is poorly understood. Plasma instabilities may severely limit energy transfer by particles.

TYPE I SUPERNOVAE

The phenomena described in the previous section for type II supernovae have generally not been observed in type I supernovae. Type I supernovae do show optical line emission at late times, but this emission is adequately explained by radioactive heating.[49,50] Radio emission has not been detected from the type I supernova SN 1981b at a luminosity limit several times lower than that observed from SN 1980k[20] (see TABLE 2).

The three historical galactic supernovae SN 1006, SN 1572, and SN 1604 can plausibly be classified as type I events. All of these supernovae created remnants that are fairly strong x-ray emitters,[51–53] which implies that they are interacting with moderately dense gas. Although there are uncertainties due to nonequilibrium ionization and heavy element overabundances, White and Long deduce an ambient hydrogen density $n_0 > 0.1\ \mathrm{cm}^{-3}$ for the SN 1604 remnant.[53] This density is higher than might be expected for the ambient interstellar medium, especially because the remnants are some distance from the galactic plane. This has led to the suggestion that the remnants are interacting with circumstellar mass loss[53] (see also Reference 54). The required amount of mass loss would be similar to that believed to occur in the progenitors of type II supernovae (although more extended). However, Chevalier argued that the remnants appear to be expanding in a constant density medium,[55] as opposed to a circumstellar medium with $\rho \propto r^{-2}$. The best evidence for this is the observed expansion rates for the SN 1572 and SN 1006 remnants. Further evidence is the x-ray and radio radial profiles of the three remnants.

YOUNG EXTRAGALACTIC REMNANTS

The radio observations of type II supernovae already show that there is considerable variation in flux from object to object. It is likely that some supernovae give rise to interactions that are even more luminous than SN 1979c. These events may be rare, but should be detectable in external galaxies. There are a number of nonthermal nonnuclear compact radio sources in external galaxies (see TABLE 2); many of these sources show time variability (τ_{decay} in TABLE 2 is the time for the flux to decrease by a factor of two). Given the additional fact that the sources are more luminous than in Cas A, their identification with supernovae interacting with circumstellar matter is plausible.

The most detailed radio observations are available on the source 41.9 + 58 in M82.[56,57] The observed radio evolution is remarkably similar to that of the supernovae SN 1979c and SN 1980k. The source showed a decreasing flux with approximately constant spectral index at high frequency and a low-frequency turnover. The turnover frequency has decreased with time. The observed properties can be fitted by a circumstellar interaction model in which the supernova occurred sometime between 1956 and 1966 (the later limit comes from the fact that the source was observed in that year) and the presupernova mass loss had $\dot{M}_{-4}v_{w1}^{-1} \approx 4$. There is again a factor of 2–3 uncertainty in the mass loss rate. While the mass loss is high, such high rates have been deduced for dust-enshrouded stars in our galaxy. Scaling from the observed radio luminosities of SN 1979c and SN 1980k gives about the correct radio luminosity for 41.9 + 58.

The main problem with the supernova model is that it is inconsistent with the very long baseline interferometry (VLBI) observations by Geldzahler *et al.*[58] The 1974 measurements by Geldzahler *et al.* showed that most of the 8 GHz flux came from a region only 0.0015 arcsec in diameter ($\sim 7 \times 10^{16}$ cm). This is too small by a factor of about 5 to be consistent with the circumstellar interaction model. More recently Jones, Sramek, and Terzian carried out 18 cm VLBI observations which indicated a substantial fraction of the flux in an elongated structure up to ~ 0.015 arcsec in position angle 56°.[59] While this size is more compatible with the model, the elongated structure is not compatible with the model in its spherically symmetric form. However, the mass loss may take an asymmetric form with most of the mass in a plane (see the section Hydrodynamic Theory of the Interaction). In this case, the interaction region would move out preferentially along the poles and an elongated emission region is possible. More definitive VLBI observations of 41.9 + 58 are crucial for testing the supernova model.

Another well-studied object is the source in NGC 4449. It is less obscured than the M82 source and it has been identified with an optical source that shows motions typical of supernovae (see Reference 60 and references therein). VLBI observations of the radio source indicate a size <0.07 arcsec and perhaps <0.04 arcsec.[61,62] When this size is combined with the observed optical velocities, an upper age limit of 200 years is implied. De Bruyn *et al.* suggest that the supernova occurred before 1925 because there is some evidence for the presence of the optical object on a plate taken in that year,[63] but this result is not secure. The optical line observations by Blair *et al.*[60] imply that the pressure in the line emitting region is somewhat more than 10 times that in Cas A. If the remnants have the same thermal energy content, this implies a size 2.3 times smaller than that of Cas A and a correspondingly smaller age. If the thermal energy is still considerably less than the kinetic energy, a smaller age is possible.

The source in Markarian 297 (NGC 6052), discovered by Heeschen, Heidmann, and Yin,[64] is unusual because of its high luminosity and because it increased in flux by a factor of 3 over a period of 27 months. Heeschen *et al.* have suggested that it is related to multiple supernovae.

It can be seen in TABLE 2 that the radio spectral indices of these young objects are larger than that of typical galactic supernova remnants, which have an average index of about 0.5. In this connection, it is of interest that the spectrum of Cas A appears to be flattening with time.[65] Another feature of the observations is that both SN 1979c[20] and the NGC 4449 source[61] show variability at the 10–20% level. The processes of particle acceleration and magnetic field amplification are not sufficiently well understood to be able to explain this phenomenon.

The existing results suggest that not all supernovae result in bright radio sources at an early age ($\lesssim 100$ years) but that a substantial fraction do. These sources should be especially prevalent in regions of massive star formation. A good example is the set of sources found in the central region of M82.[57] Once such sources are discovered, follow-up observations at x-ray and optical wavelengths will be valuable. For the brighter radio sources, VLBI observations will provide a confrontation with detailed theoretical models.

EFFECTS ON DYNAMICS AND LIGHT CURVES

The previous discussion strongly suggests that there have been varying degrees of mass loss from a massive star at the time that it explodes. The early dynamical evolution of the explosion depends on the amount of mass loss. Based on the fraction of the envelope that is lost, three types of evolution appear to be possible. I call them types A, B, and C in order to contrast them with supernova types. In type A, a small fraction of the stellar envelope has been lost by the time of the explosion. In type B, a substantial fraction

($\gtrsim$ half) of the hydrogen envelope has been lost, and in type C essentially all of the envelope has been lost.

The type A case applies to most calculations of stellar evolution and supernova explosions. The extended hydrogen envelope makes up most ($\sim$⅔) of the total stellar mass. During the supernova explosion, the mantle matter initially has most of the supernova energy, but it is decelerated by the envelope gas.[1,3] The mantle gas subsequently expands slowly and its high column density can give a photosphere to late times (several hundred days). The supernova will interact with circumstellar medium as described in the second section, but type A events will be of relatively low luminosity. The interaction only affects the outer part of the expanding supernova gas, and the deceleration and thermalization of the gas must wait for the interaction with the interstellar medium.

In type B events, an extended envelope is still present at the time of the explosion, but it is of such low mass that it cannot effectively decelerate the mantle gas. In this case, there is too little slow-moving gas to give a photosphere at late times. The interaction with circumstellar matter is likely to be bright and the supernova ejecta can be substantially decelerated by the interaction. If the surrounding interstellar medium density is low, the supernova undergoes a period of adiabatic expansion after having shocked the circumstellar gas. This phase lasts until the interaction with the interstellar medium is the dominant effect.

Case C is even more extreme in that the envelope is absent and the exploding star has a relatively small radius. Wolf-Rayet stars are probably in this class. Stars that have lost their envelopes may be hot enough to generate fast winds which can create a cavity around the star. This has the effect of delaying the interaction with the dense circumstellar gas. Otherwise the evolution would be similar to that of case B.

Cases A and B have the extended envelopes that are required to give the observed properties of type II supernovae. The difference between them is that only case A events can maintain a photosphere at late times. A possible example of a case A event is SN 1969l, which had a B luminosity observable for over 400 days.[66] A possible example of a case B event is SN 1979c, which did not appear to have significant continuum emission at late times, although it did have strong Hα emission. The estimated circumstellar mass based on the circumstellar infrared emission model is $>2\ M_{\odot}$ (see Infrared section), which is compatible with the notion that a substantial fraction of the envelope was lost in presupernova mass loss. As noted in the Infrared section, some of the type B events may have so much circumstellar dust that they are infrared supernovae rather than optical supernovae.

The type C events are probably not normal supernovae at all, but are faint because of their small radii at the time of the explosion. Chevalier argued that the Cas A explosion was such an event.[67]

These arguments suggest that there is a close correlation between the explosion events and the presupernova mass loss evolution. While the circumstellar interaction is expected to be stronger in going from type A to type C, this may not be the case if there are systematic differences between the types with regard to total mass or energy. Galactic studies of mass loss from massive stars should provide valuable information on this matter. If mass is the primary stellar parameter for single stars, it may be possible to determine a correspondence between mass ranges and the explosion types. The type will also be influenced by presence in a close binary system.

ACKNOWLEDGMENTS

I am grateful to E. Dwek, C. Fransson, M. Merrill, and K. Weiler for discussions of their work and for information in advance of publication.

REFERENCES

1. CHEVALIER, R. A. 1976. Astrophys. J. **207:** 872.
2. FALK, S. W. & W. D. ARNETT. 1977. Astrophys. J. Suppl. **33:** 515.
3. WEAVER, T. A. & S. E. WOOSLEY. 1980. Ann. N.Y. Acad. Sci. **336:** 335.
4. REIMERS, D. 1975. Mem. Soc. R. Sci. Liege Ge Ser. **8:** 369.
5. ZUCKERMAN, B. 1980. Annu. Rev. Astron. Astrophys. **18:** 263.
6. EPSTEIN, R. I. 1981. Astrophys. J. Lett. **244:** L89.
7. FRANSSON, C. 1982. Astron. Astrophys. **111:** 150.
8. CHEVALIER, R. A. 1982. Astrophys. J. **258:** 790.
9. NADYOZHIN, D. K. 1982. Institute of Theoretical and Experimental Physics. Moscow, USSR. (Preprint.)
10. JONES, E. M., B. W. SMITH & W. C. STRAKA. 1981. Astrophys. J. **249:** 185.
11. JONES, E. M. 1982. Private communication.
12. CHEVALIER, R. A. 1982. Astrophys. J. **259:** 302.
13. BRANCH, D., S. W. FALK, M. L. MCCALL, P. RYBSKI & A. K. UOMOTO. 1981. Astrophys. J. **244:** 780.
14. LIANG, E. 1982. Private communication.
15. COHEN, M. & G. D. SCHMIDT. 1982. Astrophys. J. **259:** 693.
16. BERNAT, A. P. & D. L. LAMBERT. 1976. Astrophys. J. **210:** 395.
17. ALLEN, R. J., W. M. GOSS, R. EKERS & A. DE BRUYN. 1976. Astron. Astrophys. **48:** 253.
18. WEILER, K. W., J. M. VAN DER HULST, R. A. SRAMEK & N. PANAGIA. 1981. Astrophys. J. Lett. **243:** L151.
19. WEILER, K. W., R. A. SRAMEK, J. M. VAN DER HULST & N. PANAGIA. 1982. *In* Supernovae. M. J. Rees & R. J. Stoneham, Eds.: 281. Reidel. Dordrecht, the Netherlands.
20. WEILER, K. W., R. A. SRAMEK, J. M. VAN DER HULST & N. PANAGIA. 1983. Int. Astron. Union Symp. **101.** (In press.)
21. MARSCHER, A. P. & R. L. BROWN. 1978. Astrophys. J. **220:** 474.
22. PACINI, F. & M. SALVATI. 1981. Astrophys. J. Lett. **245:** L107.
23. SHKLOVSKY, I. S. 1981. Sov. Astron. Lett. **7:** 263.
24. CHEVALIER, R. A. 1981. Astrophys. J. **251:** 259.
25. GULL, S. F. 1973. Mon. Not. R. Astron. Soc. **161:** 47.
26. GULL, S. F. 1975. Mon. Not. R. Astron. Soc. **171:** 263.
27. FRANSSON, C. 1982. Lund Observatory. Lund, Sweden. (Preprint.)
28. BROWN, R. L. & A. P. MARSCHER. 1978. Astrophys. J. **220:** 467.
29. ULMER, M. P., P. C. CRANE, R. L. BROWN & J. M. VAN DER HULST. 1980. Nature **285:** 151.
30. COWAN, J. J. & D. BRANCH. 1982. Astrophys. J. **258:** 31.
31. CANIZARES, C. R., G. A. KRISS & E. D. FEIGELSON. 1982. Astrophys. J. Lett. **253:** L17.
32. BEALL, J. H. 1979. Astrophys. J. **230:** 713.
33. PALUMBO, G. G. C., T. MACCACORO, N. PANAGIA, G. VETTOLANI & G. ZAMORANI. 1981. Astrophys. J. **247:** 484.
34. PANAGIA, N. *et al.* 1980. Mon. Not. R. Astron. Soc. **192:** 861.
35. FRANSSON, C., P. BENVENUTI, C. GORDON, K. HEMPE, G. G. C. PALUMBO, N. PANAGIA, D. REIMERS & W. WAMSTEKER. 1983. Astron. Astrophys. (In press.)
36. FRANSSON, C. 1982. Lund Observatory. Lund, Sweden. (Preprint.)
37. PANAGIA, N. *et al.* 1980. Int. Astron. Union. Circ. No. 3537.
38. MERRILL, K. M. 1980. Int. Astron. Union Circ. No. 3444. (And private communication.)
39. TELESCO, C., E. BECKLIN, R. KOEHLER & I. GATLEY. 1981. Int. Astron. Union Circ. No. 3613.

40. BODE, M. F. & A. EVANS. 1980. Mon. Not. R. Astron. Soc. **193:** 21P.
41. FALK, S. W. & J. M. SCALO. 1975. Astrophys. J. **202:** 690.
42. WRIGHT, E. L. 1980. Astrophys. J. Lett. **242:** L23.
43. DWEK, E., M. F. A'HEARN, E. E. BECKLIN, R. BROWN, R. W. CAPPS, H. L. DINERSTEIN, I. GATLEY, D. MORRISON, C. M. TELESCO, A. T. TOKUNAGA, M. W. WERNER & G. C. WYNN-WILLIAMS. 1983. Astrophys. J. (In press.)
44. DWEK, E. 1983. Astrophys. J. (In press.)
45. MORRIS, M. & M. JURA. 1983. Astrophys. J. **267:** 179.
46. THOMPSON, L. A. 1982. Astrophys. J. Lett. **257:** L63.
47. KIRSHNER, R. P. & J. KWAN. 1975. Astrophys. J. **197:** 415.
48. GORDON, C. 1978. Astrophys. J. **220:** 484.
49. MEYEROTT, R. E. 1980. Astrophys. J. **239:** 257.
50. AXELROD, T. S. 1980. *In* Proceedings, Texas Workshop on Type I Supernovae. J. C. Wheeler, Ed.: 80. University of Texas. Austin, Texas.
51. PYE, J. P., K. A. POUNDS, D. P. ROLF, F. D. SEWARD, A. SMITH & R. WILLINGALE. 1981. Mon. Not. R. Astron. Soc. **194:** 569.
52. REID, P. B., R. H. BECKER & K. S. LONG. 1982. Astrophys. J. **261:** 485.
53. WHITE, R. L. & K. S. LONG. 1983. Astrophys. J. **264:** 196.
54. FABIAN, A. C., G. C. STEWART & W. BRINKMANN. 1982. Nature **295:** 508.
55. CHEVALIER, R. A. 1982. Astrophys. J. Lett. **259:** L85.
56. KRONBERG, P. P., P. BIERMANN & F. SCHWAB. 1981. Astrophys. J. **246:** 751.
57. KRONBERG, P. P. & P. BIERMANN. 1983. Int. Astron. Union Symp. **101.** (In press.)
58. GELDZAHLER, B. J., K. I. KELLERMANN, D. B. SHAFFER & B. G. CLARK. 1977. Astrophys. J. Lett. **215:** L5.
59. JONES D. L., R. A. SRAMEK & Y. TERZIAN. 1981. Astrophys. J. **246:** 28.
60. BLAIR, W. P., R. P. KIRSHNER & P. F. WINKLER. 1983. Astrophys. J. **272:** 84.
61. BIGNELL, R. C. & E. R. SEAQUIST. 1983. Astrophys. J. **270:** 140.
62. DE BRUYN, A. G. 1983. Astron. Astrophys. **119:** 301.
63. DE BRUYN, A. G., W. M. GOSS & H. VAN WOERDEN. 1981. Astron. Astrophys. **94:** L25.
64. HEESCHEN, D. S., J. HEIDMANN & Q. F. YIN. 1983. Astrophys. J. Lett. **267:** L73.
65. DENT, W. A., H. D. ALLER & E. T. OLSEN. 1974. Astrophys. J. Lett. **188:** L11.
66. CIATTI, F., L. ROSINO & F. BERTOLA. 1971. Mem. Soc. Astron. Ital. **42:** 163.
67. CHEVALIER, R. A. 1976. Astrophys. J. **208:** 826.
68. KIRSHNER, R. P. & R. A. CHEVALIER. Unpublished data.

Interaction of Supernova Ejecta with Interstellar and Circumstellar Matter

E. P. LIANG

Lawrence Livermore National Laboratory[a]
University of California
Livermore, California 94550

R. A. CHEVALIER[b]

Department of Astronomy
University of Virginia
Charlottesville, Virginia 22903

This paper summarizes the results of a series of one-dimensional numerical simulations of supernova ejecta interacting with interstellar (ISM) and circumstellar (CSM) matter. This work was motivated by the recent publications of Chevalier,[1,2] which show that the late-time behavior of such interactions can be modeled by certain exact analytic self-similar solutions corresponding to ideal gas spheres of idealized profiles expanding into a rarefied ambient medium. A novel aspect of these solutions is that the density profile in the interaction region between the forward and reverse shocks appears to be Rayleigh-Taylor unstable. It is suspected that such instabilities may be responsible for some of the recently observed SNR features (e.g., Einstein observations by Seward).[3] Hence the goal of the current work is twofold: (a) to check the validity and limitations of the self-similar solutions, and (b) to see the change in the shock structure when electron and ion thermal conductivities are included and $T_{ion} > T_e$.

We treat two separate cases:

A. In the ISM case, the ambient density is assumed uniform with $\rho_0 = 8 \times 10^{-24}$ g/cm^3. The stellar ejecta has initial density profile $\rho = \rho_0 (r/r_s)^{-7}$ and velocity profile $v = v_0 (r/r_s)$ for $r \leqq r_s$ where $r_s = 1.8 \times 10^{18}$ cm and $v_0 = 2 \times 10^4$ km/s.
B. For the CSM case, the ambient density is assumed to be $\rho = (10^{-14}$ g/cm$^3)(r/r_s)^{-2}$ where $r_s = 2 \times 10^{14}$ cm. The ejecta has initial density profile $\rho = (10^{-13}$ g/cm$^3)(r/r_s)^{-7}$ and velocity profile $v = (2.9 \times 10^4$ km/s$)(r/r_s)$ for $r \leqq r_s$.

We use a spherical Lagrangian hydrodynamics with fine zoning concentrated in the interaction region ($\geqq$50 zones). In conducting cases the heat conduction flux is limited so that $F_e \leqq n_e\,(0.5)\,(\tfrac{3}{2}\,kT_e)\,(3\,kT_e/m_e)^{1/2}$ and $F_{ion} \leqq n_i\,(1.23)\,(\tfrac{3}{2}\,kT_i)\,(3\,kT_i/m_i)^{1/2}$ and electron-ion is coupled via coulomb scattering. We assume ideal gas with $\gamma = \tfrac{5}{3}$.

FIGURES 1 and 2 illustrate the late-time profiles of the interaction regions. The nonconducting, $T_i = T_e$, runs agree quite well with the similarity solutions (with the exception of one zone at the contact surface). The conducting, $T_i > T_e$, runs show that the density profiles are stable and that T_e is much lower. It is conceivable that in the

[a]Operated under Department of Energy Contract No. W-7405-Eng-48.
[b]To whom correspondence should be addressed.

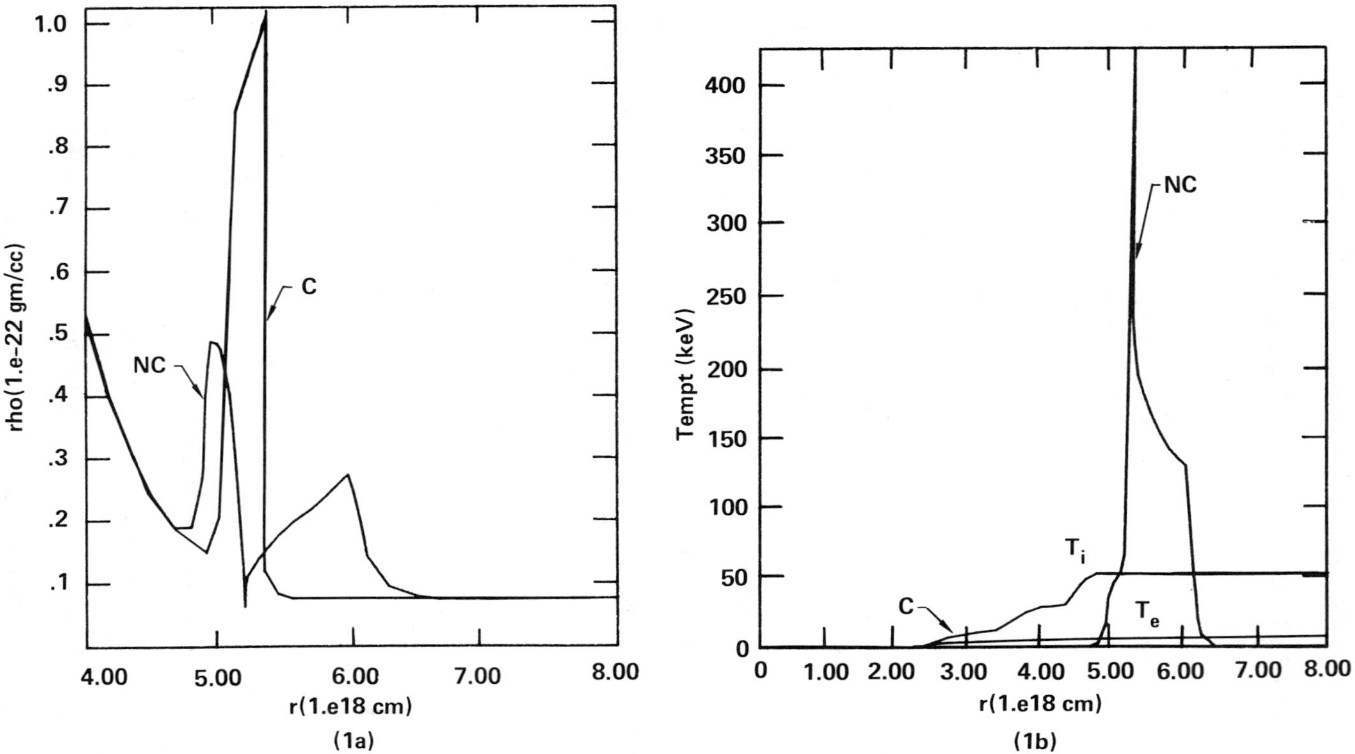

FIGURE 1. Density (a) and temperature (b) profiles at $t = 5 \times 10^9$ sec for the ISM case. C = thermally conducting, $T_i > T_e$, and NC = nonconducting, $T_i = T_e$.

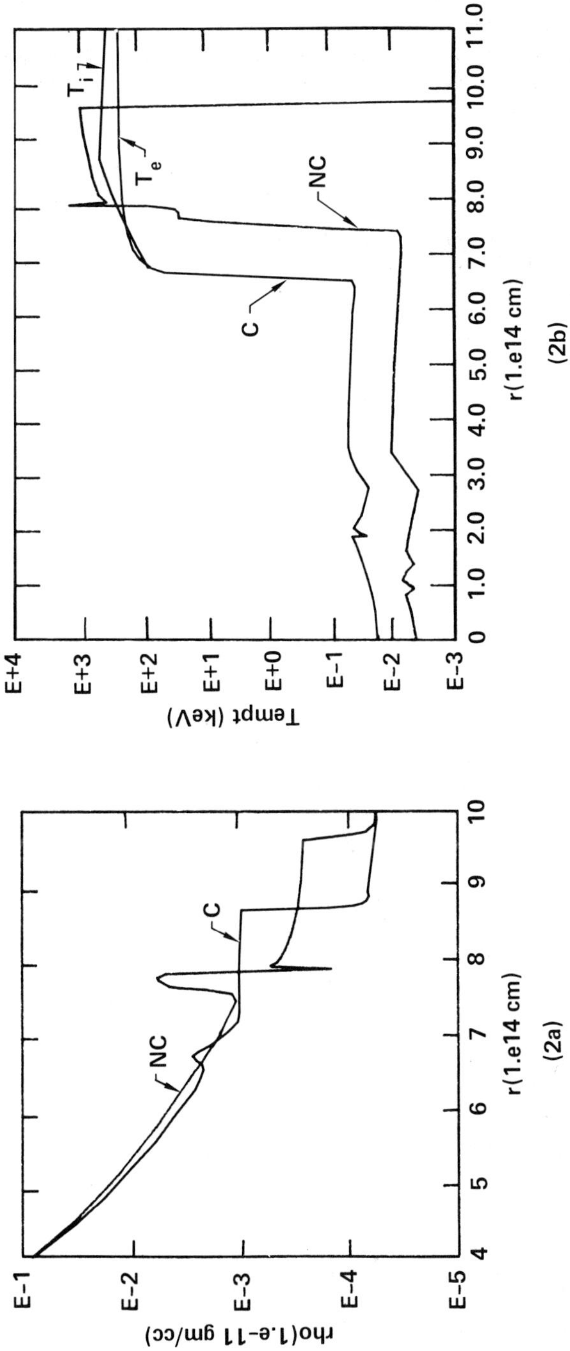

FIGURE 2. Same as FIGURE 1 for the CSM case at $t = 3.2 \times 10^5$ sec.

CSM case the thermal wave may alter the free-free absorption of early radio waves from the SNR and provide a test for their presence. Of course, in the presence of magnetic fields, the gyroradii would be so short that thermal conduction will be inhibited across field lines, but not along field lines. The manifestion of such dynamics remains to be explored.

REFERENCES

1. CHEVALIER, R. 1982. Astrophys. J. Lett. **259:** L85.
2. CHEVALIER, R. 1982. Astrophys. J. **258:** 790.
3. SEWARD, F. D. 1981. Talk at NATO Advanced Study Institute on Supernova, Cambridge, England.

Physics of Gamma-Ray Bursts[a]

D. Q. LAMB

Harvard-Smithsonian Center for Astrophysics
Cambridge, Massachusetts 02138

INTRODUCTION

The Texas Symposium on Relativistic Astrophysics was last held in Texas in 1974, a year or so after the announcement of the discovery of gamma-ray bursts.[1] In his review of burst models then, Ruderman wryly noted that there were more tenable ideas than observed bursts.[2] It is now widely believed, however, that the bursts come from strongly magnetic neutron stars. We begin by reviewing the events that have led to this changed perception. We mention current models of the bursts involving neutron stars, and comment on them in the light of the most recent observations. We then discuss the physical properties of the emission region and the radiation processes expected in strong magnetic fields. We emphasize, in particular, that the observed burst spectra require the emission region to be optically thin. This means that the energy of the emitting plasma and/or the plasma itself must be continuously replenished during a burst. It also means that the cooling time scale of the emitting plasma is much shorter than the observed durations of the bursts. This implies that the burst intensity and spectrum can vary on extremely short time scales, and that the burst duration must have a separate explanation. We also emphasize that synchrotron emission is favored as the gamma-ray production mechanism, since for moderate or strong magnetic fields it is far more efficient than bremsstrahlung, inverse Compton scattering, and pair annihilation, and it is the *only* mechanism capable of satisfying the constraint on optical thinness while producing the observed luminosity. We conclude by mentioning several important issues that any attempt to explain gamma-ray bursts must face.

For excellent observational reviews, see Hurley,[3,4] Vedrenne,[5] Mazets and Golenetskii,[6] Cline,[7] Mazets *et al.*,[8] Klebesadel *et al.*,[9] and Teegarden.[10] For recent theoretical reviews, see Lamb[11] and Rees.[12]

WHY NEUTRON STARS?

TABLE 1 lists the general properties of gamma-ray bursts. The first spacecraft instruments designed expressly for gamma-ray burst spectroscopy were flown in 1978, and ushered in a period of rapid observational development. Although some of the properties listed in TABLE 1 had been known before, many new features were discovered that suggest that gamma-ray bursts come from magnetic neutron stars. We review some of these discoveries below.

[a]This research was supported in part by National Aeronautics and Space Administration Grant NAGW-246.

5 March 1979 Event

Of central importance in focusing attention on neutron stars as the sources of gamma-ray bursts was the famous 5 March 1979 event (for a comprehensive discussion of the observations, see Cline;[13] a specific model of the event is discussed by Ramaty, Lingenfelter, and Bussard).[14] The time profile of the burst is shown in FIGURE 1.[15] The rise time of the initial spike was less than 0.25 ms, comparable to the dynamical time scale at the surface of a neutron star, and its duration was only 150 msec, again indicating a compact source.[15,16] The large peak intensity of the initial spike[16] is illustrated in FIGURE 2, which compares the peak intensity of the 5 March burst with those of other bursts. The less intense flux that followed the initial spike exhibited 8-second pulsations,[15–18] as shown in FIGURE 1, similar in character to those widely observed in accreting neutron star x-ray sources.

TABLE 1. Gamma-Ray Burst Characteristics

Property	Observed Values
Intensity	$\sim 10^{-5}$–10^{-3} ergs cm^{-2} [a]
Distance	$\sim$100–300 pc (?)
Luminosity	$\sim 10^{37}$ $(d/100\ \mathrm{pc})^2$ ergs s^{-1}
Energy	$\sim 10^{38}$ $(d/100\ \mathrm{pc})^2$ ergs
Duration	$\sim$0.1–30 s
Temporal structure	aperiodic variations $\lesssim$0.1 s
Occurrence rate	$\sim$10 yr^{-1} prior to 1979; $\sim$100 yr^{-1} since[a]
Recurrence rate	$\sim$0.1–10 yr^{-1} [a]
Spectral energy	$\gtrsim$300 keV
Spectral features	red-shifted pair annihilation line @$\sim$400 keV, cyclotron (?) features @$\sim$50 keV
Sources	typically no candidates $m_v \lesssim 22^m$, no known x-ray sources
Source directions	$\sim$isotropic
Intensity spectrum	$\sim S^{-3/2}$ for $S > 3 \times 10^{-5}$ ergs cm^{-2}; upper limits claimed below extrapolation of power law for 10^{-7} ergs cm$^{-2} \lesssim S \lesssim 3 \times 10^{-5}$ ergs cm^{-2}

[a]Dependent on detector threshold.

FIGURE 3 shows spectra taken at two different times during the burst.[15] The spectrum is relatively soft, with a characteristic temperature $kT \simeq 30$ keV. During the initial 4 seconds, the spectrum showed a feature at $\simeq$430 keV that appears to be an emission line produced by pair annihilation. If so, it is red shifted below 511 keV by just the amount expected for radiation coming from near the surface of a neutron star.

FIGURES 4 and 5 show the preliminary[19] and final[20] gamma-ray error boxes for the 5 March event, superimposed on optical and x-ray[21] images of the region. The burst source, GBS0526-66, is clearly coincident with the supernova remnant N49 in the large magellanic cloud. This has led to the supposition that the burst source is a neutron star formed in the explosion that produced the supernova remnant N49. If so, the source lies at a distance of 55 kpc and the peak intensity of the burst was $\simeq 10^{45}$ ergs s^{-1} (i.e., the source was briefly 10 times more luminous than the Galaxy!). However, if the luminosity of the 8-second pulsations in the decaying portion of the burst is similar

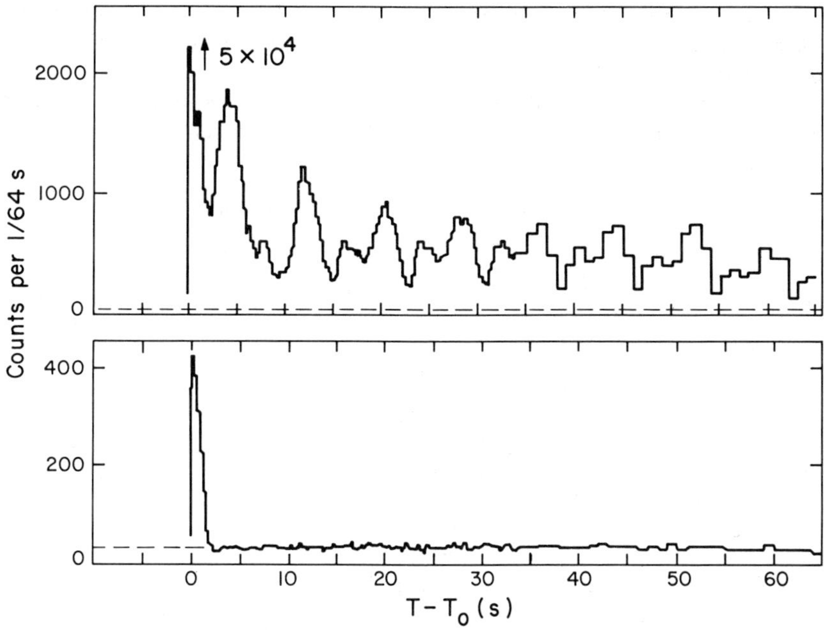

FIGURE 1. Time profiles of the 5 and 6 March 1979 bursts. Upper panel: the March 5 burst, showing the 8-second periodicity. Lower panel: the March 6 burst. (From Mazets *et al.* by permission of *Nature.*)[15]

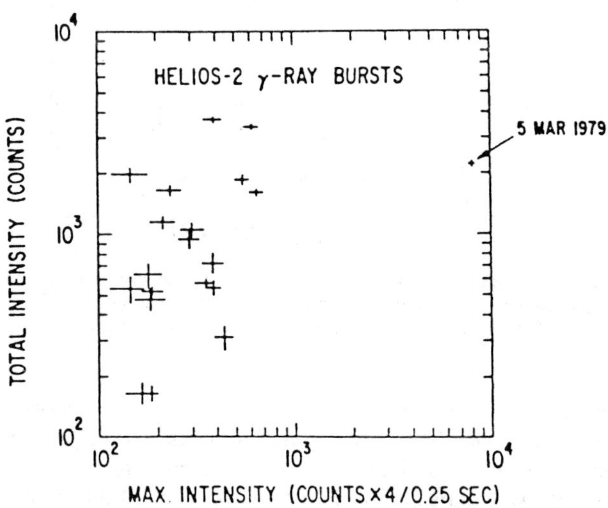

FIGURE 2. Total versus maximum intensity of the bursts observed with Helios-2 from 1976 through the 5 March 1979 burst. (From Cline *et al.* by permission of *Astrophys. J. Lett.*)[16]

to that of accreting neutron star x-ray sources, $\simeq 10^{37}$–10^{38} ergs s^{-1}, the distance is more typical of other burst sources, i.e., $\simeq$100–300 pc. Thus the identification with N49 is controversial.

During the two months following the 5 March event, 3 successively weaker bursts were detected from the same direction.[6,15] The first of these occurred only a day later,

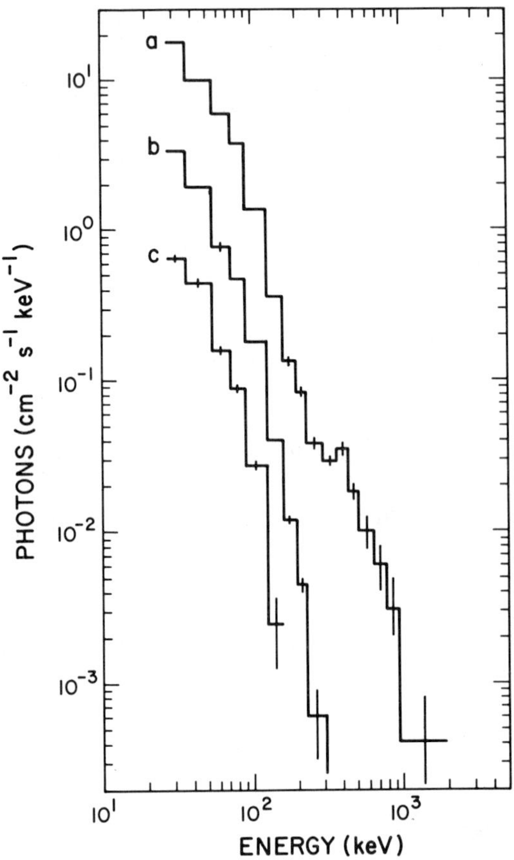

FIGURE 3. Spectra of the 5 and 6 March 1979 bursts. (**a**) Initial 4 seconds of the March 5 burst; (**b**) time-averaged spectrum of the pulsed radiation (i.e., the spectrum from 4–32 seconds) of the 5 March burst; (**c**) 6 March burst. (From Mazets *et al.* by permission of *Nature*.)[15]

on 6 March. Its time profile and spectrum are shown in FIGURES 1 and 2, for comparison with the 5 March event. All 3 weaker bursts had similar time profiles and spectra. Recently, 8 additional bursts have been detected from the same direction.[22] The first 2 of these were more intense; FIGURE 6 shows their time profiles and spectra. The Leningrad group has demonstrated that the rings of position of all 12 bursts, including the 5 March event, intersect at the same point, as shown in FIGURE 7. This

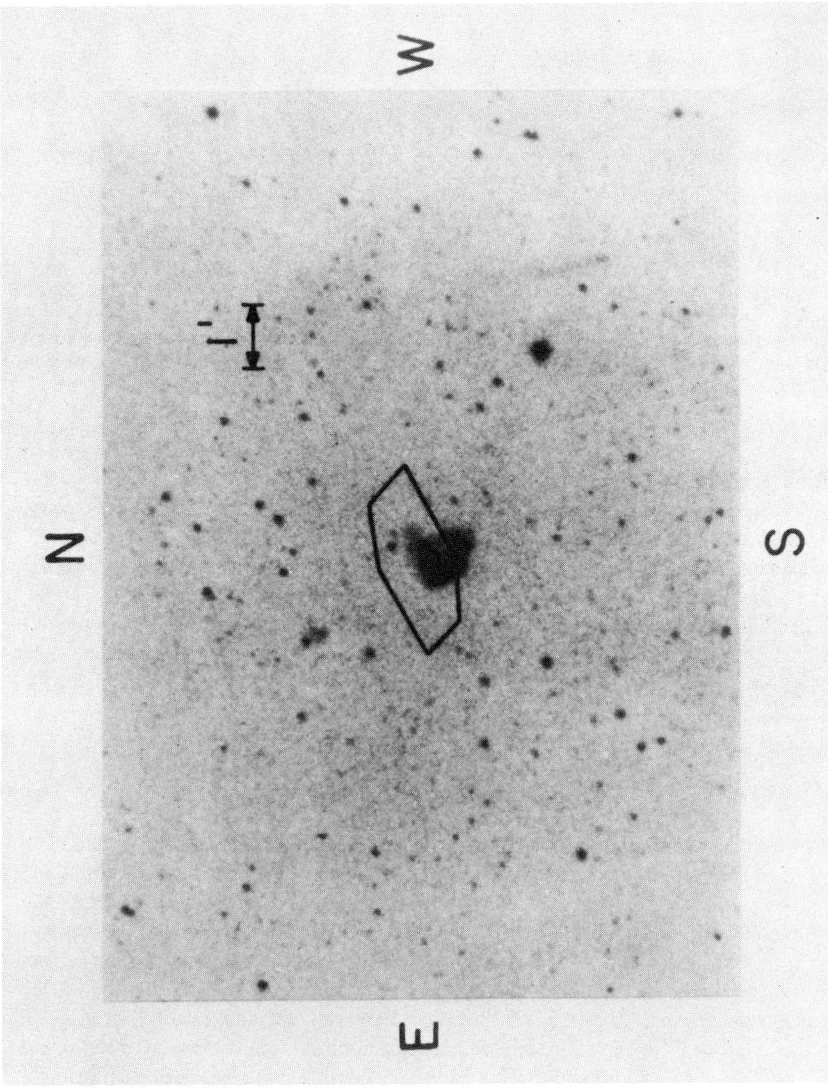

FIGURE 4. Preliminary gamma-ray error box for the 5 March 1979 burst shown on a reproduction of the optical field, in which the highest surface brightness portion of the SNR N49 in the large Magellanic cloud is visible. (From Evans *et al.* by permission of *Astrophys. J. Lett.*)[19]

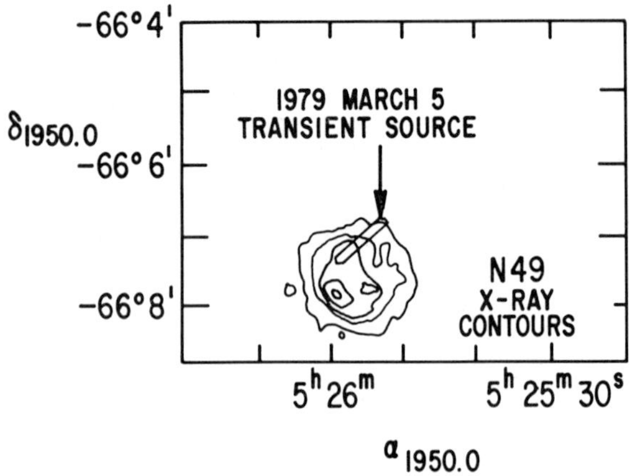

FIGURE 5. Precise source position for the 5 March 1979 burst derived from the final analysis of the observations made with the interplanetary network, plotted on the x-ray surface brightness contour map of the N49 region, as observed with the Einstein Observatory.[21] (From Cline *et al.* by permission of *Astrophys. J. Lett.*)[20]

confirms the recurrent nature of the burst source GBS0526-66. Such repetitive burst behavior is reminiscent of x-ray bursters, which are accreting neutron stars undergoing nuclear shell flashes (see Lewin and Joss and references therein).[23]

The short rise time, large peak intensity, eight-second oscillations in the decaying portion of the burst, and soft spectrum of the 5 March event are unique. These, together with the recurrence of weaker bursts, indicate that GBS0526-66 is distinct from other gamma-ray burst sources (but see Reference 24). Nevertheless, the association of this event with so many features characteristic of neutron stars has had a large impact on the thinking about gamma-ray bursts in general.

19 November 1978 Event

A second gamma-ray burst that has greatly influenced thinking about the nature of burst sources was the intense 19 November 1978 event. Two time profiles of this burst are shown in FIGURE 8. They show that the burst varied on time scales as short as 100 ms,[25,26] indicating a compact source. The spectrum of this burst is shown in FIGURE 9; it was the first to show evidence of emission lines.[26] The lines, at ≃420 and 740 keV, have been interpreted as due to pair annihilation and to the first-excited state of the ^{56}Fe nucleus, red shifted from 511 and 847 keV. It has also been suggested, alternatively, that the ≃740 keV line is due to one photon pair annihilation.[27,28] Either way, the red shifts are about what one expects for radiation coming from near the surface of a neutron star.

The 19 November event is the first gamma-ray burst for which a likely optical counterpart has been found. A search of archival plates has turned up an optical transient on a plate of the burst field taken in 1928.[29] FIGURE 10 shows the third and fourth of six archival plates taken ≃45 minutes apart, with the gamma-ray error box[30] superimposed. An optical image is clearly visible within the gamma-ray error box on the fourth (trailed) plate. There is also a soft x-ray source coincident with the gamma-ray error box and optical position,[31] as shown in FIGURE 11. The spectrum is hard, a characteristic of accretion sources. The x-ray luminosity is $\sim 1 \times 10^{30}$ ergs s^{-1} for a distance of 300 pc. If it is assumed that the x-ray source is powered by accretion onto a neutron star, the inferred accretion rate is $\sim 3 \times 10^{-16}\ M_\odot\ yr^{-1}$. This is somewhat large to be accounted for by accretion from the interstellar medium, and suggests that the burst source GBS0117–29 is a binary or is surrounded by a long-lived disk.

If all optical "flashes" are accompanied by gamma-ray bursts, the discovery of an optical "flash" in 1928 from GBS0117-29 implies that bursts recur on a time scale $\lesssim 50$

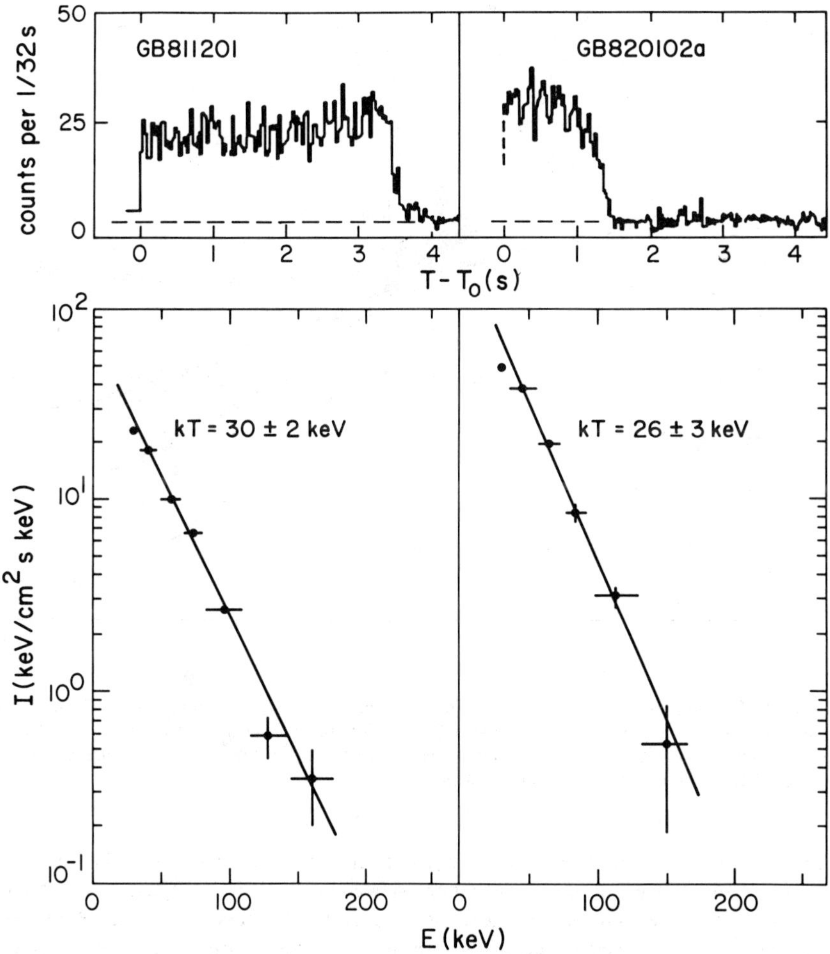

FIGURE 6. Time profiles (upper panel) and spectra (lower panel) of the bursts on 1 December 1981 and 2 January 1982. (From Golenetskii *et al.*)[22]

years. Actually, the recurrence time scale must then be much smaller. The total observing time represented by the search of archival plates for optical counterparts of the 19 November event,[29] and of two other events, is ≃ 2 years. The detection of three (and possibly as many as six) optical "flashes" therefore implies that the burst recurrence rate for these sources is ~ 1–3 yr^{-1} (Schaefer, private communication). However, such a rate may not be consistent with the observation by the Leningrad

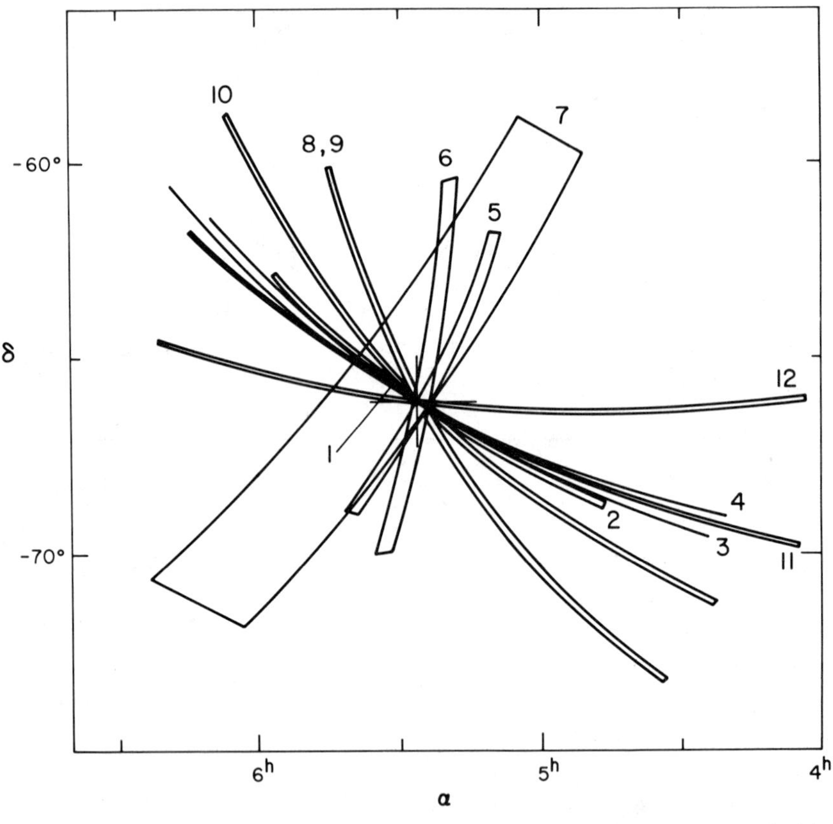

FIGURE 7. Gamma-ray error boxes for the following bursts: (1) 5 March; (2) 6 March; (3) 4 April; (4) 24 April 1979; (5) 1 December 1981; (6) 2 January; (7) 15 February; (8) and (9) 23 April a and b; (10) 10 May; (11) 30 May; (12) 8 July 1982. The cross shows the location of the source GBS0526-66. (From Golenetskii *et al.*)[22]

group of only two cases of recurrent bursts (one of which is GBS0526-66, the source of the 5 March 1979 event) in a total gamma-ray observing time ~ 2 years.[6] Of course, these statements depend on the optical and gamma-ray detector sensitivities. However, the existence of a second repeating gamma-ray burst source and several optical transients does suggest that many sources, not just GBS0526-66, probably produce

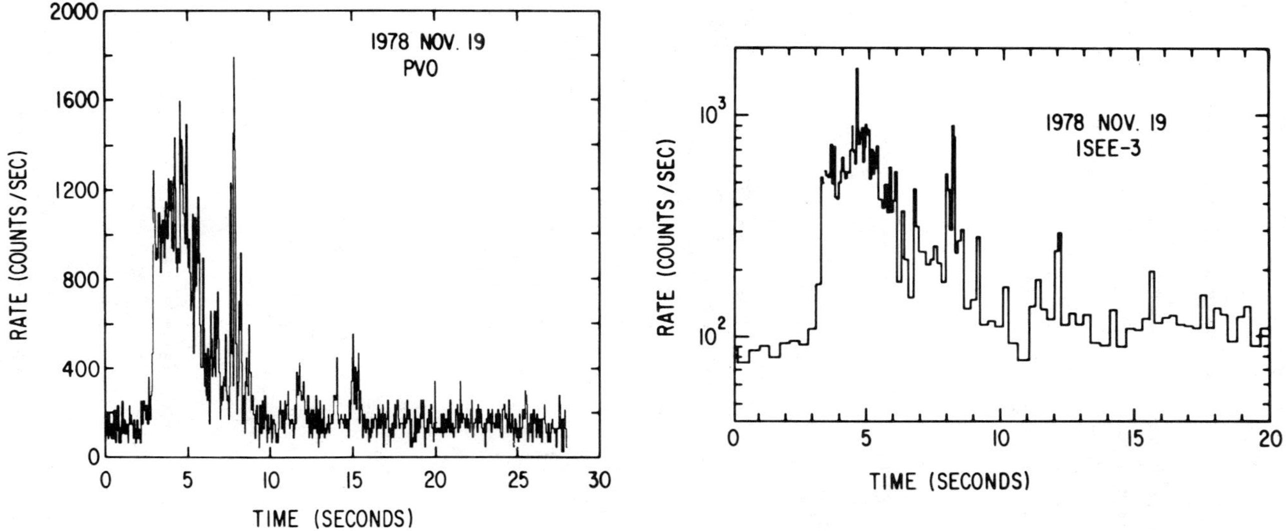

FIGURE 8. Two time profiles of the 19 November 1978 burst (of more than 12 observed with various spacecraft), showing the temporal structure of this rapidly varying and intense event. [From Evans *et al.* (PVO data)[25] and Teegarden and Cline (ISEE-3 data)[26] by permission of *Science* and *Astrophys. J. Lett.*]

recurrent bursts. As already remarked, such behavior is reminiscent of x-ray bursters, which are thought to be accreting neutron stars undergoing nuclear flashes.[23]

Annihilation and Cyclotron Spectral Features

Another important observational development has been the detection of spectroscopic features in many gamma-ray burst spectra by the Leningrad group, using the Konus experiments on Venera 11–14[6,8,32–35] (see also the review by Teegarden).[10]

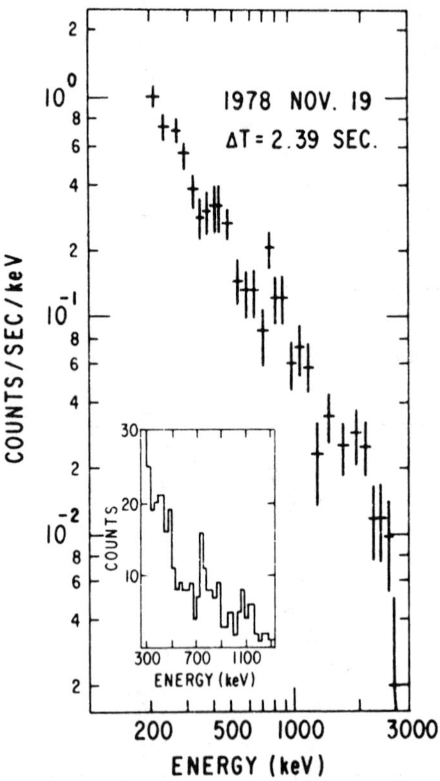

FIGURE 9. Spectrum of the 19 November 1978 burst. The insert shows the spectrum on a linear count rate scale. Evidence for a ≃420 keV feature is strong, and there is also evidence for a line at ≃700 keV. These features are interpreted as pair annihilation and first-excited iron emission lines, red shifted from 511 and 847 keV. (From Teegarden and Cline by permission of *Astrophys. J. Lett.*)[26]

FIGURE 12 shows a selected sample of such burst spectra. About 7% of the spectra show an emission feature near ≃ 400 keV that is probably red-shifted radiation from pair annihilation. The feature is ≃ 250 keV wide and contains 5–10% of the energy of the burst.

Nearly 30% of the burst spectra show low-energy-absorption features. These features have been questioned,[36,37] but their observation with two spacecraft and their variability during bursts argue for their reality. Furthermore, a similar feature has recently been reported for a burst observed with the hard x-ray detector on HEAO-1.[38]

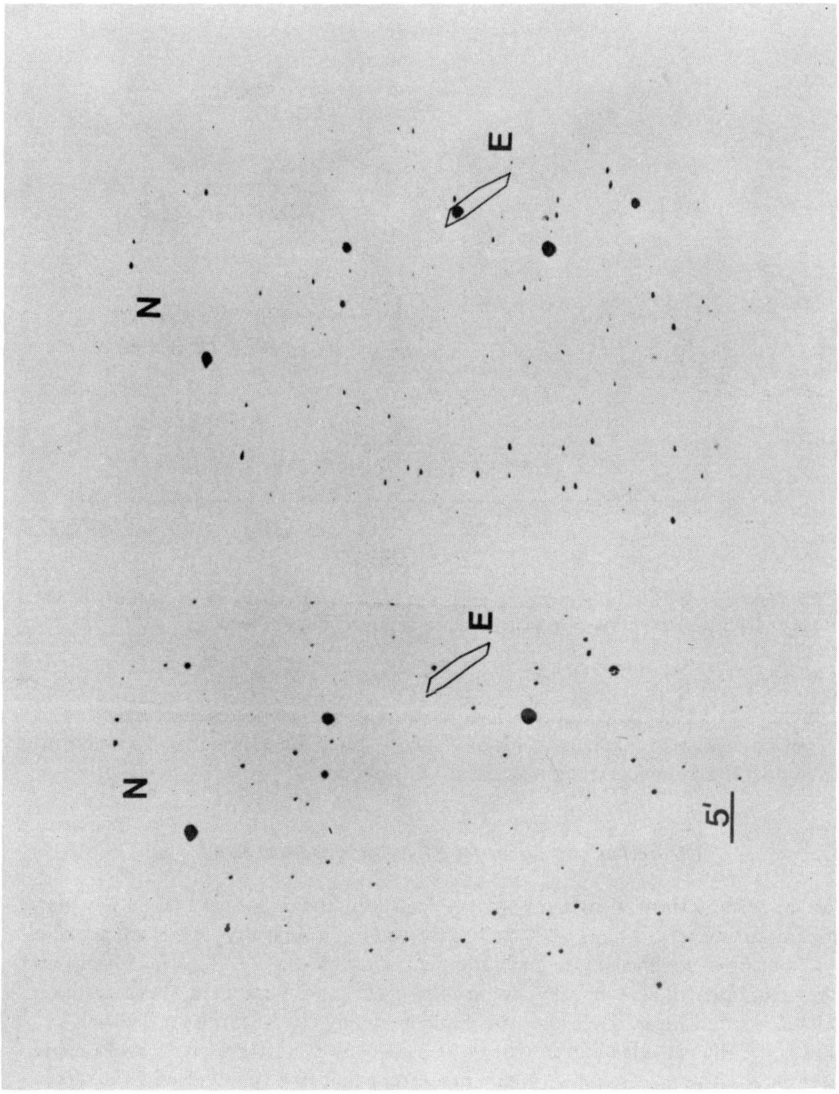

FIGURE 10. Two optical plates (taken in 1928) of the 19 November 1978 burst field, with the gamma-ray error box superimposed.[29] Note the probable burst image on the trailed plate at right, which was taken 45 minutes before the plate at left. (Courtesy of B. E. Schaefer.)

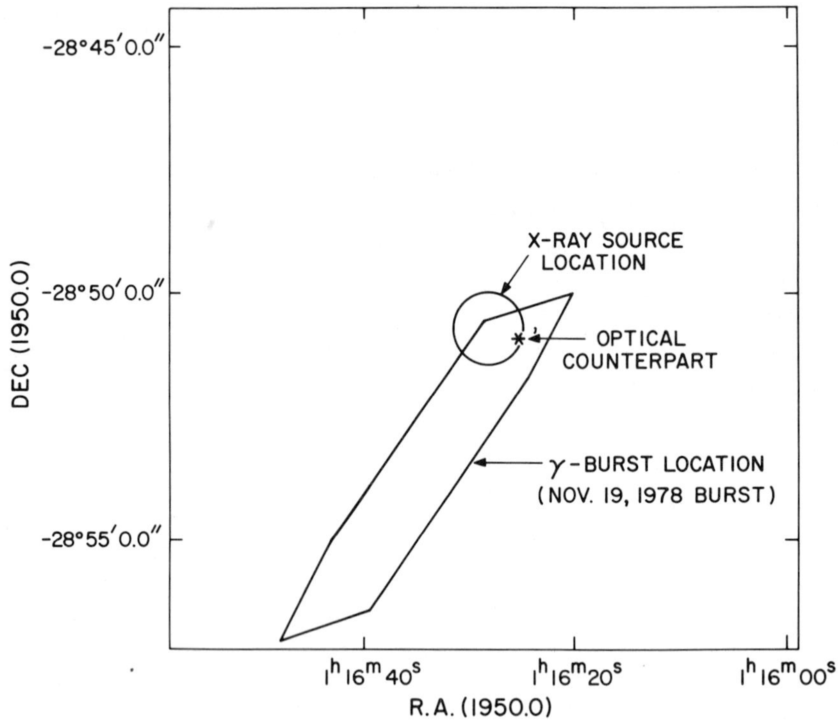

FIGURE 11. Superposition of gamma-ray and x-ray error boxes, and optical position for the burst source GBS0117-29. (From Grindlay *et al.* by permission of *Nature.*)[31]

The low-energy-absorption features have been widely interpreted as being due to cyclotron absorption in magnetic fields of $\simeq 10^{12}$ G. Such field strengths are similar to those in pulsars and in accreting neutron star x-ray sources.

Theoretical Argument for a Strong Magnetic Field

An important theoretical development has been the recognition that the high spectral temperatures, $kT_{\text{obs}} \gtrsim 200$ keV, observed in gamma-ray bursts are possible only if matter is confined by a magnetic field of order 10^{12} G. This fact has been emphasized, in particular, by Colgate and Petschek[39] and Ruderman,[40] and has been used by Ramaty, Lingenfelter, and Bussard in modeling the 5 March 1979 event.[14]

Let us briefly consider the argument, as presented by Colgate and Petschek,[39] in greater detail. The maximum surface temperature that can be achieved by a star in hydrostatic equilibrium occurs when the force of gravity just balances the radiation stress; that is,

$$\frac{a}{3}T^4 = \frac{\mu m_{\text{H}}}{\sigma_{\text{T}}}\frac{GM}{R^2} \equiv \frac{\mu m_{\text{H}}}{\sigma_{\text{T}}}g \qquad \textbf{(1)}$$

where T is the surface temperature, μ is the mean molecular weight per particle, m_H is the atomic mass of hydrogen, σ_T is the Thomson cross section, M and R are the mass and radius of the star, and we have assumed that electron scattering is the dominant opacity. This defines an Eddington temperature (corresponding to the Eddington luminosity, if the entire stellar surface radiates) which is a function only of the surface gravity of the star,

$$T_E \equiv (4\mu m_H g/\sigma_T a)^{1/4} \simeq 1.8\mu^{1/4}(M/M_\odot)^{1/4}R_6^{-1/2} \text{ keV} \quad (2)$$

where $R_6 = R/10^6$ cm (throughout the rest of this paper we will take $M = 1\ M_\odot$ and $R_6 = 1$). The Eddington temperature is therefore greatest for neutron stars. Assuming that the hot plasma producing the observed spectral temperatures $T_{obs} \simeq 200$ keV of gamma-ray bursts is optically thick, the ratio of radiation to gravitational forces is $\simeq (T_{obs}/T_E)^4 \simeq 10^8$, even for neutron stars! If the plasma is pair dominated, as may be

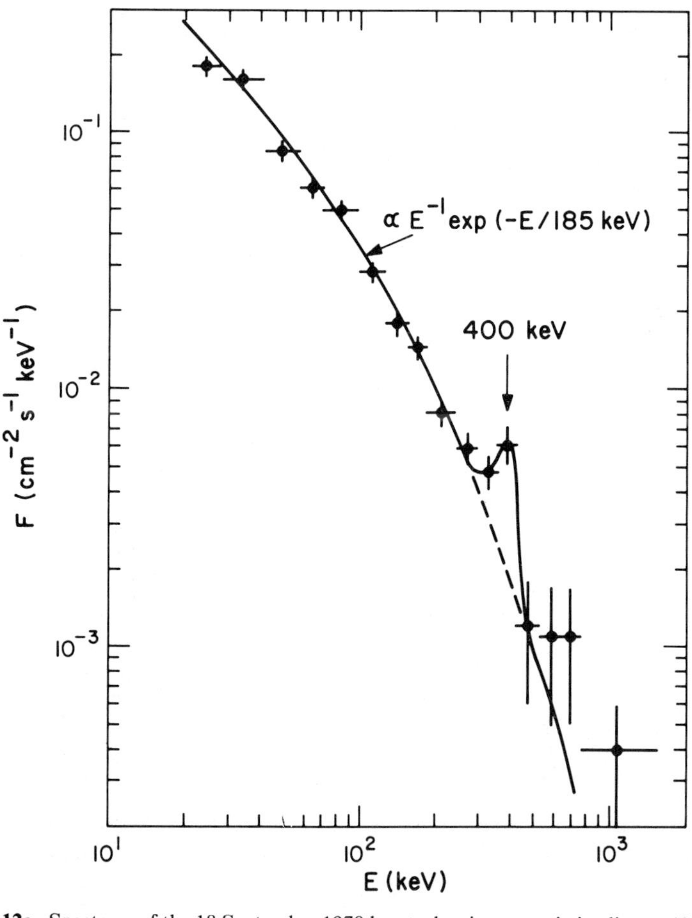

FIGURE 12a. Spectrum of the 18 September 1978 burst, showing an emission line at 400 keV.

the case for gamma-ray bursts (see below), the ratio is even larger (since $m_e \ll m_p$). Gravity cannot restrain the surface layers of the star, and they freely expand at the local sound speed, producing an explosion.

However, such explosions are inefficient sources of high-temperature radiation (for example, in supernovae only a small fraction of the explosion energy is converted into radiation; see also the discussion of explosions by Cavallo and Rees).[41] The reason is the following. As the matter expands, a rarefaction wave moves inward from the surface; at the same time, radiation is diffusing out. Matter above the rarefaction wave is cold due to adiabatic expansion, and therefore the expanding matter radiates at high temperature only until the rarefaction wave reaches the depth from which photons are diffusing out. For typical neutron star parameters, the duration and total energy of this phase are only $t \simeq 10^{-12}$ s and $E \simeq 10^{33} R_6^2 T_9^4$ ergs. These values are woefully inadequate to explain gamma-ray bursts.

Following the explosion, the photosphere lies at a radius such that Equation 2 is

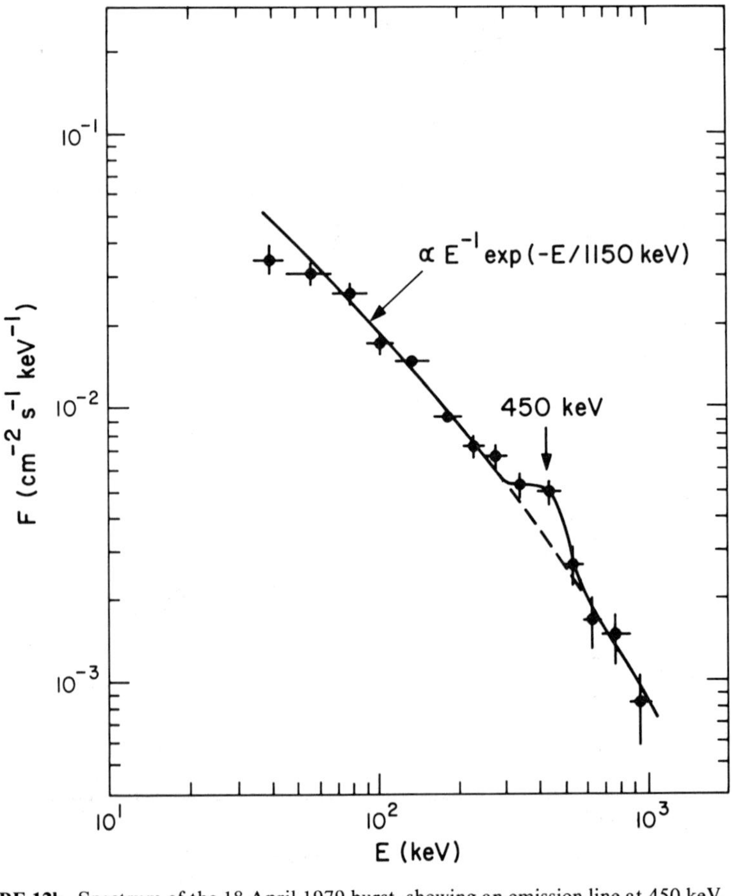

FIGURE 12b. Spectrum of the 18 April 1979 burst, showing an emission line at 450 keV.

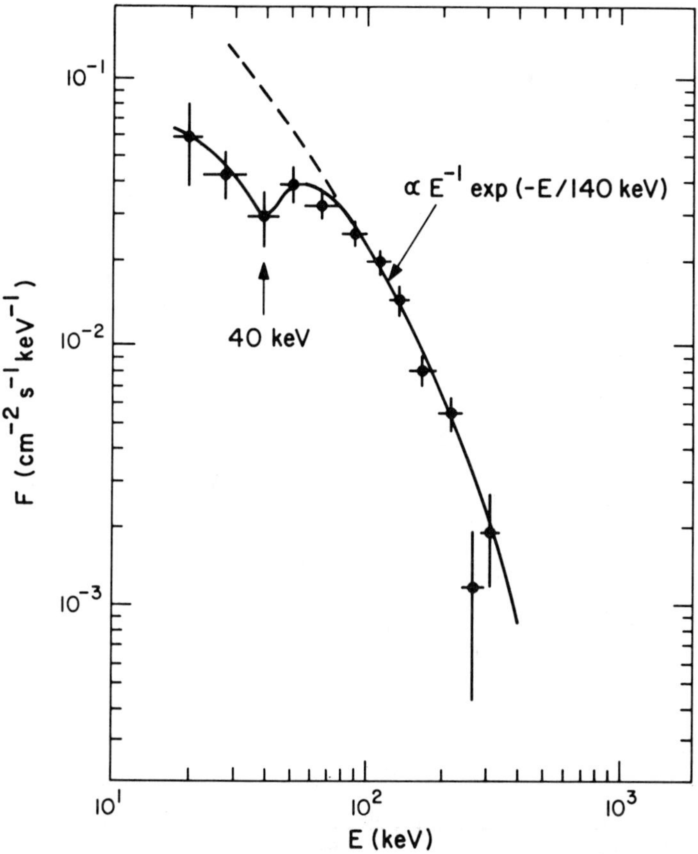

FIGURE 12c. Spectrum of the 12 June 1979 burst, showing an absorption feature at 40 keV.

satisfied. The plasma thus shrouds the star, and it radiates at a temperature T_E, much lower than that observed in gamma-ray bursts. Recent calculations of accretion onto nonmagnetic neutron stars by Newman and Cox and by Howard, Wilson, and Barton show just such behavior.[42,43]

In contrast, if a strong magnetic field is present and the plasma is optically thick and radiation dominated, it can be confined perpendicular to the field so long as $\beta_m \equiv P_r/P_m \lesssim 1$; that is, the pressure P_r of the hot plasma is less than the pressure P_m exerted by the magnetic field. Equivalently,

$$\frac{1}{3}aT^4 \lesssim \frac{B^2}{8\pi} \tag{3}$$

where B is the magnetic field. The corresponding temperature is

$$T \lesssim 170 B_{12}^{1/2}\ \text{keV} \tag{4}$$

where $B_{12} = B/10^{12}$ G. If the plasma can be confined along the field by closed field lines

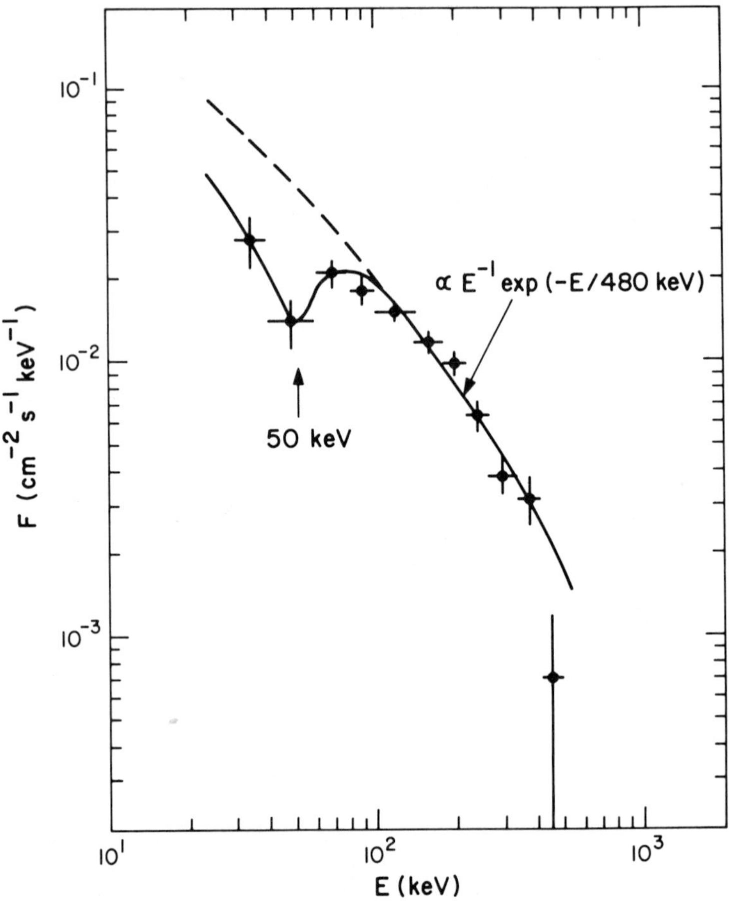

FIGURE 12d. Spectrum of the initial stage of the 12 April 1979 a burst, showing an absorption feature at 50 keV. (From Mazets *et al.* by permission of *Nature*.)[32]

or by the weight or ram pressure of overlying matter, then for magnetic field strengths of a few times 10^{12} G the plasma temperature can be similar to that observed.

These arguments, which predated the announcement of the discovery of low-energy spectral features by Mazets and his colleagues,[6,32] provide powerful indirect support for the idea that gamma-ray bursts originate from neutron stars with strong magnetic fields.

MAGNETIC NEUTRON STAR MODELS

Accretion

There are three general categories of gamma-ray burst models that invoke magnetic neutron stars. In the first category, the energy comes from the gravitational potential energy released by accreting plasma[44–47] or solid cometary material[39,48] as it impacts the magnetosphere and surface of the star. Plasma accretion models gain support from the recurrent nature of burst sources and from the binary nature of GBS0117-29 inferred from its x-ray luminosity. However, it is not clear how they can produce burst durations as short as ~0.1 seconds, although they may be able to account for bursts that last ~1–30 seconds. Furthermore, the question of why such accretion might produce gamma-rays rather than x-rays, as do the accreting neutron star x-ray sources, has not been satisfactorily answered. In contrast, the cometary accretion model appears able to produce short bursts. However, it has become less favored following recognition of the recurrent nature of burst sources. It is also hard to understand why (and perhaps how) burst sources should be binaries, if the accreting matter is supplied by a comet.

Internal Energy

In a second category of models, the energy is also derived from gravitational potential energy, but the source is internal to the neutron star. Examples include a corequake[49] or a crustquake,[50,51] which may produce a sudden rearrangement or motion of the magnetosphere.[11,51,52] Again, in these models it appears difficult to explain the recurrence rate of bursts and to understand why they should be binaries.

Nuclear Burning

In a third category of models, the energy comes from nuclear burning of accreted matter.[53–59] These models gain support from the recurrent nature of bursts and from their probable binary nature. Both properties are reminiscent of x-ray bursters,[23] as we have already remarked. However, it still remains something of a challenge in these models to account simultaneously for the recurrence rate and for the burst energy. For recent reviews of the thermonuclear flash model, see Ruderman[40] and Woosley.[60,61]

All of these models attempt to answer the question of what supplies the energy of the burst. None, so far, explains how this energy is coupled to the hot plasma that radiates and produces the gamma-rays. Contact with the continuum spectrum, and the pair annihilation lines and low-energy cyclotron features, revealed by recent spectral observations is therefore slight. We make a few conjectures concerning this difficult, but important, problem. However, we focus our attention primarily on the properties of the burst emission region and the radiation processes occurring there, as implied by the spectral observations.

Available Energy

A characteristic feature of neutron stars is the large gravitational potential at the surface. Typically,

$$\phi_{\mathrm{grav}} = \frac{GM}{R} A m_{\mathrm{H}} \simeq 0.15 A m_{\mathrm{H}} c^2 \ \langle = \rangle \ 140 \ \mathrm{MeV/nucleon} \tag{5}$$

where A is the ion atomic number. This provides a bound to the specific energy of the matter that produces the burst, whether it originates from outside the star (as in accretion) or from inside (as in a starquake or in pulsations). Given the above bound, the minimum amount of matter required to power a typical gamma-ray burst is

$$\Delta M = m_{\mathrm{H}}(E/\phi_{\mathrm{grav}}) \simeq 7.5 \times 10^{17}(E/10^{38}\ \mathrm{ergs})\ \mathrm{g} \tag{6}$$

where $E \simeq 1 \times 10^{38}(d/100\mathrm{pc})^2$ ergs is the burst energy.

If the energy is supplied by nuclear burning, the bound is about 20 times smaller, since the total energy available from burning hydrogen to iron peak nuclei is $\phi_{\mathrm{nuc}} \simeq 0.007\ m_{\mathrm{H}}c^2 \langle = \rangle$ 7 MeV/nucleon. The minimum amount of matter required to power a typical gamma-ray burst is then

$$\Delta M = m_{\mathrm{H}}(E/\phi_{\mathrm{nuc}}) \simeq 1.5 \times 10^{19}(E/10^{38}\ \mathrm{ergs})\ \mathrm{g} \tag{7}$$

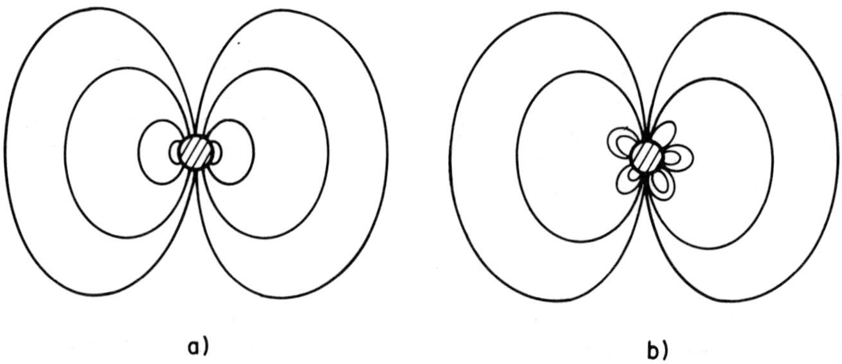

FIGURE 13. Possible configuration of the lines of force surrounding a magnetic neutron star. (**a**) Simple dipole geometry. (**b**) Complex field geometry near the neutron star and dipole field at larger distances.

PROPERTIES OF THE EMISSION REGION

Size

Little is known for certain about the detailed structure of the magnetic fields of neutron stars. Near the surface of the star, the field may be a simple dipole, as illustrated in FIGURE 13a, or the field may be more complex with concentrations of magnetic flux at several points on the surface as illustrated in FIGURE 13b. If the field is a simple dipole, then accreting matter will likely be funneled toward one or both of the magnetic poles. However, if the field is more complicated, accreted matter may arrive and accumulate at a number of places on the surface. This would certainly affect the spectra and the time variability of any gamma-ray burst; if accretion occurs, emission might come successively from locations of flux concentration on the stellar surface, or if nuclear burning occurs, a succession of flashes might occur. In either case, the series of outbursts taken together might constitute a burst.[39,55] However, the general simplicity of the pulse profiles of accreting neutron star x-ray sources at higher energies (cf. that of Vela X-1)[62] suggests that the field near the star may be a simple dipole or, at least, that accreting matter tends to be funneled toward the magnetic poles for both rapidly and slowly rotating stars.

Any concentrations of magnetic field at the stellar surface die away rapidly with distance from the star, so that at sufficiently large distances the field is dipolar. However, even the dipole field falls off as $(R/r)^3$, so that with a surface magnetic field $B = 1 \times 10^{12}$ G, the field at 10^7 cm ($\simeq 10\ R$) is 1×10^9 G while at 10^8 cm ($\simeq 100\ R$) it is only 1×10^6 G. This places important constraints on the location and volume of hot plasma that can be confined by the field. If the plasma is radiation dominated, as expected if the emission region is optically thick, then from Equation 3 the field required to confine the plasma in directions perpendicular to the field is

$$B_c \gtrsim 2.5 \times 10^{11} T_9^2 \text{ G} \tag{8}$$

and the confining radius must satisfy

$$r_c \lesssim 1.6 \times 10^6 (B_s/10^{12}\text{G})^{1/3} T_9^{-2/3} \text{ cm} \tag{9}$$

where B_s is the field strength at the stellar surface. Note that this is more stringent than the requirement that only matter at this temperature be confined, since $\beta_r \equiv P_r/P_g > 1$ where P_r and P_g are the radiation and gas pressures. The latter requirement, which may be appropriate if the emission region is optically thin, gives

$$B_c' \gtrsim 1.9 \times 10^{10} n_{26}^{1/2} T_9^{1/2} \text{ G} \tag{10}$$

and

$$r_c' \lesssim 3.7 \times 10^6 (B_s/10^{12}\text{G})^{1/3} n_{26}^{-1/6}\ T_9^{-1/6} \tag{11}$$

results that have been quoted by a number of authors (cf. Reference 28). Here n_{26} is the ion number density in units of 10^{26} cm^{-3}, which is the maximum number density likely in the emission region of gamma-ray burst sources (see below). For comparison, the scale height at the neutron star surface of matter at such temperatures is only

$$h = kT/m_H \simeq 620 T_9 \text{ cm} \tag{12}$$

Plasma flowing along the magnetic field can be channeled by the field only so long as the flow is sub-Alfvénic; that is,

$$v_p \lesssim v_A \equiv (B_p^2/4\pi\rho)^{1/2} \tag{13}$$

where v_p and B_p are the poloidal components of the bulk flow velocity and of the magnetic field. For flow at the sound speed, the confining radius is the same as that given by Equation 11. However, if the flow is supersonic, then the confining radius can be much smaller,

$$r_c'' \lesssim 1.1 \times 10^6 (B_s/10^{12}\text{G})^{2/5} (v_p/v_{ff})^{-2/5} n_{26}^{-1/5} \text{ cm} \tag{14}$$

where we have scaled the velocity by its free-fall value, $v_{ff} = (2GM/r)^{1/2}$. Beyond this radius the magnetic field is torn away by the streaming plasma.

From this discussion, we conclude that plasma as hot as that observed in gamma-ray bursts can be confined only near the star. This is consistent with the red shift $z \simeq 0.2$ and narrow width of the 511 keV annihilation line seen in a number of gamma-ray burst spectra.[6,8,26,32] We also conclude that the magnetospheres of burst sources are likely to be open, at least near the magnetic poles where the dipole field lines extend to large radii in the absence of hot plasma.

Density

If the energy of the emitting plasma or the plasma itself is not replenished during a burst, the characteristic number of particles required to power a gamma-ray burst is

$$N = E/\left(\frac{3}{2}kT\right) \simeq 4.8 \times 10^{44}(d/100\text{pc})^2T_9^{-1} \tag{15}$$

where we have taken an energy per particle of $\frac{3}{2}kT$. This number and the volume

$$V = 4\pi fR^2l = 1.3 \times 10^{19}fR_6^2l_6 \tag{16}$$

where f is again the fraction of the stellar surface and l_6 is the radial thickness of the burst emission region in units of 10^6 cm, define a characteristic maximum number density,

$$n = N/V \simeq 3.8 \times 10^{25}(d/100\text{pc})^2T_9^{-1}(fl_6)^{-1}\ \text{cm}^{-3} \tag{17}$$

for the emitting plasma. If we again assume no replenishment during a burst, but that the energy is supplied by the annihilation of pairs rather than by thermal energy, we find a somewhat smaller characteristic number density,

$$n_{\text{pair}} = (E/2m_ec^2)/V \simeq 4.9 \times 10^{24}(d/100\text{pc})^2(fl_6)^{-1}\ \text{cm}^{-3} \tag{18}$$

If replenishment occurs instead and the only requirement is that the plasma supply the observed luminosity, then the number density can be far smaller. For example, if the luminosity of the burst is due to synchrotron emission, the characteristic number density can be as small as $\simeq 10^9\ \text{cm}^{-3}$ (see below).

Parameter Regimes of a Hot Plasma in a Strong Magnetic Field

FIGURE 14 shows various parameter regimes for a hot ($T \simeq 10^6$–10^{10}K) low-density ($n \simeq 10^{12}$–$10^{30}\ \text{cm}^{-3}$) plasma in thermal equilibrium in a strong magnetic field. These regimes can give us a feel for the properties of the emission region, even though the hot plasma there may well be out of thermal equilibrium. In the example shown, the composition is taken to be hydrogen and the magnetic field is assumed to be 1×10^{12} G. The matter is nondegenerate if $\epsilon_F/kT < 1$, where ϵ_F is the Fermi energy of the electrons. In the quantum regime, this condition is satisfied as long as[63]

$$kT \gtrsim \frac{1}{2}m_ec^2\left(\frac{B_{\text{cr}}}{2B}\frac{n\lambda_C^3}{2\pi}\right)^2 \tag{19}$$

where $B_{cr} = m_e^2c^3/e\hbar = 4.4 \times 10^{13}$ G and $\lambda_C = h/m_ec$ is the electron Compton wavelength. Equivalently,

$$T \gtrsim 7.5 \times 10^4 B_{12}^{-2}\, n_{26}^2 \text{ K} = 6.4 B_{12}^{-2}\, n_{26}^2 \text{ eV} \quad \textbf{(20)}$$

In the classical regime, the usual expression for ϵ_F applies and $T \gtrsim 9.0 \times 10^6\, n_{26}^{2/3}$ K =

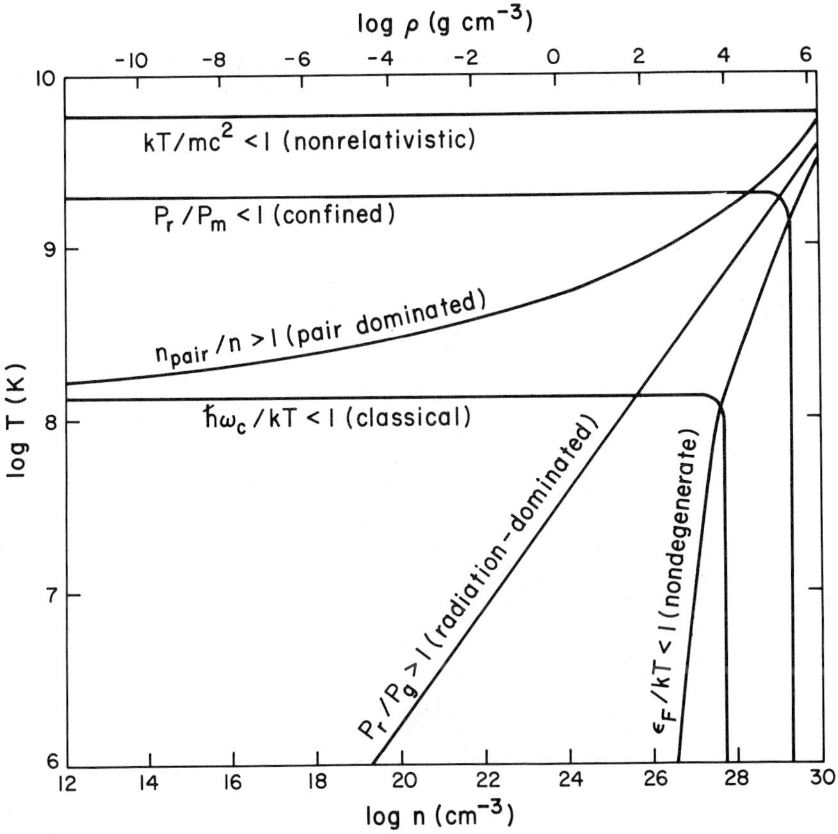

FIGURE 14. Parameter regimes of a hot plasma in a magnetic field $B = 10^{12}$ G. Above and to the left of the curve labeled $\epsilon_F/kT < 1$, the plasma is nondegenerate. To the upper left of the curve labeled $P_r/P_g > 1$, the plasma is radiation dominated. Below the curve labeled $kT/m_ec^2 < 1$, the plasma is nonrelativistic. Above the curve labeled $\hbar\omega_c/kT < 1$, a classical treatment of kinetic and radiation processes is valid. The magnetic field is able to confine the plasma for densities and temperatures below and to the left of the curve labeled $P_r/P_m < 1$.

$0.86\, n_{26}^{2/3}$ keV. The plasma is radiation dominated if $\beta_r \equiv P_r/P_g > 1$. In the nondegenerate regime, this condition is satisfied as long as

$$T \gtrsim 1.8 \times 10^8 n_{26}^{1/3} \text{ K} = 15 n_{26}^{1/3} \text{ keV} \quad \textbf{(21)}$$

The magnetic field is able to confine the plasma in the directions perpendicular to the field if $P_m > P_r + P_g$. In the nondegenerate regime, this condition is satisfied as long as $\beta_m \equiv P_r/P_m < 1$; that is, as long as Equation 4 is satisfied. When the electrons are degenerate, the condition becomes $P_m > P_e = 2n_e\epsilon_F/5$; otherwise, the magnetic field is not important in confining the electrons. The electrons in the plasma are nonrelativistic if $kT/m_ec^2 < 1$; that is, as long as

$$T \lesssim 5.9 \times 10^9 \text{ K} = 511 \text{ keV} \tag{22}$$

We note that if the emission region is optically thick, the electrons are unlikely to be relativistic since confinement of such a plasma requires a magnetic field strength $B_c \gg 1 \times 10^{13}$ G. However, if the emission region is optically thin, as expected (see below), the electrons could well be highly relativistic (see Equation 10). If $\hbar\omega_c/kT < 1$ or, equivalently,

$$T \gtrsim 1.3 \times 10^8 B_{12} \text{ K} = 12 B_{12} \text{ keV} \tag{23}$$

then the motion of the electrons is not quantized by the strong magnetic field. Quantum effects on kinetic and radiation processes[28,64] can be neglected and, in particular, synchrotron (or gyrosynchrotron) emission can be treated classically.

Pair Plasmas

One of the most fascinating and intriguing aspects of gamma-ray bursts is the implied existence in the burst emission region of plasma in which the electron-positron pair density is significant, if not dominant. The pair annihilation lines observed in many burst spectra[6,8,26,32] contain $\simeq$ 5–10% of the energy in the burst; this represents a lower limit to the fraction of the energy of the plasma that is in pairs, since the line must originate in a region that is optically thin (see below).

In equilibrium, the number density of pairs in a plasma exceeds that of the ions ($n_{pair}/n > 1$) at nonrelativistic temperatures if[65]

$$n \lesssim 2.2 \times 10^{30} (kT/m_ec^2)^{-3/2} e^{-(kT/m_ec^2)} \text{ cm}^{-3} \tag{24}$$

At relativistic temperatures, the condition becomes

$$n \lesssim 1.5 \times 10^{31} T_{10}^3 \text{ cm}^{-3} \tag{25}$$

When $n_{pair}/n > 1{,}836$, the mass density in pairs exceeds that of the ions and the plasma is less easily confined by gravity. The plasma can, however, be confined by a magnetic field. Note also that at relativistic temperatures, the energy $E_{pair} = 7aT^4/4$ in pairs exceeds the energy $E_\gamma = aT^4$ in photons; that is,

$$E_{pair}/E_\gamma = 7/4 \tag{26}$$

From the foregoing discussion, we conclude that the radiation-emitting plasma in gamma-ray burst sources is likely nondegenerate and relativistic, that the energy density of the plasma is likely dominated by radiation and possibly even by electron-positron pairs, and that a classical treatment of kinetic and radiation processes may be valid.

Stability

Hot plasma may enter the magnetosphere by accreting along magnetic flux sheets or tubes[44–47] or by being thrown up in a similar pattern from the stellar surface, either by the impact of accreting matter with the stellar surface[39,44,45,48] or by a nuclear explosion in the surface layers.[53,61] Hot plasma might also be produced *in situ* by violent motion of magnetic field lines or by field annihilation.[14,49–52] Whatever the source, such hot plasma is notoriously unstable when confined in a magnetic field, particularly if $\beta_m \gg 1$. One way in which plasma may rapidly cross field lines is by the interchange (or Rayleigh-Taylor) instability.[66,67] This instability occurs on a time scale,

$$t_{\text{inter}} \simeq (gk)^{-1/2} \simeq (\lambda/2\pi g)^{1/2} \simeq 3.5 \times 10^{-5}\lambda_6 r_6 \text{ s} \quad \textbf{(27)}$$

where $k = 2\pi/\lambda$ is the wave number of the fastest growing unstable mode and λ is the corresponding wavelength, here taken to be of order r.

RADIATION PROCESSES

A hot plasma in a strong magnetic field cools via bremsstrahlung and synchrotron (or gyrosynchrotron) emission. If pairs are present, it may also radiate via pair annihilation. If a tenuous plasma of hot electrons exists above the stellar surface, blackbody emission from the surface can be Comptonized to higher energies. Several authors, including Ramaty, Lingenfelter, and Bussard,[14] Katz,[28] Bussard and Lamb,[64] Gould,[68] Fenimore *et al.*,[69] Lamb,[11] Liang,[70–72] and Rees,[12] have recently discussed these processes in the context of gamma-ray bursts. Here we note their characteristics and compare their importance, assuming that the plasma is semirelativistic, that pairs do not dominate, and that the emission region is in thermal equilibrium but optically thin. (We consider the nonequilibrium and optically thick cases below.)

Bremsstrahlung

Cooling Rate and Time Scale

The usual electron-ion thermal bremsstrahlung cooling rate and cooling time scale are[73]

$$\Lambda_{\text{br}} = 4.4 \times 10^{29} \bar{g} Z^2 n_{26} n_{e,26} T_9^{1/2} \text{ ergs cm}^{-3}\text{ s}^{-1} \quad \textbf{(28)}$$

and

$$t_{\text{br}} = \frac{3}{2}\frac{n_e kT}{\Lambda_{\text{br}}} \simeq 4.7 \times 10^{-11} \bar{g}^{-1} Z^{-2} n_{26}^{-1}\, T_9^{1/2} \text{ s} \quad \textbf{(29)}$$

where $\bar{g}$ is the Gaunt factor, of order unity. At temperatures above 10^9 K, relativistic corrections and the contribution of electron-electron bremsstrahlung are important. These points have been carefully discussed recently by Gould.[74] We also note that if the plasma is pair dominated, the bremsstrahlung cooling rate in the nonrelativistic limit is four times larger than that above because the electron and positron radiation fields add coherently.[28]

Spectrum

At temperatures above 10^9 K, the usual electron-ion bremsstrahlung spectral emissivity[73]

$$j_\eta = 1.7 \times 10^9 \bar{g} n_{e,26}^2 T_9^{-1/2} e^{-h\nu/kT} \text{ ergs cm}^{-3}\text{ s}^{-1}\text{ sr}^{-1}\text{ Hz}^{-1} \qquad \textbf{(30)}$$

is modified by relativistic corrections and the contribution of electron-electron bremsstrahlung.[74]

Inverse Compton

Cooling Rate and Time Scale

Consider a blackbody source of soft photons with luminosity L_{bb} embedded in a tenuous hot plasma with temperature T_e. As they escape through the hot plasma, the soft photons will be shifted to higher energies by Compton scattering off hot electrons. The nonrelativistic inverse Compton cooling rate and time scale of the hot plasma are[75]

$$\Lambda_C = 4\sigma_T c U_{bb} n_e \frac{kT_e}{m_e c^2} = 3.6 \times 10^{25} L_{bb,37} n_{e,26} T_9 \text{ ergs cm}^{-3}\text{ s}^{-1} \qquad \textbf{(31)}$$

and

$$t_C = \frac{\frac{3}{2} n_e k T_e}{\Lambda_C} = 5.8 \times 10^{-7} L_{bb,37}^{-1} \qquad \textbf{(32)}$$

where $cU_{bb} = L_{bb}/4\pi R^2$. At electron temperatures above 10^9 K, there is a relativistic correction to these expressions.[76]

Spectrum

The shape of the emergent Comptonized spectrum is determined by the parameter[77–81]

$$y = \frac{4kT_e}{m_e c^2} \text{Max}(\tau_{es}, \tau_{es}^2) \qquad \textbf{(33)}$$

where τ is the electron scattering optical depth. The luminosity of the emergent radiation is also determined by y. The ratio $A = L/L_{bb}$, where L is the luminosity of the emergent radiation and L_{bb} is the luminosity of the blackbody source, is given by the relation[77]

$$y = \frac{1 - A^{-1}}{1 - \dfrac{T_\gamma}{T_e}} \qquad \textbf{(34)}$$

where T_γ is the intensity-weighted mean spectral temperature of the emergent radiation. Three regimes are apparent. When $y \ll 1$, $T_\gamma/T_e \ll 1$, $A \simeq 1$, and the emergent spectrum is essentially unchanged. When $y \simeq 1$, the spectrum remains soft, $T_\gamma/T_e \ll 1$, so A^{-1}, i.e., A becomes large. The spectrum has a power law form, $I(E) \sim E^{-\alpha}$, where $\alpha = [-\ln \tau_{es} = 2/(\theta + 3)]/\ln(12\theta^2 + 25\theta)$ and $\theta \equiv kT_e/m_ec^2$ when $\tau_{es} \lesssim 1$,[81] and $\alpha = -3/2 + (9/4 + 4/\beta)^{1/2}$ and $\beta \equiv (3/\pi^2)(kT_e/m_ec^2)(\tau_{es} + 2/3)^2$ when $\tau_{es} \gtrsim 1$.[78,80] When $y \gg 1$, A increases to $\sim T_e/T_{bb}$, so the spectrum becomes hard. In the limit of saturated Comptonization, the emergent spectrum approaches a Wien, $I(E) \sim E^3$

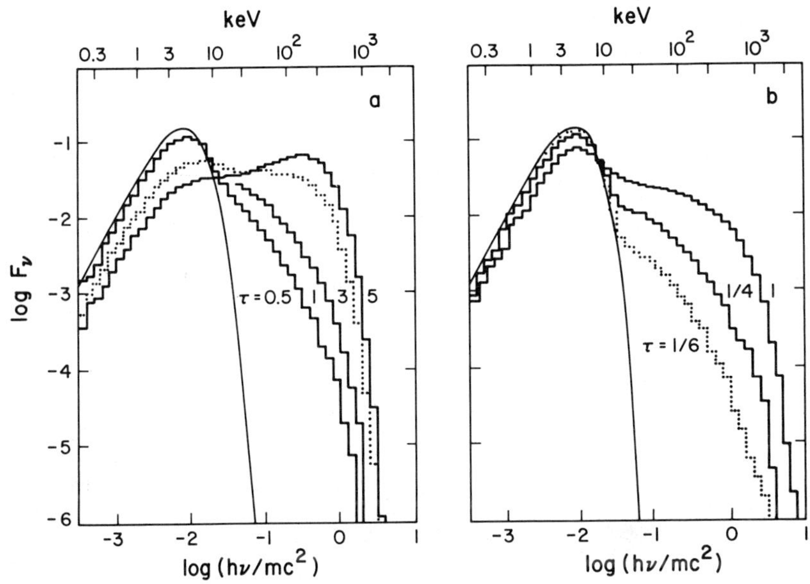

FIGURE 15. Emergent Comptonized spectrum for a blackbody central source with temperature $kT_{bb} = 1.5$ keV surrounded by a hot plasma cloud with electron temperature (a) $kT_e = 100$ keV and (b) $kT_e = 256$ keV. (From Pozdnyakov, Sobol', and Sunyaev by permission of *Sov. Astron. Lett.*)[81]

$\exp(-E/kT_e)$. FIGURE 15 shows several examples of emergent spectra, calculated for $T_e = 100$ and 256 keV and $\tau = 0.2$–5.[81]

Pair Annihilation

Cooling Rate and Time Scale

If $kT \gg 2\alpha m_ec^2 \simeq 5 \times 10^4$ K, Coulomb effects are unimportant, and the power emitted by pair annihilation and the corresponding annihilation time scale are[28]

$$\Lambda_{pair} = 2m_ec^2 n_e n_{\bar{e}} R \simeq 1.2 \times 10^{32} n_{e,26} n_{\bar{e},26} \text{ ergs cm}^{-3}\text{ s}^{-1} \qquad (35)$$

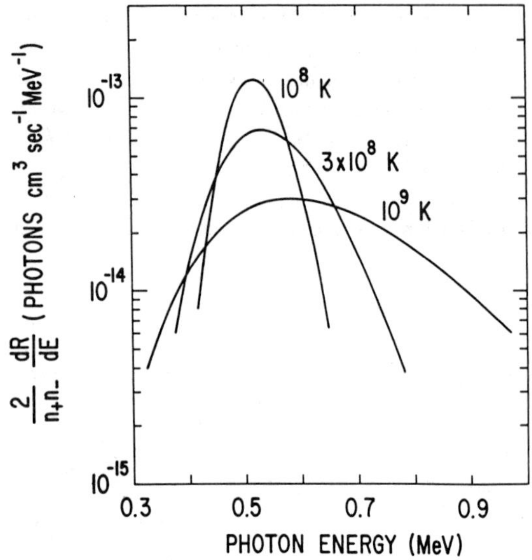

FIGURE 16. Pair annihilation spectra for various temperatures. (From Ramaty and Mészáros by permission of *Astrophys. J.*)[82]

and

$$t_{\text{pair}} = (n_e R)^{-1} \simeq 1.3 \times 10^{-12} n_{\bar{e},26}^{-1} \text{ s} \tag{36}$$

where $R = \pi r_o^2 c = 7.48 \times 10^{-15}$ cm^3 s^{-1}, $r_o = e^2/m_e c^2 = 2.82 \times 10^{-13}$ cm is the classical electron radius, and $n_{e,26}$ and $n_{\bar{e},26}$ are the electron and positron number densities in units of 10^{26} cm^{-3}.

Spectrum

At low temperatures, the pair annihilation spectrum consists of a narrow emission line centered at 511 keV. However, as the temperature approches 10^9 K and above, the annihilation spectrum becomes thermally broadened and asymmetric, with the centroid of the spectrum shifted to higher energies.[82] FIGURE 16 shows annihilation spectra for $T = 10^8$–10^9 K. FIGURE 17 shows the width and the thermally induced blue shift of the annihilation line as a function of temperature.

Synchrotron

Cooling Rate and Time Scale

The synchrotron cooling rate and cooling time scale are[83]

$$\Lambda_{\text{syn}} = 4 n_e \sigma_T c \frac{B^2}{8\pi} \frac{kT}{m_e c^2} \simeq 5.4 \times 10^{34} n_{e,26} B_{12}^2 T_9 \text{ ergs cm}^{-3} \text{ s}^{-1} \tag{37}$$

$$t_{syn} = \frac{3\pi m_e c}{\sigma_T B^2} \simeq 3.9 \times 10^{-16} B_{12}^{-2} \text{ s} \tag{38}$$

where, for simplicity, we have used the nonrelativistic expression for the synchrotron emissivity. The resulting cooling rate and cooling time scale are roughly correct, even for a semirelativistic plasma, as long as the emission region is optically thin. We note that if the plasma is pair dominated, the cooling rate is twice as large but the cooling time scale remains unchanged.

Spectrum

In gamma-ray burst sources, unlike accreting neutron star x-ray sources, a classical treatment of synchrotron emission may be valid, as discussed earlier. It is convenient to scale the thermal synchrotron emission spectrum in terms of the expression for the blackbody flux in the Rayleigh-Jeans limit,

$$F_\omega^{RJ} = \frac{kT\omega^2}{4\pi^2 c^2} \tag{39}$$

and to introduce the dimensionless size parameter

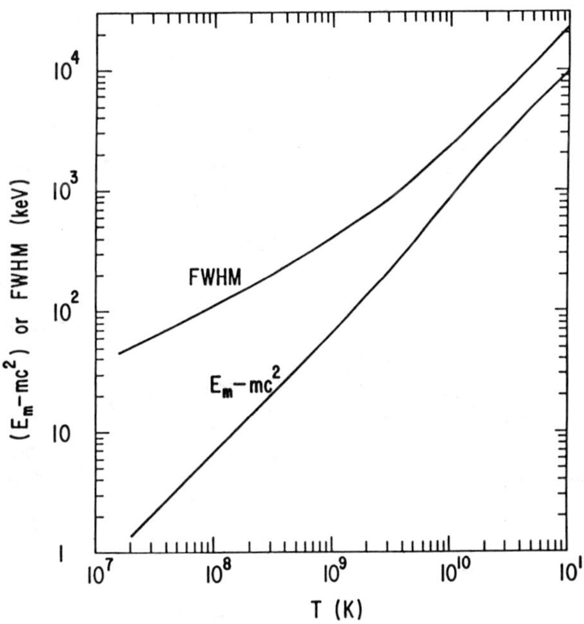

FIGURE 17. Temperature-induced blue shift, $E_m - mc^2$, and full width at half-maximum, FWHM, of the pair annihilation line as a function of temperature. (From Ramaty and Mészáros by permission of *Astrophys. J.*)[82]

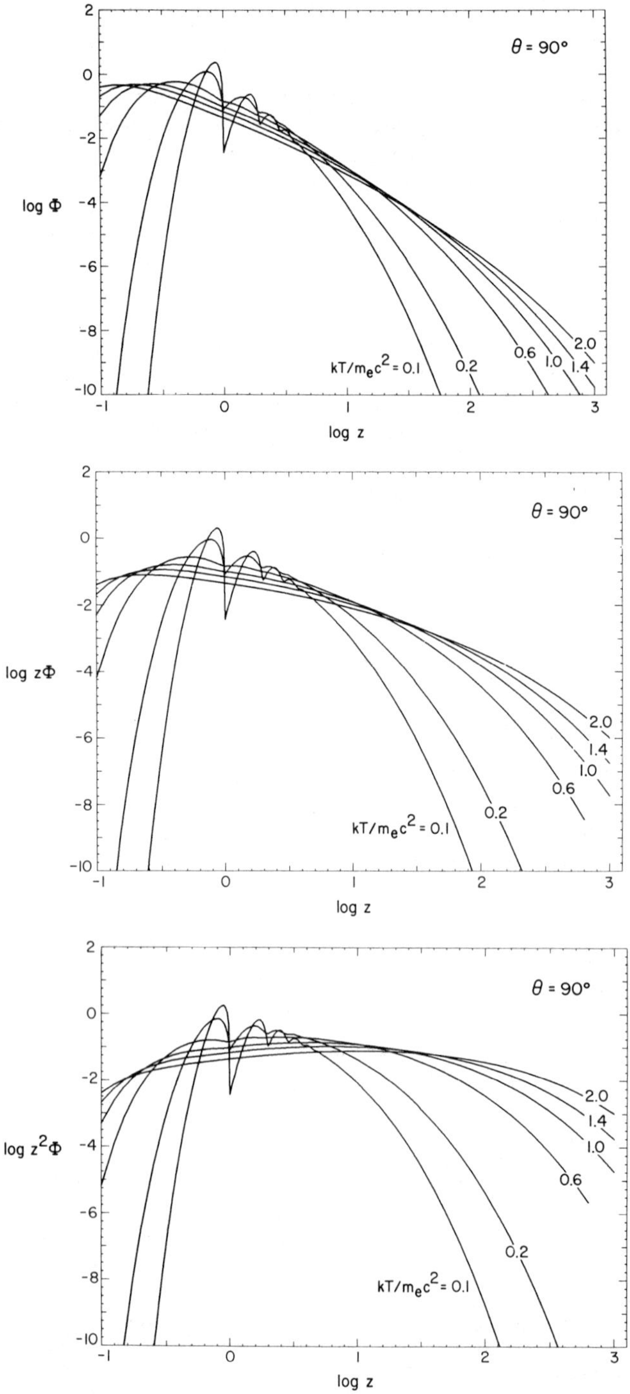
θ = 90°
log Φ
log z
kT/m_e c^2 = 0.1
0.2
0.6
1.0
1.4
2.0
θ = 90°
log zΦ
log z
kT/m_e c^2 = 0.1
0.2
0.6
1.0
1.4
2.0
θ = 90°
log z^2Φ
log z
kT/m_e c^2 = 0.1
0.2
0.6
1.0
1.4
2.0

$$\Lambda \equiv \frac{\omega_p^2 l}{\omega_c c} = \frac{3}{2}\alpha^{-1}(m_e c^2/\hbar\omega_c)n_e\sigma_T l \tag{40}$$

where ω_p is the plasma frequency, l is the thickness of the emission region, and α is the fine structure constant. The quantity Λ is related to the optical depth of the emission region (here assumed to be small). Then the angle-averaged spectral emissivity and spectral flux for thermal synchrotron emission are given by[84,85]

$$j_\omega = B_\omega^{RJ}\frac{\Lambda}{l}\Phi(\omega) \tag{41}$$

and

$$F_\omega = F_\omega^{RJ}\,\Lambda\Phi(\omega/\omega_c,T) \tag{42}$$

where $B_\omega^{RJ} = F_\omega^{RJ}/\pi$ and $\Phi(\omega/\omega_c,T)$ is a dimensionless spectral flux.

FIGURE 18a–c shows $\Phi(\omega/\omega_c,T)$, $(\omega/\omega_c)\,\Phi(\omega/\omega_c,T)$, and $(\omega/\omega_c)^2\,\Phi(\omega/\omega_c,T)$ for $T =$ 50 keV–1 MeV, from numerical calculations by Brainerd and Lamb.[86] Approximate analytic expressions for $\Phi(\omega/\omega_c,T)$ have been derived by Petrosian.[87] The curves in FIGURE 18b and c are proportional to the photon number spectral flux and the energy spectral flux, the constant of proportionality depending only on the magnetic field strength. They illustrate the spectral evolution expected as the temperature of the emitting region falls from 1 MeV to 50 keV.

At the high temperatures characteristic of the emission regions of gamma-ray bursts, the Doppler shift due to the velocity of the electrons broadens the individual synchrotron harmonics. This leads to a relatively smooth emissivity, and therefore absorption or scattering opacity, as a function of frequency. The emissivity curves shown in FIGURE 18 illustrate clearly the smoothing that occurs with increasing temperature. We note that FIGURE 18 shows the flux emitted at an angle of 90° with respect to the magnetic field, and that the broadening due to the Doppler shift is enhanced if the emission region is viewed more parallel to the field. On the other hand, the harmonics will be more pronounced if electron velocities perpendicular to the field are reduced by rapid synchrotron cooling. Both of these effects are illustrated by FIGURE 19, which shows the synchrotron opacity for a plasma with electron temperatures $T_\parallel = 300$ keV and $T_\perp \simeq 6$ keV in the directions parallel and perpendicular to the magnetic field and viewing angles $\theta = 5$–$90°$.[64]

Absorption versus Scattering

If the electron Landau levels are largely excited and deexcited by collisions, absorption and emission predominate. On the other hand, if the levels are largely excited and deexcited by radiative processes, scattering predominates and the flux at the cyclotron harmonics is reduced even if the source is isothermal. This point has been

FIGURE 18. Evolution of the thermal synchrotron emissivity of a hot plasma as the electron temperature of the plasma falls from 1 MeV to 50 keV. (Top) the dimensionless spectral flux $\Phi(z,T)$; (center) the dimensionless photon number spectral flux $z\Phi(z,T)$; and (bottom) the dimensionless energy spectral flux $z^2\,\Phi(z,T)$, where $z \equiv \omega/\omega_c$.[86]

emphasized by Bussard and Lamb,[64] and we closely follow their discussion here. Assuming only Coulomb scattering, the collisional excitation rate per electron is

$$R_{01} \simeq 6.3 \times 10^{-13} n Z^2 B_{12}^{-1} T_9^{-1/2} \ln (4kT/\hbar\omega_c)\ \mathrm{s}^{-1} \tag{43}$$

while the Einstein A coefficient for this two-level system is

$$A_{10} = \frac{4}{3}\alpha(\hbar\omega_c/m_e c^2)\omega_c \simeq 3.9 \times 10^{15} B_{12}^2\ \mathrm{s}^{-1} \tag{44}$$

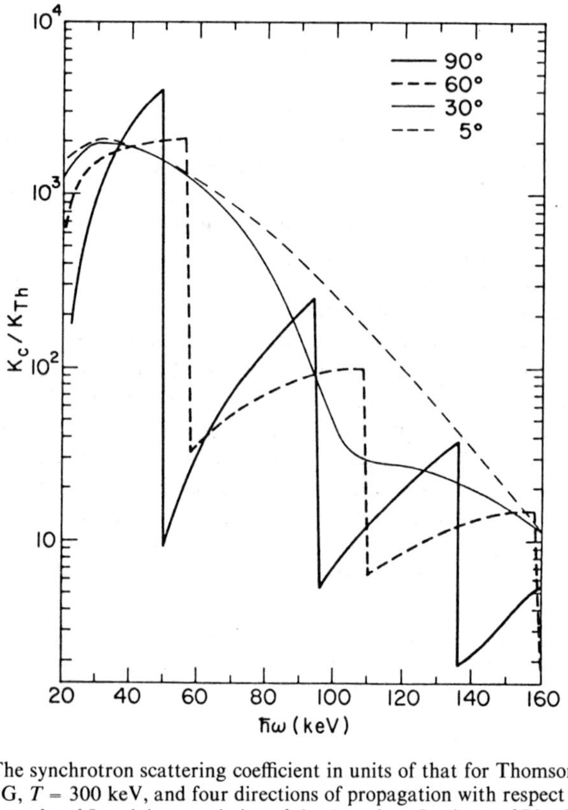

FIGURE 19. The synchrotron scattering coefficient in units of that for Thomson scattering, for $B = 4.4 \times 10^{12}$ G, $T = 300$ keV, and four directions of propagation with respect to the magnetic field. (From Bussard and Lamb by permission of the American Institute of Physics.)[64]

Cyclotron scattering predominates over absorption if $A_{10} > R_{01}$; that is, as long as

$$n \lesssim 6.2 \times 10^{27} Z^{-2} B_{12}^3 T_9^{1/2} \ln^{-1}(4kT/\hbar\omega_c)\ \mathrm{cm}^{-3} \tag{45}$$

As noted, the above analysis assumes that the Landau levels are populated by Coulomb collisions among the electrons. The results show that under these conditions only the first Landau level is significantly populated, the level distribution is determined by the radiation field, and scattering dominates. On the other hand collisionless processes, if they occur, can excite and deexcite the higher Landau levels on the much more rapid time scale ω_e^{-1} (see below). In this case, many Landau levels can be populated (a circumstance necessary for producing a hard spectrum) and absorption can dominate scattering.

Cooling Time Scale versus Burst Duration

For the densities, temperatures, and magnetic field strengths expected in the emission regions of burst sources, the cooling time scales for bremsstrahlung, inverse Compton, and synchrotron cooling are $\ll 1$ s. If the emission region is extremely optically thick, it is possible to achieve photon diffusion times,

$$t_{\rm diff} = 3lr/c \simeq 100\tau_6 l_6 \text{ s} \tag{46}$$

where τ_6 is the optical depth in units of 10^6. However, the rapid variations on a time scale $\simeq 0.1$ second seen in burst intensities and the observed burst spectra[3–9] rule out such an explanation for the duration of the bursts. We therefore conclude that the duration of gamma-ray bursts is not due to the natural cooling time of a hot $T \simeq 3 \times 10^9$ K plasma, whether in a strong magnetic field $B \simeq 10^{12}$ G or not. Rather, it requires a separate explanation.

Comparison of Radiation Processes

Comparison of Equations 28 and 37 illustrates the well-known fact that synchrotron emission in a strong magnetic field cools a hot plasma much more efficiently than does bremsstrahlung. If the spectra of some gamma-ray bursts, such as the Apollo 16 event,[88] are in fact due to bremsstrahlung, the required physical conditions are

$$B \lesssim 2.9 \times 10^9 \bar{g}^{-1/2} Z n_{26}^{1/2}\, T_9^{-1/4} \text{ G} \tag{47}$$

If the surface magnetic field is of order 10^{12} G, such a small magnetic field strength and high number density are difficult to achieve. This might occur high in the magnetosphere of a neutron star, where the magnetic field is small, or possibly in the interior of a hot plasma bubble near the stellar surface, where the field strength is reduced by expansion of the field lines.

Comparison of Equations 31 and 37 shows that synchrotron emission in a strong magnetic field also cools a hot plasma much more efficiently than does inverse Compton scattering, even for a blackbody flux equal to the typical gamma-ray burst luminosity, i.e., $L_{\rm bb} = 10^{37}$ ergs s^{-1}. If the spectra of some bursts, such as the 4 November 1978 burst,[69] are due to inverse Compton scattering, the required physical conditions are

$$B \lesssim 2.6 \times 10^7 L_{\rm bb,37}^{1/2} \text{ G} \tag{48}$$

Thus synchrotron cooling in a magnetic field $B \sim 10^{12}$ G is much more rapid than cooling due to inverse Compton scattering, even if the assumed blackbody luminosity is equal to the typical observed burst luminosity.

The annihilation line widths of $\simeq$250 keV observed in the 5 March 1979 event[13,15] and in other gamma-ray bursts[3–6,8,10,32] may be due to thermal broadening or to a strong magnetic field.[28] In either case, the observed width is an upper bound to the relative line width due to Doppler broadening, which is

$$\Delta\omega/\omega \simeq (2kT/m_e c^2)^{1/2} \tag{49}$$

The observed line widths therefore constrain the temperature of the pairs to be (see also the detailed discussion by Ramaty and Mészáros)[82]

$$T \lesssim 61(\Delta\omega/250 \text{ keV})^2 \text{ keV}. \tag{50}$$

This temperature is much smaller than that required to produce the continuum spectrum. Ramaty, Lingenfelter, and Bussard have successfully resolved this apparent contradiction by pointing out that if

$$t_{\text{syn}}/t_{\text{pair}} = \frac{9}{64}\frac{n_e m_e c^2}{(B^2/8\pi)} \lesssim 1 \tag{51}$$

synchrotron emission can cool the plasma before the pairs annihilate[14] (see also Katz).[28] Equivalently,

$$B^2/8\pi \gtrsim \frac{9}{64} n_e m_e c^2 \tag{52}$$

or

$$B \gtrsim 1.7 \times 10^{10} n_{e,26}^{1/2} \text{ G} \tag{53}$$

which is always satisfied if the magnetic field is sufficiently strong to confine the plasma. The cooling and annihilation may occur sequentially within the hot plasma itself, or they may occur in spatially distinct regions. For example, the annihilation might occur in a sheath surrounding the cooling plasma.

Evidence from Observation

We have seen that synchrotron emission in a strong magnetic field is much more efficient than bremsstrahlung, and greatly exceeds the flux produced by Comptonization of a blackbody source having a luminosity equal to that of a typical burst. Do the observed shapes of gamma-ray burst spectra point to one or another of these radiation processes?

Mazets and his colleagues have found that most of the spectra observed with the Konus experiments on Venera 11–14 can be fit by a simple exponential, apart from the pair annihilation emission lines and the low-energy cyclotron features.[6,8,32–35] Such an expression, although it has the virtue of simplicity, does not include the Gaunt factor which is required to model thermal bremsstrahlung. These fits are perhaps best regarded as a useful way of characterizing the spectral temperature of the bursts.

However, the most precisely determined spectrum of the early bursts, that of the 27 April 1972 event, is surprisingly well fit by a single-temperature (T_{obs} = 500 keV) thin bremsstrahlung spectrum with Gaunt factor over three decades in photon energy, as

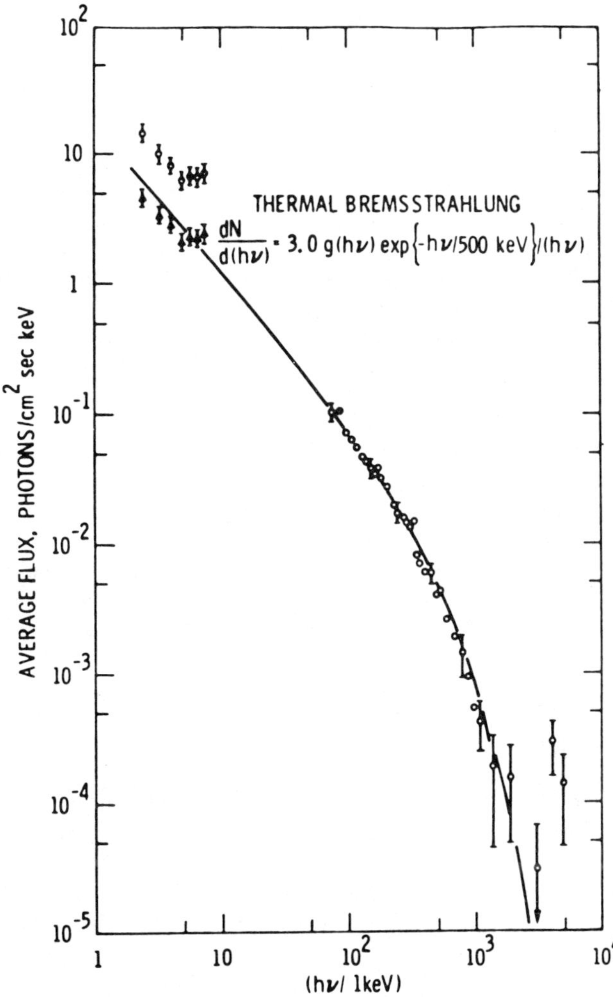

FIGURE 20. Average spectrum of the 27 April 1972 burst observed by Apollo 16, and best fit thermal bremsstrahlung spectrum. (From Gilman *et al.* by permission of *Astrophys. J.*)[88]

shown in FIGURE 20.[88] Of course, the fit did not include the contribution of electron-electron bremsstrahlung and relativistic corrections,[59] both of which are important at these energies, or effects of a strong magnetic field.[9,53] Nevertheless, it is remarkable that the fit is so good. As a result of this observation, bremsstrahlung gained some adherents.

Fenimore and his colleagues have demonstrated that the spectrum of the 4 November 1978 event is well fitted by Comptonized blackbody spectra, as shown in FIGURE 21.[69] The χ^2 per degree of freedom of the fits is much smaller than that for thermal bremsstrahlung, although this may not be surprising considering the extra fitting parameters available. The fits are characterized by an optical depth $\tau_{es} \simeq 3$, a Comptonization parameter $y \simeq 10$, and an energy enhancement factor $L/L_{bb} = A \simeq 18$. These values are characteristic of saturated Comptonization. Indeed, the energy enhancement factor A lies close to its maximum value $T_e/T_{bb} \simeq 155/2.4 \simeq 65$. Under these conditions, we expect that the spectrum should show evidence of a Wien peak. Careful examination of FIGURE 21 does reveal a broad peak centered at about $3T_e \simeq$ 450 keV. The Comptonized blackbody spectrum has the attractive feature that it links the gamma-ray burst phenomena with x-ray bursts, in the sense that the gamma-ray burst spectrum could be produced by an x-ray burst that is Comptonized as it escapes through a tenuous hot plasma surrounding the neutron star.

Synchrotron emission has been suggested as the radiation mechanism by a number of people, including Ramaty, Lingenfelter, and Bussard,[14] Katz,[28] Lamb,[11] and Liang.[70–72] Liang, in particular, has demonstrated using the approximate analytical expression derived by Petrosian[87] that the spectral data of Mazets *et al.*[32] for the 7 March, 1 November, and 6 October 1979 events are well fitted by thermal synchrotron spectra.[71] He has also demonstrated that the gamma-ray data of Gilman *et al.*[88] for the 27 April 1972 event are similarly well fitted (we note, however, that the x-ray data were omitted from the fit). These fits are shown in FIGURE 22.

What do we conclude from all of this? A reasonable conclusion would seem to be that the precision of the spectral observations is not yet sufficient to favor, or rule out, one or another of the emission mechanisms. This can be understood if we recognize that, in almost all cases, the width in photon energy of the observational window is only ~2–3 times greater than the characteristic spectral temperature of the bursts. The observations therefore tend to reveal a spectral shoulder, but do not include sufficiently low and high energies to distinguish among the spectra expected for various emission mechanisms.

In this regard, it is interesting to consider the Solar Maximum Mission (SMM) observations of the 20 March 1982 event. These observations show a burst spectrum that extends to energies higher than 25 MeV.[89] Such a hard spectrum may require a nonthermal electron or pair distribution (see below). Even so, it is difficult to account for it by bremsstrahlung or inverse Compton scattering, while synchrotron emission can easily produce such hard spectra.

THROUGH THICK AND THIN

Electron Scattering and Synchrotron Optical Depths

So far we have assumed that the emission region is optically thin. However, the electron scattering and synchrotron opacities are sufficiently large that the emission region could be extremely optically thick, as we now demonstrate. Again we assume that the electrons are at most semirelativistic, and the plasma is not pair dominated. Then the electron scattering optical depth is

$$\tau_{es} = n_e \sigma_T l \simeq 6.6 \times 10^7 \, n_{e,26} l_6 \qquad \textbf{(54)}$$

where l is the thickness of the emission region in units of 10^6 cm. For the emission

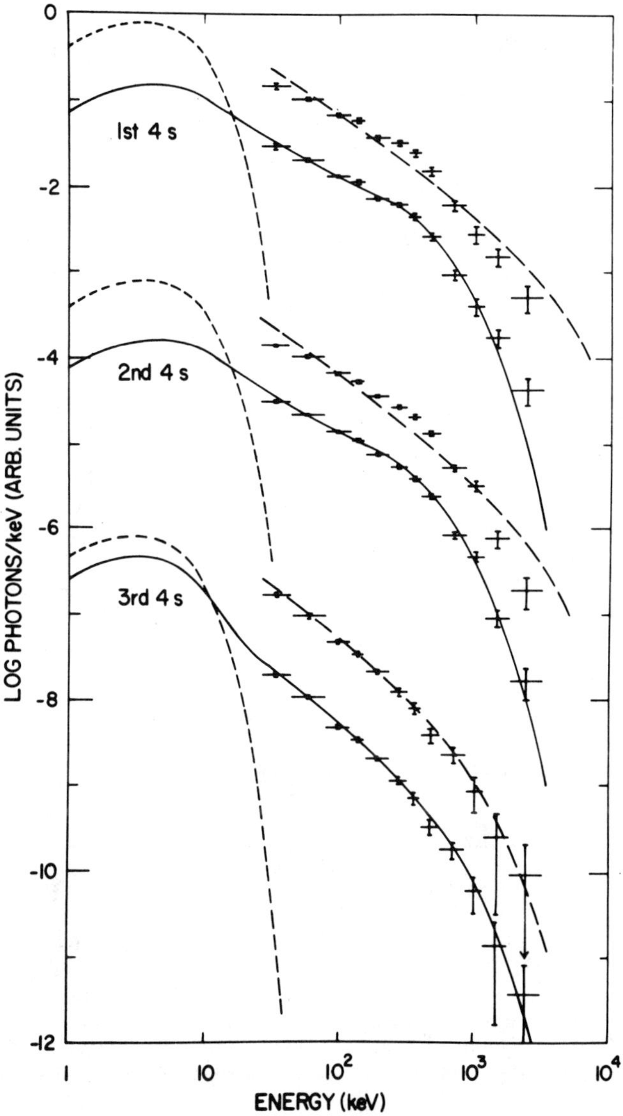

FIGURE 21. Spectra of the 4 November 1978 burst during first, second, and third 4-second intervals, and best fit thermal bremsstrahlung and Comptonized blackbody spectra for these intervals. The short-dashed curves are the initial x-ray blackbody spectra and the solid curves are the final Comptonized spectra. The long-dashed curves are the best fit thermal bremsstrahlung spectra. (From Fenimore *et al.* by permission of *Nature*.)[69]

region to be optically thin, the number density must satisfy

$$n_e \lesssim 1.5 \times 10^{18} l_6^{-1} \text{ cm}^{-3} \tag{55}$$

Similarly, the optical depth for synchrotron absorption (or scattering) at the fundamental is[90]

$$\begin{aligned} \tau_{syn} &\simeq \frac{\pi}{3} \Lambda(\Delta\omega/\omega)^{-1} \simeq \frac{\pi}{3} \Lambda(2kT/m_e c^2)^{-1/2} \\ &\simeq 1.1 \times 10^{12} B_{12}^{-1} T_9^{-1/2} n_{e,26} l_6 \end{aligned} \tag{56}$$

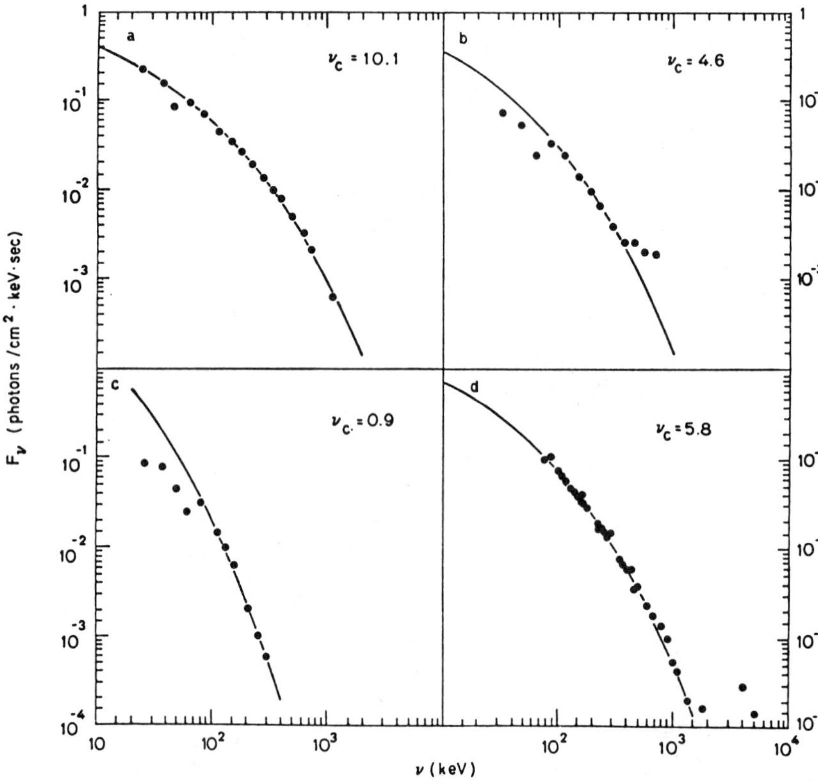

FIGURE 22. Representative burst spectra and best fit thermal synchrotron spectra. Events are (a) 7 March 1979; (b) 1 November 1979, (c) 6 October 1979 (initial stage),[32] and (d) 27 April 1972.[88] The high-energy data points have low statistical significance. (From Liang by permission of *Nature*.)[71]

where we have used Equation 49 to estimate the relative band width of the line. Then the requirement that the emission region be optically thin becomes

$$n_e \simeq 9.2 \times 10^{13} B_{12} T_9^{1/2} l_6^{-1} \text{ cm}^{-3} \tag{57}$$

From Equations 54 and 56, the ratio of electron scattering and synchrotron opacities at the cyclotron fundamental is

$$\tau_{\rm syn}/\tau_{\rm es} = \frac{\pi}{2}\alpha^{-1}(m_e c^2/2kT)^{1/2}\frac{m_e c^2}{\hbar\omega_c} \simeq 1.6 \times 10^4 B_{12}^{-1}\, T_9^{-1/2} \tag{58}$$

Equation 58 shows that, if a strong magnetic field is present, synchrotron absorption or scattering is by far the dominant source of opacity at the cyclotron fundamental. More generally, the ratio of the two opacities in the limit $h\nu \lesssim kT$ is[90]

$$\alpha_\omega^{\rm syn}/\alpha_{\rm T} = \frac{3}{2}\alpha^{-1}\frac{m_e c^2}{\hbar\omega_c}\Phi(\omega) \simeq 9.1 \times 10^3 B_{12}^{-1}\,\Phi(\omega) \tag{59}$$

where the dimensionless size parameter Λ and the dimensionless flux $\Phi(\omega)$ are those defined in Equations 40–42. FIGURE 18 shows that even for electron temperatures $T \simeq 3 \times 10^9$ K, the synchrotron emissivity and therefore the absorptivity and scattering decrease rapidly with frequency and, as a result, electron scattering may still dominate at sufficiently high frequencies. FIGURE 18a can be used, together with Equation 59, to determine the frequency at which the two opacities become equal. For example, given an electron temperature $T \simeq 100$ keV, Equation 59 shows that the two opacities are equal when $\Phi(\omega) \simeq 10^{-4}$; FIGURE 18a then shows that this occurs at $\omega/\omega_c \simeq 20$ or, assuming $B = 1 \times 10^{12}$ G, at $\simeq 200$ keV.

Earlier we derived a characteristic maximum number density $n \simeq 10^{26}$ cm^{-3} for the emission region of a gamma-ray burst (Equation 17). Comparing that result and the constraints represented by Equations 54 and 56, we conclude that the emission regions of gamma-ray bursts are either extremely optically thick, or that energy and/or matter is continuously replenished during the burst.

Evidence from Observation

All spectral observations to date suggest that the emission region of gamma-ray bursts is optically thin. None of the spectra show any indication of a Wien peak. This requires a Comptonization parameter $y \ll 10$, i.e., $\tau_{\rm es} \ll 4T_9^{-1/2}$.

A more stringent constraint comes from the pair annihilation line seen in the 5 March 1979 event[15] and in other gamma-ray bursts.[6,8,26,32] The recoil from a single Compton scattering of a 511 keV photon is sufficient to scatter it out of the line ($\Delta E/E \simeq \hbar\omega/m_e c^2$), so that observation of a line requires $\tau_{\rm es} < 1$.

Low-energy features have been observed in $\simeq 20\%$ of the burst spectra seen by the Leningrad group.[6,8,32] Even in these cases, however, the inferred optical depths are not large.[72]

We note that the low-energy features have been widely interpreted as due to emission[71–72] or absorption[3–10,32–35] at the cyclotron fundamental, and the positions of the dips have been used to infer magnetic field strengths. However, the severe thermal broadening of the cyclotron harmonics that occurs at even modest temperatures (recall FIGURE 18) rules out thermal emission at the cyclotron fundamental as an explanation of the dips, unless the temperature of the emission region is $\lesssim 50$ keV, a temperature much lower than the continuum spectra of most bursts indicate. A nonthermal distribution of electrons in the emission region could produce a deep edge at the cyclotron fundamental (recall FIGURE 19). Cold matter overlying the emission region could also produce a narrow absorption feature at the fundamental. However, in both cases strong features would then also be expected at higher harmonics, and none are observed.[6,8,32] Support for this interpretation comes from the Franco-Soviet Signe experiment on Venera 12, which allows spectra to be measured with finer time

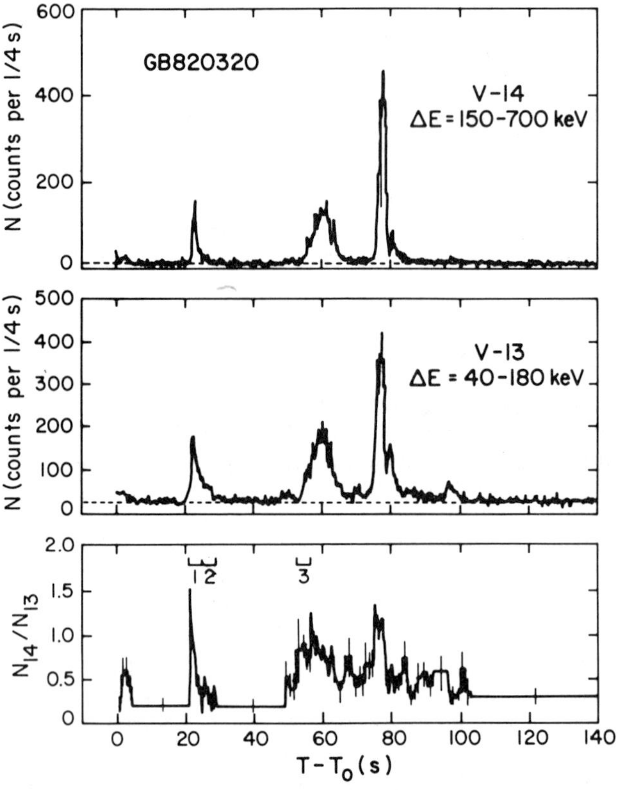

FIGURE 23. Time profiles and count rate ratio (hardness ratio) from the detectors on the Venera 13 and 14 spacecraft for the 20 March 1982 burst.[93]

resolution than the 4-second resolution of the Konus experiments. The Signe experiment shows that the low-energy cutoff of the 19 November 1978 event evolves rapidly in time.[91,92] Initially, it occurs at $\simeq$ 500 keV, but it rapidly decreases to $\lesssim$ 100 keV within about 4 seconds. More recently, the Leningrad group has presented evidence that burst spectra are rapidly varying on time scales as short as the observed intensity variations.[93] FIGURE 23 shows two time profiles of the 20 March 1982 burst from detectors on Venera 13 and 14. Also shown are spectral hardness ratios during the burst as derived from the two detectors. Clearly, the burst spectrum varies dramatically on the time scale of the intensity variations.

Given the extremely short cooling time scales of a hot plasma derived earlier, such rapid variability is expected. Therefore narrow or weak spectral features may be washed out and lost by previous and current experiments that average over time intervals $\gg 0.1$ seconds. And interpretations based on fitting theoretical spectra to the low-energy absorption features seen in some bursts (cf. References 72 and 73) should be regarded with great caution. For example, if these features are indeed the result of time averaging a rapidly varying low-energy cutoff, the magnetic field strength may not be inferred directly from the energy of the low-energy features.

In summary, observations to date suggest that the emission regions of gamma-ray bursts are relatively optically thin. As we have seen, this implies that the energy of the emitting plasma and/or the plasma itself must be continuously replenished during a burst.[11] It also implies that the cooling time scale of the emitting plasma is much shorter than the observed duration of the bursts, so that the duration must have a separate explanation.[11]

Constraints on Source Aspect Ratios and Distances

Assuming that the burst emission regions are optically thin, we may derive constraints on the number density in the emission region, and on the aspect ratio and distance of the source. The emissivity of the emitting plasma must satisfy

$$L = \Lambda_{\mathrm{br(pair,syn)}}\, V \tag{60}$$

where $L \cong 1 \times 10^{37}(d/100\mathrm{pc})^2$ ergs s^{-1} is the observed luminosity of a burst, and V is the volume of the emission region. If we parametrize V by Expression 16 and assume that bremsstrahlung is the dominant emission mechanism, then the number density in the emission region must satisfy

$$n_e \cong 1.3 \times 10^{20}(\bar{g}Z)^{-1/2}(d/100\mathrm{pc})(fl_6)^{-1/2}T_9^{-1/4}\ \mathrm{cm}^{-3} \tag{61}$$

Similarly, if instead we assume that pair annihilation or synchrotron emission is the dominant emission mechanism, the constraint on the number density becomes

$$n_{\bar{e}} \cong 8.1 \times 10^{18}(d/100\mathrm{pc})(fl_6)^{-1/2}\ \mathrm{cm}^{-3} \tag{62}$$

or

$$n_e \cong 1.5 \times 10^{9}(d/100\mathrm{pc})^2(fl_6)^{-1}B_{12}^{-2}\,T_9^{-1}\ \mathrm{cm}^{-3} \tag{63}$$

Equation 63 defines a characteristic number density that is a factor 10^{16} smaller than that given by Equation 17! Its small size reflects the high efficiency of synchrotron emission in a strong magnetic field.

By combining the above bounds on the number density and the expressions for either electron scattering or synchrotron opacities (Equations 54 and 56), as appropriate, we can obtain constraints on the aspect ratio and/or distance of the source. Katz gives a clear and comprehensive discussion of these constraints,[28] assuming bremsstrahlung or pair annihilation. Here we give them in a slightly different form and consider also synchrotron emission (see also Reference 11). We shall define the aspect ratio of the burst emission region as

$$\alpha \equiv \frac{l}{2f^{1/2}R} = \frac{l_6}{2f^{1/2}R_6} = l_6/2f^{1/2} \tag{64}$$

where in the last step we take, as usual, $R = 10^6$ cm. Assuming bremsstrahlung emission, we obtain the constraint,

$$\alpha(d/100\text{pc})^2 \lesssim 6.3 \times 10^{-5}\bar{g}ZT_9^{1/2} \tag{65}$$

Equation 65 implies that, if the source lies at a distance $\simeq$100 pc and emits via bremsstrahlung, the burst emission region must have an aspect ratio greater than 1:10^4. (If GBS0526-66, the source of the 5 March 1979 event, lies at a distance of 55 kpc, as it would if it actually is in the large Magellanic cloud,[13,19,20] the implied aspect ratio is 1:10^9!) Similar constraints can be derived from the luminosity in the observed pair annihilation lines,

$$\alpha(d/100\text{pc})^2 \lesssim 1.7 \times 10^{-2} \tag{66}$$

or assuming synchrotron emission,

$$f^{-1}(d/100\text{pc})^2 \lesssim 6.2 \times 10^4 B_{12}^3 T_9^{3/2} \tag{67}$$

Equation 67 shows that, if the burst is produced by synchrotron emission, an emission region covering only a fraction $f \simeq 1.6 \times 10^{-5}$ of the stellar surface would suffice. This is again a reflection of the high efficiency of synchrotron emission.

We note that similar arguments can be applied to inverse Compton scattering. However, the results are not very interesting, since fits to the observations indicate that the luminosity amplification factor $L/L_{bb} = A \sim 1$, i.e., the emitted luminosity L is just the assumed source luminosity L_{bb} if the emission region is assumed to be optically thin.

ENERGY TRANSPORT AND THERMAL EQUILIBRIUM

We have mentioned earlier that the way in which the energy of a burst couples to the hot plasma in the emission region is as yet unsolved. Here we permit ourselves a few conjectures about it.

Whatever the mechanism, it has a difficult challenge before it: to transfer the energy on a very brief time scale (perhaps as short as $\sim 10^{-16}$ seconds) while keeping the emission region optically thin. Coulomb collisions are hopelessly inadequate for the task. The electron-ion exchange time scale (assuming $T_i \gg T_e$) is, for example,

$$t_{e\text{-}i} = 4.0 \times 10^{-12} n_{e,26}^{-1}\, T_9^{3/2}(\ln\Lambda/20)^{-1} \tag{68}$$

However, in laboratory situations it is widely believed that collisionless plasmas can maintain near-Maxwellian distributions via collective processes that operate on a time scale of $\sim$ the inverse of the plasma frequency, i.e.,

$$\nu_p^{-1} = 1.1 \times 10^{-17} n_{e,26}^{-1/2}\ \text{s} \tag{69}$$

Such processes may therefore be faster than the synchrotron cooling time scale (Equation 38). Whether such processes are relevant, especially in strong magnetic fields, to gamma-ray burst emission regions is controversial. We merely note that if they do indeed occur, they are capable of maintaining a quasi-thermal distribution of electrons and of enhancing synchrotron absorption relative to scattering. Observation must tell us whether the burst spectra are thermal or nonthermal (see below), and therefore whether such processes are relevant.

Whether the burst energy comes from gravitational potential energy, via accretion, or an event internal to the star, or energy from nuclear burning of accreted matter, an attractive possibility for coupling this energy to the emission region is Alfvén waves.[55] Their velocity is given by Equation 13; for the low plasma densities and high magnetic fields relevant to burst-emission regions, it is essentially the speed of light, c. Alfvén waves can therefore transport energy rapidly to the emission region. Furthermore, they have the advantage of transporting energy without increasing the density of particles in the emission region, an essential requirement if the emission region is to be kept optically thin.

In the previous sections, we have described and compared various possible burst emission mechanisms assuming that the emission is thermal. This is simple and convenient because the resulting spectra are characterized by only one energy parameter, the temperature T. However, under the extreme conditions expected in the burst emission region, it is likely that the electrons are not in thermal equilibrium and their velocity distribution cannot be characterized by a Maxwell-Boltzmann distribution with temperature T. In particular, it is difficult to maintain the high-energy tail of the electron distribution. Yet it just these high-energy electrons that contribute disproportionately to the high-energy part of any thermal spectrum. For example, use of a thermal synchrotron spectrum to fit observed spectra up to only $\simeq 1$ MeV implies that a thermal distribution of electrons persists up to at least $\simeq 5$ MeV, 50 times the characteristic thermal energy! Nonthermal emission models should therefore be pursued, despite the fact that such models, by necessity, introduce additional fitting parameters.

SOME IMPORTANT ISSUES

In concluding, we mention several important issues that any theory of gamma-ray bursts must address.

What Determines the Duration of a Burst?

Observations tell us that the durations of gamma-ray bursts are as short as 0.1 sec and as long as 100 sec.[3-9] In contrast, the radiative cooling time scales for bremsstrahlung and synchrotron emission are much shorter (for example, the time scale for synchrotron cooling can be as short as $\simeq 10^{-16}$ sec; see Equation 38). Even were the emission region very optically thick, the observed intensity variations on a time scale 0.1 sec rule out an explanation for the total duration of the burst in terms of a diffusion time scale. Thus, the duration of gamma-ray bursts does not arise from the cooling time scale of the hot plasma, whether in a strong magnetic field or not. Rather, a separate explanation is needed. This issue has been faced in different ways, perhaps not fully satisfactorily, by many of the proposed models of gamma-ray bursts. For example, accretion models[39,44–48] commonly push the explanation back to an earlier stage by attributing the duration of bursts to the interval during which accreting plasma or solid cometary matter rains down on the star. Nuclear burning models[53–61] invoke the time necessary for a deflagration wave to propagate across the lake of nuclear fuel, or the interval during which nearby fuel sites successively ignite. Models that rely on an internal source of energy to power the burst invoke some other time scale, such as the gravitational damping time of neutron star oscillations.[14,49–52]

What Determines the Spectral Temperature of a Burst?

Observations tell us that the spectral temperatures of gamma-ray bursts range from $T_{obs} \simeq 20$ keV, as in the 5 March 1979 event,[13–15] to several MeV.[3,8,10,32–35] The characteristic maximum temperature of an optically thick radiation-dominated plasma confined in a magnetic field of a few times 10^{12} G (Equation 4) may indeed explain the soft gamma-ray nature of the bursts. This temperature can vary, depending on the strength of the magnetic field and the extent to which the expanding bubble of hot plasma compresses the field, yet it is not clear that this relation can give a temperature range as large as that observed. More seriously, if the emission region is optically thin, as observation indicates, it is not clear what relevance this characteristic maximum temperature may have. In addition, a complete model of the bursts must be able to answer the important question, Why are the observed spectral temperatures of bursts so much greater than those of the "steady" pulsating x-ray sources, whose emission is also due to radiation from a hot plasma at the surface of strongly magnetic neutron stars? Is the difference due to the impulsive nature of the event, or because the emission is optically thin?

How Can the Radiating Region Be Kept Optically Thin?

As we have seen, observations suggest that the spectra of gamma-ray bursts are produced by optically thin emission, whether bremsstrahlung, pair annihilation radiation, or synchrotron emission. The discussion in the last section emphasizes that if the emitting region is optically thin, the energy of the emitting plasma and/or the plasma itself must be continuously replenished during a burst. In accretion models,[39,44–48] it could occur through accretion of new matter. In models that rely on nuclear explosions,[53–61] the replenishment might occur, for example, via injection into the magnetosphere of fresh material from successive burning sites. In models deriving their energy from a source inside the neutron star,[14,50–52] the replenishment could occur via oscillations of the neutron star itself that would transport energy to the stellar surface and, possibly, even above it by rapid motions of the magnetic field lines. Whatever mechanism transports the energy, it remains a challenge in all models to do so without creating an optically thick emission region.

Are All Bursts Produced by the "Same" Mechanism?

Occam's razor instructs us to assume that all gamma-ray bursts have a single explanation. However, both their durations and their spectral temperatures encompass a dynamic range of more than 10^3. These ranges are large, even by astrophysical standards, and they raise an important question: Are all gamma-ray bursts produced by the "same" mechanism, or are such bursts the manifestations of several transient phenomena occurring in, on, and around neutron stars?[11] The latter view receives some support from indications in the burst data of the Leningrad group of at least two distinct classes of bursts; namely, short ($\lesssim$0.1 seconds) bursts with soft ($kT \cong 30$ keV) spectra, and long ($\gtrsim$10 seconds) bursts with hard ($kT \gtrsim 100$ keV) spectra.[6,93]

ACKNOWLEDGMENTS

We are grateful to Roger Bussard, Stirling Colgate, Fred Lamb, Rich Lingenfelter, Peter Mészáros, and Reuven Ramaty for valuable discussions on various aspects of the physics of gamma-ray bursts.

REFERENCES

1. Klebesadel, R. W., I. B. Strong & R. A. Olson. 1973. Astrophys. J. **182:** L85.
2. Ruderman, M. 1975. Ann. N.Y. Acad. Sci. **262:** 164.
3. Hurley, K. 1980. Adv. Space Explor. **7:** 123. (Pergamon Press. Oxford, England.)
4. Hurley, K. 1982. *In* Accreting Neutron Stars. MPE Report No. 177. W. Brinkmann & J. Trümper, Eds.: 161. Max-Planck Institute for Extraterrestrial Physics. Garching, Federal Republic of Germany.
5. Vedrenne, G. 1981. Phil. Trans. R. Soc. London Ser. A **301:** 645.
6. Mazets, E. P. & S. V. Golenetskii. 1981. Astrophys. Space Sci. **75:** 47.
7. Cline, T. L. 1981. Ann. N.Y. Acad. Sci. **375:** 314.
8. Mazets, E. P., S. V. Golenetskii, V. N. Il'inskii, Yu. A. Gur'yan, R. L. Aptekar', V. N. Panov, I. A. Sokolov, Z. Ya. Sokolova & T. V. Kharitonova. 1981. Astrophys. Space Sci. **82:** 261.
9. Klebesadel, R. W., E. E. Fenimore, J. G. Laros & J. Terrell. 1982. AIP Conf. Proc. **77:** 1.
10. Teegarden, B. J. 1982. AIP Conf. Proc. **77:** 123.
11. Lamb, D. Q. 1982. AIP Conf. Proc. **77:** 249.
12. Rees, M. J. 1982. *In* Accreting Neutron Stars. MPE Report No. 177. W. Brinkmann & J. Trümper, Eds.: 179. Max-Planck Institute for Extraterrestrial Physics. Garching, Federal Republic of Germany.
13. Cline, T. 1982. AIP Conf. Proc. **77:** 17.
14. Ramaty, R., R. E. Lingenfelter & R. W. Bussard. 1981. Astrophys. Space Sci. **75:** 193.
15. Mazets, E. P., S. V. Golenetskii, V. N. Il'inskii, R. L. Aptekar' & Yu. A. Guryan. 1979. Nature **282:** 587.
16. Cline, T. L., U. D. Desai, G. Pizzichini, B. J. Teegarden, W. D. Evans, R. W. Klebesadel, J. G. Laros, K. Hurley, M. Niel, G. Vedrenne, I. V. Estoolin, A. V. Kouznetsov, V. M. Zenchenko, D. Hovestadt & G. Gloecker. 1980. Astrophys. J. Lett. **237:** L1.
17. Barat, C., G. Chambon, K. Hurley, M. Niel, G. Vedrenne, I. V. Estulin, A. V. Kuznetsov & V. M. Zenchenko. 1979. Astron. Astrophys. **79:** L24.
18. Terrell, J., W. D. Evans, R. W. Klebesadel & J. G. Laros. 1980. Nature **285:** 383.

19. Evans, W. D., R. W. Klebesadel, J. G. Laros, T. L. Cline, U. D. Desai, G. Pizzichini, B. J. Teegarden, K. Hurley, M. Niel, G. Vedrenne, I. V. Estoolin, A. V. Kouznetsov, V. M. Zenchenko & V. G. Kurt. 1980. Astrophys. J. Lett. **237:** L7.
20. Cline, T. L., U. D. Desai, B. J. Teegarden, W. D. Evans, R. W. Klebesadel, J. G. Laros, C. Barat, K. Hurley, M. Niel, G. Vedrenne, I. V. Estulin, V. G. Kurt, G. A. Mersov, V. M. Zenchenko, M. C. Weisskopf & J. E. Grindlay. 1982. Astrophys. J. Lett. **255:** L45.
21. Helfand, D. J. & K. S. Long. 1979. Nature (London) **282:** 589.
22. Golenetskii, S. V., Yu. A. Guryan, V. N. Il'inskii & E. P. Mazets. 1983. Ioffe Fiz.-Tekh. Institute Akad. Nauk SSSR No. 813. (Submitted to Nature.)
23. Lewin, W. H. G. & P. C. Joss. 1981. Space Sci. Rev. **28:** 3.
24. Mazets, E. P., S. V. Golenetskii, Yu. A. Guryan & V. N. Il'inskii. 1982. Astrophys. Space Sci. **84:** 173.
25. Evans, W. D., J. P. Glore, R. W. Klebesadel, J. G. Laros, E. R. Tech & R. E. Spalding. 1980. Science **205:** 119.
26. Teegarden, B. J. & T. L. Cline. 1980. Astrophys. J. Lett. **236:** L67.
27. Daugherty, J. K. & R. W. Bussard. 1980. Astrophys. J. **238:** 296.
28. Katz, J. I. 1982. Astrophys. J. **260:** 371.
29. Schaefer, B. E. 1981. Nature **294:** 722.
30. Cline, T. L., U. D. Desai, G. Pizzichini, B. J. Teegarden, W. D. Evans, R. W. Klebesadel, J. G. Laros, C. Barat, K. Hurley, M. Niel, G. Vedrenne, I. V. Estulin, G. A. Mersov, V. M. Zenchenko & V. G. Kurt. 1981. Astrophys. J. Lett. **246:** L133.
31. Grindlay, J. E., T. Cline, U. D. Desai, B. J. Teegarden, G. Pizzichini, W. D. Evans, R. W. Klebesadel, J. G. Laros, K. C. Hurley, M. Niel & G. Vedrenne. 1982. Nature **300:** 730.
32. Mazets, E. P., S. V. Golenetskii, R. L. Aptekar', Yu. A. Gur'yan & V. N. Il'inskii. 1981. Nature **290:** 378.
33. Mazets, E. P., S. V. Golenetskii, V. N. Il'inskii, V. N. Panov, R. L. Aptekar', Yu. A. Gur'yan, I. A. Sokolov, Z. Ya. Sokolova & T. V. Kharitonova. 1981. Astrophys. Space Sci. **80:** 3.
34. Mazets, E. P., S. V. Golenetskii, V. N. Il'inskii, V. N. Panov, R. L. Aptekar', Yu. A. Gur'yan, I. A. Sokolov, Z. Ya. Sokolova & T. V. Kharitonova. 1981. Astrophys. Space Sci. **80:** 85.
35. Mazets, E. P., S. V. Golenetskii, V. N. Il'inskii, V. N. Panov, R. L. Aptekar', Yu. A. Gur'yan, A. V. Dyatchkov, M. P. Proskura, I. A. Sokolov, Z. Ya. Sokolova, N. G. Khavenson & T. V. Kharitonova. 1981. Astrophys. Space Sci. **80:** 119.
36. Fenimore, E. E., R. W. Klebesadel, J. C. Laros, R. E. Stockdale & S. Kane. 1982. AIP Conf. Proc. **77:** 201.
37. Fenimore, E. E., R. W. Klebesadel & J. C. Laros. 1982. Adv. Space Explor. (In press.)
38. Hueter, G. J. & D. E. Gruber. 1982. *In* Accreting Neutron Stars. MPE Report No. 177. W. Brinkmann & J. Trümper, Eds.: 213. Max-Planck Institute for Extraterrestrial Physics. Garching, Federal Republic of Germany.
39. Colgate, S. A. & A. G. Petschek. 1981. Astrophys. J. **248:** 771.
40. Ruderman, M. 1981. Prog. Part. Nucl. Phys. **6:** 215.
41. Cavallo, G. & M. J. Rees. 1978. Mon. Not. R. Astron. Soc. **183:** 359.
42. Newman, M. J. & A. N. Cox. 1980. Astrophys. J. **242:** 319.
43. Howard, W. M., J. R. Wilson & R. T. Barton. 1981. Astrophys. J. **249:** 302.
44. Lamb, D. Q., F. K. Lamb & D. Pines. 1973. Nature Phys. Sci. **246:** 52.
45. Anzer, U., G. Boerner & P. Mészáros. 1976. Astron. Astrophys. **58:** 305.
46. Hoshi, R. 1976. Prog. Theor. Phys. **56:** 542.
47. Takahara, F. & R. Hoshi. 1978. Prog. Theor. Phys. **59:** 425.
48. Harwit, M. & E. E. Salpeter. 1973. Astrophys. J. Lett. **187:** L97.
49. Tsygan, A. I. 1975. Astrophys. Space Sci. **44:** 21; **49:** 159.
50. Fabian, A. C., V. Icke & J. E. Pringle. 1976. Astrophys. Space Sci. **42:** 77.
51. Bisnovatyi-Kogan, G. S. & V. M. Chechetkin. 1980. Preprint. Institute for Space Research. Acad. Nauk SSSR No. 561.

52. Pacini, F. & M. Ruderman. 1974. Nature **251:** 399.
53. Woosley, S. E. & R. E. Taam. 1976. Nature **262:** 101.
54. Tsygan, A. I. 1980. Astron. Astrophys. **87:** 224.
55. Mitrofanov, I. G. & V. M. Ostryakov. 1981. Astrophys. Space Sci. **77:** 469.
56. Woosley, S. E. & R. K. Wallace. 1982. Astrophys. J. **258:** 716.
57. Fryxell, B. A. & S. E. Woosley. 1982. Astrophys. J. **258:** 733.
58. Fryxell, B. A. & S. E. Woosley. 1982. AIP Conf. Proc. **77:** 299.
59. Hameury, J. M. S. Bonazzola, J. Heyvaerts & J. Ventura. 1982. Astron. Astrophys. **111:** 242.
60. Woosley, S. E. 1982. AIP Conf. Proc. **77:** 273.
61. Woosley, S. E. 1982. *In* Accreting Neutron Stars. MPE Report No. 177. W. Brinkmann & J. T. Trümper, Eds.: 189. Max-Planck Institute for Extraterrestrial Physics. Garching, Federal Republic of Germany.
62. McClintock, J. E., S. Rappaport, P. C. Joss, H. Bradt, J. Buff, G. W. Clark, D. Hearn, W. H. G. Lewin, T. Matilsky, W. Mayer & F. Primini. 1976. Astrophys. J. Lett. **206:** L99.
63. Bussard, R. W. 1981. Private communication.
64. Bussard, R. W. & F. K. Lamb. 1982. AIP Conf. Proc. **77:** 189.
65. Chiu, H.-Y. 1968. Stellar Physics. Blaisdell Publishing Co. Waltham, Mass.
66. Arons, J. & S. M. Lea. 1976. Astrophys. J. **207:** 914.
67. Elsner, R. F., P. Ghosh & F. K. Lamb. 1980. Astrophys. J. **241:** L155.
68. Gould, R. J. 1982. AIP Conf. Proc. **77:** 169.
69. Fenimore, E. E., R. W. Klebesadel, J. G. Laros, R. E. Stockdale & S. R. Kane. 1982. Nature **297:** 665.
70. Liang, E. P. 1981. Nature **292:** 319.
71. Liang, E. P. 1982. Nature **299:** 321.
72. Liang, E. P. 1983. Astrophys. J. Lett. **268:** L89.
73. Allen, C. W. 1973. Astrophysical Quantities. 3rd edit.: 103. Athlone Press. London, England.
74. Gould, R. J. 1980. Astrophys. J. **238:** 1026.
75. Weymann, R. 1965. Phys. Fluids **8:** 2112.
76. Shapiro, S. L. & E. E. Salpeter. 1975. Astrophys. J. **198:** 671.
77. Illarionov, A. F. & R. A. Sunyaev. 1972. Sov. Astron. **16:** 45.
78. Shapiro, S. L., A. P. Lightman & D. M. Eardley. 1976. Astrophys. J. **204:** 187.
79. Katz, J. I. 1976. Astrophys. J. **206:** 916.
80. Sunyaev, R. A. & L. G. Titarchuk. 1980. Astron. Astrophys. **86:** 121.
81. Pozdnyakov, L. A., I. M. Sobol & R. A. Sunyaev. 1979. Sov. Astron. Lett. **5:** 149. (1979. Pis'ma Astron. Zh. **5:** 279.)
82. Ramaty, R. & P. Mészáros. 1981. Astrophys. J. **250:** 384.
83. Trubnikov, B. A. 1958. Ph.D. Thesis. Moscow Institute of Engineering and Physics. (AEC-tr-4073. Office of Technical Services. Washington, D.C.)
84. Hirschfield, J. L., D. E. Baldwin & S. C. Brown. 1961. Phys. Fluids **4:** 198.
85. Lamb, D. Q. & A. R. Masters. 1979. Astrophys. J. Lett. **229:** 652.
86. Brainerd, J. J. & D. Q. Lamb. 1983. (In preparation.)
87. Petrosian, V. 1981. Astrophys. J. **251:** 727.
88. Gilman, D., A. E. Metzer, R. H. Parker, L. G. Evans & J. I. Trombda. 1980. Astrophys. J. **236:** 951.
89. Rieger, E., C. Reppin, G. Kanbach, D. J. Forrest, E. L. Chupp & G. H. Share. 1982. *In* Accreting Neutron Stars. MPE Report No. 177. W. Brinkmann & J. Trümper, Eds.: 229. Max-Planck Institute for Extraterrestrial Physics. Garching, Federal Republic of Germany.
90. Masters, A. R. 1978. Ph.D. Thesis. University of Illinois. Urbana, Ill.
91. Knight, F. K., J. L. Matteson & L. E. Peterson. 1981. Astrophys. Space Sci. **75:** 21.
92. Barat, C., G. Chambon, K. Hurley, M. Niel, G. Vedrenne, I. V. Estulin, A. V. Kuznetsov & V. M. Zenchenko. 1981. Astrophys. Space Sci. **75:** 83.
93. Golenetskii, S. V., Yu. A. Guryan, V. N. Il'inskii & E. P. Mazets. 1983. Ioffe Fiz.-Tekh. Inst. Akad. Nauk SSSR No. 819.

History of X-Ray Transients as Seen by Vela, 1969–1979[a]

J. TERRELL, W. C. PRIEDHORSKY, R. D. BELIAN,
J. P. CONNER, AND W. D. EVANS

Los Alamos National Laboratory
Los Alamos, New Mexico 87545

Vela spacecraft 5A and 5B, launched into orbit in May 1969, were among the first to be capable of x-ray astronomy. The x-ray counters aboard Vela 5B operated for an unprecedented length of time, observing the entire x-ray sky for 10 years, until June 1979. These spacecraft, monitoring for nuclear tests in space with various detectors, were put into very high orbits at 118,000 km radius. Collimated NaI detectors, sensitive to 3–12 keV x-rays, scanned the sky at 90° to the earth-spacecraft axis over a 6.1° FWHM (full width at half-maximum) square field of view. Any given source, when in view, was scanned every 64 s as the spacecraft rotated, with 1-second count accumulations in two energy channels. The entire sky was observed every 56 hours, half of the orbital period.

During the 10-year lifetime of the x-ray detectors, the Vela spacecraft provided data on many new types of sources. Cen X-4, an exceedingly bright x-ray transient, was observed in the summer of 1969, shortly after launch,[3] and did not return to the x-ray sky until May 1979, when it was again observed by Vela 5B. This source also produced the first of many x-ray bursts detected by Vela, the bright, hard, precursor observed on 7 July 1969.[1] The Vela gamma-ray detectors revealed the new phenomenon of gamma-ray bursts.[5] It has recently been found that at least two gamma-ray bursts were also detected by the x-ray counters.[10]

The data produced by Vela 5B over the 7-year period May 1969–June 1976, during which real-time tracking was reasonably complete, have been put into the form of a series of 10-day x-ray sky maps. Although the Vela x-ray detectors were relatively small (27 cm^2 effective area), they were sensitive enough to yield good data on Cen A (NGC 5128) during the period 1973–1975 when it was unusually bright and active.[2] The Vela data show that Cen A often changed dramatically in intensity in less than 10 days.[8]

This series of sky maps has recently been made into a movie (*The X-ray Sky, 1969–1976*), showing in time-lapse form the appearance and reappearance of transient sources and the continual fluctuations of other sources. This period includes times when no other spacecraft were monitoring the sky. Sources as weak as Her X-1, Per X-1, and SMC X-1 are often quite apparent in these sky maps, as are transient outbursts not previously known, such as 4U1145-61 in April 1973, Aql X-1 in May 1974, and 4U0115+63 in August 1974. An otherwise unknown transient x-ray source (VO332+53) was observed by Vela in Camelopardalis during June–August 1973,[9,11] at galactic coordinates 146.1°, −2.2°, with the position known to ~0.2°. This hard

[a]This work was supported by the U.S. Department of Energy.

x-ray transient was brighter than the Crab in July 1973, has not been seen since, and was not observed by other spacecraft, to our knowledge. Its time history is different from that of other bright transients in its gradual rise to peak brightness. There is evidence of rapid fluctuations in the source, but no clear periodicity has yet been found.

Cyg X-1 is a source of considerable interest in the Vela data. This black-hole candidate exhibits chaotic fluctuations on a time scale of seconds, which appear to be nonperiodic shot noise.[7] On longer time scales it is modulated by the 5.6-day orbital period of the binary system,[4] and has high and low states at irregular intervals over the years. The long complete record available in the Vela data now makes it obvious that Cyg X-1 fluctuates in intensity with a period of 294 ± 4 days.[6] The origin of this long period is of considerable interest, whether it is due to precession of the supergiant companion star or of the accretion disc, or to some other cause.

Much other evidence of outbursts, fluctuations, and periodicity is present in the Vela sky-map data base, and apparent in the movie made from it. It is clear that even a small detector, scanning the entire x-ray sky for many years, can yield data of great value to x-ray astronomy.

REFERENCES

1. BELIAN, R. D., J. P. CONNER & W. D. EVANS. 1972. Astrophys. J. **171:** L87–L90.
2. BEALL, J. H., *et al.* 1978. Astrophys. J. **219:** 836–844.
3. CONNER, J. P., W. D. EVANS & R. D. BELIAN. 1969. Astrophys. J. **157:** L157–L159.
4. HOLT, S. S., L. J. KALUZIENSKY, E. A. BOLDT & P. J. SERLEMITSOS. 1979. Astrophys. J. **233:** 344–349.
5. KLEBESADEL, R. W., I. B. STRONG & R. A. OLSON. 1973. Astrophys. J. **182:** L85–L88.
6. PRIEDHORSKY, W. C., J. TERRELL & S. S. HOLT. 1983. Astrophys. J. **270:** 233–238.
7. TERRELL, N. J. 1972. Astrophys. J. **174:** L35–L41.
8. TERRELL, J. 1982. Int. Astron. Union Symp. **97:** 117–118.
9. TERRELL, J., R. D. BELIAN, J. P. CONNER, W. D. EVANS & W. C. PRIEDHORSKY. 1982. Am. Astron. Soc. Bull. **14:** 619.
10. TERRELL, J., E. E. FENIMORE, R. W. KLEBESADEL & U. D. DESAI. 1982. Astrophys. J. **254:** 279–286.
11. TERRELL, J., W. C. PRIEDHORSKY, R. D. BELIAN, J. P. CONNER & W. D. EVANS. 1982. *In* Accreting Neutron Stars. W. Brinkmann & J. Trümper, Eds.: 210–212. Max-Planck Institute for Extraterrestial Physics. Garching, Federal Republic of Germany.

The Emission-Line Regions of Quasars

JOHN KWAN

Department of Physics and Astronomy
University of Massachusetts
Amherst, Massachusetts 01003

INTRODUCTION

Quasars and active galactic nuclei have always aroused considerable interest among astronomers. The recent accumulation of x-ray and ultraviolet data has further spurred theoretical activity on this topic. An ultimate goal of this research is to uncover the origin of the powerful continuum radiation. One area of study that bears information on this problem is the properties of the matter (at $\sim$1 pc in the case of quasars) around the continuum source. Analysis of the emitted line intensities permits a determination of the physical conditions and abundance of elements in that region. More interestingly, knowledge of the cloud kinematics provides direct information about the continuum source. If the clouds were infalling matter, an estimate of the rate of mass input to the central engine, as well as its total mass, could be obtained. If the clouds were outflowing matter, their mass and momenta would place constraints on acceleration mechanisms. The possibility also exists that the emission-line region might affect the continuum radiation through gas absorption and dust reddening, and that it might itself contribute some continuum emission. The wealth of observational data—namely, the intensities of the many UV, optical, and infrared lines and the highly resolved line profiles—provide exacting tests for any theoretical model, and will lead to a definitive understanding of the emission-line region.

I shall summarize the present understanding of the physical conditions of the emission-line gas in the next section, and address the question of the cloud kinematics in the third.

PHYSICAL CONDITIONS OF THE BROAD-LINE EMISSION REGION

The emission spectra of quasars consist prominently of broad permitted and semiforbidden lines (e.g., Lyα, Hα, OVIλ1034, CIVλ1549, CIII]λ1909, MgIIλ2798) of full width at half-maximum (FWHM) $\geqq 10^4$ km s^{-1}, and narrow forbidden lines (e.g., [OIII]$\lambda\lambda$5007, 4959) of FWHM $\leqq$ 500 km s^{-1}. It is generally believed that the emission lines are produced as a result of photoionization of surrounding gas by the strong central continuum[1,9,21] (for a review see Reference 10). The power-law UV continuum not only provides a convenient energy source for the line emission, but also creates naturally the wide range of observed high and low ionization species; and theoretical photoionization models are fairly successful in reproducing the observed line intensities. The different dependences of the forbidden, semiforbidden, and permitted line excitation on density, and the absence of broad forbidden lines, suggest that the permitted and semiforbidden lines are primarily formed in a region of high density, such that collisional deexcitation suppresses forbidden line emission. A typical nucleon density N_0 of $\sim 10^9$–10^{10} cm^{-3} is derived from analysis of those lines whose intensities are density sensitive (e.g., CIII]λ1909, HeIλ5876). To produce the observed ionization species (e.g., OVI, NV) an ionization parameter Γ^i (which is the ratio of the incident ionizing [$h\nu > 13.6$ eV] photon density to cloud nucleon density) of $\sim$0.03 is needed. This corresponds to placing the broad-line emission clouds at a distance from the continuum source of $\sim$1 pc $(L/10^{46}\ \mathrm{erg\ s^{-1}})^{1/2}\ (3 \times 10^9\ \mathrm{cm^{-3}}/N_0)^{1/2}\ (0.03/\Gamma^i)^{1/2}$, where L is the ionizing UV luminosity. Since the spectra among quasars and Seyfert 1 nuclei are quite similar, despite their very different luminosities, the size of the broad-line region of each object scales approximately as the square root of its luminosity. Similar considerations place the lower density ($\sim 10^3$ cm^{-3}) narrow-line gas at a distance of $\sim$1 kpc. Henceforth I shall be concerned only with the broad-line region, because of its proximity to the central source.

There has been a recent upsurge in interest in the study of photoionization models and theoretical line intensities (e.g., References 15, 18, 19, 25, 5, and 37). It is prompted by the observed intensities of UV lines which, together with known optical data, indicate that the ratio of Lyα flux to Hβ flux is at least several times smaller than that expected.[2,8,40] These recent theoretical studies all attempt to improve upon the earlier photoionization models by extending the calculations of ionization and thermal equilibria to depths far beyond the fully ionized zone. Although differing considerably in details, they reach the general conclusion that, owing to the strong x-ray ionization heating, the region beyond the HII zone produces a large fraction of the hydrogen Balmer and higher series line and continuum emission, as well as the line emission (e.g., OIλλ1304, 8446, MgIIλ2798, FeII optical bands) of lowly ionized heavy elements (see the following subsection for some details). Besides the Lyα/Hβ ratio, the relative intensities of the more than 20 broad emission lines can be reproduced reasonably well by such photoionization calculations for very optically thick clouds.[19] The relative abundances of the elements, within a factor of a few, are solar.[10,19,11]

Besides the indication from photoionization models, observations also point to very optically thick clouds. First, the profiles of lines that are produced at appreciable Lyman continuum optical depths (e.g., MgIIλ2798) are usually quite similar to those of lines of highly ionized elements.[10,14] Second, those quasars whose Lyman continua are observable in the optical do not show a sharp Lyman edge absorption attributed to the broad-line clouds,[28] indicating that the absorption of at least a few percent of the Lyman continuum (in order to produce the Lyα flux) is accomplished, not by optically thin clouds covering the whole continuum source, but by optically thick clouds covering a small fraction of the continuum area.

In order to be very optically thick to the Lyman continuum, the linear dimension of a cloud need not be very large. Knowing that the density within a cloud is about 10^9–10^{10} cm^{-3} and that the distance between a cloud and the central continuum source is $\sim$1 pc, it is found that a column density of $\sim 3 \times 10^{22}$ cm^{-2} suffices to produce a Lyman edge continuum optical depth of $\sim 10^5$. The linear cloud size is then only $\sim 10^{13}$ cm, much smaller than the overall size of the region wherein the clouds are found. Thus the broad-line clouds fill only a miniscule fraction of the volume.

The total mass of the broad-line gas can be determined. Since a Lyα photon is emitted following each proton-electron recombination in the HII zone, the total rate of emission of Lyα is equal to the total rate of recombinations. From the latter rate the total mass of the broad-line gas can be derived. It is

$$M \geqq 30\, M_\odot \left(\frac{L_{\mathrm{Ly}\alpha}}{3 \times 10^{44}\ \mathrm{erg\ s^{-1}}}\right)\left(\frac{3 \times 10^{9}\ \mathrm{cm^{-3}}}{N_0}\right)\left(\frac{\text{column density}}{5 \times 10^{21}\ \mathrm{cm^{-2}}}\right)$$

where $L_{\mathrm{Ly}\alpha}$ represents the luminosity in the Lyα line. The above estimate assumes a cloud column density of 5×10^{21} cm^{-2}, which is the minimum value needed to absorb all the UV continuum between 13.6 and 54.4 eV. Since the clouds are considerably thicker, the total mass M increases proportionally. The mass within a spherical cloud 10^{13} cm in radius is only $\sim 10^{-8}\ M_\odot$. Even if the clouds are highly flattened in shape, a very large number of them is expected. Together they cover a few percent or more of the area of the continuum source, and their different velocities produce the broad-line profiles.

The small physical thickness of a cloud (at least toward the continuum source) has an important implication on the cloud dynamics. The sound travel time (temperature $\sim 10^4$ K) across 10^{13} cm is less than a year, which is much shorter than the time ($\sim$100 years) for the cloud to traverse the broad-line region. Thus the clouds must be confined, generally believed to be by a very hot intercloud medium. It is found that such a two-phase equilibrium can occur only over a narrow range of the cloud ionization parameter.[16] Remarkably, this range is the same as that deduced from photoionization models. The hot ambient medium may affect the line and continuum emission via electron scattering,[34] and may accelerate the broad-line clouds.[38]

A Few Details of the Photoionization Calculations

Two conditions make the photoionization models of the quasar emission-line clouds different from standard nebular calculations. The first, mentioned earlier, is the strong x-ray luminosity of the quasar. Of the continuum photons, those with energies between 13.6 and $\sim$300 eV are absorbed within the fully ionized zone. The more energetic ones, however, cause ionization and heating in the region beyond. The typical quasar continuum from UV to hard x-ray indicates that as much energy can be deposited in the region beyond the HII zone as within. Compared to standard nebular calculations, then, the gas emission from the HI zone is much more significant. The second condition is the relatively high density ($\sim 10^9$–10^{10} cm^{-3}) in the quasar emission-line cloud. Collisional excitation, together with the strong trapping of Lyα when hydrogen becomes partially neutral, will sustain a fraction of the hydrogen atoms in the $n = 2$ level. This fraction is small, $\leqq 10^{-5}$, but significant in that it can be photoionized by the Balmer continuum. Furthermore, once population is built up into $n = 2$, it is quickly built up into higher excited levels, whence collisions can produce more ionizations. Detailed calculations indeed show that the electron fraction, although dropping sharply once the Lyman continuum is significantly depleted, remains $\geqq 0.2$ up to a depth almost 10 times the thickness of the HII zone, and that the temperature stays at $\sim$8,000 K there.[19,37] In this extended partially ionized zone, collisional excitations produce Balmer and higher series photons but few Lyα photons, as the latter are so strongly trapped that few of them escape. This provides an explanation for the observed ratios of Lyα to the Balmer lines which are much lower than the values expected when only the fluxes from the HII zone were considered.

Many improvements of the present photoionization calculations are possible. First, FeII emission is strong in the extended ionized zone, and it consists of many UV and optical multiplets. Because these multiplets have a wide range of line strengths, their relative intensities depend sensitively on the line opacities. A careful modeling of the FeII emission will provide a measure of the cloud column density, as well as a check of the physical conditions in the extended ionized zone. An obstacle to this study, unfortunately, is the uncertain FeII collisional rates. Second, the broad feature (so-called 3,000 Å bump) in the spectrum needs to be satisfactorily explained. One suggestion is that it consists of Balmer continuum and FeII emission.[12,31] Another is that it is thermal emission from an accretion disk.[23] It is important that the flux attributed to the broad feature be accurately determined observationally, since the Balmer continuum and FeII emission produced in a photoionization model is limited by the available x-ray heating. Third, most of the present calculations assume that the line photons escape from the illuminated surface only, i.e., they pertain to infinite slab geometries. For those lines that are produced over a considerable portion of the emission-line cloud (e.g., Hα, Hβ, FeII optical multiplets), their emitted fluxes are better modeled by finite slab calculations. Fourth, the present models are compared with the mean quasar spectrum or the spectrum of an individual object to derive typical values of the nucleon density N_0 and ionization parameter Γ^i. Since the line emission from a given quasar is contributed by an ensemble of clouds, models need to be calculated for different parameters and averaged in order to determine over what ranges of N_0 and Γ^i the average spectrum will resemble that obtained with the typical values.

Dust Reddening of the Quasar Radiation?

It is important to know if the observed optical and UV radiation from a quasar or active galactic nucleus represents the intrinsic energy distribution. McKee and Petrosian first evaluated the possibility of dust reddening of the quasar continuum by measuring the 2,200 Å feature.[26] They concluded that there is little dust reddening. This question was reopened recently when the Lyα/Hβ "problem" was discovered. Besides Lyα/Hβ, there are several other ratios that are possible indicators of dust reddening, namely, OIλ1304/λ8446 and HeIIλ1640/λ4686.[27,22]

As mentioned earlier, photoionization calculations for very optically thick clouds can account for the low Lyα/Hβ ratio. The same calculations also obtain values of Pα/Hα that are lower than given by standard recombination theory. This result is just opposite to the effect of dust reddening. Together, then, the observed Lα/Hβ and Pα/Hα ratios provide a measure of the relative importance of collisional excitation in a very optically thick cloud and dust reddening. They indicate that for quasars, dust reddening is not significant,[32,35] but for Seyfert 1s, both dust reddening and collisions are needed.[20]

The ratio OIλ1304/λ8446 is an ideal indicator of dust reddening. The relevant energy level structure of OI is such that each λ8446 photon that is produced, as a result of Bowen fluorescence following absorption of a Lyβ photon,[4] also produces a λ1304 photon during the cascade down to the ground state. The ratio OIλ1304/λ8446 is then given simply by the ratio of the energies of the two photons or 6.5. When the excitation of these two lines is included in photoionization calculations for very optically thick clouds, it is found that the above ratio is reduced somewhat, to ~4.9 in one particular model,[19] owing to a small amount of collisional excitation of λ8446 and Balmer continuum absorption of λ1304. Recently, it was noted that the upper state of λ1304 has also a semiforbidden transition, which can become important at large λ1304 optical depths.[13] When this aspect is included in the above photoionization model, the OIλ1304/λ8446 ratio is further reduced to ~3.3. It is so significantly smaller than the "expected" value that the use of OIλ1304/λ8446 as a reddening indicator might be compromised.

The HeIIλ1640/λ4686 ratio is a better indicator of reddening than $Ly\alpha/H\beta$, because the corresponding effects of optical depth and collision are much weaker. Its observed values have led to a conclusion of significant dust reddening for quasars.[22] A practical problem in the use of this ratio is the uncertainty involved in isolating λ4686 from the broad FeIIλ4570 band. The possibility that HeIIλ1640/λ4686 might be affected by $Ly\alpha$ fluorescence is also suggested,[13] but has to be definitively demonstrated.

Judging by the overall available observational and theoretical evidence, I believe that dust reddening is not significant for quasars.

KINEMATICS OF THE BROAD-LINE CLOUDS

Among the proposed kinematic models of the broad-line gas are outflow of matter being accelerated by radiation pressure[3,24] or by a hot wind;[38] ballistic outflow of matter;[6] matter in circular orbits;[29] and matter in parabolic orbits.[17]

Each of the above models has reasons for its promotion. The process of acceleration by radiation pressure produces a logarithmic line shape most naturally (which, except at line center and at the far wings, fits observed profiles well), and this result is independent of the optical thickness of the clouds. The additional acceleration supplied by a thermal wind takes into account the dynamical effect of the hot intercloud medium. The process of ballistic outflow avoids the problem of cloud formation, and has more freedom to produce the extended high-velocity wings that are observed (e.g., Reference 41). All these models share the knowledge that outflows have been observed in some quasars, as evidenced by the presence of P Cygni profiles and broad blue-ward absorption troughs (e.g., Reference 36). The models of rotation and parabolic-orbital motion both involve a gravitational origin for the cloud velocities. The former has the appeal that a rotating disk is an expected mode of accretion onto a compact object, while the latter reckons that infall of matter from far away will have a range of impact parameters. Both models provide a means of mass input to the central source.

The theoretical questions associated with the above kinematic models are many and formidable, but will not be dealt with here. Ultimately the merit of a particular model must depend on how it fares when compared with observations. In this respect, it is unfortunate that the different models can all produce, given the freedom of choice of parameters, broad-line profiles that agree reasonably well with observations.

One property of the broad-line clouds has an important implication for kinematic models. It is that each cloud is very optically thick to the Lyman continuum. The $Ly\alpha$ emission from such a cloud is highly anisotropic, i.e., most of the $Ly\alpha$ radiation emerges from the face of the cloud illuminated by the continuum source. This effect then places a constraint on kinematic models. For example, in cases of purely radial motion (either infall or outflow), the Lyα profile is expected to be very asymmetric. If the broad-line clouds are all very optically thick, the observed line symmetry (e.g., Reference 39) is then an argument in favor of the rotation and parabolic-orbital models. But, by itself, it is hardly definitive enough to decide such an important issue.

Another test is also available and probably more conclusive, as it is a direct consequence of the cloud kinematics. Besides producing permitted line emission, the clouds, during their course of motion, also produce forbidden line emission, as the density within a cloud decreases (following the decreasing external pressure) with increasing distance from the central source. The two classes of kinematic models (orbital versus outflow) predict very differently, both in the relative amount and in the line profile, for that associated forbidden line emission. The orbiting models produce a significant amount of it, with line widths that are smaller than those of the broad permitted lines but greater than those of the narrow cores.[33,17] For example, in the parabolic-orbiting model, the associated [OIII]λ5007 emission is about 0.3 the total broad Hβ emission, and has a full width at zero intensity (FWZI) of ~4,000 km s^{-1}. Also, different forbidden lines are predicted to have different line widths. In order of increasing line width, they are [OII]λ3727, [SII]$\lambda\lambda$6716, 6731, [NII]λ6583, [OIII]λ5007, [OI]λ6300, [NeIII]λ3869, [NeV]λ3426, [OIII]λ4363, and [SII]$\lambda\lambda$4069, 4076.[7] Outflow models, on the other hand, produce considerably less associated forbidden line emission. Compared to the infall case, the outflowing gas, by the time it expands to a density low enough to favor forbidden line emission, is moving at a much higher speed, making the time interval of that emission comparatively shorter. The outflowing gas also becomes optically thin to the Lyman continuum then, and the fraction of ionizing photons absorbed is reduced. The profile of the forbidden line emission is flat topped and has a broad width equal to the FWZI of the permitted lines, making it more difficult to discern. Thus the two classes of kinematic models predict very differently for the associated forbidden line emission. These lines, although heretofore not prominent in studies of the broad-line region, may contain the clues to the cloud kinematics. High-resolution observations of the strong [OIII]λ5007 line (e.g., Reference 30), and more importantly other weaker forbidden lines, will reveal whether the different lines, besides their narrow cores, have broader components of different widths. They will severely test models with a gravitational origin for the cloud velocities.

It may also be possible to distinguish observationally between the case of circular rotation and the case of infall or parabolic-orbital motion. In the former case it is expected that in roughly 10% of the quasars (deduced from the typical covering factor of ~0.1), the rotating disk will be observed edge on. For these quasars, the broad-line emission region completely covers the central continuum source along the line of sight. Since the permitted resonance lines (e.g., Lyα, CIVλ1549, MgIIλ2798) have very large optical depths, and since the emission gas is moving transversely to the line of sight, scattering of the continuum radiation occurs at the center of each optically thick line. Those line profiles should then have absorption cores with an extent at least several thermal Doppler widths. Unfortunately the line emission from the narrow-line region somewhat complicates the situation. Nevertheless, it may still be possible to determine, by comparison, if a fraction of the quasars show line center absorptions.

The observations of time variation in the emission-line intensities and profiles can also probe the velocity structure of the broad-line gas. These, for Seyfert galaxies, are discussed by Dr. Ulrich (this volume).

REFERENCES

1. BAHCALL, J. N. & B. KOZLOVSKY. 1969. Astrophys. J. **155:** 1077.
2. BALDWIN, J. A. 1977. Mon. Not. R. Astron. Soc. **178:** 67P.
3. BLUMENTHAL, G. R. & W. G. MATHEWS. 1975. Astrophys. J. **198:** 517.
4. BOWEN, I. S. 1947. Publ. Astron. Soc. Pac. **59:** 196.
5. CANFIELD, R. C. & R. C. PUETTER. 1981. Astrophys. J. **243:** 390.
6. CAPRIOTTI, E., C. FOLTZ & P. BYARD. 1980. Astrophys. J. **241:** 903.
7. CARROLL, T. J. & J. KWAN. 1983. Astrophys. J. **274.** (In press.)
8. DAVIDSON, A. F., G. F. HARTIG & W. G. FASTIE. 1977. Nature **269:** 203.
9. DAVIDSON, K. 1972. Astrophys. J. **171:** 213.
10. DAVIDSON, K. & H. NETZER. 1979. Rev. Mod. Phys. **51:** 715.
11. GASKELL, C. M., G. A. SHIELDS & E. J. WAMPLER. 1981. Astrophys. J. **249:** 443.
12. GRANDI, S. A. 1982. Astrophys. J. **255:** 25.
13. GRANDI, S. A. 1983. Astrophys. J. **268:** 591.
14. GRANDI, S. A. & M. M. PHILLIPS. 1979. Astrophys. J. **232:** 659.
15. KROLIK, J. H. & C. F. MCKEE. 1978. Astrophys. J. Suppl. **37:** 459.
16. KROLIK, J. H., C. F. MCKEE & C. B. TARTER. 1981. Astrophys. J. **249:** 422.
17. KWAN, J. & T. J. CARROLL. 1982. Astrophys. J. **261:** 25.
18. KWAN, J. & J. H. KROLIK. 1979. Astrophys. J. Lett. **233:** L91.
19. KWAN, J. & J. H. KROLIK. 1981. Astrophys. J. **250:** 478.
20. LACY, J. H., B. T. SOIFER, G. NEUGEBAUER, K. MATTHEWS, M. MALKAN, E. E. BECKLIN, C. C. WU, A. BOGGESS & T. R. GULL. 1982. Astrophys. J. **256:** 75.
21. MACALPINE, G. M. 1972. Astrophys. J. **175:** 11.
22. MACALPINE, G. M. 1981. Astrophys. J. **251:** 465.
23. MALKAN, M. A. 1983. Astrophys. J. **268:** 582.
24. MATHEWS, W. G. 1982. Astrophys. J. **252:** 39.
25. MATHEWS, W. G., G. R. BLUMENTHAL & S. A. GRANDI. 1980. Astrophys. J. **235:** 971.
26. MCKEE, C. F. & V. PETROSIAN. 1974. Astrophys. J. **189:** 17.
27. NETZER, H. & K. DAVIDSON. 1979. Mon. Not. R. Astron. Soc. **187:** 871.
28. OKE, J. B. & D. G. KORYCANSKY. 1982. Astrophys. J. **255:** 11.
29. OSTERBROCK, D. E. 1978. Proc. Nat. Acad. Sci. USA **75:** 540.
30. PELAT, D. & D. ALLOIN. 1982. Astron. Astrophys. **105:** 335.
31. PUETTER, R. C. & P. D. LEVAN. 1982. Astrophys. J. **260:** 44.
32. PUETTER, R. C., H. E. SMITH, S. P. WILLNER & J. L. PIPHER. 1981. Astrophys. J. **243:** 345.
33. SHIELDS, G. A. 1978. *In* Pittsburgh Conference on BL Lacertae Objects. A. M. Wolfe, Ed.: 257. University of Pittsburgh Press. Pittsburgh, Pa.
34. SHIELDS, G. A. & C. F. MCKEE. 1981. Astrophys. J. Lett. **246:** L57.
35. SOIFER, B. T., G. NEUGEBAUER, J. B. OKE & K. MATTHEWS. 1981. Astrophys. J. **243:** 369.
36. TURNSHEK, D. A., R. J. WEYMANN, J. W. LEIBERT, R. E. WILLIAMS & P. A. STRITTMATTER. 1980. Astrophys. J. **238:** 488.
37. WEISHEIT, J. C., G. A. SHIELDS & C. B. TARTER. 1981. Astrophys. J. **245:** 406.
38. WEYMANN, R. J., J. S. SCOTT, A. V. R. SCHIANO & W. A. CHRISTIANSEN. 1982. Astrophys. J. **262:** 497.
39. WILKES, B. J. & R. F. CARSWELL. 1982. Mon. Not. R. Astron. Soc. **201:** 645.
40. WU, C. C. 1977. Astrophys. J. Lett. **217:** L117.
41. WU, C. C., A. BOGGESS & T. R. GULL. 1981. Astrophys. J. **247:** 449.

Line Variability in Active Nuclei and the Structure of the Broad-Line Region

M. H. ULRICH

European Southern Observatory
D-8046 Garching, Federal Republic of Germany

INTRODUCTION

Given the fact that the typical linear dimension of the broad-line region of an active nucleus is 10^{16} to 10^{18} cm and that the closest example is at about 10 Mpc, there is little hope that a resolved image or a map of a broad-line region can be obtained in the near future. The variability of the profile and of the intensity of the emission lines thus provides the only way to obtain information on the structure of the broad-line region, i.e., on the spatial distribution of the gas and the velocities.

Several groups of observers are currently making repeated observations of Seyfert nuclei of type 1 with the aim of studying the continuum and the line variability. In this talk I would like to review new data that have recently become available. I shall cover in some detail the variations of the ultraviolet spectrum of NGC 4151 since this is the galaxy that has been observed the most extensively.

RESULTS

The first reports of variability of the Balmer lines in Seyfert 1 nuclei were published several years ago.[1,10] Since then it has been realized that line variability is not an uncommon phenomenon among Seyfert nuclei. For example, Osterbrock and Shuder find variations of the Balmer lines in one-third of the nuclei that have been observed repeatedly at Lick Observatory over a period of several years.[18] No variations of the permitted lines in quasars have yet been reported. This is not unexpected because intrinsically bright broad-line regions like those of quasars have larger linear dimensions and therefore must vary on a longer time scale than the intrinsically weaker broad-line regions in the nearby Seyfert galaxies.

In the classical model of active nuclei, the permitted lines are emitted by dense ($N_e \geqq 10^9$ cm^{-3}) small gas clouds close to the central continuum source. It is generally believed that the large width of the emission lines is due to the motions of the gas clouds (3×10^3 to 20×10^3 km s^{-1}) and that the intensity variations of the lines are caused by changes in the degree of ionization which are themselves caused by the variations of the continuum source.

Variations of the Balmer decrement have been observed in several galaxies.[8,18,21,24] Particularly interesting are the variations reported by de Bruyn, who found several instances where the intensity ratios Hβ/Hγ and Hα/Hβ became shallower when the continuum increased.[7] This suggests that, as the continuum increased, large opacity effects reached transitions that were higher and higher in the Balmer series.

Models of the response of a Balmer line intensity and profile to the increase or decrease of the continuum source have been made by Capriotti *et al.*[9] in the case of a collection of clouds moving ballistically outward. This model reproduces rather well the Hβ profiles observed at different epochs in Arakelian 120. However, the agreement between profiles calculated with a model and profiles observed at isolated epochs is a necessary but not a sufficient test of the model. A nearly continuous time coverage of the line variations for a certain period is necessary to make further progress in this area.

I would like to turn now to results obtained with the satellite IUE.[5] Given the fact the the ultraviolet spectrum is known for quasars from numerous ground-based observations, the IUE observations allow one to make a comparison of the UV spectra of active nuclei over a wide range of absolute luminosity.

There are many well-known advantages in observing with a satellite, e.g., repeatability of the observing procedure, possibility of continuous monitoring without interference of the moon and the sun. But the most interesting characteristic of the UV range, 1,200 to 3,000 Å, is that it contains intense lines produced in regions of very different physical conditions: CIVλ1550 comes from a region that is highly ionized, and could therefore be very close to the continuum source; CIII]λ1909 comes from a region where $N_e \lesssim 10^{9.5}$ cm^{-3}. Finally MgIIλ2800 comes from a region that is partially ionized and where the main portion of the Balmer lines are also believed to be emitted.

The active nucleus that is the nearest to us, the brightest in apparent magnitude and the most active, is NGC 4151. It has been extensively monitored with IUE by a consortium of astronomers from different groups in the framework of a collaboration which I organized in 1978. The results of this collaborative program are being published in a series of articles in the *Monthly Notices*.[19,20,26] Here I shall restrict myself to an outline of the results concerning the variations of the emission lines and their implications on the structure of the broad-line region. In addition, I discuss the evidence provided by the May 1979 data that the wings and the center of the CIV line behave differently indicating that the highest velocity clouds are the closest to the continuumn source.

Line variations in other galaxies also observed with IUE but at a few isolated epochs have been reported recently.[3,11,13,25] They appear to be consistent with our results on NGC 4151.[26]

The galaxy NGC 4151 was observed at 30 epochs between July 1978 and May 1980; at each epoch we obtained several spectra in each of the two wavelength ranges of IUE 1,200–2,000 and 2,000–3,000. The repeatability of consecutive measurements is excellent with an rms of 5% in 50 Å bands.[20] The results on the main emission lines of CIVλ1550, MgIIλ2800, and CIII]λ1909 are summarized below. (The Lyα line is not studied because it is heavily blended with the intense geocoronal Lyα.)

THE STRUCTURE OF THE BROAD-LINE REGION INFERRED FROM THE VARIATIONS OF THE EMISSION LINES IN NGC 4151

The permitted lines of CIVλ1550, SiIV1400, HeIIλ1640, and MgIIλ2800 vary in intensity and profiles in response to the variations of the UV continuum (FIGURES 1, 2

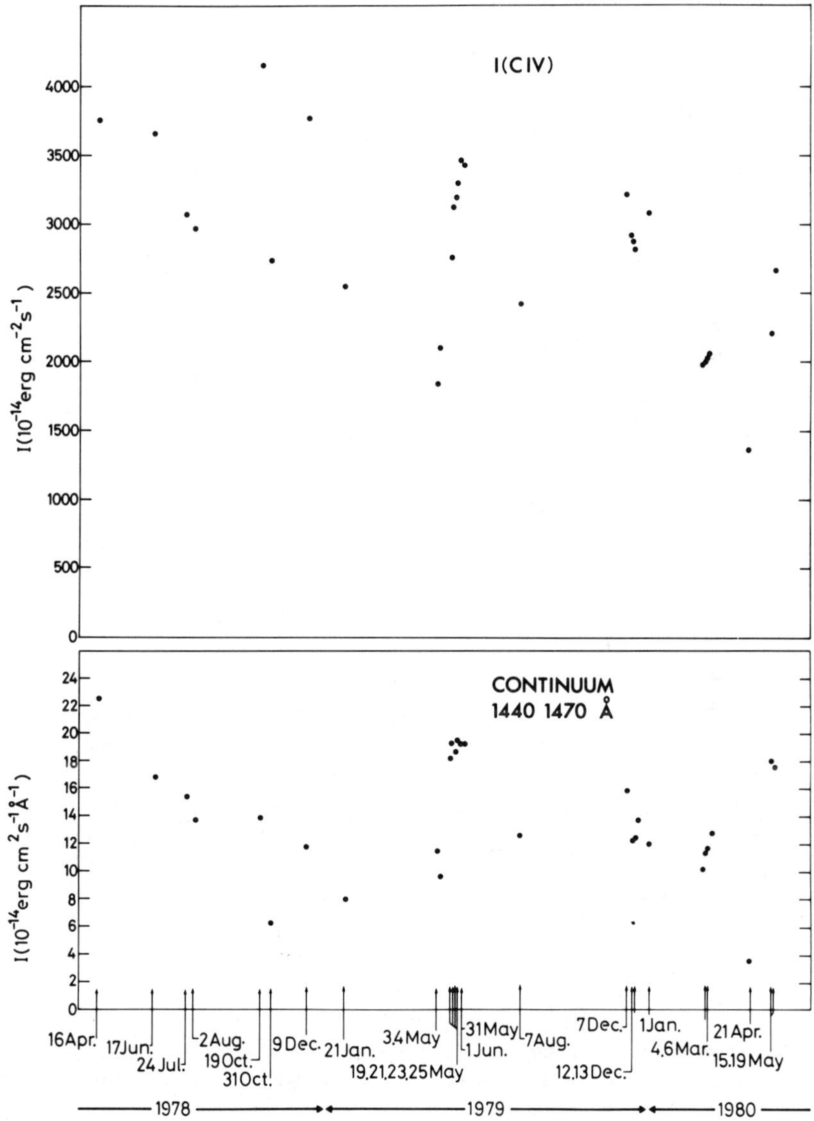

FIGURE 1. Time variations of the intensity of CIVλ1550 and of the continuum in the range 1,440–1,470 Å in NGC 4151.

and 3). The variations of the flux of photons incident on the broad-line region clouds change the degree of ionization and the emissivity of the clouds in the various lines. The most significant characteristics of the lines can be summarized as follows.

Differences between Line Profiles

The full width at zero intensity of CIVλ1550 is at most epochs larger than that of MgIIλ2800 (FIGURES 3 and 4). The clouds with the largest velocities are therefore

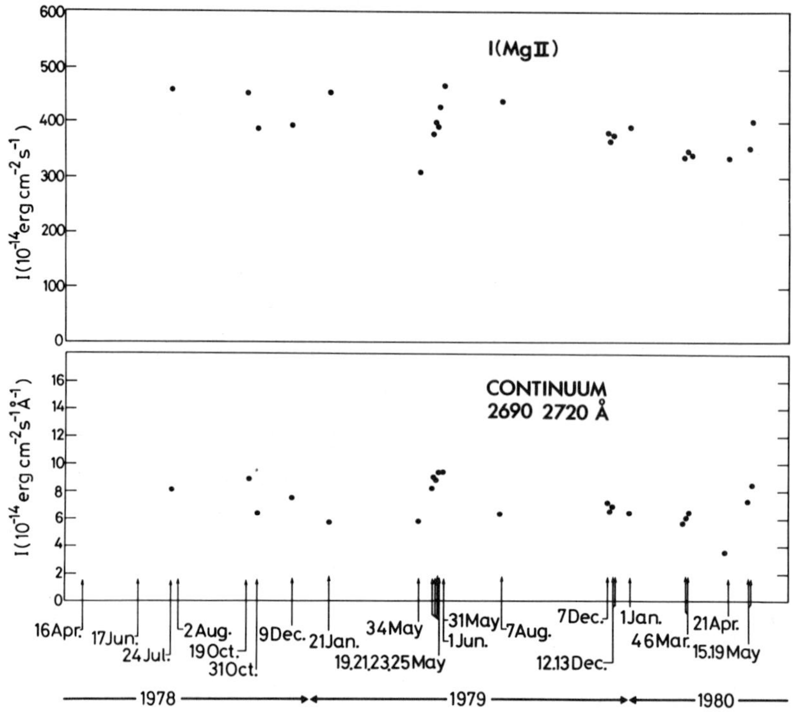

FIGURE 2. Time variations of the intensity of MgIIλ2800 and of the continuum in the range 2,690–2,720 Å in NGC 4151.

optically thin or have a smaller optical depth than those at lower velocities. This is in contrast with quasars where in most cases the line profiles are similar.[2,27] Differences in the line profiles indicating inhomogeneities in the broad-line region have also been found in the optical and in the UV range in other galaxies.[6,18,23,25]

Minima

NGC 4151 passed through periods of pronounced minima during which most of the ionized gas in the broad-line region recombines. Such a minimum occurred on 21 April

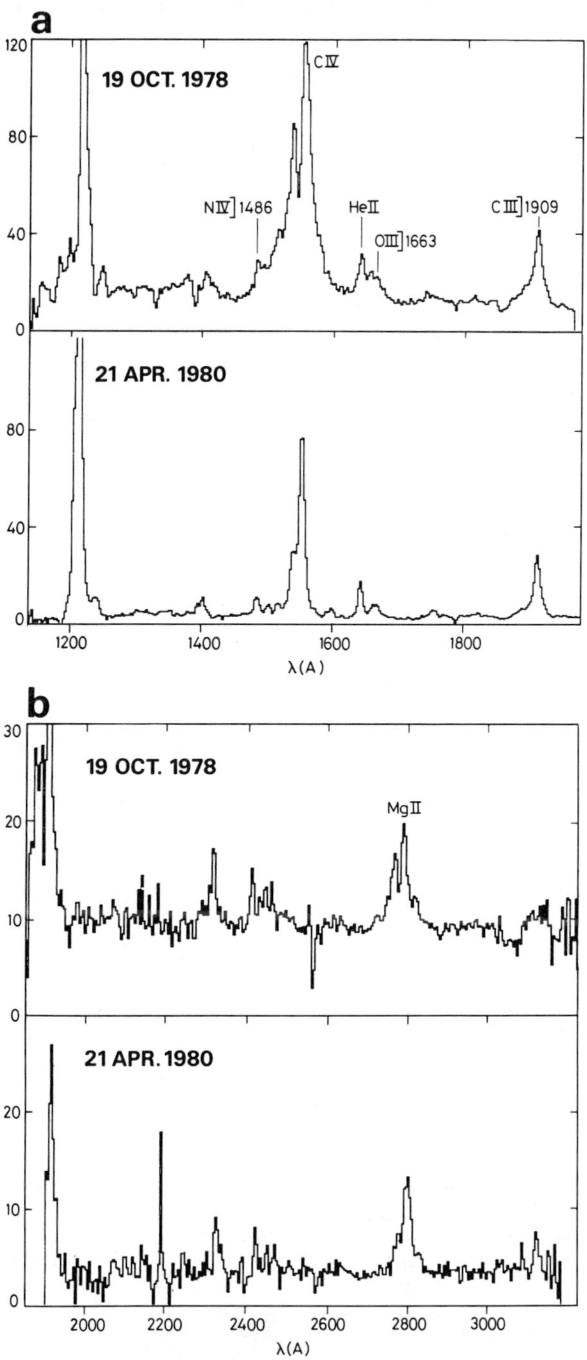

FIGURE 3. UV spectra in the intervals 1,200–2,000 Å and 2,000–3,000 Å taken with IUE when NGC 4151 was in a maximum phase (19 October 1978) and a minimum phase (21 April 1980). Vertical axis: intensity per unit wavelength in 10^{-14} erg cm^{-2} s^{-1} A^{-1}.

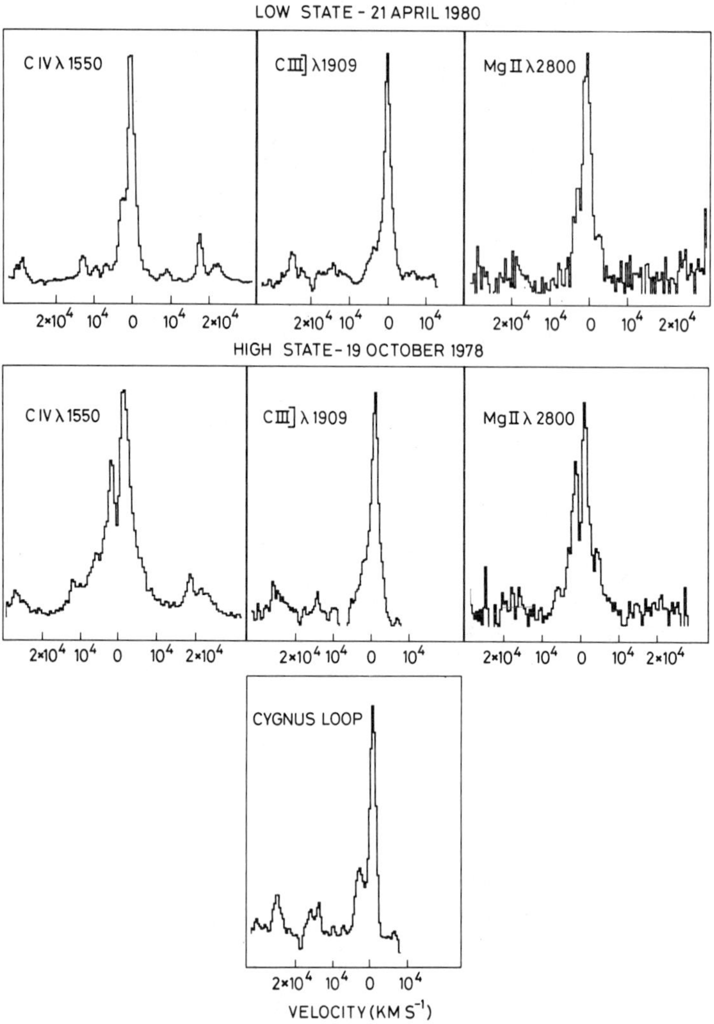

FIGURE 4. Line profiles for the three strong emission lines in NGC 4151 plotted on the same velocity scale and normalized to the same peak intensity.[25] The blue wing is on the left. The top row shows the lines when the nucleus is at a minimum: they are all relatively narrow (HWZI ~6,000 km s^{-1}) and similar in width. The second row shows the same lines when the continuum is bright; the lines have markedly different widths demonstrating that the physical conditions in the gas clouds correlate with the clouds' velocities. To illustrate the IUE resolution at λ1909, the CIII] line profile of the Cygnus loop[4] is shown at the bottom.

1980 and over the period April–July 1981. During these minima the broad components of CIV, MgII and other permitted lines nearly vanish and the continuum at 1,450 Å is less than 3×10^{-14} erg cm^{-2} s^{-1} A^{-1}. The "narrow" components that remain have a width of 6,000 km s^{-1} and must come from a region that does not vary on a time scale of several months and that is therefore larger than the region emitting the broad components of the lines. Most of the CIII]λ1909 line comes from this region. At any epoch, and even at maximum, the intensity of the broad component of this line is not larger than 0.3 the intensity of the "narrow" component, causing the overall profile to be strikingly different from the profile of the permitted lines.

Time Delay between the Variations of the CIV Line and the Variations of the Continuum

In May 1979, the period during which the time resolution of the observations is best, we observed a time delay between the variation of the CIV line intensity and of the continuum: FIGURE 1 shows that the continuum increases between May 3 and May 19, then stays at a nearly constant level between May 19 and June 1; in contrast, the CIV line (and to a lesser extent MgII) increases at a constant rate between May 3 and June 1. This difference between the light curves of CIV and of the continuum is interpreted as a delay due to the light travel time between the continuum source and the broad-line region gas. This time delay indicates a *lower* limit for the dimension of the broad line region of 10 to 15 light days.

The variations are consistent with the model sketched in FIGURE 5. The radiation emitted by the central source at a given instant reaches the points A and B at the same time, but the signals carrying the information about the effect of the radiation from the central sources at A and B arrive at the observer with a time spread. Similarly the ray from C carrying a signal from a point farther out in the broad-line region also arrives later than the signal from A. The delays are caused by the finite light travel time across the nebulosity and between the central source and the nebulosity.

With this model the observations of the profile immediately after a minimum, when one can follow the effect of the increase in the continuum as it spreads through the nebulosity, can in principle provide an answer as to whether the gas is inflowing or outflowing. In the first case the CIV line would first appear with a blue wing, and in the second case with a red wing. If turbulent, isotropic, or rotation motions dominate, the broad component would be symmetrical.

Variations of the CIV Profile in May 1979: Relationship between Velocity and Distance from Center

Let us turn now to the interpretation of the variations of the profiles of the CIV line in May 1979 in terms of this model. It is assumed here that the increase in the continuum between May 3 and May 19 is the continuation of an increase that started some time before May 3.

FIGURE 6 provides evidence that the wings and the line center behave differently: the blue wing of CIV remains nearly constant through the whole period May 3 to June

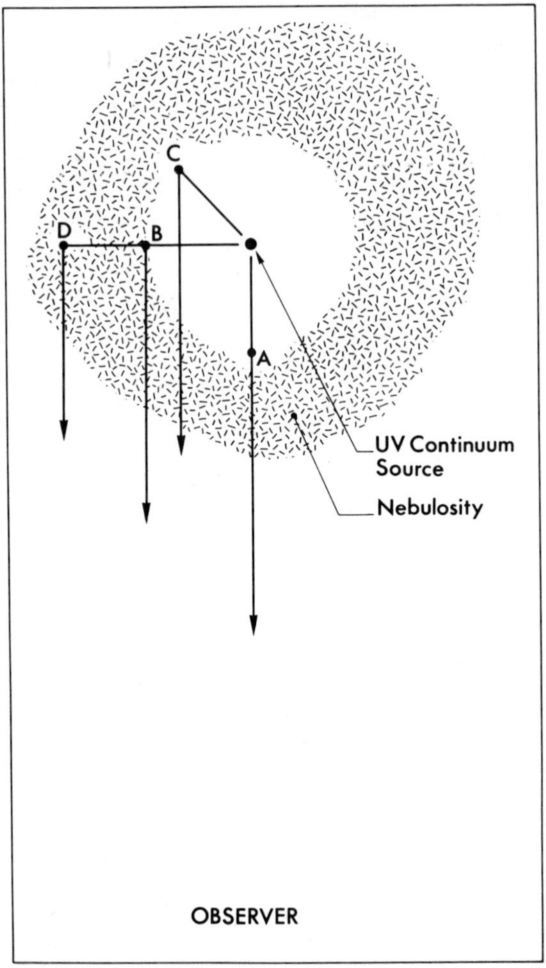

FIGURE 5. Sketch of a model for the broad-line region consistent with the observations of NGC 4151 available at the present time. The radiation emitted by the points A, B, C, and D reaches the observer with a time spread that is caused by the finite light travel time through the broad-line region and between the central continuum source and the nebulosity. The data obtained in May 1979 suggest that the clouds closer to the center have higher velocities and a smaller optical thickness than the clouds farther out.

1, while the total intensity of the line increases at a practically constant rate between May 3 and June 1.[a] Most of the increase in the total intensity of CIV between May 3

[a] Regarding the red wing, FIGURE 6 gives some hint that it appears after the blue wing, which is as expected if the gas is outflowing. There is however some ambiguity in this interpretation as the red wing of CIV is blended with the blue wing of HeIIλ1640.

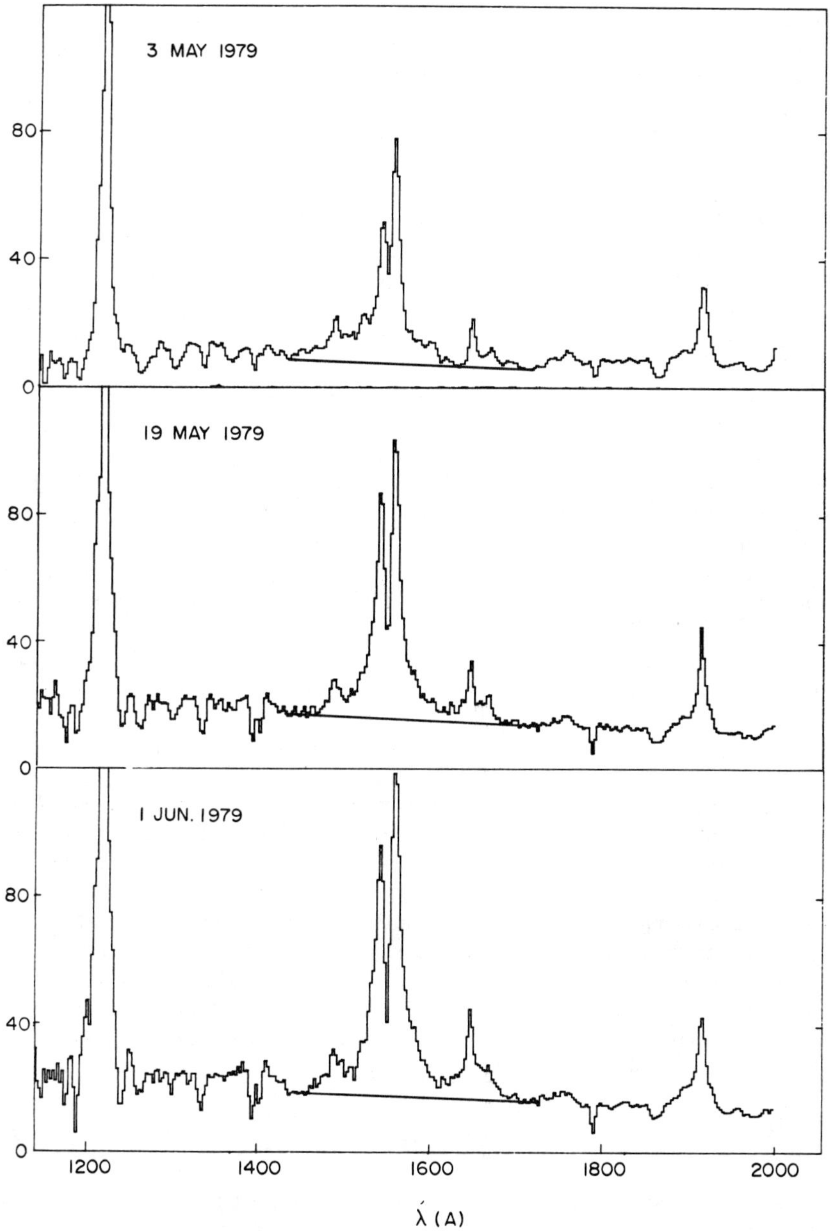

FIGURE 6. Short wavelength spectra of NGC 4151 during the outburst of May 1979. The straight line is the local continuum under the CIV line as defined in Reference 26. The blue wing of CIVλ1550 did not vary significantly during this outburst, most of the extra intensity appearing near the line center. This suggests that the high velocity clouds were already illuminated on May 3 and therefore that they are the closest to the central continuum source.

and June 1 occurred near the line center and therefore represents an increase in the emissivity of the lower-velocity clouds. On May 3 the high-velocity clouds emitting the blue wing were already illuminated, indicating that they are *closer* to the continuum source than are the clouds at lower velocities whose emission contributes to the line center. The time delay deduced from the light curves above in fact represents the light travel time between the continuum source and the relatively low-velocity clouds producing the line center.

Regarding the nonvariation of the CIV blue wing between May 3 and June 1, it is useful to recall that beyond a certain value of the flux of ionizing photons, C^{+3} is ionized into C^{+4} and consequently I(CIV) stops increasing, remains constant, and then decreases when the continuum increases above this value. This is illustrated in Figure 7 in the review article of Davidson and Netzer.[12] Such a situation seems to apply to the region producing the blue wing of CIV. The constancy of this wing after May 3 is explained if on that date the ionizing flux incident on the region producing the blue wing had reached the value beyond which C^{+3} starts being ionized into C^{+4}.

Finally it is interesting to note that the clouds producing the CIV wings and that, from the argument above, are the closest to the central source have a ratio I(CIV)/I(MgII) larger than at the line center, indicating that they are optically thin or have a smaller optical thickness than the clouds emitting the bulk of the CIV line.

In summary, the differences between the line profiles, the variations of the line intensities, and the variations of the CIVλ1550 profile lead us to propose a model for the broad-line region where the optical thickness, the distance from the central continuum source, and the velocity of the clouds forming the broad-line region are correlated, the clouds with the highest velocities and smaller optical thickness being closer to the center.

Other models for the broad-line region of Seyfert nuclei and quasars have been proposed. Their characteristic ingredients are (1) the atmosphere of a supermassive object;[22] (2) a binary black hole;[15] (3) changes in the physical conditions of the clouds as they approach the continuum source along nearly parabolic orbits;[17] and (4) the possibility that the clouds form two dynamical systems, a disk where the optically thick clouds are located and a spherical or conical system of optically thin clouds.[16] These models have been proposed to explain the main features of the broad-line region and in particular the symmetry of Lyα. It is not our purpose to discuss these models here, but it is worthwhile to point out that a collection of optically thin clouds responsible for the wings of CIV and the bulk of Lyα could produce symmetric Lyα lines. Moreover, the asymmetry produced by the finite light travel time through the broad-line region occurs only in nuclei that are variable, and even there it occurs only occasionally.

Origin of the Absorption Lines

NGC 3516 and NGC 4151 are the only two Seyfert nuclei, among the ~35 observed with IUE, where there is a prominent CIVλ1550 absorption line. It has been argued that this line could be caused by the clouds forming the broad-line system.[14] There are however several arguments in favor of another origin for the absorption. Firstly, the absorption lines have a depth that is much larger than the height of the continuum at 1,500 Å, indicating that a large fraction of the broad-line region is covered by the absorber.[25] Secondly the absorber occupies a velocity range of $\sim 10^3$ km s^{-1} compared to more than 1.5×10^4 km s^{-1} for the emission clouds. Thirdly, in NGC 3516 and in NGC 4151 the absorption line of CIV seems to be present even during a minimum when the CIV line is relatively narrow. The narrow components of CIV in NGC 3516 and NGC 4151 are constant on a time scale of several months and must be one year or more away from the nucleus. These three points together suggest that the absorber is in a sheet of gas nearly covering the entire broad-line region, and located at more than one light year from the center.

FUTURE WORK ON NGC 4151

The monitoring of NGC 4151 that has been done so far has shown that measurable variations of the CIV line intensity (and of the continuum) can occur in 4 days, and also that the line intensity and line profile observed on a given day do not depend on the continuum intensity on that day but on the continuum intensity in the previous 5 to 30 days. This shows the way in which further observations of NGC 4151 should be undertaken: monitoring of NGC 4151 every 4 or 5 days during a sufficiently long time so that the nucleus passes by at least one clearly defined maximum and minimum would allow us to confirm and refine the model proposed above and to obtain quantitative data on the run of the velocity and of the physical conditions of the clouds in the broad-line region vs. the distance to the continuum source. Such a monitoring was undertaken in the period April–July 1980, but the nucleus remained at a minimum during the whole campaign. A similar experiment will be attempted again in the coming months simultaneously with x-ray observations with the satellite Exosat.

SUMMARY

We review the observations of the time variations of the emission lines in Seyfert nuclei and in particular in NGC 4151, the galaxy that has been the most extensively studied in the ultraviolet. In this galaxy, the variations of the line intensities and of the line profiles allow us to gain an insight into the structure of the broad-line region. The variations are consistent with a model where the optical thickness, the distance from the central continuum source, and the velocity of the clouds forming the broad-line region are correlated, the clouds with the highest velocities and smaller optical thickness being closer to the center. It is argued that the prominent CIV 1550 absorption lines in NGC 3516 and NGC 4151 are not caused by clouds in the broad-line region. Instead the absorber is likely to be an outflowing sheet of gas covering a large fraction of the broad-line region, and located more than one light year away from the continuum source.

REFERENCES

1. ANDRILLAT, Y. & S. SOUFFRIN. 1968. Astrophys. Lett. **1:** 111.
2. BALDWIN, J. A. 1979. *In* Active Galactic Nuclei. C. Hazard & S. Mitton, Eds.: 51. Cambridge University Press. Cambridge, England.

3. Barr, P., A. J. Willis & R. Wilson. 1983. Mon. Not. R. Astron. Soc. **203:** 201.
4. Benvenuti, P., M. Dopita & S. D'Odorico. 1980. Astrophy. J. **238:** 601.
5. Boggess, A., *et al.* 1978. Nature **275:** 372.
6. Boksenberg, A., K. Shortridge, D. A. Allen, R. A. E. Fosbury, M. V. Penston & A. Savage. 1975. Mon. Not. R. Astron. Soc. **173:** 381.
7. de Bruyn, A. G. 1980. Highlights Astron. **5:** 631.
8. de Bruyn, A. G. & W. L. W. Sargent. 1978. Astron. J. **83:** 1257.
9. Capriotti, E. R., C. B. Foltz & B. M. Peterson. 1982. Astrophys. J. **261:** 35.
10. Cherepashchuk, A. M. & V. M. Lyutyi. 1973. Astrophys. Lett. **13:** 165.
11. Clavel, J. 1983. Mon. Not. R. Astron. Soc. **204:** 189.
12. Davidson, K. & H. Netzer. 1979. Rev. Mod. Phys. **51:** 715.
13. Elvius, A., B. A. M. Westin & J. Lind. 1982. *In* Proceedings of the Third European IUE Conference, Madrid, May 1982: 543. ESA. Nordwijk, the Netherlands.
14. Ferland, G. J. & R. F. Mushotzky. 1982. Astrophys. J. **262:** 564.
15. Gaskell, C. M. Ann. N.Y. Acad. Sci. (This volume.)
16. Gordon, C., S. Collin-Souffrin & D. Dultzin-Hacyan. 1981. Astron. Astrophys. **103:** 69.
17. Kwan, J. & T. J. Carroll. 1982. Astrophys. J. **261:** 25.
18. Osterbrock, D. E. & J. M. Shuder. 1982. Astrophys. J. Suppl. **49:** 149.
19. Penston, M. V., A. Boksenberg, G. E. Bromage, J. Clavel, A. Elvius, P. M. Gonhalekar, C. Jordan, J. Lind, L. Lindengren, G. C. Perola, M. Pettini, M. A. J. Snijders, E. G. Tanzi, M. Tarenghi & M. H. Ulrich. 1981. Mon. Not. R. Astron. Soc. **196:** 857.
20. Perola, G. C., A. Boksenberg, G. E. Bromage, J. Clavel, M. Elvis, A. Elvius, P. M. Gondhalekar, J. Lind, C. Lloyd, M. V. Penston, M. Pettini, M. A. J. Snijders, E. G. Tanzi, M. Tarenghi, M. H. Ulrich & R. S. Warwick. 1982. Mon. Not. R. Astron. Soc. **200:** 293.
21. Peterson, B. M., C. B. Foltz, P. L. Byard & R. M. Wagner. 1982. Astrophys. J. Suppl. **49:** 469.
22. Shields, G. A. 1983. University of Texas. Austin, Tex. (Preprint.).
23. Shuder, J. M. 1982. Astrophys. **259:** 48.
24. Tohline, J. E. & D. E. Osterbrock. 1976. Astrophys. J. **210:** L117.
25. Ulrich, M. H. & C. Boisson. 1983. Astrophys. J. **266:** 515.
26. Ulrich, M. H., A. Boksenberg, G. E. Bromage, J. Clavel, A. Elvius, M. V. Penston, G. C. Perola, M. Pettini, M. A. J. Snijders, E. G. Tanzi & M. Tarenghi. 1983. Mon. Not. R. Astron. Soc. (In press.)
27. Wilkes, B. J. & R. F. Carswell. 1982. Mon. Not. R. Astron. Soc. **201:** 645.

Models of Active Galactic Nuclei[a]

R. D. BLANDFORD

Theoretical Astrophysics
California Institute of Technology
Pasadena, California 91125

INTRODUCTION

Roughly 40 years have elapsed since the discovery of Seyfert galaxies,[1] 30 years since extragalactic radio sources were shown to be double,[2] 20 years since it was realized that quasars were at cosmological distances,[3] and 10 years since BL Lac objects were recognized as an important class of active galaxy.[4] Over this period we have learned much about the properties of these objects. Observations extend from the lowest radio frequencies to hard gamma rays. Variability has been shown to be quite common, jet features have been discovered in abundance and evolutionary effects revealed.

There has been less solid progress on the theoretical side. We have been most successful in interpreting "secondary" phenomena, and these are generally discussed within a reasonably coherent framework. For example, we are arguing about the densities and kinematics of emission-line clouds, the speeds and confinement of jets, and the cosmological distribution of absorption-line clouds rather than questioning their existence at all. However, when we turn to "primary" phenomena, we find that there is still no such consensus as to the identity of the "prime mover." While it is true that black hole models have gained widespread acceptance in recent years,[5,6] we are still justly skeptical of them because there is not a single observation that demands the existence of massive black holes, although many observed properties (especially the central light cusp in M87, rapid x-ray variability, and the persistence of jets) are naturally interpreted in this context. The principal arguments in favor of black holes, which in my view are highly persuasive, are essentially theoretical. We are unlikely to make much further theoretical progress until we are surer about their observational status.

Most theoretical studies of black hole models of active galactic nuclei have been highly idealized; for example a lot of work has assumed spherical symmetry. Furthermore it seems that many theoreticians are engaged in trying to explain an abstraction—an amalgam of the better publicized features of the totality of active nuclei rather than accounting for the considerable diversity which most observers take for granted. Correspondingly, many observers are unduly spectrally chauvinist and appear reluctant to interpret their observations in the context of the total electromagnetic spectrum which is the instinctive approach of the theorist. In this talk I wish to argue that black hole models should account for both the diversity of extragalactic phenomena and the correlations between different wave bands. I shall summarize a particular unified scheme. This is not with the intent of trying to promulgate a

[a] I thank the National Science Foundation for support under grant AST 80-17752 and the Alfred P. Sloan Foundation for financial assistance.

preferred ideology but more to give an example of this approach and perhaps to provide a target for further observational studies. Of course alternative schemes are possible and have been discussed.[7]

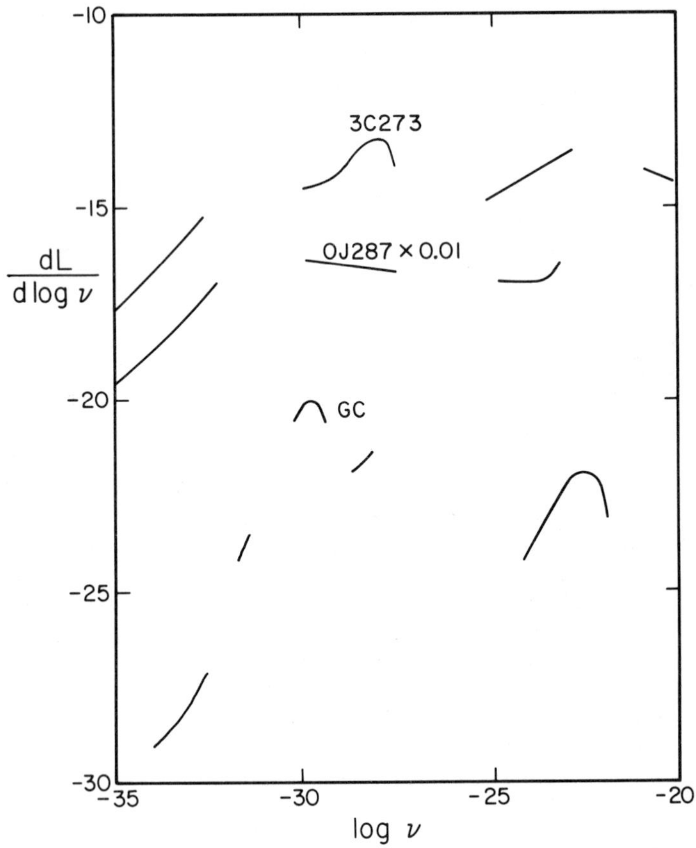

FIGURE 1. Continuum luminosity for a selection of galactic nuclei. The continuum power per $\log_{10}\nu$ in gravitational units. We show 3C273, a QSR. QSOs and Seyferts appear to have similar spectral shapes except that their radio emission is far weaker than in QSRs. OJ287 is a relatively powerful BL Lac object. It has a flat radio spectrum, a featureless optical continuum, and a two-component x-ray spectrum. Our galactic center (GC) is presumably not untypical of the nuclei of ordinary spiral galaxies. It too exhibits activity from radio to γ-ray energies.

TAXONOMY OF ACTIVE GALACTIC NUCLEI

The classification of active galactic nuclei is still in a very confused state. For the purposes of the present talk, I shall need the following general properties of active objects which gloss over a wealth of observational detail and ignore many exceptions and qualifications.

QSOs and type I Seyfert galaxies are believed to be respectively the high and low luminosity ends of a continuous range of optical powers. We assume that properties of one class are typical of both. They appear to be radio quiet, have "bumpy" power law continua with spectral index $\alpha = -\mathrm{d} \ln L_\nu / \mathrm{d} \ln \nu \sim 1\text{–}1.5$, extending from the infrared to the ultraviolet punctuated by broad $\Delta V \sim 10{,}000$ km s^{-1} and narrow $\Delta V \lesssim 1{,}000$ km s^{-1} emission lines.[8,9] They are generally not rapidly variable, and only exhibit small polarization. Fe II emission is common.[11] QSOs have a UV excess probably containing most of the bolometric power which has been modeled as black-body radiation with temperature $T_{uv} \sim 25{,}000$ K.[12,13] Their luminosities range from $\sim 10^{43}\text{–}10^{47}$ergs^{-1}. Their ~ 3 keV x-ray powers are typically a few percent of their UV powers, but their x-ray spectra have spectral index $\alpha \sim 0.6$ and extend up to and perhaps beyond 100 keV.[14] Rapid variability at x-ray energies on time scales of minutes has been reported in several objects,[15] but is not prevalent. Seyfert galaxies and (more controversially) QSOs appear to be associated with spiral galaxies[16] of which they represent a fraction, ~0.01. *Type II Seyfert galaxies* are similar except that they do not show broad emission lines.

Roughly 10% of quasars are radioloud or QSRs.[17] Their radio powers contain up to a few percent of the total luminosity and may either be dominated by a "compact," "flat spectrum"[18] component or an "extended" "steep spectrum" component.[17] The radio emission is believed to be synchrotron radiation, and the compact sources are thought to owe their flat spectra to a superposition of self-absorbed components. Both compact and extended components are much larger than the central continuum source, and they appear to be energized by jets emerging from the nucleus. Optically, and at x-ray energies, they are similar to the QSOs except perhaps in FeII emission.

A subset (about 15%) of QSRs are optically violently variable (OVV).[10] They also exhibit high polarization and are associated with low-frequency radio variability and steep, smooth optical continua. Nevertheless, they display normal broad emission spectra in contrast to *BL Lac* objects, which are probably an extreme case of OVVs in which the continuum dominates the lines. They can vary on time scales as short as a day and appear to be associated with giant elliptical galaxies. Their x-ray spectra are different from those of the Seyferts[19] and display both hard and soft components.

Radio galaxies are always elliptical and are also divided into compact and extended sources. They often display comparatively weak nuclear emission lines and are classified in a similar fashion to Seyferts as broad-line or narrow-line objects.

Low levels of activity may be present in the nuclei of most normal galaxies. The best-studied example is of course our own galaxy. There is an unusual compact radio source[20] within the galactic center located near a strong infrared source and identified with a prominent hard x-ray object.[21] A high-velocity width (1,500 km s^{-1}) infrared helium emission line has been reported.[22] A strong, narrow, variable position annihilation line is also observed.[23] There is no reason to believe that our galaxy is atypical of spirals.

ENERGY PRODUCTION BY BLACK HOLES

Active galactic nuclei are popularly supposed to be powered by accretion onto black holes. It seems likely that both the hole and the accreting gas possess significant

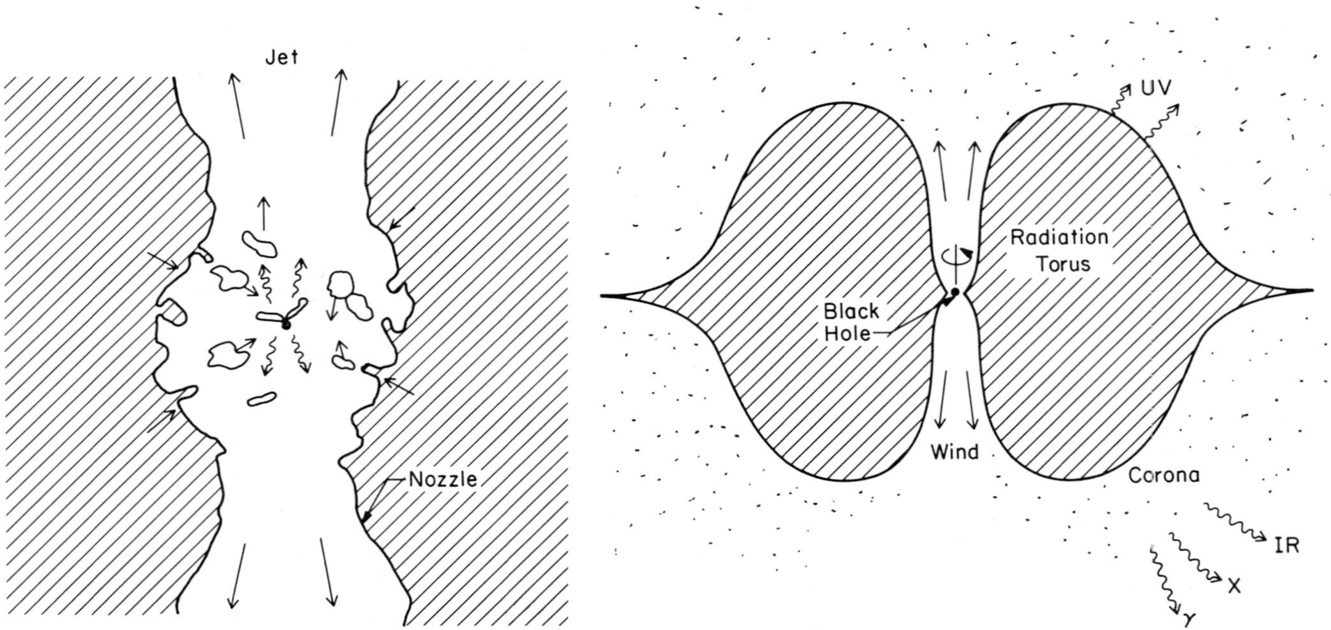

FIGURE 2. The observed properties of an active galactic nucleus may be controlled by the ratio of the mass accretion rate to the mass of the hole, $\dot{M}/M$. FIGURE 2a (left). When the accretion rate is much larger than critical, radiation will be trapped by the inflowing gas until it reaches a central "cauldron" within which there is a high dissipation rate through instabilities, shocks, gas cloud collisions, etc. A substantial fraction of the internal energy created in this way may escape through two nozzles as radiation-dominated, though inhomogeneous, jets. FIGURE 2b (right). When the time-averaged accretion rate is only somewhat larger than critical, a radiation torus may form. This quasi-spherical structure will probably have an effective temperature in the UV. It may be surrounded by a tenuous, active corona, perhaps dominated by electron-positron pairs and heated by a radiation-driven wind flowing outward through the central funnels. The nonthermal components of the spectrum may arise in the corona through synchrotron radiation and inverse Compton scattering.

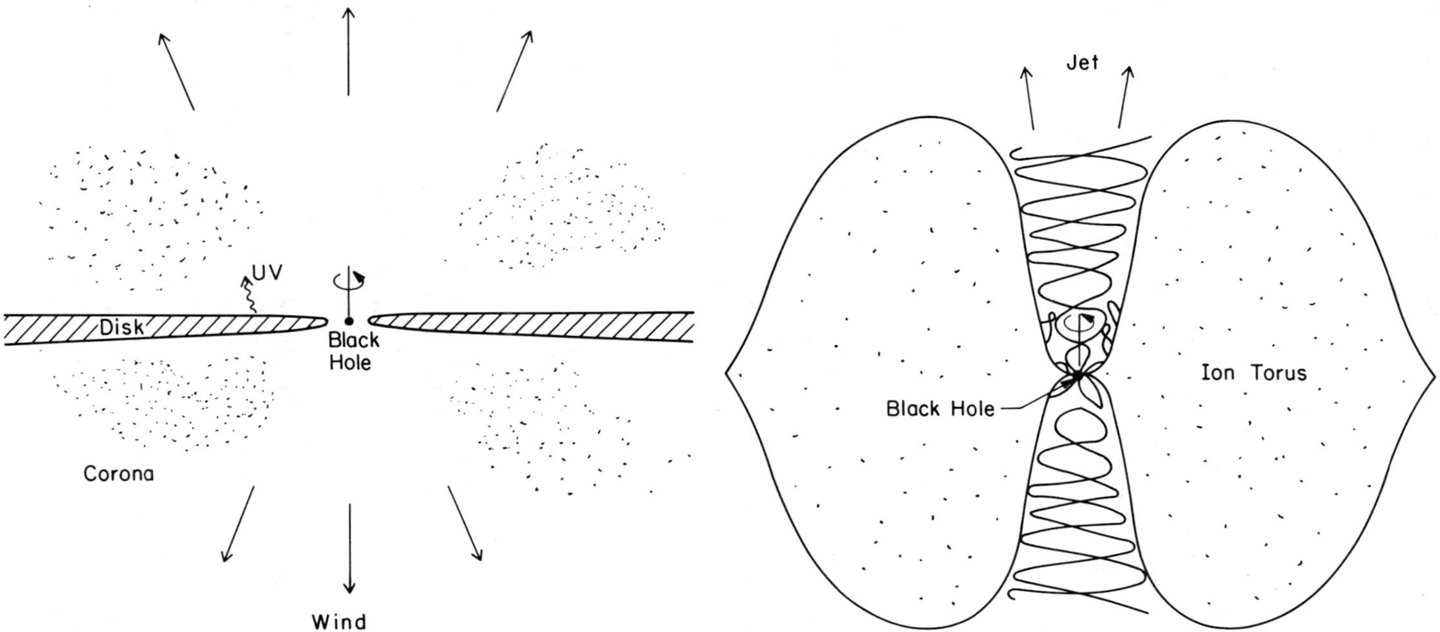

FIGURE 2c. (left). At subcritical accretion rates, the disk will flatten though much of the dissipation may still occur within a tenuous corona from which a poorly collimated relativistic wind may escape. FIGURE 2d (right). At low accretion rates and for large effective viscosity, the inflowing gas may have insufficient time to cool and instead may inflate a large torus supported by ion pressure. Under these conditions, most of the power may be released by magnetic torques braking the spinning hole.

angular momentum. In a standard model of a thin accretion disk, the inner parts are radiation-pressure supported.[24] Such disks are subject to both secular and thermal instability, and much of the energy liberated may be dissipated in a tenuous corona.[25]

If the infalling gas is unable to cool effectively, the inner disk may become thick and toroidal. This can happen either at high accretion rates when a radiation-supported torus will be formed or at low accretion rates when the ions may be unable to cool and form an ion-pressure-supported torus. These two fundamentally different structures may be reflected in the types of active nucleus that we observe. In particular, QSOs may be identified with the former case and radio galaxies with the latter.

Radiation tori[26–28] have generally been modeled as stationary flows. The integrated luminosity from the surface is of order several times the Eddington limit for the mass of the hole although the accretion rate can be arbitrarily high with constant loss of radiative efficiency. A more realistic approach is to assume that the gas is supplied intermittently and that the gas is usually not flowing into the hole. Instead the surface luminosity is supplied at the expense of the orbital and internal energy of the torus, which slowly deflates until the next episode of gas supply.[29]

The outer radius of the torus depends upon the size of the hole and its history. If it is much larger than the Schwarzschild radius, then two antiparallel funnels will be defined along the spin axis of the torus (presumed to coincide with the spin axis of the hole). Several authors have suggested that these be the origins of radio jets.[31,32] However, there are two severe difficulties with this idea. Firstly it may be difficult to achieve the relativistic speeds necessitated by observations of superluminal sources because of the radiation drag on the outflowing plasma. Secondly the nuclear powers associated with most of the jets that we actually observe are very much lower than the Eddington value if we assume that the mass of the hole is at least comparable with the energy content of the radio components.

A slightly different structure is possible when the accretion rate is very high and there is also a high dissipation rate within a central cavity or "cauldron."[33] If the optical thickness of the cauldron is sufficiently large, then radiation will be trapped within it and the energy carried off in two radiation-dominated jets through a pair of nozzles. It can then be released at large distances from the hole where the jet optical depth has decreased sufficiently to allow the photons to escape. Jets formed in this way may correspond to the more powerful QSRs like 3C273.

In an ion torus,[34] the main source of power may be the rotational energy of the hole, which has presumably been spun up in an earlier epoch of high mass accretion. Up to 29% of the rest mass energy of the hole can be stored in this way. An attractive way to extract this energy is through the use of large-scale magnetic fields, supported by external currents which pass through the horizon of the hole.[35] Charged particles have to be created continuously in the region outside the horizon (most plausibly as electron-positron pairs) for this mechanism to operate. Fortunately the gamma rays created by the inflowing plasma are probably adequate for this purpose. Recent studies[36] of electromagnetic energy extraction from black holes have greatly clarified its relativistic basis and shown how perfect MHD can be incorporated into the description.[36] It now appears that no more than 3% of the rest mass energy is likely to be extracted. This still means that black holes no more massive than $10^8\ M_\odot$ can supply the energy requirements of the largest double radio sources.

EMISSION MECHANISMS

Radio

The sizes of compact radio sources are at least a hundred times larger than those of the associated black holes. They can therefore be construed as secondary phenomena, and the details of their emission (self-absorbed synchrotron radiation) are not directly relevant to the problem of the central energy source. What is important is that these sources usually have one-sided core-jet structure[37] and that in several sources features have been reported to be moving with apparent superluminal speed.[38] This is now conventionally interpreted as implying that the radio-emitting material moves with an outflow speed of Lorentz factor $\gamma \sim v_o/c$, with v_o being the superluminal speed, at an angle $\sim c/v_o$ to the line of sight. Jets that are beamed toward us will be strongly Doppler boosted, and this may be true of the majority of strong compact sources. Jets should also be able to radiate at optical and x-ray energies through synchrotron and inverse Compton emission. BL Lac objects and OVV quasars can be interpreted as sources in which the Doppler-boosted nonthermal continuum from the jet outshines the normal continuum from the nucleus.[39]

Power-Law Continuum

The most simple (and common) explanation for the power-law continuum is that it is synchrotron radiation by a nonthermal distribution of relativistic electrons. However this is hard to reconcile with the low observed degree of polarization. For this reason, some authors have advocated Comptonization models in which infrared photons (probably created as synchrotron or high harmonic cyclotron emission) are Fermi accelerated by hot electrons to produce a power law.[40–46] If the Thomson optical depth of the source region is $\tau_T \gtrsim 1$ then the spectral index is given by

$$\alpha(\alpha + 3) = \frac{m_e c^2}{\tau_T^2 kT} \tag{1}$$

where T is the electron temperature. Furthermore, if as is probably the case the magnetic field strength exceeds $B \gtrsim 1$ G then an electron plasma will be able to depolarize optical radiation through Faraday rotation. As it stands, this mechanism requires a coincidence between τ_T and T in order to give a standard power law. However, if we couple the heating of the plasma directly to the emission of the seed infrared photons and stipulate that similar powers flow through the two channels, then we constrain the spectral index to be $\alpha \gtrsim 1$ as observed.

Ultraviolet

If the UV excess really is approximately black body with a range of effective temperatures 20,000 K $\lesssim T_e \lesssim$ 30,000 K, then it seems likely that it is formed in a photosphere. One possibility is that it be formed at the surface of a thin accretion disk.

Now if the radiation is produced at radius r, the effective temperature is given by

$$T_{\text{eff}} \sim 2 \times 10^5 \left[\frac{L}{L_{\text{edd}}}\right]^{1/4} \left[\frac{r}{10r_s}\right]^{-1/2} M_8^{-1/4}\ \text{K} \tag{2}$$

However, a disk would be expected to give a range of effective temperatures and unless the hole were very massive only a small fraction of the total binding energy liberated by the infalling gas will be radiated at 20,000 K. Both of these drawbacks can be avoided if the emission is from the surface of a radiation torus. The gas in the interior of the torus will be fully ionized and sufficiently tenuous that electron-scattering opacity dominates. However, if the temperature of the surface layers is allowed to fall below ~25,000 K, then helium will start to recombine. The opacity will then decrease through the increase in mean molecular weight per electron and the surface layers become hard to support. In this way, a radiation torus can radiate a full Eddington luminosity from a surface whose radius is limited to $\sim 10^{16} M_8^{1/2}$ cm but whose mass can steadily increase as the average specific entropy of the torus decreases.

Emission Lines

The broad-emission-line region[47] is believed to be located at a distance $\sim L_{\text{uv}46}^{1/2}$ pc from the far smaller central continuum source. It involves a comparatively small quantity of gas and can be regarded as secondary emission. The narrow-line region is at even greater distances.

X-Ray

The comparatively standard ratio of the x-ray to optical fluxes suggests that the two fluxes are intimately related although the difference in the spectral indices implies that they are not produced by the same emission mechanism.[48] If the x-rays are produced by inverse Compton scattering of the UV radiation,[49] then an electron distribution function of the form $N_\gamma \propto \gamma^{-2.3}$ is called for. Now the relativistic electron cooling time is much shorter than the dynamical time scale. Therefore if the electrons are accelerated impulsively to an energy $\gamma_{\max}$ an average distribution function of the form $N_\gamma \propto \gamma^{-2}$ for $\gamma \lesssim \gamma_{\max}$ will result. A range of values of $\gamma_{\max}$ will steepen the distribution function slightly as required.

This is not the only possibility. Synchrotron[49,50] and thermal Compton[51,52] models of the x-ray emission have also been proposed.

γ-Ray

There are several reasons why we should consider γ-ray emission. Firstly there is an excess in the background radiation around 1 MeV which has been attributed to the superposition of active galactic nuclei.[53] Secondly, we do not know where the upper cutoff is in the x-ray spectrum. It has been shown to exceed 50 keV in several Seyferts. Thirdly, 3C273 is positionally identified with a 100 MeV γ-ray source.[54] (It cannot be typical; otherwise the γ-ray background would be exceeded by the superposition of quasars. The γ-rays are probably produced in the jet.) Finally, Comptonization models require the electron temperature to be almost relativistic and this therefore raises the possibility that the plasma comprises electrons and positrons which can be freely produced at relativistic energies.

There has been considerable recent work on the properties of relativistic plasmas. Equilibration,[55] pair-production, and annihilation[56] time scales have been carefully calculated. Equilibria have been computed in which electron-positron pair production is balanced by annihilation, and it has been shown that there is a maximum temperature, $T_{max} \sim 25$ MeV, above which an equilibrium is impossible.[57] If the Thomson optical depth of the plasma exceeds unity, then γ-rays will be trapped and have further opportunity to bring about pair production, thereby lowering T_{max}. If we also add magnetic field, allowing the electrons and positrons to cool through synchrotron radiation, then T_{max} is lowered even further.[58] It can be demonstrated that quasar emission cannot be attributed to radiation from a uniform, equilibrium, relativistic plasma.[59]

The possibilities increase when we allow the plasma to be inhomogeneous and out of thermal equilibrium.[60] Nonthermal radiation from impulsively accelerated electrons may extend up to γ-ray energies $\gtrsim 1$ MeV. If the γ-ray luminosity exceeds $\sim(m_e/m_p)(r/r_s)_{L_{edd}}$, then the opacity to pair production exceeds unity, giving rise to a pair plasma that equilibrates to a mildly relativistic temperature and has a Thomson optical depth $\tau_T \sim 1$. This plasma can reprocess lower frequency photons and may be responsible for the nonthermal spectrum that we observe.

In fact, it is possible to construct a unified model of QSO emission as follows. Suppose that there is an active corona above a radiation torus with effective temperature $T_e \sim 25{,}000$ K. Suppose further that the corona is heated by sound or Alfvén waves produced by subphotospheric convection or rotation-driven instability and that these waves steepen to form shocks as they travel down the density gradient into the corona. (As an alternative it is possible that the corona is powered by a radiation-driven outflow through the central funnels defined by the torus.) The shocks are capable of accelerating electrons to energies $\sim$200 MeV,[61] and these electrons can then:

1. Inverse Compton scatter the UV photons to give pair-producing MeV γ-rays.
2. Synchrotron radiate in the ambient magnetic field to give marginally optically thin infrared radiation.
3. Cool through processes 1 and 2 to establish a steady particle spectrum that produces the x-ray emission as described above.

The pair plasma produced by the γ-rays will equilibrate to a temperature $\sim 1/4\, h\bar{\nu}$, typically 200 keV. The Thomson optical depth of the corona is given by balancing pair creation by annihilation and is roughly $\tau_T \sim [(m_p/m_e)(r_s/r)]^{1/2} \sim 4\, M_8^{1/4}$. This is roughly the value necessary from Equation 1 to give the infrared optical power law. Occasional rapid variability can be associated with flares in this corona. The plasma would probably be hottest and densest on the closed field lines where it would be retained for many dynamical times. A radiation-pressure-driven electron-positron wind might be expected to emanate from the open field lines. More detailed calculations are imperative to see if this model really is self-consistent.

BEAMING

We have argued that beaming contributes to the observed properties of BL Lacs, OVV quasars, and compact radio sources and that it is the near coincidence of the line of sight with the source symmetry axis that distinguishes these objects. It is natural to identify this symmetry axis with the spin of the black hole. There are then two distinct ways that beaming may occur. Either we are looking down the accretion funnel defined by a radiation torus at much hotter plasma somewhat closer to the black hole or there is a relativistic jet and the emission is beamed in our direction. Some rather severe constraints are posed by the former possibility and in any case it is irrelevant to the radio emission. We therefore confine our attention to relativistic beaming.

We list several arguments in favor of relativistic beaming:

1. Superluminal expansion in compact radio sources indicates speeds $v_o \sim$ 5–10[38].
2. Jets in compact sources appear to be more bent and misaligned than those associated with extended sources, which can be explained as a projection effect if the jets are beamed toward us.[37]
3. In several sources, there would be catastrophic inverse Compton losses in the absence of relativistic expansion;[62] indeed this has been used as a predictor of superluminal expansion.[63]
4. Low-frequency radio variability, which is independently associated with relativistic expansion, is particularly common among the OVV quasars.[64]
5. Compact radio sources are often surrounded by faint halos, presumably the isotropically radiating lobes of weak double radio sources seen along their axes.[68]
6. The space densities of BL Lacs and OVVs do not appear to exhibit strong density evolution,[65] just like the weaker QSRs (both steep spectrum[66] and flat spectrum[67]) and QSOs.[69]
7. The hydrogen line profiles are narrower and simpler in the nuclei of compact radio sources than in the extended radio sources, as might be expected if the emission line clouds have a rotationally flattened space distribution.[70]

The beaming hypothesis is both vulnerable, in the sense that it can be cast in a simple form and then shown to be inconsistent with observations, and durable, in that it is not at all hard to create more complex models that have sufficient flexibility to survive more or less any feasible statistical test![72,73] The simple view is that compact radio sources comprise similar synchrotron-emitting clouds shot out at an angle θ to the line of sight. The emission observed from them is therefore boosted by a factor $[\gamma(1 - \beta \cos \theta)]^{-(3+\alpha)}$. This implies that the number of sources observed in a complete sample down to a limiting flux density S is roughly $N(>S) \propto S^{-1/(3+\alpha)}$ as long as the flux is boosted by a factor $\lesssim\gamma^{(3+\alpha)}$. However, real sources must be far more complex than this (FIGURE 3). With differing assumptions about the source kinematics, the exponent in the boost factor can range from 2 to 6. The intrinsic powers of the unboosted source components and their relativistic speeds must vary substantially from source to source and within individual sources. In particular jets may contain thick boundary layers and shocks in which the dissipation is far higher and the speed lower than in most of the jet. The most important effect is probably that the total width of the radiation beam

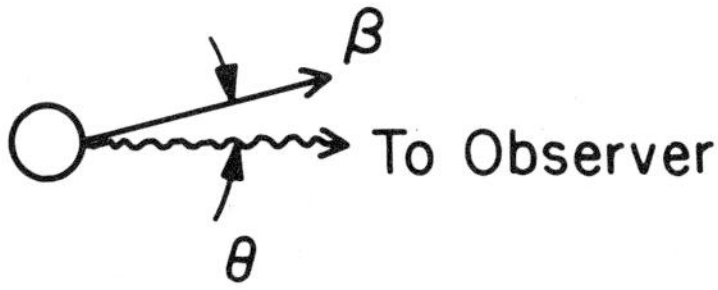

FIGURE 3a. Simple model of compact radio source used in elementary discussions of the statistics of beaming. The observed flux is enhanced by a factor $\propto(1 - \beta \cos \theta)^{-(3+\alpha)}$. The cloud will appear to the observer to move transversely with speed $V_0 = \beta c \sin \theta/(1 - \beta \cos \theta)$.

greatly exceeds γ^{-2}. We observe those features that are highlighted by moving in our direction.

BL Lacs are identified with giant elliptical galaxies, of which they make up a fraction roughly 1% (comparing objects of the same galaxy luminosity). If we compare objects of the same total magnitude then we must allow for the fact that the beamed sources are drawn from a larger sample of intrinsically fainter galaxies. In view of the current uncertainty in the local space density of BL Lacs there seems no statistical objection to their properties being attributed to relativistic beaming. It is even harder to make a quantitative comparison for the OVVs. These constitute about 15% of QSRs at the same limiting total magnitude. (We know very little about the nature of the underlying galaxies.) Most QSRs observed down to a given flux density at short wavelengths are compact and therefore perhaps beamed themselves. The emission lines are probably a better measure of the intrinsic luminosity as they are manifestly not Doppler enhanced (although they could radiate anisotropically if they are surrounded by an obscuring ring of gas). One difficulty that a beaming explanation must overcome is that the wavelength independence of the optical polarization implies

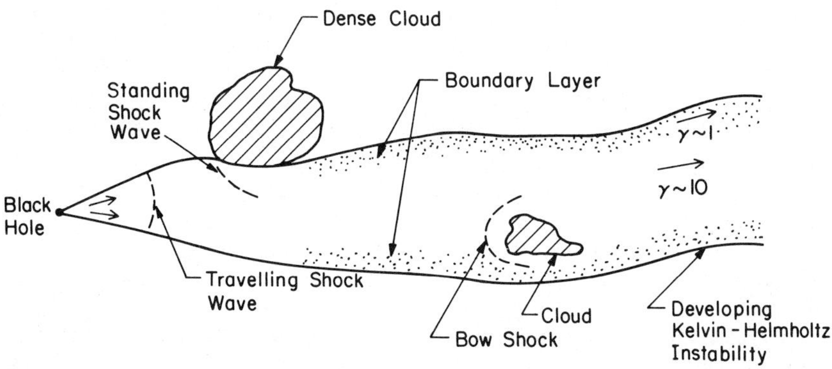

FIGURE 3b. Schematic illustration of "realistic" model of compact radio jet. Shear flows and shocks abound when the outflow interacts strongly with the surrounding medium. These are probably the main sites of particle acceleration and field amplification. As the radio emission is associated with material moving with a wide range of speeds and directions, the beam pattern will be quite complex and will vary from source to source. In particular features moving superluminally with speeds in excess of V_0 may be seen within a solid angle much larger than $2\pi(c/V_0)^2$.

that the beamed continuum outshines the unpolarized and unbeamed continuum. However the emission-line equivalent widths are not correspondingly smaller in the OVVs than in the unpolarized QSRs.

Most compact radio sources observed in high-frequency radio surveys are quasars. On the assumption that these are beamed, the unbeamed counterparts have been separately identified with (1) the QSOs of which they make up about 5% of an optically selected sample;[72] and (2) the steep spectrum extended quasars.[73,77,78] The former hypothesis seems to be in fairly severe difficulty because deep radio surveys have failed to increase the fraction of radio identifications as would be expected.[74,75] Furthermore the radio halos that are being seen increasingly around the compact QSRs and are presumably unbeamed are not being found around the QSOs. The second hypothesis also has problems but these are so far not as severe. One problem is with the quasar 3C179, which was selected as being a normal, double-lobed, steep-spectrum quasar and therefore was presumably not beamed toward us.[76] However when the central compact component was studied using VLBI, it exhibited superluminal expansion. It has been shown that the frequency and radio flux limit dependence of the fraction of compact sources in radio surveys can be reproduced if the compact jets comprise simple jets with Lorentz factor $\gamma \sim 5$. However it has not been demonstrated that this can be done with quasars of similar emission-line luminosity. Both difficulties can be alleviated if the opening angle of the outflow is made large, $\gtrsim 30°$,[79] as suggested by recent low-frequency, high-resolution maps of compact radio sources that show low brightness features with no particular preferred orientation.[80]

SUMMARY

In this review, I have used the organizational device of constructing a unified model of active galactic nuclei to summarize recent work on a wide variety of relevant physical mechanisms. Only the grossest characteristics can be accommodated within this scheme, and there are already some inconsistencies. It is suggested that the primary properties of active galactic nuclei are basically controlled by three parameters, which can be chosen to be the black hole mass, the accretion rate, and the viewing angle (FIGURE 4). Of course the spin angular momentum and the immediate environment may also be influential, particularly on the evolution of active galaxies. Seyferts and QSOs are associated with radiation tori orbiting black holes whose masses are such as to make the observed UV flux equal to the Eddington flux. The remaining luminosity extending from the infrared to gamma-ray energies is attributed to a pair-dominated corona heated by the underlying torus. As radiation tori are essentially spherical structures, they give low polarization.

The radio galaxies are associated with holes that are accreting at levels far below the Eddington rate and in which the power is derived from the hole's residual spin energy. Conversely, the black holes powering radio galaxies (associated with ellipticals) are more massive than those in QSOs (mostly associated with spirals). This correlation fits in with the increasingly attractive hypothesis that general elliptical galaxies (not just the giant cDs) are merger products.[81] Central black holes more massive than $\sim 3 \times 10^5\ M_\odot$ residing in the nuclei of the component spirals should be able to fall into the center of the compound galaxy under dynamical friction. In this scheme, our galaxy contains a small ($\lesssim 10^6 M_\odot$) hole in its nucleus surrounded by an ion torus.

These two structures represent the extremes of what is more likely to be a continuous progression as $\dot{M}/M$ is increased. The ratio of thermal to nonthermal coronal power increases with the thickness of the accretion disk, and the importance of radiation pressure support also increases. QSRs can be associated with the intermediate case when the disk is thin enough for a corona to extend down to the event horizon of the hole. The degree of collimation of the outflowing relativistic plasma may be

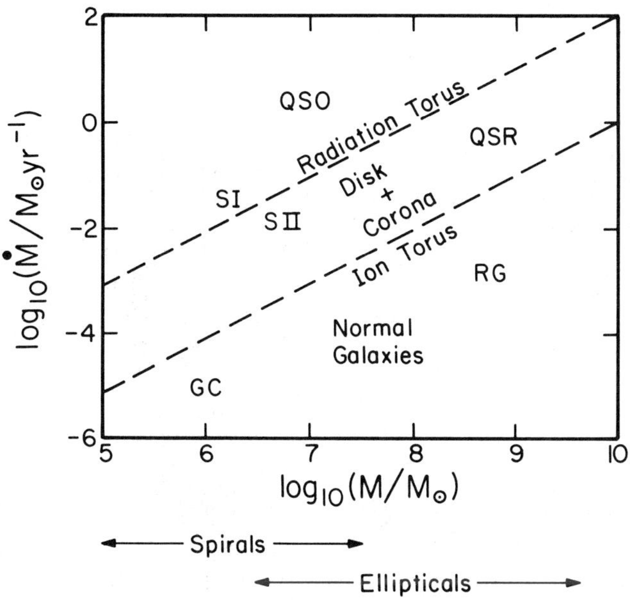

FIGURE 4. Simple unified scheme for different types of extragalactic object. In this scheme, the observed properties are assumed to be mainly controlled by the hole mass M, the accretion rate $\dot{M}$, and the viewing angle. SI and SII are type I and type II Seyfert galaxies. GC is the galactic center, and RG are radio galaxies. Beamed QSRs (generally of lower intrinsic power) are classified as OVV. Compact radio sources are associated with beamed radio galaxies and quasars. When the line of sight is very well aligned with the jet axis, the optical continuum will be strongly beamed and the object will be classified as a BL Lac.

inversely correlated with $\dot{M}/M$. QSRs may be expected in both ellipticals and spirals.

We identify OVVs with beamed QSRs and BL Lacs with beamed giant ellipticals. In this reviewer's opinion, the circumstantial evidence in favor of the existence of relativistic outflows exceeds the difficulties associated with implementing beaming in a general scheme such as that discussed above. Perhaps beaming is important in some of the more radio-prominent Seyfert galaxies also.

The observational implications of this scheme are not particularly well defined, but the space telescope should be able to determine if the QSOs really are mainly associated with spirals and whether or not those associated with ellipticals have larger UV luminosities and hence more massive black holes, as is suggested above. Future VLBI studies, in particular those with the projected array, should be especially useful in probing the importance of beaming in compact radio sources.

ACKNOWLEDGMENTS

I am indebted to my collaborators M. Begelman, D. Payne, S. Phinney, and M. Rees as well as several observational colleagues for invaluable discussions on active galactic nuclei.

REFERENCES

1. Seyfert, C. K. 1942. Astrophys. J. **97:** 28.
2. Jennison, R. C. & M. K. das Gupta. 1953. Nature **172:** 996.
3. Schmidt, M. 1963. Nature. **197:** 1040.
4. Blake, G. M. 1970. Astrophys. Lett. **6:** 201.
5. Rees, M. J. 1977. Ann. N.Y. Acad. Sci. **302:** 613.
6. Rees, M. J., M. C. Begelman & R. D. Blandford. 1981. Ann. N.Y. Acad. Sci. **375:** 254.
7. Leiter, D. & E. Boldt. 1982. Astrophys. J. **260:** 1.
8. Rieke, G. H. & M. J. Lebofsky. 1979. Annu. Rev. Astron. Astrophys. **17:** 477.
9. Neugebauer, G., J. B. Oke, E. E. Becklin & K. Matthews. 1979. Astrophys. J. **230:** 79.
10. Angel, J. R. P. & H. S. Stockman. 1980. Annu. Rev. Astron. Astrophys. **18:** 321.
11. Peterson, B. M., C. B. Foltz, & P. L. Byard. 1981. Astrophys. J. **251:** 4.
12. Richstone, D. O. & M. Schmidt. 1980. Astrophys. J. **235:** 361.
13. Malkan, M. A. & W. L. W. Sargent. 1982. Astrophys. J. **254:** 22.
14. Mushotzky, R. F. 1982. Paper presented at the Eleventh Texas Symposium on Relativistic Astrophysics, Austin, Tex., December 12–17.
15. Tennant, A. F., *et al.* 1981. Astrophys. J. **251:** 15.
16. Boroson, T. A., J. B. Oke & R. F. Green. 1982. Astrophys. J. **263:** 32.
17. Miley, G. 1980. Annu. Rev. Astron. Astrophys. **18:** 165.
18. Kellermann, K. K. & I. I. K. Pauliny-Toth. Annu. Rev. Astron. Astrophys.
19. Miley, G. K. & J. S. Miller. 1979. Astrophys. J. **228:** L55.
20. Lo, K. Y. 1982. *In* The Galactic Center. G. R. Riegler & R. D. Blandford, Eds. American Institute of Physics. New York, N.Y.
21. Matteson, J. L. 1982. *In* The Galactic Center. G. R. Riegler & R. D. Blandford, Eds. American Institute of Physics. New York, N.Y.
22. Hall, D. N. B., S. G. Kleinmann & N. Z. Scoville. 1982. (Preprint.)
23. Jacobson, A. S. 1982. *In* The Galactic Center. G. R. Riegler & R. D. Blandford, Eds. American Institute of Physics. New York, N.Y.
24. Pringle, J. E. 1981. Annu. Rev. Astron, Astrophys. **19:** 137.
25. Galeev, A. A., R. Rosner & G. S. Vaiana. 1979. Astrophys. J. **229:** 318.
26. Jaroszyński, M., M. A. Abramowicz & B. Paczyński. 1980. Acta. Astron. **30:** 1.
27. Paczyński, B. & P. Wiita. Astron. Astrophys. **88:** 23.
28. Wiita, P. J. 1982. Astrophys. J. **256:** 666.
29. Blandford, R. D. 1983. *In* Numerical Astrophysics. J. Centrella & L. L. Smarr, Eds. (In press.)
30. Lynden-Bell, D. 1979. *In* Active Galactic Nuclei. C. Hazard & S. A. Mitton, Eds. Cambridge University Press. Cambridge, England.
31. Abramowicz, M. & T. Piran. 1980. Astrophys. J. Lett. **241:** L7.

32. SIKORA, M. & D. B. WILSON. 1981. Mon. Not. R. Astron. Soc. **197:** 529.
33. BEGELMAN, M. C. & M. J. REES. 1982. *In* Proceedings, Turin Conference on Astrophysical Jets. (In press.)
34. REES, M. J., M. C. BEGELMAN, R. D. BLANDFORD & E. S. PHINNEY. 1982. Nature **295:** 17.
35. MACDONALD, D. & K. S. THORNE. 1982. Mon. Not. R. Astron. Soc. **198:** 345.
36. PHINNEY, E. S. 1982. *In* Proceedings, Turin Conference on Astrophysical Jets. (In press.)
37. READHEAD, A. C. S. & T. PEARSON. 1981. *In* Extragalactic Radio Sources. D. S. Heeschen & C. M. Wade, Eds. Reidel. Dordrecht, the Netherlands.
38. COHEN, M. H. & S. C. UNWIN. 1982. *In* Extragalactic Radio Sources. D. S. Heeschen & C. M. Wade, Eds. Reidel. Dordrecht, the Netherlands.
39. KONIGL, A. 1981. Astrophys. J. **243:** 700.
40. BLANDFORD, R. D. & M. J. REES. 1978. *In* Pittsburgh Conference on BL Lac Objects. A. M. Wolfe, Ed. Pittsburgh, Pa.
41. ZUREK, W. H. 1981. Astron. Astrophys. **91:** 90.
42. IPSER, J. & R. M. PRICE. 1982. Astrophys. J. **255:** 654.
43. KATZ, J. I. 1976. Astrophys. J. **206:** 910.
44. SUNYAEV, R. A. & L. G. TITARCHUK. 1980. Astron. Astrophys. **86:** 121.
45. TAKAHARA, F., S. TSURUTA & S. ICHIMARU, 1981. Astrophys. J. **251:** 26.
46. SHAPIRO, S. L., A. P. LIGHTMAN & D. M. EARDLEY. 1976. Astrophys. J. **204:** 87.
47. KWAN, J. Ann. N.Y. Acad. Sci. (This volume.)
48. MUSHOTZKY, R. 1982. Paper presented at the Eleventh Texas Symposium on Relativistic Astrophysics, Austin, Tex., December 12–17.
49. LIGHTMAN, A. P. 1982. Space Sci. Rev. (In press.)
50. TAKAHARA, F. 1980. Prog. Theor. Phys. **613:** 1551.
51. GUILBERT, P. W., A. C. FABIAN & R. R. ROSS. 1982. Mon. Not. R. Astr. Soc. **199:** 763.
52. PAYNE, D. G. 1980. Astrophys. J. **237:** 951.
53. FABIAN, A. C. 1981. Ann. N.Y. Acad. Sci. **375:** 235.
54. SWANENBURG, B. N. 1981. Ann. N.Y. Acad. Sci. **375:** 381.
55. GOULD, R. J. 1982. Astrophys. J. **254:** 755.
56. LIGHTMAN, A. P. 1982. Astrophys. J. **253:** 842.
57. SVENSSON, R. 1982. Astrophys. J. **258:** 335.
58. ARAKI, S. & A. P. LIGHTMAN. 1983. Astrophy. J. (In press.)
59. LIGHTMAN, A. P. & D. BAND. 1981. Astrophys. J. **251:** 713.
60. FABIAN, A. C. & M. J. REES. (In preparation.)
61. AXFORD, W. I. 1981. Ann. N.Y. Acad. Sci. **375:** 297.
62. JONES, T. W. & G. R. BURBRIDGE. 1973. Astrophys. J. **186:** 791.
63. MARSCHER, A. P. & J. J. BRODERICK. 1981. Astrophys. J. **249:** 406.
64. FANTI, R., L. PADRIELLI & M. SALVATI. 1982. *In* Extragalactic Radio Sources. D. S. Heeschen & C. M. Wade, Eds. Reidel. Dordrecht, the Netherlands.
65. MOORE, R. L. 1983. Astrophys. J. (In press.)
66. WALL, J. V. 1980. Phil. Trans. R. Soc. London Ser. A **A296:** 367.
67. PEACOCK, J. A. & S. F. GULL. 1981. Mon. Not. R. Astron. Soc. **196:** 611.
68. PERLEY, R. A., E. B. FONALONT & K. J. JOHNSON. 1980. Astron. J. **85:** 649.
69. SCHMIDT, M. & R. F. GREEN. 1982. (Preprint.)
70. MILEY, G. & J. S. MILLER. 1979. Astrophys. J. **228:** 655.
71. WOLTJER, L. & G. SETTI. 1982. *In* Cosmology and Fundamental Physics.
72. SCHEUER, P. A. G. & A. C. S. READHEAD. 1979. Nature London **277:** 182.
73. BLANDFORD, R. D. & A. KONIGL. 1979. Astrophys. J. **232:** 34.
74. CONDON, J. J., S. L. O'DELL, J. J. PUSCHELL & W. A. STEIN. 1981. Astrophys. J. **246:** 624.
75. STRITTMATTER, P. A. 1980. Astron. Astrophys. **88:** L12.
76. PORCAS, R. W. 1982. *In* Extragalactic Radio Sources. D. S. Heeschen & C. M. Wade, Eds. Reidel. Dordrecht, the Netherlands.
77. ORR, M. J. L. & W. A. BROWNE. 1982. Mon. Not. R. Astron. Soc. **200:** 1067.
78. BROWNE, I. W. A., M. J. L. ORR, R. J. DAVIS, A. FOLEY, T. W. B. MUXLOW & T. H. O. MASSON. 1982. Mon. Not. R. Astron. Soc. **198:** 673.

79. REES, M. J. 1982. *In* Extragalactic Radio Sources. D. S. Heeschen & C. M. Wade. Eds. Reidel. Dordrecht, the Netherlands.
80. SIMON, R. S. 1982. Doctoral Thesis. California Insititute of Technology. Pasadena, Calif.
81. FALL, S. M. 1979. Nature **281:** 200.
82. SILK, J. & C. A. NORMAN. 1981. Astrophys. J. **281:** 200.
83. BRADT, H. V. 1980. Ann. N.Y. Acad. Sci. **336:** 59.

Supernovae, Neutrino Rest Mass, and the Middle-Energy Neutrino Background in the Universe

GENNADYI S. BISNOVATYI-KOGAN

Space Research Institute
Moscow 117810, Union of Soviet Socialist Republics

ZAKIR F. SEIDOV

Shemakha Astrophysical Observatory
Shemakha, 373243 Azerbaijan,
Union of Soviet Socialist Republics

INTRODUCTION

The neutrinos emitted during the formation of the neutron stars and black holes form a background in the present universe, together with the relict microwave radiation and relict neutrinos. About 10% Mc^2 is emitted in neutrinos during the formation of the neutron stars[10] and maybe more during the formation of the black hole. The energy of one such neutrino, ~3–30 MeV, is much greater than that of relict neutrinos and much smaller than that of the cosmic ray neutrinos, so we shall call them middle-energy neutrinos (MEN). We show here that MEN background has the density 2–10 × 10^{-33} g cm^{-3}, which is larger than the density of relict microwave radiation $\rho_r = 4 \times 10^{-34}$ g cm^{-3} and less than the matter density. We calculate the MEN spectra and show that they give 2–8 × 10^{-3} solar neutrino units (SNU) in the solar chlorine-argon detector.

The effects of the possible neutrino rest mass are discussed: the propagation in the expanding universe, influence on the spatial structure of MEN background. The horizon of the particle with rest mass is calculated for the expanding universe and it is shown that the relict neutrino horizon is of the order of the characteristic distance between rich clusters of the galaxies in a neutrino-dominated universe when $m_\nu \simeq 30$ eV.

ENERGY DENSITY AND SPECTRUM OF MEN WITHOUT GALAXY EVOLUTION

Integral Characteristics

The total energy flux of MEN from supernovae (SN) in all galaxies with red shifts z from 0 to z_F at present is (see details in Reference 4):

$$\varphi_\epsilon(z_F) = \frac{W_0 \tau_1 c n_0}{\tau_2 H} \int_0^{z_F} \frac{dz}{(1+z)^3 \sqrt{1 + \Omega z}} \tag{1}$$

and the total number flux of MEN is

$$\varphi_n(z_F) = \frac{N_0\tau_1 c n_0}{\tau_2 H}\int_0^{z_F}\frac{dz}{(1+z)^2\sqrt{1+\Omega z}} \tag{2}$$

Here W_0 and N_0 are the total energy and the total number of neutrinos emitted in one second in an SN explosion, τ_1 is the SN neutrino flash duration, τ_2 is the present time interval between two SN explosions in one galaxy, n_0 is the present spatial density of galaxies, H and Ω are the present values of the Hubble constant and the ratio of the universe mean density to the critical one, and c is the light velocity.

According to Nadezhin and Otroshenko,[12] we assume $\tau_1 \simeq 50$ s, and the spectral density of neutrino (ν_e) flash to be

$$\left.\begin{aligned} P(E) &= \frac{Ax^3\exp(-\alpha x^2)}{1+\exp(x)}, \\ x = E/kT, &\quad W_0 = 18.4\,A, \quad N_0 = 1.69\,A \end{aligned}\right\} \tag{3}$$

here E is the neutrino energy in MeV, $A = 8.92 \times 10^{55}$ s^{-1}, $\alpha = 5.09 \times 10^{-3}$, $kT = 3.59$.

From (**1–3**) we get (for $z_F \rightarrow \infty$) the mean particle energy of MEN:

$$\frac{\varphi_\epsilon}{\varphi_n} = \begin{cases} \dfrac{1}{2}\dfrac{W_0}{N_0} = 5.46 \text{ MeV}, \Omega = 0 \\ \dfrac{3}{5}\dfrac{W_0}{N_0} = 6.55 \text{ MeV}, \Omega = 1 \end{cases} \tag{4}$$

and the mass density of MEN:

$$\rho_\nu = 4.10^{-33}\left(\frac{75 \text{ km s}^{-1} \times \text{Mpc}}{H}\right)\left(\frac{n_0}{0.6 \text{ Mpc}^{-3}}\right)\left(\frac{30 \text{ yr}}{\tau_2}\right) \text{g cm}^{-3} \times \begin{cases} 1/2, \Omega = 0 \\ 2/5, \Omega = 1 \end{cases} \tag{5}$$

The Galaxy mass is taken to be $10^{10}\ M_\odot$, the matter density 5×10^{-31} g cm^{-3}, and we introduce in (**5**) an additional factor of 4 because in SN explosions, four sorts of neutrinos (ν_e, $\tilde{\nu}_e$, ν_μ, $\tilde{\nu}_\mu$) with the same total energy are emitted. We conclude that SN explosions provide the middle-energy neutrino background in the present universe with the energy density about one order of magnitude higher than that of the relict microwave radiation.

The Energy Spectrum of MEN

The resulting spectrum of MEN from SN in all galaxies from 0 to z_F is:

$$\left.\begin{aligned} N_{z_F}(E) &= \frac{A\tau_1 n_0 c}{\tau_2 H (kT)^3} E^2 \varphi_{z_F}(E) \\ \varphi_{z_F}(E) &= \int_0^{z_F} \frac{(1+z)\exp\{-\alpha[E(1+z)/kT]^2\}}{\sqrt{1+\Omega z}\,\{1+\exp[E(1+z)/kT]\}}\,dz \end{aligned}\right\} \tag{6}$$

For finite z_F and small E we get $N_{z_F} \propto E^2$; if however $z_F = \infty$ then for small E, $N_\infty \propto E^{1/2}$. The function $E^2 \varphi_{z_F}(E)$ is shown in FIGURE 1 for various values of Ω and z_F.

GALAXY EVOLUTION AND SN RATES

It is rather probable that in the past the SN rate was higher than at present. To take into account this evolution factor, we assume the simple power law, $\tilde{\tau}_2 = \tau_2 (1 + z)^{-\gamma}$, and confine ourselves to $z < z_B \lesssim 10$, z_B being the galaxy formation epoch.

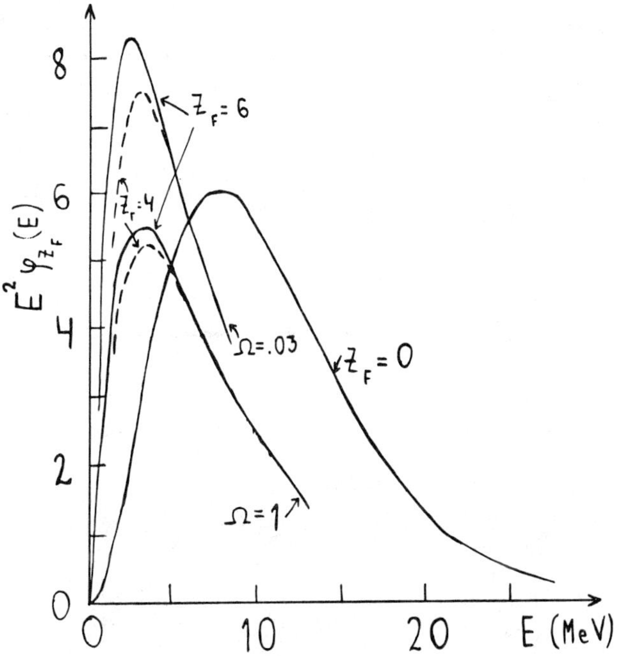

FIGURE 1. The energy spectrum of neutrinos emitted in SN, according to Nadezhin and Otroshenko ($z_F = 0$),[12] and integrated energy spectra of MeV in our neighborhood for various values of Ω and z_F.

Chemical Constitution and Constraints on γ

The present concentration of heavy elements $z_{he} \lesssim 1.5\%$ is presumably provided by the mass ejections from SN. Let every SN give $\eta M_\odot$ of heavy elements, then:

$$z_{he} = \frac{\eta M_\odot}{M_G \tau_2 H} \int_0^{z_F} \frac{(1 + z)^{\gamma - 2}}{\sqrt{1 + \Omega z}} dz \tag{7}$$

Assuming $\eta = 0.1$, $M_G = 10^{10}\, M_\odot$, $H = 75$ km s^{-1} × Mps, $\tau_2 = 30$ yr, and $\Omega = 0, 1$, we calculated from (7) the function $z_F(\gamma)$, shown for $z_{he} = 0.015$ and 0.0075 in FIGURE 2. For $z_F < 10$ we have $\gamma > 1.3$ (0.7) (if $\Omega = 0$) and $\gamma > 2.3$ (1.7) (if $\Omega = 1$) for $z_{he} = 0.015$

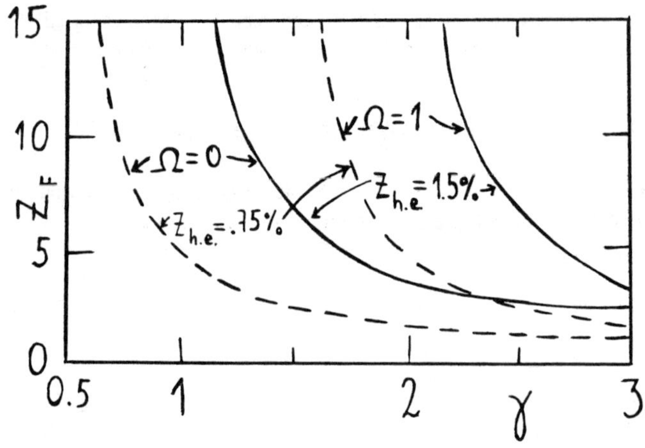

FIGURE 2. The galaxy formation epoch z_p as a function of the power of the galaxy evolution law γ provided that the present concentration z_{he} of heavy elements is due to SN explosions. Solid lines: $z_{he} = 0.015$. Dashed lines: $z_{he} = 0.0075$.

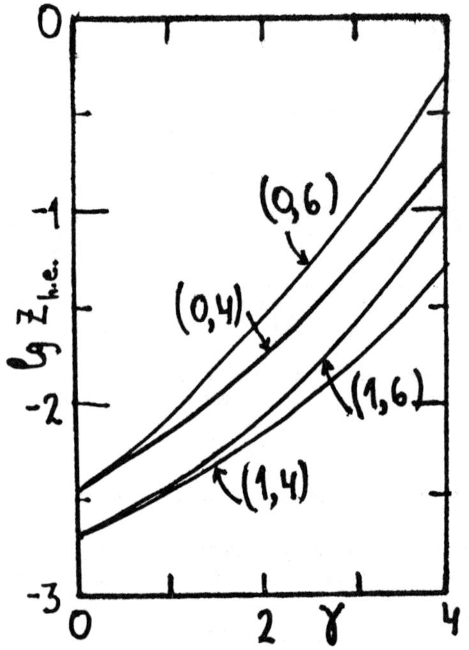

FIGURE 3. The present concentration of heavy elements z_{he} due to SN explosion as a function of the power of the galaxy evolution law γ, in open ($\Omega = 0$) and flat ($\Omega = 1$) universes for two values of z_F—galaxy formation epoch. The curves are labeled by values of Ω and z_F.

(0.0075), which indicates the possibility of evident evolution of galaxies in the past. For the case $z_{he} = 0.0015$, the needed evolution is negligible; moreover for our adopted modern rate of heavy element production, the values of z_F are equal to 0.5 (0.8) for $\Omega = 0$ (1). This is not consistent with the observations of the element abundances in distant quasars.[2] In FIGURE 3 we show the function $z_{he}(\gamma)$ from (**7**). For given z_F and γ we have $z_{he}(\Omega = 0) > z_{he}(\Omega = 1)$, and z_{he} increases with γ faster when $\Omega = 0$ than when $\Omega = 1$.

MEN BACKGROUND IN SOLAR NEUTRINO EXPERIMENT

No Galaxy Evolution

The energy spectrum of MEN found above allows us to calculate the rate of neutrino capture reactions in one second per one target nucleus. In radiochemical solar neutrino experiments, the reaction $^{37}Cl\,(\nu_e, e^-)\,Ar^{37}$ is used. For this reaction we use the cross section by Bahcall,[1] keeping in mind that it is the lower boundary (see Reference 6). The results of the calculation are for $z_F = 4$–6:

$$n_{capt} = \left.\begin{matrix} 1.7 \\ \\ 1.5 \end{matrix}\right\} 10^{-3}\ \text{SNU} \quad \begin{matrix} \Omega = 0 \\ \\ \Omega = 1 \end{matrix} \qquad (\mathbf{8})$$

That is, the MEN background in the radiochemical experiment is comparable with other backgrounds in chlorine-argon experiments (see Reference 8).

The Evolution Effect

We recalculated n_{capt} with the additional factor $(1 + z)^\gamma$ in the integrand (**6**). The value of n_{capt} is almost independent of z_F for $z_F \gtrsim 3$ and increases with γ much slower than z_{he} does because of the rapid fall of the neutrino-capture reaction cross section with decreasing (due to red shift) neutrino energy.

THE EFFECTS OF NEUTRINO FINITE REST MASS

The Propagation of Particles with Nonzero Rest Mass in the Expanding Universe

The Lagrange path covered by the nonzero rest-mass particle in the Friedmann universe in a time interval between t_1 and t_2 is:[4]

$$\left.\begin{aligned} r &= \int_{t_1}^{t_2} \frac{v(t)}{a(t)}\,dt, \quad v(t) = c\left[1 + \frac{a^2(t)}{B^2}\right]^{-1/2} \\ B &= [\gamma^2(t) - 1]^{1/2} a(t) = \text{const.}, \quad \gamma = (1 - v^2/c^2)^{-1/2} \end{aligned}\right\} \qquad (\mathbf{9})$$

Here v is the velocity of particles in the comoving coordinates, that is, the velocity with respect to the matter taking part only in Hubble expansion, and $a(t)$ is "radius of the universe."

The Open Universe, Ω < 1

In this case:

$$r = \int_{\eta_1}^{\eta_2} \left[1 + \left(\frac{ch\eta - 1}{A}\right)^2\right]^{-1/2}, A = (ch\eta - 1)(\gamma^2 - 1)^{1/2} = \text{const.} \qquad \textbf{(10)}$$

Here η is the parameter of Friedmann solution for the open universe (see Reference 11).

The maximal possible path of the particle, $R = r(\eta_1 = 0, \eta_2 = \infty)$, in two extreme limits is equal to:

$$R = (2A)^{1/2}K(1/\sqrt{2}),\ A \ll 1$$

$$R = \ln(4A),\ A \gg 1 \qquad \textbf{(11)}$$

here $K(k)$ is the complete elliptical integral of first kind.

We conclude that the Lagrange path of free particles with finite rest mass during the whole evolution of the open universe is finite (for photon $A \to \infty$ and $R \to \infty$).

From (**10**) one sees that for given η_2 and γ_2 the function $r(\eta_1)$ is monotonic and the neutrino horizon r_0 corresponds to $\eta_1 = 0$.

However for given η_2 and γ_1 the function $r(\eta_1)$ has a maximum, $r_{max} = r\,(\eta_{1,max})$:

$$\eta_{1,max} = \frac{\eta_2}{2}, \quad r_{max} = \frac{v_1}{c}\, th\, \frac{\eta_2}{4}, \quad \frac{v_1}{c} \ll 1 \qquad \textbf{(12)}$$

$$\left.\begin{aligned} \eta_{1,max} &= \left[\frac{8}{\gamma_1^2}\left(\frac{1}{4} sh2\eta_2 + \frac{3}{2}\eta_2 - 2sh\eta_2\right)\right]^{1/5} \\ r_{max} &= \eta_2 - \frac{5}{4}\eta_{1,max} \end{aligned}\right\} \gamma_2 \gg 1 \qquad \textbf{(13)}$$

The functions $\eta_{1,max}(\gamma_1)$, $z_{max}(\gamma_1)$, and $r_{max}(\gamma_1)$ for the value of $\eta_2 = 4.877$ $(\Omega = 0.03)$ are shown in FIGURE 4.

Comparison with Light

For photons $v = c$ and $r = \eta_{2,ph} - \eta_{1,ph}$, where $\eta_{2,ph}$ and $\eta_{1,ph}$ are connected with Ω (at the moment $\eta_{2,ph}$) and with the red shift of the photon source (galaxy) z, emitting at the moment $\eta_{1,ph}$:

$$\eta_{2,ph} = \ln\frac{1 + \sqrt{1 - \Omega}}{1 - \sqrt{1 - \Omega}}, \quad \eta_{1,ph} = \ln\frac{\sqrt{1 + \Omega z} + \sqrt{1 - \Omega}}{\sqrt{1 + \Omega z} - \sqrt{1 - \Omega}} \qquad \textbf{(14)}$$

The photon horizon $r_{0,ph}$ at the moment η_2 corresponds to $\eta_1 = 0$ $(z = \infty)$ and $r_{0,ph} = \eta_2$. If a neutrino is emitted by a galaxy simultaneously with light at the moment η_1 and the light of the galaxy is observed at moment $\eta_{2,ph}$ with red shift z, then the neutrino with finite rest mass and $\gamma_2 \gg 1$ comes to the observer after the time delay:

$$\Delta t(\text{sec}) = t_{2,\nu} - t_{2,ph} = \frac{1}{2H\gamma_2^2}\int_0^z \frac{dz}{(1 + z)^3(1 + \Omega z)^{1/2}} \qquad \textbf{(15)}$$

The Flat Universe, Ω = 1

In this case it is convenient to pass from r to red shift z:

$$1 - \frac{1}{\sqrt{1+z}} = \int_{x_1}^{1} \left[1 + \frac{x^4}{x_1^4(\gamma_1^2 - 1)}\right]^{-1/2} dx, \; x = (t/t_2)^{1/3} \tag{16}$$

The neutrino horizon (maximal $z = z_{max}$) for given γ_1 is:

$$x_{max} = ((2/5)\gamma_1^{-2})^{1/5}, \; z_{max} = \left(\frac{512}{625}\gamma_1^2\right)^{2/5}, \; \gamma_2 = \left(\frac{4\gamma_1}{25}\right)^{1/5}, \; \gamma_2 \gg 1 \tag{17}$$

$$x_{max} = \frac{1}{2}, \; z_{max} = \frac{1}{2}\frac{v_1}{c}, \; v_2 = \frac{1}{2}v_1, \; v_1 \ll c \tag{18}$$

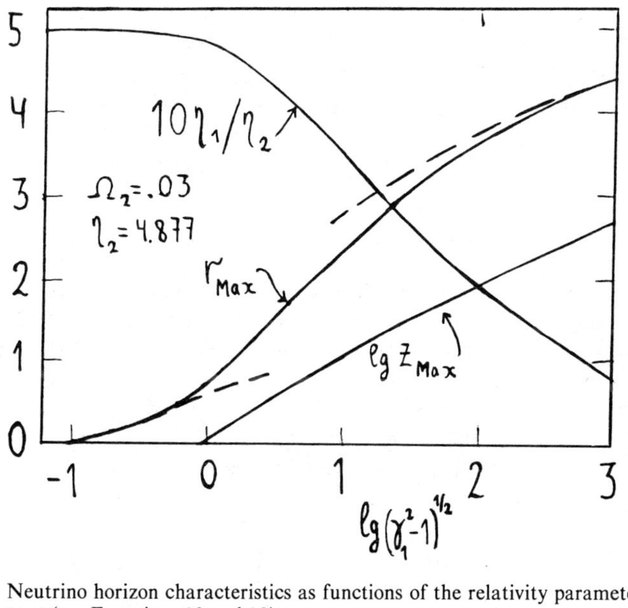

FIGURE 4. Neutrino horizon characteristics as functions of the relativity parameters γ_1 at the emission moment (see Equations 12 and 13).

We conclude (see Equations 12, 13, 17, and 18) that neutrinos emitted with the same energy come to the observer from the region smaller than the light horizon, and the neutrino horizon corresponds to the nonzero moment of emission.

For fixed γ_2 from (**16**), the neutrino horizon corresponds to $t_1 = 0$:

$$\left.\begin{aligned} z_0 &= 100\,\gamma_2^4, \; \gamma_2 \gg 1 \\ z_0 &= 3.71\,(v_2/c)^{1/2}, \; v_1 \ll c \end{aligned}\right\} \tag{19}$$

The Cosmological Neutrino Background Horizon

Let us briefly consider the question,[5] What is the dimension of the region of the universe that takes part in formation of the cosmological neutrino background (CNB)? In other words, what is the maximal distance to which it is possible to get information from CNB observations?

For massive neutrinos of the uniform CNB the present mean velocity is $v_2 \approx 6$ km s^{-1}.[3] From (**17**) we get the distance to the horizon of CNB, $R = cz_0/H = 66$ Mpc, the distance to the Coma cluster. The mass inside the present CNB horizon is of the order of the characteristic mass of a condensation in the neutrino-dominated universe.

The Temporal Structure of the Observed Neutrino Signal from SN

Scintillation detectors are often proposed for registration of antineutrino ($\tilde{\nu}_e$) radiation during the relativistic collapse (or SN explosion). The characteristics of $\tilde{\nu}_e$ radiation from SN are close to that of ν_e radiation,[12] so it is possible to calculate the temporal structure of the response of the scintillation detector [based on the $p(\tilde{\nu}_e, e^+)n$ reaction] to the SN signal.[9] It is shown that the shape of this response is rather sensitive both to SN neutrino radiation characteristics and to the neutrino rest mass. The very short and even instantaneous neutrino flashes are lengthened in the detector to several seconds for $m_\nu = 30$ eV, so discrimination between quiet relativistic collapse (short neutrino burst) and SN explosion ($\tau_1 \simeq 50$ s) may be rather difficult.

The Temporal Structure of the MEN Background

The MEN background is formed by all galaxies up to $z \lesssim 10$ (galaxy birth epoch), but the input of galaxies with $z > 1$ is rather small. In fact, from (**2**) it follows that the number fluxes $\varphi_n(1)/\varphi_n(\infty) = 0.5$–$0.65$ and the mean energies $\bar{\epsilon}\,(1)/\bar{\epsilon}\,(\infty) = 1.5$–$1.2$ for $\Omega = 0$–1. So we confine ourselves to $z < 1$; then with $m_\nu = 0$ the neutrino flash duration from farthest galaxies ($z = 1$) is two times larger than from the nearest galaxies ($z = 0$). With a nonzero rest mass of the neutrino, the duration of neutrino pulses from SN in galaxies at $z = 1$ is (see Equation 15):

$$\Delta t = 10^7\ \text{s} \times \begin{cases} 3/4, \Omega = 0 \\ 2/3, \Omega = 1 \end{cases} \quad \mathbf{(20)}$$

where we assume $H = 75$ km s^{-1} Mpc, $\gamma_2 = 10^5$ ($E_\nu = 3$ MeV, $m_\nu = 30$ eV).[7] In the volume $z < 1$, the mean duration of neutrino pulses is 10^7 s.

The expected number of SN explosions in all galaxies up to $z_F = 1$ is:[4]

$$\frac{dN_{SN}}{dt} = (37\text{–}21)\ s^{-1}, \Omega = 0\text{–}1 \quad \mathbf{(21)}$$

We conclude that the observed SN rate in the volume up to $z = 1$ is $\sim 20\ s^{-1}$, the mean duration of neutrino pulses is 10^2–10^7 s for $m_\nu = 0$–30 eV, and the relative amplitude of MEN background fluctuations is 10^{-2}–10^{-4}.

The observations of MEN with such characteristics with the help of the scintillation detectors will be of great import to the universe models, formation of the universe structure, star evolution theory, and elementary particle physics.

SUMMARY

We have calculated the density and spectrum of neutrinos with energy 3–30 MeV (MEN) due to SN explosions and their influence on the solar neutrino detector. The density of these neutrinos in the universe at present is 2–10 × 10^{-33} g cm^{-3}, that is, larger than the density of the relict microwave radiation. Propagation of a nonzero rest-mass neutrino in the Friedmann universe is considered, and the temporal characteristics of the MEN background are calculated. In the volume $z \lesssim 1$ the SN neutrino pulse frequency is ~ 20 s^{-1}, the single pulse duration is 10^2–10^7 s for m_ν = 0–30 eV, and the relative amplitude of MEN background fluctuation is 10^{-2}–10^{-4}.

ACKNOWLEDGMENT

The authors are grateful to I. D. Novikov for useful discussions.

REFERENCES

1. BAHCALL, J. N. 1979. Space Sci. Rev. **24:** 227.
2. BIERMAN, P. 1978. Astron. Astrophys. **62:** 225.
3. BISNOVATYI-KOGAN, G. S. & I. D. NOVIKOV. 1980. Astron. Zh. **57:** 899.
4. BISNOVATYI-KOGAN, G. S. & Z. F. SEIDOV. 1982. Astron. Zh. **59:** 213.
5. BISNOVATYI-KOGAN, G. S. & Z. F. SEIDOV. 1982. Astron. Zh. **60:** 220.
6. BUGAEV, E. V., M. A. RUDZSKYI, G. S. BISNOVATYI-KOGAN & Z. F. SEIDOV. 1979. Nucl. Phys. **234A:** 350.
7. LYUBIMOV, V. A., E. G. NOVIKOV, V. Z. NOZIK, E. F. TRETYAKOV & V. S. KOSIK. 1980. J. Nucl. Phys. **32:** 301.
8. RUDZSKYI, M. A. & Z. F. SEIDOV. 1979. J. Nucl. Phys. **30:** 1063.
9. SEIDOV, Z. F. 1982. Astrophys. Space Sci. **81:** 483.
10. ZELDOVICH, YA. B. & I. D. NOVIKOV. 1972. Gravitation Theory and Star Evolution. Nauka. Moscow, USSR.
11. ZELDOVICH, YA. B. & I. D. NOVIKOV. 1975. Structure and Evolution of the Universe. Nauka. Moscow, USSR.
12. NADEZHIN, D. K. & I. V. OTROSHENKO. 1980. Astron. Zh. **57:** 78.

On the Magnetization and Origin of the Millisecond Pulsar PSR 1937+214[a]

JONATHAN ARONS

Astronomy Department, Physics Department, and Space Sciences Laboratory University of California Berkeley, California 94720

and

Institute of Geophysics and Planetary Physics Lawrence Livermore Laboratory Livermore, California 94550

Lack of an analogue of the Crab nebula around PRS 1937+214 is shown to be consistent with the theory of pulsar magnetospheres only if the dipole component of the surface magnetic field is unusually weak, 10^8 gauss $< B_d < 3 \times 10^9$ gauss. This requirement is in satisfactory agreement with observations of this pulsar's spindown rate, $dP/dt \sim 10^{-19}$ s/s. Predictions are made for observations of weak nebular emission around this object. It is suggested that the neutron star was born as a weakly magnetized, rapidly rotating pulsar, and now is 10^5–10^6 years old.

[a]The substance of this paper was submitted to *Nature* on 30 November 1982, and was published in 1983. *Nature* **302:** 301.

Pulsar Magnetospheres

E. ASSÉO, D. BEAUFILS, AND R. PELLAT

Center for Theoretical Physics
Polytechnical School
91128 Palaiseau Cedex, France

Starting from the aligned rotator model introduced by Goldreich and Julian,[1] we first study a cold, force-free magnetospheric pulsar plasma, finite in extent. From Pilipp's theorem,[2] we know that such a plasma cannot be in corotation with the neutron star. We investigate the possibility for a noncorotating magnetospheric plasma to be separated from a corotating plasma (and neutron star) by a vacuum gap limited by two force-free surfaces. We find that axisymmetry together with the force-free condition implies the presence of a cusp in the equipotential surfaces which prevents the existence of the vacuum gap. Consequently we looked next for the possible existence of charged beams above the pulsar polar caps which propagate up and down along the magnetic dipole field lines, between the neutron star surface and a non-force-free surface, located at a finite distance, beyond which vacuum exists. We find that there is no vacuum potential solution that can be linked to the potential that describes the beam zone, and we conclude that there exists no self-consistent solution. We then investigate an alternative configuration in which particles are exchanged between the stellar surface and a force-free magnetospheric plasma. We obtain a physical situation reminiscent of the Goldreich and Julian charge-separated magnetosphere.[1]

REFERENCES

1. GOLDREICH, P. & W. H. JULIAN. 1969. Astrophys. J. **157:** 869.
2. PILIPP, W. G. 1974. Astrophys. J. **190:** 391.

Relativistic Stars with Anisotropic Pressure

SELÇUK Ş. BAYIN[a]

Department of Physics
Middle East Technical University
Ankara, Turkey

In this article we present various analytic solutions for anisotropic fluid spheres in general relativity.[1] First we consider generalizations of the $p = \alpha\rho$ solution to the case where pressure is anisotropic and study the effects of anisotropy on the structure of neutron stars. Next we study radiating anisotropic fluid spheres and present three classes of analytic solutions. We also study slowly rotating anisotropic fluid spheres and present two analytic solutions corresponding to the nonradiating case. One of these solutions corresponds to uniform rotation, while the other corresponds to differential rotation. We also present differential equations to be solved for slowly rotating and radiating anisotropic fluids.

Anisotropic pressure could be introduced by the existence of a solid core, by the presence of type-p superfluid, by the complexity of interactions, or by the existence of an external field. Recently it has been suggested that cooling of neutron stars might be accompanied by a phase transition from one anisotropic superfluid to another with significantly different properties.[1] Besides these, an interesting way to generate anisotropic pressure is through the presence of two perfect fluids with the energy-momentum tensor given as

$$T^{\mu\nu} = (P_1 + \rho_1)U^\mu U^\nu - P_1 g_\alpha^{\mu\nu} + (P_2 + \rho_2)W^\mu W^\nu - P_2 g_\alpha^{\mu\nu}$$

where $U^\mu U_\mu = W^\mu W_\mu = 1$, $W^\mu U_\mu = 1 + a; a \neq 0$.

This energy-momentum tensor can be cast into the standard form for anisotropic fluids by a suitable transformation.[1]

In general one expects these two fluids to reach equilibrium through dissipative mechanisms, and hence $U^\mu \longrightarrow W^\mu$. However in some cases it is possible that they are completely decoupled, or at least weakly interacting, throughout the object or in some portion of it. We have shown that in extremely relativistic objects these two fluids do not have to vibrate violently with respect to each other in order to introduce a significant amount of anisotropy. We have also argued that such vibrations could actually exist in neutron stars between the neutron and the electron components of the fluid.

Anisotropic pressure introduced through the presence of two perfect fluids does also have very interesting applications to cosmology; this is currently under investigation.

REFERENCE

1. BAYIN, S. S. 1982. Phys. Rev. **D26:** 1262.

[a]Present affiliation: Materials and Metallurgical Engineering Department, University of Michigan, Ann Arbor, Mich. 48109.

Trace Anomaly in the "Massive" Gowdy T^3 Cosmology[a]

BEVERLY K. BERGER

Physics Department
Oakland University
Rochester, Michigan 48063

The addition of a "mass" term to the scalar wave equation for the single polarization of gravitational waves in the Gowdy T^3 cosmology is explored. Previous semiclassical analysis[1]—a spatially homogeneous background cosmology driven by the quantized (scalar) gravitons—has relied on the method of adiabatic regularization[2,3] definition of positive frequency and construction of stress-energy tensor expectation values. The "mass" term alters the asymptotics of the equation for the time dependence of the mode amplitude. (Solutions previously in terms of zero-order Hankel functions are expressed in terms of Hankel function of order $i\kappa$ where $\kappa^2 = VM^2$ is the mass term.) The t-t and θ-θ components of the stress-energy tensor expectation values are no longer identical so that regularization may be expected to lead to an anomalous trace. Calculation of this trace anomaly is in progress with preliminary results presented. Direct comparison may be made with calculations by Bunch *et al.*[4] on a massive scalar field in two dimensions by constructing a fictitious two-dimensional metric which yields the "massive" Gowdy T^3 wave equation. (Symmetries of the model cause the effective stress-energy tensor to have only two nonvanishing components.) A comparison with the work of Hu[5] on a massive scalar field in a Kasner space time is also made.

REFERENCES

1. BERGER, B. K. 1982. Phys. Lett. **108B:** 394.
2. PARKER, L. & S. A. FULLING. 1974. Phys. Rev. **D9:** 341.
3. FULLING, S. A., L. PARKER & B. L. HU. 1974. Phys. Rev. **D12:** 3905.
4. BUNCH, T. S., S. M. CHRISTENSEN & S. A. FULLING. 1978. Phys. Rev. **D18:** 4435.
5. HU, B. L. 1978. Phys. Rev. **D18:** 4460.

[a]Supported in part by National Science Foundation Grant No. PHY-81-07387 to the University of Maryland.

Cosmology in General Relativity with Higher Derivatives

KENNETH BRECHER

Department of Astronomy
Boston University
Boston, Massachusetts 02215

The "standard" hot big bang model accounts for the expansion of the universe, the existence of the microwave background radiation, and the mass fraction of the light elements up to ^{4}He. It does not naturally account for the high degree of isotropy and homogeneity of the universe in the large, nor for the existence of structure (galaxies, clusters) on smaller scales. Other problems, such as the lepton to baryon ratio, the preponderance of matter over antimatter, and the "coincidences" of dimensionless ratios of several fundamental physical and cosmological "constants," also lie outside the "standard" model at present.

A solution of the large-scale homogeneity and isotropy problems, as well as the so-called flatness problem, has been proposed recently involving extreme supercooling during a phase transition associated with spontaneous symmetry breaking in grand unified theories (GUTs).[1] This and related solutions are based on a direct application of the Einstein field equations for general relativity $G_{\mu\nu} = -8\pi G T_{\mu\nu} - \Lambda g_{\mu\nu}$. The new feature of such theories is due to the introduction of a cosmological constant Λ or stress tensor with $p = -\rho$ at high densities due to the GUTs. Such theories do not, however, alleviate the problem of the origin of galaxies.

We propose a solution to both the large- and small-scale structure problems based on a modification of the left-hand side of the field equations.[2] By starting with a gravitational field Lagrangian $R + AR^2$ (where A is a new dimensional constant), we derive a new set of field equations, consistent with all current weak field gravitational tests, which modify the cosmological solutions to general relativity at early times when the high-order terms predominate. Such solutions are without a particle horizon [with scale factor $a(t) \propto t$ as $t \rightarrow 0$]. Therefore, large-scale fluctuations present at an early epoch could in principle be damped. On the other hand, small-scale perturbations in both the density and metric tend to grow much more rapidly than in unmodified general relativity, easing the problem of the formation of galaxies later in the evolution of the universe.

REFERENCES

1. GUTH, A. H. 1981. Phys. Rev. **D23:** 347.
2. FRENKEL, A. & K. BRECHER. 1982. Phys. Rev. **D26:** 368.

Neutrino Transport and Shock Propagation in the Core Collapse of a Massive Star

STEPHEN W. BRUENN

Department of Physics
Florida Atlantic University
Boca Raton, Florida 33431

The collapse of the 1.398 $M_\odot$ core of a 10 $M_\odot$ star evolved from the main sequence by Weaver and Woosley is followed numerically. General relativistic hydrodynamics was used to evolve the matter, and a multigroup, flux-limited diffusion scheme was used to independently evolve ν_e's, $\bar{\nu}_e$'s, and ν_μ and ν_τ pairs. The equation of state of Lamb *et al.* was used up to nuclear densities; a soft equation of state modeled after the hot nucleon EOS of Friedman and Pandharipande was used above nuclear density. The zero and first Lengendre moments of all important neutrino interactions were calculated from exact expressions (numerically integrated when necessary) obtained from the standard model of electroweak interactions with $\sin \theta_W = 0.23$. The initial evolution was characterized by the approximately homologous collapse of an inner core of mass 0.55 $M_\odot$ and a neutrino trapping density of between 6×10^{11} g cm^{-3} and 10^{12} g cm^{-3} for all neutrino types. The central density at bounce was 9.7×10^{14} g cm^{-3}, setting quickly afterwards to 6.6×10^{14} g cm^{-3}. Shortly before bounce a quasi-stationary shock formed near the edge of the inner core at a radius of 25 km and a density of 6×10^{12} g cm^{-3}. On rebound the inner core transferred about 6×10^{51} ergs to the shock which thereupon began a rapid outward propagation ($v_{\text{shock}} \sim 10^{10}$ cm s^{-1}). The location of the neutrinosphere was neutrino energy dependent, and about 2 ms were required for the shock to propagate from the 8 MeV to the 23 MeV neutrinosphere. The shock weakened and eventually ceased to propagate outward at all, becoming an accretion shock; no supernovalike explosion occurred and no matter was expelled. A total of $\sim 3.3 \times 10^{51}$ ergs, $\sim 2.1 \times 10^{50}$, and $\sim 1.2 \times 10^{51}$ ergs were radiated as ν_e's, $\bar{\nu}_e$'s, and ν_μ and ν_τ pairs by the time the shock reached maximum radius. The ν_e luminosity rose gradually to a maximum of about 2.2×10^{53} ergs s^{-1}. The luminosities of the $\bar{\nu}_e$'s and the ν_μ and ν_τ pairs rose abruptly (rise time 1 ms) after shock breakout and were still slowly rising after maximum shock radius, with values at this time of 2.0×10^{52} and 8.8×10^{52} ergs s^{-1}, respectively. Mean emergent neutrino energies at maximum shock radius were 8.0 MeV for the ν_e's, 10.5 MeV for the $\bar{\nu}_e$'s, and 14.4 MeV for the ν_μ and ν_τ pairs.

The Radio Structure of the X-Ray Jet in Centaurus A

J. O. BURNS

Department of Physics and Astronomy
University of New Mexico
Albuquerque, New Mexico 87131

E. D. FEIGELSON

Department of Astronomy
Pennsylvania State University
University Park, Pennsylvania 16802

E. J. SCHREIER

Space Telescope Science Institute
Homewood Campus
Johns Hopkins University
Baltimore, Maryland 21218

Multifrequency (1.4 and 4.9 GHz) observations of the jet and inner lobes of the nearby radio galaxy Centaurus A (=NGC5128) were made with all four configurations of the very large array radio telescope (VLA). These data were used to construct maps of the total intensity and polarization at resolutions of 31 × 10 arcsec and 3.6 × 1.1 arcsec. Among the results of the analysis of these maps are as follows. (1) The distributions of surface brightness and pressure in the jet combined with the apparent x-ray emission from the interstellar medium (ISM) of NGC5128 indicate that the radio jet is thermally confined. (2) Six different model-dependent assumptions were used to estimate the velocity of the jet; our best guess based primarily on energetics requirements is 5,000 km s^{-1}. (3) A comparison of the radio structure and the optical galaxy shows that the jet in Cen A emerges nearly along the major axis of the elliptical stellar component, which is parallel to the rotational velocity vector of the dust lane; the outer radio structure bends toward the galaxy minor axis possibly as a result of the complex gas dynamics in the galaxy. (4) A detailed comparison of the x-ray and radio emission in the jet demonstrates that the knots coincide precisely at the two energies. This fact, combined with the radio to x-ray spectrum and the limits on the thermal density in the jet set by the radio depolarization, argues in favor of a common synchrotron radiation origin for the full-spectrum jet emission.

Manifestations of a Cosmological Density of Compact Objects in Quasar Light

CLAUDE R. CANIZARES[a]

Department of Physics and Center for Space Research
Massachusetts Institute of Technology
Cambridge, Massachusetts 02139

and

Institute of Astronomy
University of Cambridge
Cambridge CP3 0HA, England

A cosmological density of compact objects with masses in the range $0.01 < M < 10^5 M_\odot$ will cause observable manifestations in the properties of statistical samples of quasars. The manifestations are caused by amplification of the quasar continuum through the gravitational lens effect [image splitting is detectable by very long baseline interferometry (VLBI) techniques only for lens masses $>10^5 M_\odot$, as shown by Press and Gunn].[1] The effects are computed using an analytic probability distribution for amplification that accounts for single and, approximately, for multiple lens amplification and that incorporates flux conservation and the diminution caused by light propagation in an inhomogeneous universe. The size of the emitting region is taken into account and is very important in determining the observable effects for a given range of lens masses. The results are easily scaled to different values of quasar red shift and mean lens density.

If the continuum emitting regions in quasars are small ($\lesssim 10^{-3}$ pc) then continuum flux limited samples should contain a large fraction of lensed quasars. Observable manifestations include time variability, higher mean continuum to line ratios for high z versus low z quasars, and large root-mean-square scatter in continuum to line ratios for quasars with $z > 1$–2. The effects on the ratios result from the difference in the sizes of the continuum and line emitting regions—the former can be strongly lensed whereas the latter cannot (for $M < 10^5 M_\odot$). Examination of a variety of quasar data suggests that these effects are not seen, so the density of compact objects in the 0.01–$10^5 M_\odot$ range is not sufficient to close the universe (this conclusion is invalidated if most quasar continuum emission regions are $\gtrsim 1$ pc). This would exclude nuclear burning stars and their remnants. The conclusion is based on several scant and heterogeneous data sets, but it can be refined and strengthened with further data. As gravitational lensing predicts a minimum scatter in various observed quantities, upper limits of the

[a]Alfred P. Sloan Research Fellow; supported in part by National Aeronautics and Space Administration contract NAS-8-30752.

cosmological density of compact objects are not invalidated by the unknown evolution of intrinsic quasar properties.

A full report of this work appears in the *Astrophysical Journal.*[2]

REFERENCES

1. PRESS, W. H. & J. E. GUNN. 1973. Astrophys. J. **185:** 397.
2. CANIZARES, C. R. 1982. Astrophys. J. **263:** 508.

Dwarf Nova Instabilities

J. K. CANNIZZO AND J. C. WHEELER

Department of Astronomy
University of Texas
Austin, Texas 78712

Dwarf novae are binary systems consisting of a late-type star overflowing its Roche lobe and dumping matter onto a white dwarf. These systems are observed to undergo outbursts in which the visual output increases by three to four magnitudes. These bursts last for one to two weeks and recur every few months. In outburst one sees an optically thick accretion disk, but not in quiescence. There are currently two ideas as to the burst mechanism. It is possible that because of a pulsational instability in the secondary, its mass outflow periodically increases. However one does not see a brightening of the accretion disk—mass flow stream "hotspot" interaction at the burst onset. Also, this mass increase must be five orders of magnitude. The second idea is that the cause of the instability lies in the disk itself. Meyer and Meyer-Hofmeister have proposed that the peak in Rosseland opacity at $T \sim 20{,}000$ K induces an instability,[1] while Cannizzo, Ghosh, and Wheeler have proposed a mechanism related to the turn-on of convection at $T \sim 2{,}000$ K.[2] It appears the former mechanism cannot satisfy self-consistency checks. The CGW mechanism can also explain recurrent x-ray transients if scaled up to their larger orbital radii.

REFERENCES

1. MEYER, F. & E. MEYER-HOFMEISTER. 1981. Astron. Astrophys. **104:** L10.
2. CANNIZZO, J. K., P. GHOSH & J. C. WHEELER. 1982. Astrophys. J. **260:** L83.

FRW Models as Exact Viscous Magnetohydrodynamical Cosmologies

A. A. COLEY[a]

Department of Applied Mathematics
University of Waterloo
Waterloo, Ontario, Canada N2L 3G1

B. O. J. TUPPER

Department of Mathematics and Statistics
University of New Brunswick
Fredericton, New Brunswick, Canada E3B 5A3

It can be shown[1-5] that Friedmann-Robertson-Walker (FRW) cosmologies, and in particular the zero-curvature models, do not necessarily represent perfect fluid solutions, but also can be exact solutions of the Einstein field equations for a viscous fluid, with or without an electromagnetic field. We present an exact viscous magnetohydrodynamical solution of the field equations with Einstein–de Sitter metric. In the solution there is a "tilted" fluid velocity vector with spacelike component in the z direction, and the magnetic field and heat conduction vector are also both axially directed. All the physical quantities appearing in the solution (that is, the density, pressure, shear viscosity, magnetic field, temperature, baryon particle density, entropy density, and thermal conductivity) are well behaved and the Einstein-Maxwell equations and the standard thermodynamic relations are all satisfied; therefore we conclude that the exact viscous magnetohydrodynamical model presented is physically satisfactory. The model is both homogeneous and isotropic and the matter fields depend on t only. As $t \rightarrow \infty$, the model approaches the usual perfect fluid Einstein–de Sitter cosmology (in that all quantities become negligible compared with the density). However, the behavior of the model differs considerably as one extrapolates back to t equals zero, which suggests that small amounts of viscosity (etc.) in the universe at present would lead to a history for the model that is quite different from, and perhaps more physically realistic than, that of the standard perfect fluid model.

REFERENCES

1. COLEY, A. A. & B. O. J. TUPPER. 1983. Gen. Rel. Grav. (In press.)
2. COLEY, A. A. & B. O. J. TUPPER. 1983. Astrophys. J. **271:** 1.
3. TUPPER, B. O. J. 1981. J. Math. Phys. **22:** 2666.
4. TUPPER, B. O. J. 1983. Gen. Rel. Grav. **15:** 849.
5. TUPPER, B. O. J. 1983. Gen. Rel. Grav. **15.** (In press.)

[a]Present affiliation: Department of Mathematics, Statistics and Computing Science, Dalhousie University, Halifax, Nova Scotia, Canada B3H 4H8.

A Single Expression for the General Solution of Einstein's Vacuum and Electrovac Field Equations with Cosmological Constant for Petrov Type D Admitting a Nonsingular Aligned Maxwell Field[a]

R. DEBEVER

Department of Mathematics
Free University of Brussels
B-1050 Brussels, Belgium

N. KAMRAM AND R. G. McLENAGHAN

Department of Applied Mathematics
University of Waterloo
Waterloo, Ontario, Canada N2L 3G1

A single expression is given for the general solution of Einstein's vacuum and electrovac field equations with cosmological constant for Petrov type D admitting a nonsingular aligned Maxwell field satisfying the generalized Goldberg-Sachs theorem from which all the particular cases until the present known in partial versions may be deduced.

[a]To appear in *Physics Letters A.*

Killing Tensors and the Geodesic Deviation Equation

P. DOLAN

Department of Mathematics
Imperial College
London SW7 2AZ, England

J. L. SAFKO

Department of Astronomy
University of South Carolina
Columbia, South Carolina 29208

A recently derived formula of Caviglia, Zordan, and Salmistraro (1982) relates Killing tensors of order ρ generating ρ^{th}-fold integrals of the geodesic equations to solutions of the geodesic deviation equation. Applications are made of this to quadratic Killing tensors on the triaxial ellipsoid and on the Kerr space-time to give some solutions for geodesic deviation not previously stated in the literature.

When geodesics are obtained from a Hamilton-Jacobi equation, a reformulation of the above CZS formula gives a new solution for geodesic deviation in terms of the action S and the ρ-fold Killing tensor.

Gravitational Lenses Producing a Large Number of Images[a]

CHARLES C. DYER

Department of Astronomy
Scarborough College
University of Toronto
Toronto, Ontario, Canada M1C 1A4

The problem of finding all the images produced by a gravitational lens consisting of many spherical deflectors with distributed masses is considered. A modification is described for Muller's zero finding algorithm for complex functions to allow the consideration of complex nonanalytic functions as arise in this problem. This is applied to specific cases of these lenses, and the effects of changing parameters for these lenses are considered, for example where there are five deflectors giving rise to 21 distinct images.

[a]Dyer, C. C. A modified Muller routine for finding the zeroes of a non-analytic complex function. J. Comput. Phys. (In press.)

Implications of an Alternative Relativistic Transport Theory

JOHN J. DYKLA

Department of Physics
Loyola University
Chicago, Illinois 60626

ANTHONY E. HWANG

Bell Laboratories
Holmdel, New Jersey 07733

We suggest some implications of an alternative to standard general relativistic transport theory based on some recent work in differential geometry which shows a natural way to make the phase space of kinetic theory fully Riemannian. While preserving all the successes of orthodox equilibrium statistical mechanics, this approach produces a new Boltzmann equation for evolving distributions of particles. Physical consequences of the new model include the possibility of a crucial experiment based upon supernovae light curves. Still more speculatively, application to conditions in the very early universe should lead to predictions quite distinct from those of other approaches that have been considered.

Loss of Angular Momentum in Binary System Due to Collisionless Particles as Monopoles, Gravitinos, or Mirror Particles

Does It Exceed the Gravitational Radiation Emission in Binary System PSR 1913 +16?

DANIELE FARGION

National Institute of Nuclear Physics—Rome Section
G. Marconi Institute of Physics
University of Rome
Rome, Italy

Dimopoulos *et al.*[1] have recently shown that the Cabrera magnetic monopole detection[2] may be consistent with the Hooft-Polyakov monopole masses[3,4] as well as with the present known cosmological[5,6] and astrophysical[7] limits, only in terms of an orbiting cloud of monopoles in the solar system, which is replenished by a small monopole wind from the sun or from the Galaxy.

The solar source of magnetic monopoles implies a solar monopole number density of $n \sim 10^{-7}$ cm^{-3}, corresponding to a number density on Earth of $n \sim 10^{-15}$ cm^{-3} for a slow flux of monopoles with typical velocities ~30 km sec^{-1}.[1]

Here we consider the gravitational drag[8,9] for a massive object moving in the (almost) collisionless fluid made by magnetic monopoles: the drag mechanism may be relevant to the loss of angular momentum in binary systems;[10] in particular for a known binary system in our galaxy associated with pulsar PSR 1913 + 16,[11] we show that a monopole number density larger than 10^{-7} cm^{-3} (as the solar one) may induce an energy loss comparable to or even larger than the one due to the gravitational radiation of the system.[12]

ACKNOWLEDGMENTS

The author wishes to thank Prof. M. Virasoro for very helpful discussions and useful suggestions, and Dr. G. Ponte for help in numerical calculation.

REFERENCES

1. Dimopoulos S., S. L. Glashow & E. M. Purcell. 1982. Wilczek Nature **298:** 824.
2. Cabrera, B. 1982. Phys. Rev. Lett. **48:** 1378–1381.
3. t'Hooft, G. 1974. Nucl. Phys. **B79:** 276–284; 1976. BIOS: 538–547.
4. Polyakov, A. M. 1974. Pis'ma Exp. Teor. Fiz. **20:** 430–433; 1974. JEPT Lett. **20:** 194–195.
5. Preskill, J. 1979. Phys. Rev. Lett. **43:** 1365.
6. Lazarides, G., Q. Shafi & T. F. Walsh. 1981. Phys. Lett. **100B:** 21–24.

7. Parker, E. N. 1964. Astrophys. J. **139:** 951–958.
8. Ruderman, M. A. & E. A. Siegel. 1971. Astrophys. J. **165:** 1.
9. Rephaeli, Y. & E. Salpeter. 1980. Astrophys J. **240:** 20.
10. Fargion, D. 1981. Lett. Nuovo Cimento **31:** 2, 49.
11. Kolb, E. W., S. Colgate & J. Harvey. 1982. (Preprint.)
12. Fargion, D. 1983. Lett. Nuovo Cimento **36:** 449.

A Camera for Very High Energy Gamma-Ray Astronomy

D. J. FEGAN, D. McLAUGHLIN, J. CLEAR,
M. F. CAWLEY, AND N. A. PORTER

Department of Physics
University College
Dublin 4, Ireland

K. GIBBS AND T. C. WEEKES

Harvard-Smithsonian Center for Astrophysics
Cambridge, Massachusetts 02138

R. C. LAMB

Department of Astronomy
Iowa State University
Ames, Iowa 50011

P. GORHAM AND V. J. STENGER

Department of Astronomy
University of Hawaii
Honolulu, Hawaii 96822

In the last dozen years, very high energy gamma-ray astronomy has achieved a sensitivity sufficient to observe four sources emitting TeV photons. These sources, the Crab and Vela pulsars, Cen A with its active galactic nucleus, and Cygnus X-3, already show an interesting diversity, and suggest that this range of the electromagnetic spectrum will have much to teach us about the universe's high-energy processes as the sensitivity of instrumentation is improved.

Detectors in this energy regime sense the Cerenkov light emitted by the air showers created by the primary gamma rays. The principal limit to sensitivity is the background from cosmic ray air showers, which is generally many times the signal rate. One promising and previously untested way to reduce this background is to image each Cerenkov air shower event. Calculations indicate that with two-dimensional imaging, the angular resolution (and the sensitivity) of TeV photon detectors can be improved substantially.

At the Fred Whipple Observatory, Mt. Hopkins, Arizona, the focal plane of the 10-m optical reflector has recently been fitted with an array of high-speed phototubes acting as a Cerenkov image camera. Fast digitization techniques are used to give a 19 pixel image of each Cerenkov light air shower (full field of 2.5°). Initial operation of the camera has demonstrated the feasibility of this approach, and the camera will be expanded to 37 phototubes (3.5° full field) in 1983.

REFERENCE

1. FEGAN, D. J., D. MCLAUGHLIN, J. CLEAR, M. F. CAWLEY & N. A. PORTER. 1983. Nucl. Instrum. Methods **211:** 179–191.

Program and Results of the Homestake Muon, Neutrino, and Proton Decay Experiment[a]

ERVIN J. FENYVES

Physics Department
University of Texas at Dallas
Richardson, Texas 75080

The research program of the Homestake Muon, Neutrino, and Proton Decay Experiment is the following:

1. Search for nucleon decay with the deep underground 300-ton water Cerenkov detector and with a planned 1,400 ton liquid scintillation detector.
2. Search for massive magnetic monopoles with the 140-ton liquid scintillation detector.
3. Search for neutron-antineutron conversion by detecting neutron-antineutron annihilation in nuclei.
4. Search for bursts of neutrinos and antineutrinos emitted in the final gravitational collapse of massive stars into neutron stars or black holes.
5. Measurement of the angular distribution, multiplicity, and transverse momentum distribution of cosmic ray muons deep underground.
6. Study the oscillation of neutrinos produced in the atmosphere by cosmic rays as they pass through the earth (10^2 to 10^4 km).
7. Study the composition and energy spectrum of extensive air shower generating cosmic ray primaries in the 10^{14}–10^{16} eV energy range with the surface-underground telescope consisting of the 130 m^2 liquid scintillation detector located at 4,200 m.w.e. depth and an extensive air shower electron array extending to about 1 km^2 at the surface.
8. Search for localized sources of cosmic rays utilizing the very fine angular resolution of the surface-underground telescope.

The initial results of the Homestake collaboration are the following:

1. The nucleon total lifetime was found to be $T_N > 3 \times 10^{31}\ B_\mu$ yr where B_μ is the model-dependent probability of producing and detecting a stopping and decaying muon following the nucleon decay event. In the SU(5) grand unified model, calculations on B_μ vary from ~5% to 10%. Thus the nucleon lifetime limits are $T_N > (1.5–3) \times 10^{30}$ yr.
2. The 300-ton water Cerenkov detector has sensitivity sufficiently high to observe 10 MeV neutrinos and antineutrinos from collapsing stars releasing 10^{53} erg

[a]Brookhaven National Laboratory–University of Pennsylvania–University of Texas at Dallas–University of Leeds–South Dakota School of Mines and Technology collaboration.

neutrino energy at a distance of 10 kpc. No neutrino bursts have been seen over a running time of about 400 days.

3. The decoherence curve of muons and multiple muon rates were measured at a depth of 4,200 m.w.e. The average transverse momentum of muons in the range $\sqrt{s} = 400$–600 GeV was found to be consistent with $\langle p_t \rangle = 500$ MeV/c.
4. Preliminary Monte Carlo simulations of extensive air showers to be detected by the surface-underground telescope showed that for primary energies above 10^{15} eV there is a good chance to clearly distinguish between proton and iron primaries by measuring the number of electrons and high-energy muons in the shower simultaneously.

Quasars as Supermassive Binaries

C. MARTIN GASKELL[a]

Institute of Astronomy
Cambridge, CB3 0HA, England

The peaks of the broad and narrow emission lines are shown to differ significantly in red shift in almost all active galactic nuclei. It is argued that the narrow lines are yielding the "true" red shift to much better than 100 km s^{-1} in almost all cases. No narrow forbidden line emission has been detected at the broad emission line red shift. Relative blue shiftings and red shiftings of the broad lines appear to be equally common, thus presenting serious difficulties for previous explanations involving anisotropic emission from radially moving clouds. I propose instead that the displaced broad-line peaks are the result of orbital motion. Since the observed velocities are very high (~3,000 km s^{-1}) and the sizes of the broad-line regions are known roughly (~1 pc), the masses involved must be very large (~10^9 $M_\odot$). A number of objects show *two* broad-line peaks (always one blue shifted and one red shifted), but there is no known object with *more* than two broad-line peaks. The observed velocities of the peaks are consistent with the expected evolution of supermassive binaries. The binary will be sped up by encounters with stars until the orbital velocity is significantly greater than the stellar velocity dispersion (~300 km s^{-1}). On the other hand the orbital velocity has to be significantly less than the speed of light or else gravitational radiation losses will cause the orbit to decay extremely rapidly. If the displaced broad-line emission peaks are due to dense gas associated with one or both of the massive objects the orbital period is sufficiently short that radial velocity changes should be noticeable in a few decades. It is interesting to note that the most extreme examples of displaced broad emission lines tend to be extended radio structure objects.

[a]Present affiliation: Department of Astronomy, University of Texas, Austin, Texas 78712.

Broad-Line Clouds in Active Galactic Nuclei

Physical Conditions and Motions

C. MARTIN GASKELL[a]

Institute of Astronomy
Cambridge, CB3 0HA, England.

The equivalent width (i.e., relative strength) of the broad Hβ emission line in Seyfert I galaxies is shown to depend on the velocity width of the lines, being very weak for objects with narrow lines. The equivalent width of the FeII λ4570 blend, on the other hand, is independent of velocity width. Simple calculations show that the decrease in equivalent width of Hβ is due to *either* slightly higher densities *or* slightly larger optical depths in the partially ionized zones of the broad-line clouds of the narrower-line Seyfert I galaxies. Hβ is collisionally suppressed at much lower densities or optical depths than the optical FeII lines are. The correlation of Hβ emissivity with cloud velocity provides strong support for the radiative acceleration models developed by Mathews and Blumenthal. It is suggested that the recent IUE satellite observations of low CIII]λ1909/CIVλ1549 ratios in Seyfert I galaxies (compared with high red shift quasars) are due to higher densities rather than high ionization parameters. It is predicted that this ratio should be higher in objects lacking strong FeII emission, such as radio galaxies.

[a]Present affiliation: Department of Astronomy, University of Texas, Austin, Texas 78712.

Curvature Coordinates in Cosmology

RONALD GAUTREAU

Department of Physics
New Jersey Institute of Technology
Newark, New Jersey 07102

We show how to develop cosmological theory from first principles starting with curvature coordinates (R,T) in terms of which the metric for the universe takes the form[1]

$$ds^2(R,T) = \frac{dR^2}{A(R,T)} + R^2 d\Omega^2 - B(R,T)dT^2 \quad \textbf{(1)}$$

We solve the timelike geodesic equations of (**1**) for radially moving particles, and show how to replace T with a new time coordinate τ that is equal to the time measured by radially moving geodesic clocks. We then solve the field equations with cosmological constant Λ in the (R, τ) coordinate system for a source of the universe that is a perfect fluid composed of radially moving geodesic particles to get the metric coefficients in terms of the pressure P and density ρ of the fluid. The end result is a metric for the universe having the form

$$ds^2(R, \tau) = [dR - m(1 - A)^{1/2}d\tau]^2 + R^2 d\Omega^2 - d\tau^2, \quad m = \pm 1 \quad \textbf{(2)}$$

where $m = +1$ or -1 depending respectively on whether the cosmological fluid particles are moving in the sense of increasing or decreasing R; A is given by

$$A(R, \tau) = A(R,T) = 1 - (8\pi\rho/3)R^2 - (\Lambda/3)R^2 \quad \textbf{(3)}$$

and P and ρ are related by

$$P + \rho + m(24\pi\rho + \Lambda)^{-1/2} d\rho/d\tau = 0 \quad \textbf{(4)}$$

From (**2**) it is seen that the metric on the subspace τ = constant is equal to $dR^2 + R^2 d\Omega^2$, and so is flat, with R having the physical significance that it is a measure of proper distance in this subspace. We demonstrate how to apply the curvature formalism to the de Sitter and Einstein–de Sitter universes. We also show how to use the curvature formalism to develop a method for imbedding a Schwarzschild mass into cosmology.[2] In the imbedding procedure, we solve the field equations for a stress-energy tensor that corresponds to a radially moving geodesic fluid outside the boundary R_b of the imbedded mass, while inside R_b the stress-energy tensor is composed of a perfect fluid part and a Schwarzschild matter part. We give specific examples of imbedding a Schwarzschild mass into a de Sitter universe and an Einstein–de Sitter universe. A consequence is that there will be spiraling of planetary orbits around a mass imbedded in a nonstatic universe. We relate this work to recent work done by Dirac with regard to his large-numbers hypothesis.

REFERENCES

1. GAUTREAU, R. 1984. Phys. Rev. **D29:** 181.
2. GAUTREAU, R. 1984. Phys. Rev. **D29:** 193.

Variability of SS433 and Extragalactic Compact Radio Sources

B. J. GELDZAHLER, J. H. SPENCER, K. J. JOHNSTON,
AND E. B. WALTMAN

E. O. Hulburt Center for Space Research
Naval Research Laboratory
Washington, D.C. 20375

W. J. KLEPCZYNSKI, P. E. ANGERHO FER,
D. R. FLORKOWSKI, D. D. McCARTHY,
AND D. N. MATSAKIS

United States Naval Observatory
Washington, D.C.

We present 3.5 years worth of results from our on-going program of monitoring daily flux densities of selected galactic and extragalactic compact radio sources at 2695 and 8085 MHz (11.1 and 3.7 cm, respectively). SS433 is rarely found to be in a "quiescent" state. It varies sporadically and dramatically, yet its spectral index remains fairly constant at ($S \propto \nu^{\alpha}$), $\alpha \sim -0.6$. Harmonic analysis on data obtained during one quiescent period reveals a periodicity of 6.5 days. No other periodicity was seen at any time. BL Lac also varies aperiodically although during times of both rise and fall of major outbursts the spectral index exhibits flickers on a time scale of 50–100 days. The rapid variations suggest a Doppler beaming factor of 10–40 to keep the brightness temperature below 10^{12} K and, thus, avoid Compton cooling.

Other sources on our observing program that were not displayed were discussed privately.

The Gravitationally Bound Current Loop Model for Quasars and Galactic Nuclei

HOWARD D. GREYBER
Naval Intelligence Support Center
Washington, D.C. 20390

This is a distinct physical model for the "engine" that powers quasars and active galactic nuclei. A large-scale dipole magnetic field (whose axis is the rotation axis) is produced by a very intense relativistic current loop around the center of the galaxy or quasar. The toroidal plasma surrounding the loop is bound to it by the Maxwell "frozen-in" field condition, and, in turn, the toroidal mass is gravitationally bound to the central core object, probably a black hole. The bursting force of the strong dipole magnetic field, plus the effect of rotation, balances the force of gravity and produces an equilibrium state. (References 1 to 4 elucidate features of the model.)

The extreme alignment of jets and blobs in radio galaxies, and the acceleration of plasma are explained simply and directly by the "toothpaste tube" effect of an increasing dipole magnetic field on plasma. The double-lobe radio sources are explained by the morphology of the dipole magnetic field anchored in the optical galaxy. The very rapid time variations observed in some quasars and galactic nuclei, such as the hundred-second variability in the x-ray flux of quasar 1525+227,[5] are much more easily explained than in the conventional black hole accretion disk model. In the latter, the characteristic time for a significant variation is a few times the light travel time across the diameter of the black hole. In the GBCL model, it is a few times the *small* diameter of the current loop times the velocity of light, a much smaller quantity.

Some other evidence that the GBCL model helps to explain includes the magnetic field configuration in galactic spiral arms, the formation of stars in the galactic plane, the well-known fact that spiral galaxies dominate in loose clusters of galaxies while ordinary ellipticals dominate the tight clusters, and the puzzling rapid structural variations on the milliarcsecond scale of 3C 120 (see Reference 6).

In our model accretion may be occurring, of course, but the dynamics are dominated, in general, by the gravitationally bound current loop and its galactic dipole magnetic field. In the GBCL model, the general morphology of objects of galactic dimension is determined by the ratio of magnetic field energy to gravitational energy, with rotational energy secondary. It is important to note the conclusion of W. G. Mathews regarding quasars: "Rotation is a most unlikely source of emission line broadening. . . . In fact, there is no direct and unambiguous evidence for rotation in the observed spectra of quasars at the present time."[7] If Mathews is correct, the rotating accretion disk model of Rees, Blandford, *et al.* must be rejected but our GBCL model is unaffected and wholly viable.

REFERENCES

1. GREYBER, H. D. 1967. Publ. Astron. Soc. Pac. **79:** 341.
2. GREYBER, H. D. 1966. *In* Proceedings of the 14th International Astrophysics Symposium: 189, 197. Astrophysics Institute. Cointe-Sclessin, Belgium.
3. GREYBER, H. D. 1965. *In* Proceedings of the First Texas Symposium on Relativistic Astrophysics: 389. University of Chicago Press. Chicago, Ill.
4. GREYBER, H. D. 1962. On the Steady State Dynamics of Spiral Galaxies. Report No. 2958. U.S. Air Force Office of Scientific Research. Washington, D.C.
5. MATILSKY, T., C. SHRADER & H. TANANBAUM. 1982. Astrophys. J. Lett. **258:** L1.
6. WALKER, R. C., G. SEIELSTAD *et al.* 1982. Astrophy. J. **257:** 56.
7. MATHEWS, W. G. 1982. Astrophys. J. **258:** 425.

Possibility of Detecting the Magnetic Field of a Rotating Neutron Star with SQUID[a]

J. GRYGAR, M. ODEHNAL, AND V. PETŘÍČEK

Institute of Physics
Czechoslovak Academy of Sciences
CS-250 68 Řež, Czechoslovakia

K. PRIKNER

Geophysical Institute
Czechoslovak Academy of Sciences
CS-141 31 Praha, Czechoslovakia

Ultra-low-frequency (below 3 Hz) magnetic dipole radiation of a highly magnetized rotating neutron star produces magnetic induction of the order 10^{-14} to 10^{-15} T in the distance of 50 to 500 parsecs from the source. Although this ULF radiation has a frequency well below that of the estimated interstellar plasma, it may propagate in a whistler mode along suitably aligned magnetic lines of force in the interstellar medium. Moreover, the earth's ionosphere does not attenuate signals with frequencies lower than about 5 Hz. Thus the magnetic field of a neutron star with rotational period longer than about 0.3 seconds might well be detected with the sensitive SQUID (for superconducting quantum interference device) magnetometers. Radio pulsars or gamma-ray bursters are the most obvious candidates for the detection.

[a]Full paper submitted to the *Bulletin of the Astronomical Institutes of Czechoslovakia.*

High-Energy Gamma Rays from Pulsars

S. K. GUPTA, P. V. RAMANA MURTHY,
B. V. SREEKANTAN, S. C. TONWAR,
AND P. R. VISWANATH
Tata Institute of Fundamental Research
Bombay-400 005, India

Pulsed γ-ray emission at high energies, 10^7–10^9 eV, has been observed with satellite-borne detectors for Crab and Vela pulsars. γ-Ray luminosity of these pulsars is higher than in all other spectral ranges. There is considerable interest in determining the γ-ray energy spectra for these pulsars at ultra high energies, 10^{11}–10^{13} eV, since these high-energy photons carry a lot of information on the radiation mechanism and the pulsar magnetosphere. The ground-based atmospheric Cerenkov technique is ideally suited for the study of ultra-high-energy γ-rays since the effective γ-ray detection area is many orders of magnitude larger than the physical area of the Cerenkov radiation detector itself. The Cerenkov radiation telescope, CRT, operating at Ooty, 11.4° N and 2.2 km altitude, in India since 1977 has detected[1,2] pulsed emission at energies $\sim 10^{12}$ eV from the Crab and Vela pulsars with light curves showing a double peak structure as at lower energies. The Ooty CRT, composed of 18 reflectors, has been modified to a "distributed mode" for observations in early 1981 to increase its sensitivity through a significant improvement in its angular resolution using a relative timing method. New observations in 1981 have been analyzed recently and again show positive signals for both Crab and Vela pulsars. It is seen that the pulsed flux from both pulsars is highly variable over tens of minutes. Our observations show that the measured pulsed flux of $2.2 \pm 0.5 \times 10^{-11}$ photons cm^{-2} s^{-1} at energies $>9 \times 10^{11}$ eV for the Crab pulsar during the active phase agrees well with the extrapolation of the COS-B spectra, 2.9×10^{-8} $E_{GeV}^{-1.17}$ photons cm^{-2} s^{-1}, both in amplitude and spectral slope. The time-averaged flux is however considerably smaller. On the other hand the time-averaged pulsed flux of $1.0 \pm 0.4 \times 10^{-11}$ photons cm^{-2} s^{-1} at energies $>2.3 \times 10^{12}$ eV for the Vela pulsar is about 2 orders of magnitude smaller than the extrapolated value from the COS-B spectrum, 1.8×10^{-7} $E_{GeV}^{-1.11}$ photons cm^{-2} s^{-1}. Also the energy spectrum for the Vela pulsar at energies ~ 2–5×10^{12} eV with a slope of -2.2 ± 0.2 is much steeper than at lower energies. These observations suggest significant differences between γ-ray emission from these two pulsars. More observations are continuing to confirm these interesting results with better statistics.

REFERENCES

1. GUPTA, S. K., *et al.* 1978. Astrophys. J. **221:** 268.
2. BHAT, P. N., *et al.* 1980. Astron. Astrophys. **81:** L3.

The Evolution of Voids in the Expanding Universe

MARK A. HAUSMAN,[a,b] DONALD W. OLSON,[a,c]
AND BERNARD D. ROTH[d]

We examine the growth and nonlinear evolution of spherically symmetric, negative density perturbations in an otherwise uniform Friedmann cosmology. In general, the cavity expands and its density relative to the background decreases. In certain cases, the cavity comes to be surrounded by a narrow shell of matter with a density substantially greater than the mean cosmological value; we show that such an overdense shell can form even when no overdense region is present anywhere in the initial density profile. These "walled-in" cavities may be identified with recently observed, large-scale holes in the galaxy distribution, and the shells themselves with the superclusters that define the boundaries of such holes.

However, we also show that there exist choices for the initial density profile that lead to the formation of deep cavities which are not surrounded by any overdense shells.

We examine the behavior of the shell-growth phenomenon under various assumptions about the background universe and the initial density perturbation, and find that shell growth is enhanced for high values of the density parameter Ω, positive (repulsive) cosmological constant, steep density profile of the initial perturbation, large-amplitude perturbation, and/or early origin of the perturbation. We derive analytic criteria for classifying arbitrary initially negative density profiles to determine (a) whether shell crossing with formally infinite density will eventually occur for any given comoving mass coordinate, and (b) whether matter at adjacent mass coordinates will form a shell of finite overdensity which persists indefinitely.

[a]Department of Astronomy, University of Texas, Austin, Texas 78712.
[b]Address correspondence to Mission Research Corporation, P.O. Drawer 719, Santa Barbara, California 93102.
[c]Department of Physics, Southwest Texas State University, San Marcos, Texas 78666.
[d]Department of Physics, University of Texas, Austin, Texas 78712.

Active Galaxies and the Diffuse Gamma-Ray Background

D. KAZANAS[a] AND R. J. PROTHEROE[a,b]

Laboratory for High Energy Astrophysics
NASA/Goddard Space Flight Center
Greenbelt, Maryland 20771

A new model for the origin of relativistic particles and gamma-rays in active galactic nuclei and quasars,[1] together with recent HEAO-1 observations of the spectra of active galaxies from 2 to 165 keV,[2] provides the basis for a reexamination of the nature of the extragalactic gamma-ray background. We find that active galaxies could account for the observed background if their x-ray spectra steepen to $E^{-2.1}$ above $\sim$100 keV, as has been observed in Cen-A,[3] together with a further steepening to $E^{-2.7}$ as a result of absorption of gamma-rays by photon-photon pair production interactions with x-ray photons. The compactness of active galaxies required to give this steepening is consistent with current estimates of their typical luminosity and radius.

REFERENCES

1. PROTHEROE, R. J. & D. KAZANAS. 1983. Astrophys. J. **265:** 620.
2. ROTHSCHILD, R. E., *et al.* 1983. Astrophys. J. (In press.)
3. BAITY, W. A., *et al.* 1981. Astrophys. J. **244:** 429.

[a]Also affiliated with the Department of Physics and Astronomy, University of Maryland.
[b]Present address: Department of Physics, University of Adelaide, Adelaide, South Australia 5001, Australia.

Astrophysical Boosters

W. KUNDT

Institute for Astrophysics
University of Bonn
5300 Bonn, Federal Republic of Germany

INTRODUCTION

Our cosmic neighborhood is capable of accelerating protons and/or ions to energies in excess of 10^{20} eV—corresponding to Lorentz factors γ larger than 10^{11}—and of accelerating electrons to Lorentz factors in excess of 10^5, perhaps even 10^8. The former are observed directly as "cosmic rays" by particle detectors, or indirectly via scintillation events and air showers. The latter are inferred from the upper cutoff frequency ν_u in nonthermal spectra:

$$\nu_u = eB_\perp \gamma^2/\pi m_e c = 5.6 \times 10^2 \gamma^2 B_{-4}\,\text{Hz} \tag{1}$$

where $B_\perp$ is the transverse magnetic field strength, with $B_{-4} := B/10^{-4}$ Gauss. ν_u reaches optical frequencies in the jets of some extragalactic radio sources, like Cen A, M 87, 3C 273, NGC 1097, and several others, and also in young pulsars. What engines can achieve such high energies at reasonable efficiencies?

In a microscopic description, a charge e can only gain energy ΔE w.r.t. an inertial frame by falling through an electric field $\vec{E}$:

$$\Delta E = e \int \vec{E} \times d\vec{x} \tag{2}$$

We are not aware of large electrostatic fields in the interplanetary, interstellar, or intergalactic medium, and do not expect such to exist because of the high prevailing conductivities. But $\vec{E}$ may belong to the outgoing wave radiated by a rotating magnet, as envisaged for pulsars, or $\vec{E}$ may be the electric field $-\vec{\beta} \times \vec{B}$ seen by a charge when a magnetic field $\vec{B}$ is convected across it at velocity $\vec{v} = c\,\vec{\beta}$:

$$\Delta E = -e \int (\vec{\beta} \times \vec{B}) \times d\vec{x} \tag{3}$$

For pulsars, the highest available voltages are probably those across a certain fraction of the speed-of-light cylinder, of radius c/Ω, whence (cf. Reference 10)

$$E_{\rm PSR} \lesssim e\,B_s\,R\,(\Omega R/c)^2 = 10^{17}\,\text{eV}\,B_{13}\,\Omega_2^2 \tag{4}$$

For interplanetary shocks, field strengths $B_\perp$ of $\lesssim 10^{-4}$ G and velocities $\lesssim 10^3$ km s^{-1} lead to maximal energy gains per particle

$$E_{\rm max}^{\rm shock} \approx e\beta B_\perp\,\Delta x \approx 10^6\,\text{eV}\,(\beta B)_{-7}\,(\Delta x)_{10} \tag{5}$$

in agreement with measurements of both electrons and protons near the earth's bowshock $[(\Delta x)_{10} \lesssim 1]$ and in the corotating interaction regions of the solar wind

$[(\Delta x)_{10} \lesssim 10]$ (cf. Reference 14). In both cases, the spectra peak at energies ~ 50 times below the maximum and have exponential tails. Interstellar shocks can be more than 10 times faster; they may or may not extend through much larger distances Δx, depending on the internal geometry of a supernova shell.[15] Even in the most optimistic case, however, Equation 5 does not yield energies anywhere near the maximal ones of cosmic rays, not even anywhere near the maximal energy for galactic containment ($= 10^{19}$ eV).

It has been suggested that the highest-energy cosmic rays are of extragalactic origin, perhaps from the nuclei of active galaxies (or AGNs).[20] Difficulties of this interpretation are (1) the huge overall energy requirement, unless one restricts the claim to energies above 10^{19} eV, which are comparatively rare; (2) a monotonic increase and reorientation of the anisotropy in arrival directions,[5] even though one would sample over volumes that extend far beyond the Virgo cluster, and whose sizes change rapidly with energy; and (3) a likely rareness of high-energy ions in AGNs: the total energy radiated by high-energy electrons is already so large that ions cannot store a much larger fraction thereof, the jets may predominantly consist of electrons and positrons.[16,12] They may share this property with pulsars whose high-energy wind is thought to be mainly leptonic.[17,9]

So far, the above discussion has not implicated any astrophysical source as the booster to the highest observed energies. This lack of a clue has led theorists to revive Fermi's idea of accelerating charges in many small steps, by bouncing them off approaching walls.[1–4,7] Such walls, or shocks, are believed to exist in the form of supernova shells, strong stellar winds, and the like. The idea of tapping strong shocks—often called "*in situ* acceleration"—has become so widespread that researchers speak of "little synchrotrons drifting down the jets of extragalactic radio sources," which are conceived capable of upgrading the bulk energy of beam motion into extremely relativistic electrons (cf. Reference 18). In this way, the kinetic energy of marginally relativistic or even nonrelativistic protons is assumed to be converted into a power-law distribution of electrons, with Lorentz factors exceeding the value of 10^5, with the electron power distributed almost evenly up to the high-energy cutoff, at efficiencies reaching 30%, and possibly even in an anisotropic fashion such that we can only see the approaching jet. The existence of similar mechanisms has been claimed for laboratory plasmas.

If such were true, we should wonder why engineers build intricate machines to convert 1 MW of electric power into $\lesssim 1$ kW of electrons with Lorentz factors $\gamma \lesssim 10^4$. All they should do is blow a plasma beam into plasma in order to transform the kinetic energy of ordered proton motion into a power-law distribution of relativistic electrons, with part of the electron power stored at several hundred times the streaming energy.

ENERGY ESTIMATES

Let us look in somewhat more detail at the constraints imposed upon a strong shock in the head of an extragalactic radio source. We can estimate the injected power and spectrum $\dot{N}_{e,E}\, dE$ of the relativistic electrons from their synchrotron radiation. This power is typically $\gtrsim 1\%$ of the total radiation from the active nucleus, i.e., it corresponds to an efficiency $\eta \gtrsim 1\%$ of the central engine which generates the beams.[16] Clearly, there is not much room in the overall power budget for energetic protons or ions: their power $\dot{N}_p E_p$ should be at most comparable. In order not to be dogmatic, however, let us assume that they store $10\,\eta_{-1}$ times the electron power (with $\eta_{-1} \leqq 1$). If $c\beta_{\mp}$ is their bulk velocity (before/after) the shock, conservation of energy implies (approximately):

$$\dot{N}_p\,[m_p c^2\,(\gamma_- - 1) + 3\,kT_-] \approx \dot{N}_p[m_p\,c^2\,(\gamma_+ - 1) + 3\,kT_+/2] + \int E\dot{N}_{e,E}\mathrm{d}E \quad (6)$$

Here it has been assumed for simplicity that before the shock, electrons and protons have the same temperature T_-. Charge neutrality, in the absence of positrons, would imply $\dot{N}_p = \dot{N}_e$, hence for $3\,kT_- \ll m_p c^2\,(\gamma_- - 1)$:

$$\dot{N}\,m_p c^2\,(\gamma_- - \gamma_+ - 3\,kT_+/2\,m_p c^2) \approx \int E\,\dot{N}_E\,\mathrm{d}E \quad (7)$$

with

$$\dot{N} = \int_{E_{\min}}^{E_{\max}} \dot{N}_E \mathrm{d}E$$

Observations give $\dot{N}_E \sim E^{-g}$ for $E_{\min} \leqq E \leqq E_{\max}$ with $g = 2.2 \pm 0.2$, $E_{\min} \approx 10^2$ MeV, $E_{\max} \gtrsim 10^2$ GeV, so that the power $(E^2 \dot{N}_E)$ peaks at the lower cutoff energy, and:

$$\frac{\int E\,\dot{N}_E \mathrm{d}E}{\dot{N}} = E_{\min}\frac{g-1}{g-2}\frac{1 - (E_{\min}/E_{\max})^{g-2}}{1 - (E_{\min}/E_{\max})^{g-1}} \approx \frac{E_{\min}}{g-2} \quad (8)$$

whence from (7):

$$E_{\min} \approx (g-2)\,m_p c^2\,(\gamma_- - 1)\left[1 - \frac{\gamma_+ - 1 + 3\,kT_+/2\,m_p c^2}{\gamma_- - 1}\right] \quad (9)$$

For a nonrelativistic incoming bulk velocity $c\beta_-$, this expression yields

$$E_{\min} \lesssim \left(\frac{g}{2} - 1\right) m_p c^2 \beta_-^2 \approx 10^2\ \mathrm{MeV}\ \beta_-^{\,2} \quad (10)$$

Comparison with the "observed" $E_{\min}$ implies that either $\beta_- \approx 1$, in mild conflict with the assumption of a nonrelativistic incoming flow, or else there must be a large number of low-energy electrons in addition to the high-energy power law. In both cases, most of the electrons have lower energies than the (shocked) protons, which causes problems with the second law of thermodynamics.

ENTROPY ESTIMATES

In order to see this, remember that the μ-space entropy S of a homogeneous system of N classical particles reads

$$S/Nk = 1 - \int f \ln f\,\mathrm{d}^3p/nh^3 \quad (11)$$

where the phase space density $f = f(\vec{p})$ is normalized according to $\int f\, d^3p = nh^3$, and $n := N/\text{volume}$. Call $s := S/Nk$ the (dimensionless) "entropy per particle." A nonrelativistic Maxwell distribution f (of temperature T) has

$$s_{\rm th} = 5/2 - \ln (n\,\lambda^3) \qquad \text{with} \qquad \lambda := h/(2\pi mkT)^{1/2} \tag{12}$$

a supersonic bi-Maxwellian beam has

$$s_{\rm b} = 5/2 - \ln (n\,\lambda_\parallel\,\lambda_\perp^2) \qquad \text{with} \qquad \lambda_{\rm j} := h/(2\pi\, mkT_{\rm j})^{1/2} \tag{13}$$

and an isotropic truncated relativistic power-law distribution with $f \sim p^{-g-2}$ for $p_{\min} \leqq p \leqq p_{\max}$, $E = pc$, $g > 1$:

$$s_{\rm p} \approx (2g + 1)/(g - 1) - \ln (n\, l^3\, (g - 1)/4\pi) \qquad \text{with} \qquad l := hc/E_{\min} \tag{14}$$

The particle density n in the beam of an extragalactic radio source is of order $n = L/A\beta cE \approx 10^{-7}\ \text{cm}^{-3}/\beta$ where L = power, A = beam cross section, and E = typical particle energy; $\lambda_{\rm p} = 10^{-13}\ \text{cm}/T_{12.5}^{1/2}$, $l = 10^{-12}\ \text{cm}\ (10^2\ \text{meV}/E)$. We thus get

$$\left.\begin{aligned} s_{\rm th} &= \ln(10^{46}\, T_{12.5}^{3/2}) + 5/2 = 108 \\ s_{\rm b} &= \ln(10^{43}\, T_{10.5}^{3/2}) + 5/2 = 102 \\ s_{\rm p} &= \ln(10^{42}\, E_{-4}^{3}) + 9/2 = 101 \end{aligned}\right\} \tag{15}$$

for the respective entropies per particle. These numbers depend somewhat on one's preferred values for the densities, temperatures, and velocities before and after shocking, but satisfy the strict inequalities $s_{\rm b}$, $s_{\rm p} < s_{\rm th}$. The comparatively low entropy of a beam is due to its low temperature, and that of a soft ($g > 2$) power-law distribution (of comparable total power, Equations 9 and 10) is due to the preponderance of cold particles ($E \approx E_{\min}$).

For the joint system of protons and electrons, the average entropy per particle $s_{\mp}$ (before/after) shocking depends on the relative numbers ν of thermalized ($\nu_{\rm th}$), power-law ($\nu_{\rm p}$), and cold ($\nu_{\rm c}$) particles:

$$\left.\begin{aligned} 2\,s_- &= (1 + \nu_{\rm th}^{\rm e})\, s_{\rm b} + (1 - \nu_{\rm th}^{\rm e})\, s_{\rm p} \\ 2\,s_+ &= \nu_{\rm th}\, s_{\rm th} + (1 - \nu_{\rm th} + \nu_{\rm p}^{\rm e})\, s_{\rm p} + (1 - \nu_{\rm p}^{\rm e})\, s_{\rm c} \end{aligned}\right\} \tag{16}$$

with $\nu_{\rm p}^{\rm e} := \nu_{\rm p}$ for electrons, $0 < \nu_{\rm j} < 1$, $s_{\rm c} \lesssim s_{\rm p}$. Proponents of cosmic-ray production via shocks want $\nu_{\rm p}^{\rm p} \approx \nu_{\rm p}^{\rm e}$. The power budget in extragalactic radio sources wants $\nu_{\rm th}$ to be small. The second law of thermodynamics demands $s_- < s_+$ and is not automatically satisfied: it needs a sufficient number of thermalized protons ($\nu_{\rm th}$ not too small) and/or a sufficiently low temperature T_- ($\approx T_\parallel$) of the incoming beam.

However, a growth of entropy is not the only condition imposed by nonequilibrium thermodynamics. In unconstrained situations, the entropy approaches its thermal equilibrium value $S_{\rm th}$ in a few response times of the system; exceptions are open systems and systems with hurdles ("forbidden" reactions). In the present case, s_+ (in Equation 16) is much smaller than $s_{\rm th}$. Most electrons would have to stay colder than most of the protons whereas few of them would have to undergo very many interactions, in order to cascade in energy up to excessive values. A single fall through a potential drop would not lead to a power-law distribution, as demonstrated by the earth's bow shock. Many falls are expected to achieve thermalization. I cannot see the constraints in the thin shock layer of an extragalactic radio source that would largely remodel the distribution functions but arrest the entropy halfway before its thermal equilibrium value, implying a conversion of mechanical energy into extremely relativistic charges at an efficiency of 10% or higher.

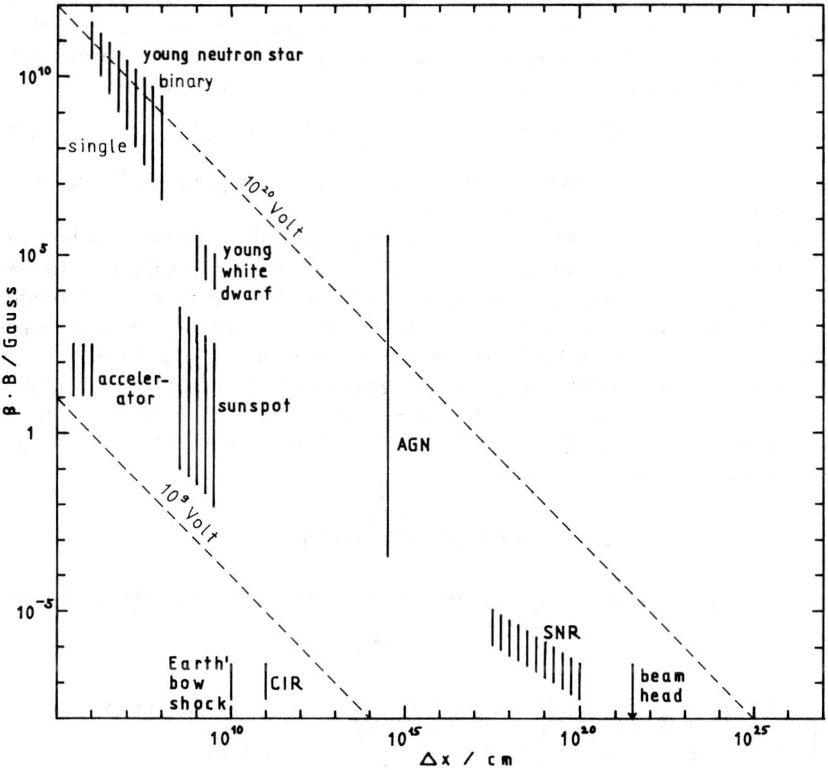

FIGURE 1. Estimated βB/Gauss versus Δx for astrophysical boosters (see Equations 3–5 and 16). $\beta = v/c$ ranges from 1 (for wavelike phenomena) through $\Omega R/c$ (for rotators) down to some 10^{-3} (for old SNRs), and is often quite uncertain (e.g., for sunspots). The largest uncertainty rests in the distance Δx that particles can traverse during the acceleration event; for instance, the entries "SNR" and "beam head" may have to be moved to the left by four and seven scale units respectively (factors 10^4 and 10^7 in Δx). The broken lines show voltages $\Delta E/e$. A similar diagram is contained in Hillas.[5]

NEUTRON STARS AND AGNs

It is my conviction that Nature has found different solutions to the relativistic acceleration problem (cf. FIGURE 1). We know of the existence of fast-spinning, magnetized neutron stars. Their strong outgoing waves involve potential drops which are a factor of order 10^{10} above those of interstellar shocks (cf. Equations 4 and 5 above) and which may well be responsible for most of the high-energy cosmic-ray electrons and positrons. Quite likely, the central engines in the active galactic nuclei are likewise magnetized rotators and act like giant pulsars.[8,12] If they can generate the observed extremely relativistic $e^{\pm}$-flows on the innermost scale of $10^{15\pm1}$ cm, there would be no need for any downstream postacceleration.

We also know of the existence of magnetized neutron stars in binary systems where the weak but nevertheless heavy wind of a companion star can quench the pulsar mechanism, impinge upon the corotating magnetosphere, and be flung out like sparks from a grindstone. When confining the magnetosphere deep inside the speed-of-light cylinder, the falling matter can (in principle) generate voltages of order

$$E_{\text{grind}} \leqq 2e\,Br = 2e\,B_s\,R^3/r^2 = 10^{20}\text{ eV } B_{13}/r_7^2 \tag{17}$$

which reach high enough to explain the observed cosmic-ray spectrum. This potential cosmic-ray booster has been recently discussed.[14]

A final word concerns SS 433. Its UV output is difficult to estimate, but both the observed fluxes and the Eddington limit suggest that the central engine radiates less than 10^{39} erg s^{-1}. At an efficiency η of order 1%, this central engine should not be able to produce beams more powerful than some 10^{37} erg s^{-1}. This consideration stands against the hydrogen beam model, in favor of extremely relativistic $e^{\pm}$-beams.[11] I even maintain that details of Hjellming and Johnston's radio spiral[6] and Seward *et al.*'s x-ray map[19] can be better understood in terms of the "soft-beam" model (hair drier)[13] than in terms of the "hard-beam" model (lawn sprinkler). If so, SS 433 would follow a pattern similar to that of the AGNs.

ACKNOWLEDGMENTS

I thank Axel Jessner for a discussion and Reinhard Schlickeiser for critical remarks.

REFERENCES

1. Axford, W. I., E. Leer & G. Skadron. 1977. Proc. Int. Cosmic Ray Conf. **11:** 132.
2. Bell, A. R. 1978. Mon. Not. R. Astron. Soc. **182:** 147.
3. Blandford, R. D. & J. P. Ostriker. 1978. Astrophys. J. **221:** L29.
4. Drury, L. O'C. 1983. Rep. Prog. Phys. **46:** 973.
5. Hillas, A. M. 1982. *In* Composition and Origin of Cosmic Rays. M. M. Shapiro, Ed.: 125. Reidel. Dordrecht, the Netherlands.
6. Hjellming, R. M. & K. J. Johnston. 1981. Astrophys. J. **246:** L141.
7. Krymsky, G. F. 1977. Dokl. Akad. Nauk USSR **234:** 1306.
8. Kundt, W. 1979. Astrophys. Space Sci. **62:** 335.
9. Kundt, W. 1980. Ann. N.Y. Acad. Sci. **336:** 429.
10. Kundt, W. 1981. Astron. Astrophys. **98:** 207.
11. Kundt, W. 1981. Vistas Astron. **25:** 153.
12. Kundt, W. 1982. Int. Astron. Union Symp. **97:** 265.
13. Kundt, W. 1982. Mitt. Astron. Gesell. **57:** 65.
14. Kundt, W. 1983. Astrophys. Space Sci. **90:** 59.
15. Kundt, W. 1983. Astron. Astrophys. Lett. **121:** L15.
16. Kundt, W. & Gopal-Krishna. 1980. Astrophys. Space Sci. **75:** 257.
17. Kundt, W. & E. Krotscheck. 1980. Astron. Astrophys. **83:** 1.
18. Rees, M. J. 1982. Int. Astron. Union Symp. **97:** 211.
19. Seward, F., J. Grindlay, E. Seaquist & W. Gilmore. 1980. Nature **287:** 806.
20. Shapiro, M. M. & R. Silberberg. 1983. Astrophys. J. **265:** 570.

A Solar Breeder to Explain Constant Luminosity and Lack of Neutrinos

L. H. LEVINE AND L. M. LIBBY

Department of Astronomy
University of California
Los Angeles, California 90024

A uranium breeder supplying energy in the sun runs for billions of years. Instead of solar neutrinos, it produces antineutrinos from fission products and actinide decays. It gives the sun constant luminosity as required by archaen biology. This follows because reactors have negative temperature coefficients. Uranium is the decay product of superheavy elements.

The existence of a superheavy of 0.5 b.y. mean life is indicated by the decay of remanent magnetism in the returned moon rocks. Thus a superheavy of 0.5 b.y. life powered the moon's juvenile magnetic field. If this radioactivity existed also in the sun, the uranium breeder would have come into existence slowly over about 2 b.y. A shell of hydrogen from neutron decay and helium from actinide decay would have collected slowly about the uranium core and begun to produce energy from fusion. Today both neutrinos and antineutrinos would be emitted. Thus the two mechanisms of fusion and fission to power the sun are symbiotic.

ACKNOWLEDGMENTS

We wish to acknowledge the assistance of C. Alexander and A. M. Perry of the Oak Ridge National Laboratory in performing some calculations related to breeding.

Three-Millimeter Anisotropy Measurement[a]

PHILIP M. LUBIN,[b,c] GERALD L. EPSTEIN,[b]
AND GEORGE F. SMOOT[b]

[b]*Space Sciences Laboratory and Lawrence Berkeley Laboratory*
University of California
Berkeley, California 94720

[c]*Joseph Henry Laboratories*
Physics Department
Princeton University
Princeton, New Jersey 08544

We have mapped the large-scale anisotropy in the cosmic background radiation at 3-mm wavelength using a liquid-helium-cooled balloon-borne radiometer sensitive enough to detect the dipole in one gondola rotation (1 minute). Statistical errors on the dipole and quadrupole components are below 0.1 mK with less than 0.1 mK galactic contribution. We find a dipole consistent with previous measurements and find no evidence for a quadrupole anisotropy with a 90% confidence level of 0.2 mK. The measurement is also useful in searching for spectral distortions.

[a]1983. *Physics Review Letters* **50:** 616.

X-Ray Spectral Variability of BL Lac Objects

DARIO MACCAGNI

Institute of Cosmic Physics, CNR
20133 Milan, Italy

TOMMASO MACCACARO

Harvard-Smithsonian Center for Astrophysics
Cambridge, Massachusetts 02138

MASSIMO TARENGHI

European Southern Observatory
D-8046 Garching/Munich, Federal Republic of Germany

The energy spectra of the BL Lac objects PKS 0548-322 and 3C66A between 0.2 and 10 KeV have been obtained at several epochs in 1979 and 1980 by combining the simultaneous imaging proportional counter and monitoring proportional counter data of the Einstein Observatory. It is found that within our data set, PKS 0548-322 shows a constant behavior and its x-ray spectrum can be represented by a power law with energy cutoff given by a hydrogen column density of 4.5×10^{20} cm^{-2}. A comparison with results from other experiments suggests a spectral variation between 1977 and 1979. In contrast, 3C66A shows an intensity decrease of about 30% on a time scale of 54 hours accompanied by a spectral steepening: the energy index changes from 2.1 ± 0.6 to 4.2 ± 0.3 while the low-energy cutoff is still represented by a hydrogen column density of about 10^{21} cm^{-2}.

An Accretion Disk Origin for the Ultraviolet Excess of Luminous Quasars

The Confrontation of Theory with Observation

MATTHEW A. MALKAN

Palomar Observatory
California Institute of Technology
Pasadena, California 91125

We have analyzed the ultraviolet, optical, and infrared spectra of luminous quasars. They are very well fitted by a combination of a power law with slope -1.1 ± 0.1, and the emission predicted from an optically thick, geometrically thin, steady-state accretion disk. The effects of general relativity on the disk structure and the propogation of light rays have been included, as they influence the results significantly. The two fitting parameters, the mass of the accreting black hole and the accretion rate, are determined by the data with a formal accuracy of 20%. The masses range from 0.2–0.5 billion $M_\odot$ for 3C 273, to 1–3 billion $M_\odot$ for the brightest quasars. The values depend on the assumed disk inclination and the angular momentum of the hole. The high luminosities of all the brightest quasars studied are not far from their Eddington limits. The ultraviolet excess in the energy distributions of Seyfert 1 galaxies can be produced by disks accreting at well below their Eddington rates.

The Unobservability of Higher-Order Moments in the Cosmic Background Radiation

DENNIS W. MARKS[a]

Physics Department
Iowa State University
Ames, Iowa 50011

The general relativistic formulation of the equation of radiative transfer is given for arbitrary velocity flows, temperature gradients, and gravitational fields. The moments of the transfer equation are used to establish that inside an optically thick gas, each part of which is in local thermodynamic equilibrium, an observer can detect only zeroth, first, and second moments. The monopole moment shows departures of the spectrum from the Planck function, while the dipole and quadrupole moments show smaller departures of the radiation from isotropy. The octopole, hexadecapole, and higher moments are essentially zero. Even if different parts of the early universe had had different local temperatures, the cosmic background radiation, which comes from the surface where the universe becomes opaque, would not show higher-order anisotropy. From the observed (near) isotropy of the cosmic background radiation, one cannot infer the (near) homogeneity of the early universe.

[a]On leave from the Department of Physics, Astronomy, and Geology, Valdosta State College, Valdosta, Georgia 36198.

A Complete Sample of Quasars at $B = 19.80$[a]

H. L. MARSHALL,[b] Y. AVNI,[b,c] A. BRACCESI,[d]
J. P. HUCHRA,[b] H. TANANBAUM,[b] G. ZAMORANI,[b,e]
AND V. ZITELLI[d]

We have completed spectroscopy for quasar candidates in a UV excess survey with $B <$ 19.80. Including the 10 previously known quasars, we find that there are 35 quasars in a 1.72 deg^2 region. None of the quasar candidates fainter than $B = 19.2$ are stars, and about 95% are quasars. About 50% of the sample is detected in x-rays by deep exposures with the Einstein Observatory. Using the x-ray data for the 35 quasars, we compute that quasars with $B < 19.80$ and $z < 2.2$ contribute $24^{+10}_{-8}\%$ of the observed x-ray background at 2 keV.

By combining the present sample with a previous sample of 22 quasars having $B < 18.25$, we find that the data can be described by a pure luminosity evolution model of $e^{7\tau}$ or $(1 + z)^{3.8}$ and a power law luminosity function $P(>L) \propto L^{-2.5}$ with very tight limits on the model parameters. These evolution models produce very good fits to the observed $N(<B)$ data and can be used to predict that quasars could comprise all of the observed 2 keV x-ray background.

[a]Research reported here used the Multiple Mirror Telescope Observatory, a joint facility of the University of Arizona and the Smithsonian Institution.

[b]Harvard/Smithsonian Center for Astrophysics, Cambridge, Mass. 02138.
[c]Weizmann Institute of Science, Rehovot, Israel.
[d]Institute of Astronomy and University Observatory, Bologna, Italy.
[e]Institute of Radioastronomy, CNR, Bologna, Italy.

Radiation from Test Charges in Yang-Mills Theories

THOMAS A. MORGAN

Physics Department
University of Nebraska
Lincoln, Nebraska 68588

It is well known that a test charge placed in a background Yang-Mills field will precess if its *iso*-charge has a component perpendicular to the *iso*-direction of the field. It is shown here that a precessing test charge radiates energy and *iso*-charge away to infinity at a constant rate. It is also shown that the test charge has a component of *iso*-charge increasing in the *iso*-direction opposite to the *iso*-direction of the background field and at such a rate that the resultant negative interaction energy exactly accounts for the energy that is being lost to infinity by radiation.

Nonlinear Density Fluctuations in a Two-Component Fluid during the Recombination Era of the Universe[a]

EKKEHARD NOWOTNY[b]

Physics Department
Autonomous Metropolitan University of Iztapalapa
CP 09340, Mexico 13, Federal District

On the basis of the hydrodynamical equations of a two-component gas (photons and hydrogen with coupling via Thomson scattering) in the recombination era of the universe (standard model), the evolution of the density perturbations up to second order are calculated. It is shown that the generated second-order amplitudes of the density fluctuations of the matter reach values of the same order as the first-order amplitudes within only one-tenth of the expansion time for fluctuations with wavelength corresponding to 10^7 $M_{\odot}$. Upper limits in the density fluctuations (for the gravitationally unstable modes) up to which first-order calculations are valid are given. These calculations indicate that the linear perturbation analysis is very restricted, especially at wavelengths near the lower limit of the Jeans length.

The linear analysis would be a good approximation only for density fluctuations of matter with a density contrast less than 10^{-5}–10^{-4} at the recombination era. Therefore a nonlinear analysis is required for making galaxies, because for this we need a density contrast of 10^{-2}–10^{-3} at the end of the recombination era.

[a]The full text of this paper has been submitted to *Astrophysics and Space Science* under the title "The evolution of nonlinear hydrodynamical density fluctuations of the photon-plasma gas during the recombination era of the universe."

[b]On leave from the Physics Faculty, University of Konstanz, Post Office Box 5560, D-7750 Konstanz, Federal Republic of Germany.

Black Hole Nuclear Cluster Model of Quasars and Active Galactic Nuclei

A. G. PACHOLCZYK[a] AND W. R. STOEGER

Vatican Observatory
Castel Gandolfo
Vatican City State

An active galactic nucleus is modeled by a cluster of black holes.[1-3] All disks are standard subcritical disks[4-6] randomly oriented to the line of sight; the accreted material comes from the galaxy in which the cluster is embedded. The mass spectrum of black holes is of the power-law type with the index β, resulting in a power-law radiation spectrum with an index $\alpha = -9 + 4\beta$. Parameters characterizing such clusters of black holes, determined from the observations,[7] are given in TABLE 1 (for the accretion rate equal to a third of the critical value).

The evolution of a cluster of black holes was estimated under the assumption that small holes (of the order of 10 solar masses with disks of 10 Schwarzschild radii) are dominated by a central large hole of mass of the order of 0.1 of the total cluster mass. The dominant processes affecting the evolution of such a cluster are the consumption of small holes by the central black hole and the evaporation of small holes from the cluster.[8,9] The computed rates of consumption (dotted lines) and evaporation (dashed lines) are summarized in FIGURE 1. The values of the total mass for this cluster model for different objects are plotted as horizontal lines since the other coordinate, the time scale of variability, is not very reliably determined at present. Velocity dispersion v_c is displayed as a solid line.

TABLE 1

Object	Observed Luminosity at 2500A (log)	Observed Spectral Index (α)	Mass Spectrum Index (β)	Number of Holes (N)	Total Mass of Cluster ($M_{tot}/M_\odot$)
PG1351+64	30.02	1.87	1.78	4.2×10^3	1.5×10^6
3CR 47	30.32	1.00	2.00	5.8×10^5	2.1×10^7
3C 279	30.56	0.97	2.01	1.0×10^6	2.0×10^8
3C 273	31.41	1.22	1.95	1.6×10^6	1.5×10^8
3CR 175	31.69	1.67	1.83	5.2×10^5	1.0×10^8
0537–286	31.83	1.00	2.00	1.9×10^7	6.7×10^8
1054–034	32.03	1.71	1.82	3.7×10^5	2.2×10^8
0207–396	32.57	1.58	1.86	3.8×10^6	5.6×10^8
2126–150	32.94	1.22	1.95	5.5×10^7	5.1×10^9

[a]Address correspondence to Steward Observatory, University of Arizona, Tucson, Ariz. 85721.

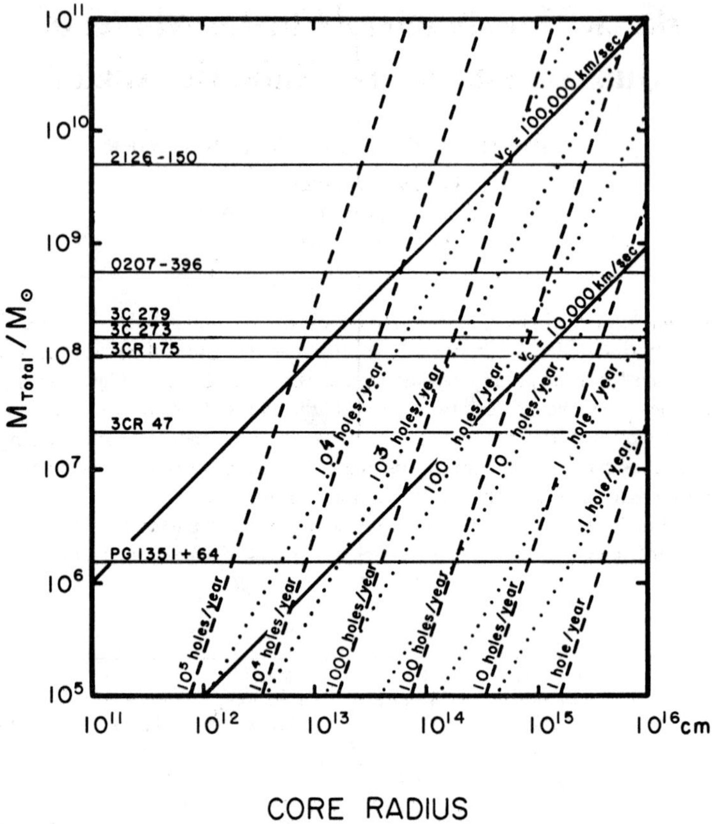

FIGURE 1. The consumption rates (dotted lines) and evaporation rates (dashed lines) of 10 solar masses black holes in a cluster dominated by a central mass of the order of 0.1 of the total cluster mass M_{total}.

REFERENCES

1. PACHOLCZYK, A. G. & W. R. STOEGER. 1982. Bull. Am. Astron. Soc. **14:** 648.
2. PACHOLCZYK, A. G. & W. R. STOEGER. 1982. Mem. Soc. Astron. Ital. **53:** 1033.
3. STOEGER, W. R. & A. G. PACHOLCZYK. 1984. Recent Dev. Gen. Relativity. (In press.)
4. SHAKURA, N. I. & R. A. SUNYAEV. 1973. Astron. Astrophys. **24:** 337.
5. NOVIKOV, I. D. & K. S. THORNE. 1973. *In* Black Holes, Les Houches. C. De Vitt & B. S. De Vitt, Eds. Gordon & Breech. New York, N.Y.
6. PRINGLE, J. E., M. J. REES & A. G. PACHOLCZYK. 1973. Astron. Astrophys. **29:** 279.
7. ZAMORANI, G., *et al.* 1981. Astrophys. J. **245:** 357.
8. FRANK, J. & M. J. REES. 1976. Mon. Not. R. Astron. Soc. **176:** 633.
9. LIGHTMAN, A. P. & S. L. SHAPIRO. 1978. Rev. Mod. Phys. **50:** 437.

A Gravitation Theory with G Determined by Retarded Cosmic Potential

KARL PHILBERTH[a]

Department of Theoretical Physics
University of New South Wales
Kensington, NSW 2033, Australia

The application of Einstein's contracted equation $-R^n_n G^{-1} = 8\pi c^{-4}\, T^n_n$ to the spheric Friedmann-Robertson-Walker metric with the "arc parameter" radial coordinate χ and the cosmic radius A, results in

$$G \int \mathrm{d}m_r / r = 3\, c^2 \qquad \textbf{(1)}$$

where G is the Newtonian gravitation factor, $\mathrm{d}m_r \equiv 4\pi\rho_r A^3 \sin^2\chi \mathrm{d}\chi$ is the "retarded" rest mass increment contained in the spherical shell of rest mass density ρ_r and of thickness $A\mathrm{d}\chi$ at the time when the now arriving information left the shell, and where the radius r is identical to sin χ times A_0, i.e., the "now-value" of A. The integration is extended from 0 ("here") to χ_h ("horizon"). With $\mathrm{d}m_r \equiv 4\pi\rho_r A^3 sh^2\chi \mathrm{d}\chi$ and $r \equiv A_0 sh\chi$, Equation 1 applies to the hyperbolic cosmos.

Since $G\mathrm{d}m_r/r$ means the "retarded gravitational potential" increment, Equation 1 represents a fundamental "Machian" cosmic relation. Therefore I postulate that the cosmic application of a realistic gravitation theory has to keep Equation 1, even if G varies.[1-3] This postulate is violated by the Brans-Dicke theory. But it is not violated by an integrable-Weyl-space two-metrics theory,[4] if its gauge function β is proportional to G^{-1}. A reasonable scalar equation that satisfies this postulate is

$$-R^n_n G^{-1} + 6\Box(G^{-1}) = 8\pi c^{-4}\, T^n_n, \quad \text{with } \Box(\ ) \equiv (\ )^{;n}_{;n} \qquad \textbf{(2)}$$

Gravitation theories with covariant equations, the contraction of which is Equation 2, deserve special interest. I propose the equations

$$(R^k_i - \frac{1}{2}\delta^k_i R^n_n)G^{-1} - 2(G^{-1})^{;k}_{;i} + 2\delta^k_i \Box(G^{-1}) + 4\, G(G^{-1})_{;i}\,(G^{-1})^{;k}$$

$$-\, \delta^k_i\, G(G^{-1})_{;n}\,(G^{-1})^{;n} = 8\pi c^{-4}\,(T^k_i + P^k_i) \text{ with } T^k_{i;k} = 0$$

where P^k_i is a gravitational energy-momentum-density tensor with zero-trace.[1,2]

[a]Address correspondence to Peter-Rosegger-Str. 6, D-8039 Puchheim, Federal Republic of Germany.

REFERENCES

1. PHILBERTH, K. 1982. Speculations Sci. Technol. **5**(2): 189–212.
2. PHILBERTH, B. & K. PHILBERTH. 1982. Das All. Christiana-Verlag. CH-8260 Stein/Rhein, Federal Republic of Germany. (ISBN No. 3-7171-0821-2.)
3. PHILBERTH, K. 1977. *In* Cosmology, History, and Theology. W. Yourgrau & A. D. Breck, Eds.: 113–129. Plenum Press. New York, N.Y.
4. CANUTO, V., P. J. ADAMS, S.-H. HSIEH & E. TSIANG. 1977. Phys. Rev. **D16**(6): 1643–1663.

Dynamics of Accretion Disks around Compact Objects

A. R. PRASANNA

Physical Research Laboratory
Navrangpura
Ahmedabad-380009, India

D. K. CHAKRABORTY

Department of Physics
Government College
Jagdalpur-494001, M.P. India

We have obtained a class of steady-state solutions for thick disks around compact objects, both in Newtonian and general relativistic formulations, and analyzed their stability under axisymmetric perturbations using normal mode analysis. The steady-state equations are solved explicitly for various types of density distributions and corresponding velocity and pressure profiles are obtained. The meridional structure is obtained by applying proper boundary conditions. In the case of background Schwarzschild geometry it has been found that the requirement of positivity of pressure leads to constraint on the inner edge that it cannot be within 4 MG/c^2, the marginally bound orbit. The condition for neutral stability ($\sigma^2 = 0$) leads to the critical value of the adiabatic index γ_c to lie within the range 0.83 and 1.18, thus showing that perfect fluid disks ($\gamma \geqq 1.33$) are always stable.

Spinning Fluid Energy-Momentum Tensors in the Einstein-Cartan Theory

JOHN R. RAY[a] AND LARRY L. SMALLEY[b]

Space Science Laboratory
National Aeronautics and Space Administration
George C. Marshall Space Flight Center
Marshall Space Flight Center, Alabama 35812

A Eulerian variational principle for a perfect fluid in general relativity was presented some years ago.[1] Recently we have generalized this variational principle to deal with fluids having intrinsic spin.[2] Recent papers by Bailey and Israel also deal with spinning fluids in general relativity using a different approach.[3–5] These papers by Bailey and Israel may be consulted for earlier references to this subject.

It is thought that the spin of "particles," protogalaxies, turbulent eddies, or primeval black holes could play an important role in the dynamics of the early universe.[3–5] Israel, in a brief study, found a solution for a spinning fluid in an anisotropic (Bianchi I) cosmological model.[4] In this example the spin induces a Lense-Thirring rotation of the local inertial axis relative to the directions along the spin of the fluid. Further such studies should be carried out to determine how spin can affect the dynamics of the early universe. The general relativistic energy-momentum tensor for a spinning fluid derived in Reference 2 would be an appropriate starting point for such studies.

Our main goal in this paper is to generalize our spinning fluid variational principle of Reference 2 to the Einstein-Cartan (EC) theory. In the EC theory spin takes on an important role as the source of the torsion part of the gravitational field. Thus, what we derive in this paper is a fundamental theory, based on a Lagrangian, which introduces spin into the EC theory for a macroscopic spinning fluid. The torsion-spin equation of our theory leads to the Weyssenhoff convective form for the canonical spin tensor, and the energy-momentum tensor has explicit spin-dependent terms. There have been several recent studies of spinning matter in the EC theory;[6–8] however, none of these studies deduces the field equations of the theory from a Eulerian variational principle such as in this paper.

The theories that we studied in Reference 2 and in this paper are based on the work of Halbwachs, who formulated a Eulerian variational principle for a spinning fluid in special relativity.[9] Reference 1 is a generalization of Halbwachs' theory to general relativity for fluids without intrinsic spin, while Reference 2 is for fluids with intrinsic spin.

[a]NASA/ASEE Summer Faculty Fellow. Permanent address: Department of Physics and Astronomy, Clemson University, Clemson, S.C. 29631.

[b]Also Department of Physics, University of Alabama, Huntsville, Ala. 35899.

REFERENCES

1. Ray, J. R. 1972. J. Math. Phys. **13:** 1451.
2. Ray, J. R. & L. L. Smalley. 1982. Phys. Rev. **D26:** 2619.
3. Bailey, I. 1979. Ann. Phys. **119:** 76.
4. Israel, W. 1973. Lett. Nuovo Cim. **7:** 860.
5. Bailey, I. & W. Israel. 1975. Commun. Math. Phys. **42:** 65.
6. Cognola, G., R. Soldati, L. Vanzo & S. Zerbini. 1982. Phys. Rev. **D25:** 3109.
7. Yasskin, P. B. & W. R. Stoeger. 1980. Phys. Rev. **D21:** 2081.
8. Hojman, S. 1978. Phys. Rev. **D18:** 2741.
9. Halbwachs, F. 1960. Theorie Relativiste des Fluides a Spin. Gauthier-Villars. Paris, France.

United Kingdom Schmidt Telescope Objective Prism Surveys for QSOs

ANN SAVAGE

Royal Observatory
Blackford Hill
Edinburgh EH9 3HJ, Scotland

The low dispersion objective prism was first used on the United Kingdom Schmidt telescope (UKST) in 1976. Its performance is described by Nandy *et al.*[5] It has a dispersion of 240 nm/mm at 430 nm with a resolution of 6 nm. Some 450 plates have now been taken through this prism, and of these 80 are limiting exposures taken in seeing $\leq$2 arcsec. The latter plates are excellent for optical QSO searches. Each plate yields about 200 obvious QSO candidates, of which 95% will be confirmed. A further 800 objects are usually selected as possible QSO candidates, 10 to 50% of which may be confirmed as QSOs. However the dispersion in the red for this prism is very low so that plates are normally taken on blue-sensitive IIIa-J emulsion and it is extremely difficult to select QSOs with $z > 3.0$.

These selected candidates have a 95% chance of being confirmed as QSOs. Of the red shifts estimated from the objective prism, 75% are correct to within 5%. For the objects confirmed as QSOs, 30% have red shifts < 1.7, in contrast to other prism surveys which find most objects with $1.8 < z < 3.5$. Although there are many selection effects operating in objective prism searches, the UKST samples are ideal for studies of QSO clustering on the 100 Mpc scale at $z = 2.0$ (scales of 1° on the sky). The Automatic QSO Detection Programme (AQD) developed by Clowes *et al.*[3] for use with the ROE automatic measuring machine COSMOS makes it possible to quantitatively select a well-defined subset of candidates with which to make isotropy studies from plate to plate.

Fuller details of the prisms, the searches, and the results can be found in the listed references.

REFERENCES

1. Clowes, R. G., M. G. Smith, A. Savage, R. D. Cannon, A. Boksenberg & J. V. Wall. 1979. Mon. Not. R. Astron. Soc. **189:** 175.
2. Clowes, R. G. & A. Savage. 1983. Mon. Not. R. Astron. Soc. **204:** 365–375.
3. Clowes, R. G., J. A. Cooke & S. M. Beard. 1982. Workshop on astronomical measuring machines. Occas. Rep. ROE No. 10: 253.
4. Morton, D. C. & K. P. Tritton. 1982. Mon. Not. R. Astron. Soc. **198:** 669.
5. Nandy, K., V. C. Reddish, K. P. Tritton, J. A. Cooke & D. Emerson. 1977. Mon. Not. R. Astron. Soc. **178:** 63p.
6. Peterson, B. A., A. Savage, D. L. Jauncey & A. E. Wright. 1982. Astrophys. J. Lett. **260:** L27.
7. Savage, A. & J. G. Bolton. 1978. Int. Astron. Union Symp. **79:** 295.
8. Savage, A. & J. G. Bolton. 1979. Mon. Not. R. Astron. Soc. **188:** 599.

X-Ray Constraints on Relativistic Beams in BL Lacertae Objects

D. A. SCHWARTZ AND G. M. MADEJSKI

Harvard-Smithsonian Center for Astrophysics
Cambridge, Massachusetts 02138

Relativistic beaming has been invoked in several astrophysical contexts, e.g., superluminal expansion, avoidance of the Compton catastrophe, and the absence of emission lines in BL Lac objects. However, the x-ray data rule out the possibility that BL Lac objects are simply those quasars whose relativistic optical or radio beam ($\Gamma \sim 5$ to 10) points toward us. This is because their local volume density is of the same order of magnitude, 10^{-7} Mpc^{-3}, as quasars, when both are selected by x-ray luminosities greater than 10^{44} ergs s^{-1}. Furthermore, their x-ray power law spectral indices span a wide range, so they are not merely the Lorentz transformation of quasars or Seyferts for which the index clusters around $\alpha = 0.7$. If BL Lac objects do have relativistic beams then it must be the x-ray emission that is beamed so that the parent population which is $2\,\Gamma^2$ more numerous really only has rest frame x-ray luminosity $\Gamma^{3+\alpha}$ less. The allowed parent population might thus have densities 10^{-5} Mpc^{-3} and x-ray luminosities a few times 10^{40} ergs s^{-1}, for $\Gamma \sim 10$. However, if the synchrotron self-Compton mechanism is responsible for the observed x-ray flux, then by using very long baseline interferometry sizes for a set of 16 BL Lac type objects we estimate that typical values of Γ are in the range of 2 to 5.

Results from the Durham-AAT Red Shift Survey

T. SHANKS

Department of Physics
University of Durham
Durham DH1 3LE, England

The analysis of the dynamics of 340 galaxies in the Durham/AAT red shift survey gives an estimate for Ω_0 of 0.15 ± 0.05 on the assumption that galaxies are good tracers of the distribution of mass in the universe. The two-point galaxy correlation function, $\xi(r)$, has been derived and shows a general level of clustering in agreement with that expected from correlation analyses of projected catalogues with a break detected at r = 5 h^{-1} Mpc. The correlation function also shows some evidence of a shoulder at 2 h^{-1} Mpc which may have been too weak to be detected in projection. At scales between 25 and 100 h^{-1} Mpc our analysis shows that the galaxy distribution is consistent with being Poissonian with no evidence of large-scale clustering. On scales between 10 and 25 h^{-1} Mpc, ξ shows some evidence of anticorrelation, although the errors in this range are still large.

Spectrum of the Cosmic Background Radiation[a]

G. SIRONI

National Research Council (CNR)
20133 Milan, Italy

N. MANDOLESI

National Research Council (CNR)
40126 Bologna, Italy

G. DE AMICI, S. FRIEDMAN, G. SMOOT,
AND C. WITEBSKY

Space Science Laboratory and
Lawrence Berkeley Laboratory
University of California
Berkeley, California 94720

R. B. PARTRIDGE

Department of Astronomy
Haverford College
Haverford, Pennsylvania 19041

L. DANESE

Astronomical Observatory
35100 Padova, Italy

We present preliminary results of measurements of the low-frequency spectrum of the cosmic background radiation. The measurements have been made at five wavelengths (12, 6.3, 3, 0.9, and 0.33 cm) using radiometers with scaled corrugated horn antennas having very low side lobes. A single large-mouth (0.7 m diameter) cold-load calibrator cooled by liquid helium was used as the absolute reference for all five wavelengths. The measurements were made during the first week of July 1982 (1983. Phys. Rev. Lett. **51:** 1099), and we expect to make additional and repeated measurements in the summer of 1983.

[a]Partially supported by the National Science Foundation (United States) and the CNR.

Gamma-Ray and X-Ray Telescope Using Variable-Metric Diffraction Crystals

ROBERT K. SMITHER

Argonne National Laboratory
Argonne, Illinois 60439

A new method has been devised for focusing and imaging the radiation from sharp-line sources of x-rays and gamma-rays, which makes use of variable-metric diffraction crystals (see R. K. Smither, Rev. Sci. Instrum., Feb. 1982). A variable-metric diffraction crystal is one in which the spacing between the crystalline planes is varied as a function of position in the crystal by either the application of a thermal gradient or by changing the composition of a two-component or multiple-component crystal. This change in planar spacing changes the Bragg diffraction angle for monochromatic radiation as a function of position in the crystal and makes it possible to obtain focusing and in some cases imaging of a sharp-line point source or parallel beam source (see R. K. Smither, NASA Technical Memorandum No. 83843, Nov. 1981). This new approach to focusing x-rays and gamma-rays is used to design a number of gamma-ray telescopes suitable for focusing the 511 keV annihilation radiation from the strong source of the center of our galaxy. The new designs are surprisingly efficient with approximately 20% of the radiation incident of the variable-metric diffraction crystals being focused on the image spot. Crystals of Ge, Ge + Si, Si, and quartz are used with mosaic widths of 10 arcsec. The size of the telescope can be scaled up or down without affecting the angular resolution or the energy resolution. The largest model described is 50 m long and has 10 crystal diffraction ring assemblies with radii between 71 cm and 200 cm. The total area of the diffraction crystal is 24,610 cm^2, and the effective area (total $\times$ diffraction coefficient $\times$ transmission) is 4,107 cm^2. This means that the average value for the photons cm^{-2} s^{-1} at the detector spot will be 4,745 times the incident photons cm^{-2} s^{-1}. For a flux of 2×10^{-3} photon cm^{-2} s^{-1} from the center of our galaxy, this would result in a flux of 9 counts per second in the 1 cm diameter focal spot. The peak intensity of the center of the focal spot is three times brighter. An example of a smaller telescope is also given that is only 12.5 m long and has an effective area of 297 cm^2. A modified version of the smaller model with poorer resolution (160 arcsec) but with a larger effective area (606 cm^2) is also described. A design for focusing 50 keV radiation is also given. This 50 keV telescope uses the same type of diffraction crystals and the same size crystal rings as the 511 keV model, the only difference being that the crystals are much thinner and the Bragg angles are much larger—thus the focal length, 5.6 m, is much shorter. The effective area for the large model (71 cm to 200 cm radiation for crystal rings) with a monochromatic source is 8,660 cm^2. This design can also be scaled up or down if more or less gain in the system is needed.

A Closed Pulsating Universe with Inner Dimensionality

F. R. TANGHERLINI

Department of Physics
College of the Holy Cross
Worcester, Massachusetts 01610

At present it is still unknown empirically whether or not the universe is open or closed. Nevertheless, the closed (FRW) universe has well-known attractive features which make it interesting to study theoretically: the volume, the number of particles (and hence galaxies), the mass, and the lifetime are all finite. It is usually simpler in physics to deal with finite quantities than with divergent ones. On the other hand, the FRW closed universe starts out from a point singularity at the big bang and returns to a point singularity at the big crunch. These singularities present serious mathematical difficulties in extrapolating to times before the big bang and after the big crunch. Clearly such situations violate Einstein's singularity-free field theory. In a different vein, John Wheeler has emphasized that in gravitational collapse there is no escape from an ultimate singularity in which the present laws of nature are transcended.[1,2]

A possible solution to the crisis posed by gravitational collapse and the singularity problem is to regard Einstein's field equations for the problem as analogous to a dynamical system that needs stabilization. There are two principal ways to produce stabilization; one through statics, the other through dynamics. For example, one might regard Einstein's original closed universe as a system that was stabilized in a static manner by means of the cosmological term. However, as Lemaître showed, the equilibrium achieved is inherently unstable. Nevertheless, one might wonder whether there exists a dynamical analogue to the method based on the cosmological term which would yield a stable universe, in the sense that small perturbations would not run away. One might also hope that other solutions associated with this dynamical stabilization would prevent gravitational collapse to a singularity and thereby lead to the possibility of a pulsating universe where the collapse rebounds at some minimum value of the expansion parameter $R(t)$.

The present work describes such a model, which has the "bounce" property for open as well as closed universes. One starts from the usual FRW metric

$$ds^2 = c^2 dt^2 - R^2\left(1 + \frac{kr^2}{4}\right)^{-2} \delta_{ij} dx^i dx^j$$

for which the Einstein field equations are

$$\dot{R}^2 - \frac{8\pi}{3} G\rho R^2 + kc^2 = 0 \tag{1}$$

$$2R\ddot{R} + \dot{R}^2 + kc^2 = 0 \tag{2}$$

in the absence of a pressure term, since it is well known that one cannot halt collapse with positive pressure. Since $M = 4\pi R^3\rho/3$ is a constant of the motion under these circumstances, it is convenient to rewrite the field equations in the following form:

$$\frac{\dot{R}^2}{2} - \frac{GM}{R} + \frac{kc^2}{2} = 0 \tag{3}$$

$$\ddot{R} = -GM/R^2 \tag{4}$$

Thus, the field equations in the FRW coordinate system have the dynamical image of a Newtonian particle undergoing radial motion in the gravitational field of a mass point M located at $R = 0$, although this is not the standard interpretation. The cycloidal singularity of the big bang and big crunch is an obvious consequence of this image. It is clear that there could be a halt to the collapse if there were some hidden angular momentum that belongs in the field equation for $R(t)$. A way to achieve this is to increase the dimensionality of the Newtonian image space. This is done by replacing the one-dimensional kinetic energy term $\dot{R}^2/2$ by a kinetic energy in a space of three dimensions corresponding to the $1/R$ behavior of the potential. Thus one sets $\dot{R}^2 \rightarrow \dot{\boldsymbol{R}} \cdot \dot{\boldsymbol{R}}$, where $\boldsymbol{R}$ is now a vector to the test particle in the Newtonian image space. Correspondingly, the equation of motion becomes a vector equation in $\boldsymbol{R}$.

The dimensionality-boosted field equations are therefore

$$\frac{\dot{\boldsymbol{R}} \cdot \dot{\boldsymbol{R}}}{2} - \frac{GM}{R} + \frac{kc^2}{2} = 0 \tag{5}$$

$$\ddot{\boldsymbol{R}} = -\vec{\nabla}(-GM/R) \tag{6}$$

On the other hand, in the FRW line element we need a one-dimensional quantity; hence, by standard arguments for the Keplerian problem, one now reduces the dimensionality of these equations back to a one-dimensional radial form, but in which a centrifugal potential is now present corresponding to the angular momentum per unit mass A

$$\frac{\dot{R}^2}{2} + \frac{A^2}{2R^2} - \frac{GM}{R} + \frac{kc^2}{2} = 0 \tag{7}$$

$$\ddot{R} = \frac{A^2}{R^3} - \frac{GM}{R^2} \tag{8}$$

These are now the stabilized form of Einstein's field equations for $R(t)$, which for a closed universe traces out a Keplerian ellipse about the focus point $R = 0$ in the Newtonian space. In the limit $A \rightarrow 0$, one of course recaptures the original field equations (3) and (4).

Thus, instead of gravitational collapse to a point, there is a minimum radius of approach, the "pericosmon," given by

$$R_{pc} = \frac{GM}{kc^2}(1 - \epsilon) \tag{9}$$

where ϵ is the "cosmological eccentricity"

$$\epsilon = \left(1 - \frac{kA^2C^2}{G^2M^2}\right)^{1/2} \tag{10}$$

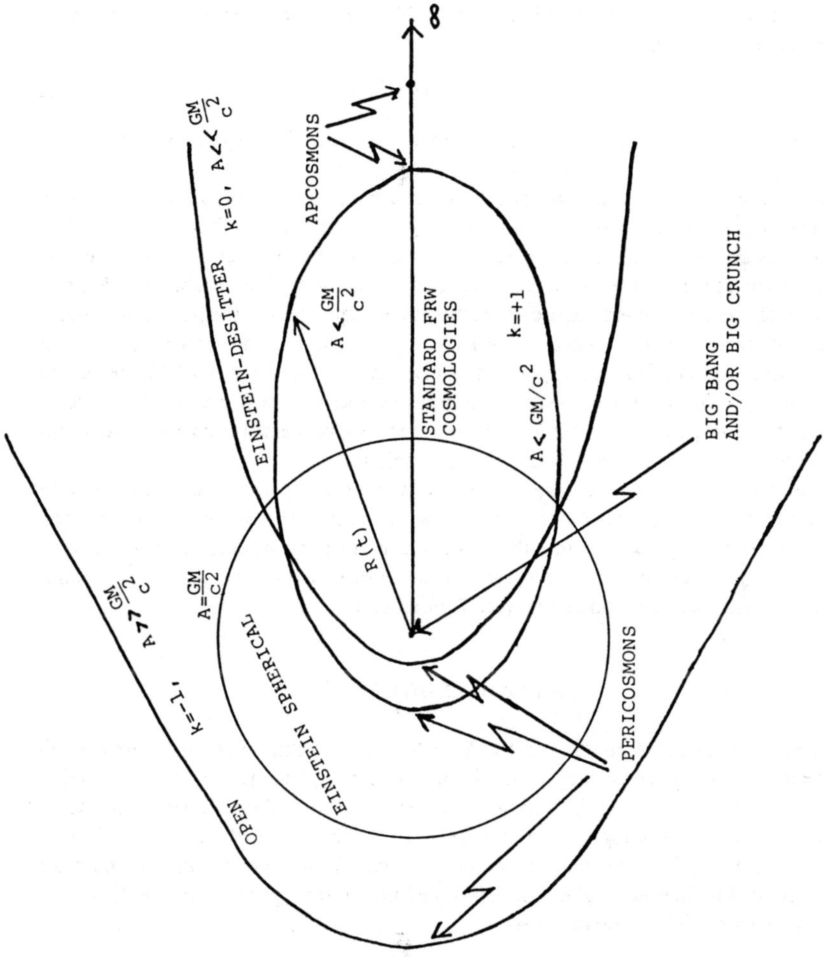

FIGURE 1. Various models of the universe.

For A small, $R_{pc} = A^2/2GM$ and is independent of k. This latter value for R_{pc} holds exactly for a "flat" Einstein–de Sitter universe.

If the universe is closed, the "apcosmon" is

$$R_{AC} = \frac{GM}{c^2}(1 + \epsilon) \tag{11}$$

and the major axis of the image ellipse is $2a = 2GM/c^2$. Since the period is the same as for the Keplerian problem,

$$T = 2\,GM/c^3 \tag{12}$$

Because T is independent of A, the above holds for the limiting cycloidal case as well. Assuming $T > 10^{10}$ years, one gets $M > 2 \times 10^{55}$ g.

The missing mass problem is not ameliorated by the approach followed here, since the density parameter required for a closed universe must satisfy $\Omega > 1 + (A^2/R^2H^2)$ where H is Hubble's constant at epoch t. On the other hand, since $\dot{R} \simeq 0$ in the early stages of the universe, one has a possible solution to the horizon problem of Rindler[3] (for a recent review, see Reference 4). Although the equations of the pulsating universe are not yet derivable from first principles (e.g., general covariance), they do give some new insight into the Hubble expansion. According to the present FRW model, the expansion is to be thought of as due to an initial velocity, $\dot{R} > 0$, acquired in a violent explosion at a singularity, rather than from a geodesic process. In contrast, the motion of the ideal pulsating universe is geodesic throughout.

The results could have bearing on understanding the explosive processes associated with supernovae and quasars, if these were associated in part with an outward geodesic motion. However, application to pulsars is unclear. It should be emphasized that A is a parameter that depends on the system and perhaps some new aspects of dimensionality[5–9] and angular momentum (see Reference 10).[a]

ACKNOWLEDGMENTS

The author is indebted to Prof. John A. Wheeler for rekindling his interest in the dimensionality of space problem and stimulating correspondence. He is also indebted to Prof. Shelley Glashow for initiating him into the problem of gravitation collapse and lending an ear to cosmology. I am also indebted to Prof. John R. Ray for calling my attention to Prof. Trautman's work at the Eleventh Texas Symposium. I would also like to thank Ms. Jane Kjems for editorial assistance and preparation of the illustration of the various models of the universe.

REFERENCES

1. MISNER, C. W., K. S. THORNE & J. A. WHEELER, Eds. 1973. Gravitation: 1196–1202. W. H. Freeman. San Francisco, Calif.

[a] This work describes another approach to averting the singularity based on spin and torsion in the Einstein-Cartan theory of gravitation.

2. WHEELER, J. A. 1980. *In* Some Strangeness in the Proportion. H. Woolf, Ed.: 50. Addison-Wesley. Reading, Pa.
3. RINDLER, W. 1956. Mon. Not. R. Astron. Soc. **116:** 662.
4. CALVANI, M., S. MATARESSE & M. D. POLLOCH. 1982. Lett. Nuovo Cimento **34:** 183.
5. TANGHERLINI, F. R. 1984. Lett. Nuovo Cimento. (In press.)
6. TANGHERLINI, F. R. 1963. Nuovo Cimento **27:** 636.
7. EHRENFEST, P. 1917. Proc. Amsterdam Acad. **20:** 200.
8. EHRENFEST, P. 1920. Ann. Physik. **61:** 440.
9. WHITROW, G. J. 1959. The Structure and Evolution of the Universe. Harper & Bros. New York, N.Y.
10. TRAUTMAN, A. 1973. Nature London Phys. Sci. **242:** 7.

Does Gravity Exhibit a $1/r$ Force on the Scale of Galaxies?

JOEL E. TOHLINE

Department of Physics and Astronomy
Louisiana State University
Baton Rouge, Louisiana 70803

Several major features of galaxy dynamics (e.g., flat rotation curves and the stability of cold stellar disks) can be modeled without invoking the presence of dark halos if one assumes that the force law of gravity has the form $F = -GMmr^{-2}(1 + r/d)$, where $d \sim 1$ to 5 kpc. In this form, the gravitational force would fall off as $1/r$ instead of as $1/r^2$ on scales $r \gg d$. The *dynamical* effect of such a deviation from a pure Newtonian force on planetary orbits in our solar system is not measurable, but the *secular* effect is probably detectable. The precession $\Delta\psi$ of the perihelion of a planetary orbit whose semimajor axis in A.U. is "a" and whose orbital eccentricity is e would be

$$\Delta\psi = -0.628\, a^{-1/2}\, d_{\text{kpc}}^{-1}\, [e^2 - 1 + (1 - e^2)^{1/2}]\, e^{-2}$$

arcseconds/century if d_{kpc} is the scale length "d" measured in kiloparsecs. For example, Mercury's orbit should exhibit $|\Delta\psi| \sim 0.5$–0.1 arcsecs/century for the chosen range $d_{\text{kpc}} \sim 1$–5. Precise observations of the motions of planets in our solar system can therefore be used to make a strong, useful statement regarding the long-range nature of the force of gravity on scales that are important to galaxy dynamics.

Optically Thick, Time-Dependent Accretion onto Black Holes

PETER VITELLO

Plasma Physics Division
Science Applications, Inc.
P.O. Box 1303
McLean, Virginia 22102

Radiation emitted from spherical black hole accretion must be generated in flight as matter flows toward the horizon. Thus for optically thick accretion, the question of what the resulting luminosity is can be quite complex and dynamics dependent. Studies of accretion flows that become optically thick to Thomson scattering between their accretion radius and the horizon have been carried out by a number of authors, who have treated the steady-state fluid equations. All studies have met with divergence problems with the luminosity. We find that a time-dependent approach avoids this divergence behavior and can be used to quickly relax the flow to steady state. We have solved the time-dependent, fully relativistic fluid equations with a simple magnetic dissipation model, assuming radiation diffusion with the Thomson opacity. Using an implicit differencing scheme on a Eulerian spatial mesh allows us to avoid time-scale, and hence radial, limitations. Flows are treated from their radius of optical depth unity, until they are well within the horizon. For no magnetic dissipation, the efficiency is quite small and luminosities well below L_E result even for $\dot{M} \gg \dot{M}_E$. Magnetic dissipative heating, however, leads to luminosities of the order of L_E independent of the accretion rate. The physical nature of dissipation in accretion flows must be examined more thoroughly in order to determine when and for what objects it does in fact provide significant heating.

Just How Fast Are Tachyons?

T. A. WEBER

Physics Department
Iowa State University
Ames, Iowa 50011

The interaction of tachyons with gravity leads to results of interest in cosmology and black hole physics.[1,2] Tachyons have the dispersion rule

$$E^2 = p^2 - m_0^2$$

in which the sign of the mass-squared term is opposite to that of ordinary particles so that their group velocity is greater than the velocity of light, which is taken to be one. However, some people claim that tachyons do not travel with their group velocity and purport to show that no tachyonic signal can travel faster than one.[3,4] Their proofs require all values of momentum to construct strictly localized packets. But momenta in the range $|p| < m_0$ lead to imaginary energies and nonnormalizable wave functions. Excluding such momenta, well-defined wave packets with velocities greater than one can be constructed provided the peak in the momentum distribution occurs for $p \gg m_0$, i.e., the expected momentum of the tachyon is much greater than m_0. But the interesting phenomena associated with tachyons occur for momenta at or near m_0. Rapid spreading of the packet would happen for such momenta, casting doubt on some of the interesting results associated with tachyons.

REFERENCES

1. Narlikar, J. V. & E. C. G. Sudarshan. 1976. Mon. Not. R. Astron. Soc. **175:** 105.
2. Dhurandhar, S. V. & J. V. Narlikar. 1978. J. Gen. Relativity Gravity **9:** 1089.
3. Fox, R., C. G. Kuper & S. G. Lipson. 1969. Nature **223:** 597.
4. Robinett, L. 1978. Phys. Rev. **D18:** 3610.

Improved Collimation for Radio Sources

PAUL J. WIITA AND M. JAVAD SIAH

Department of Astronomy and Astrophysics
University of Pennsylvania
Philadelphia, Pennsylvania 19104

Models for the formation and evolution of beams in radio galaxies have been computed using a Kompaneets-like approximation. We consider the initially spherically symmetric injection of relativistic plasma from a central engine into a confining cloud of hot gas in the nucleus of an elliptical galaxy. If the gas lies in a potential well produced by a slowly rotating stellar distribution, its isobars will take on a dimpled appearance, with partial evacuation along the rotation axes. In contrast to earlier calculations, which used a spheroidal pressure distribution, narrower jets are produced.

The key parameters are the luminosity of the central source (L), the temperature (T), central pressure (p_0), scale height (l), and eccentricity (e) of the confining cloud, and the scale height (b) of the stellar distribution. Typical parameter ranges are $L = 0.1 - 3 \times 10^{46}$ erg s^{-1}, $T = 1$–30×10^7 K, $p_0 = 3$–50×10^{-6} dyn cm^{-2}, $l = 20$–150 pc, $b = 50$–350 pc, $e = 0.7$–0.999, yielding reduced luminosities L' $(=L/l^2 p_0 c_s)$ of between 0.32 and 18,000. As in our previous work, we find that a high L' tends to produce jets directly, while lower values yield bubbles. Increasing e improves the collimation. But we now find that the ratio b/l, or equivalently, the azimuthal Mach number of the cloud, is also important. High values of this quantity favor direct jet formation and also produce better collimation.

Our models can be computed out to greater distances ($\gtrsim$10 kpc) and for longer times (equivalent to $\sim 10^6$ years) and produce narrower jets than more sophisticated and time consuming two-dimensional hydrodynamic computations, although they agree quite well in their areas of overlap. Thus they can serve as a useful link in the construction of realistic models for extended radio sources.

Index of Contributors

ics/Compu